Dynamically Reconfigurable Systems

Marco Platzner · Jürgen Teich · Norbert Wehn
Editors

Dynamically Reconfigurable Systems

Architectures, Design Methods and Applications

Springer

Editors
Marco Platzner
Department of Computer Science
University of Paderborn
Warburger Str. 100
33098 Paderborn
Germany
platzner@upb.de

Jürgen Teich
Hardware/Software Co-Design
Department of Computer Science
University of Erlangen-Nuremberg
Am Weichselgarten 3
91058 Erlangen
Germany
teich@cs.fau.de

Norbert Wehn
TU Kaiserslautern
Erwin-Schrödinger-Str.
67663 Kaiserslautern
Germany
wehn@eit.uni-kl.de

ISBN 978-90-481-3484-7 e-ISBN 978-90-481-3485-4
DOI 10.1007/978-90-481-3485-4
Springer Dordrecht Heidelberg London New York

Library of Congress Control Number: 2009944064

Printed on acid-free paper

Springer is part of Springer Science+Business Media (www.springer.com)

Foreword

After six years of exciting and intensive research, one of the world's largest and longest-running programs of research into Reconfigurable Computing is concluding with the publication of the research papers in this volume. The research was launched in 2003 by Deutsche Forschungsgemeinschaft, the German Research Foundation, as "Schwerpunktprogramm (SPP 1148)" under the title "Rekonfigurierbare Rechensysteme". This translates as the Priority Programme in Reconfigurable Computing Systems. For convenience, we will refer to it here as SPP 1148.

Several aspects of SPP 1148 made it noteworthy from its inception. Beneath the umbrella of reconfigurable computing systems, several key topics were identified. These included theoretical aspects, modeling, languages, analysis, design methodologies, computer-aided design (CAD) tools, architectures and applications. From the onset, the scope of the research was deliberately formulated to be as multidisciplinary as possible. A national network of excellence was established with the aim of encouraging participation from as many academics with novel individual contributions as possible while simultaneously emphasizing and promoting interdisciplinary collaboration.

The six-year program was organized as three successive funding periods of two-years each. At each interval, new projects were proposed and existing projects were reviewed. The two largest project sub-themes were design methodologies and tools, and architectures and applications. Feedback from graduate students who participated in SPP 1148 indicates that they felt especially privileged to have been able to conduct their studies within the framework of this program. In particular, they benefited from the highly collaborative environment that was nurtured by the participating institutions and the continuity afforded by such a sustained period of research funding.

Throughout the conduct of SPP 1148, we have had the privilege of interacting and collaborated directly with many of the research teams. We have become familiar with the majority of the research projects albeit at different levels of detail. We have had the opportunity to observe the progress of SPP 1148 as a whole from our complementary positions in academia and industry. There can be no doubt that the most immediate impact of the DFG's (German Research Foundation) foresight was to establish German academia as the foremost group in the world for advanced research into reconfigurable computing over the last six years.

The legacy of the SPP 1148 in Reconfigurable Computing Systems is that Germany has developed an impressive national capability in precisely those areas that will be of greatest relevance for the next decade. As we reach for 40 nm integrated circuits and beyond, it is clear that fewer companies and fewer architectures will be viable at these ever more advanced process nodes. The successful architectures will be highly concurrent and will combine programmability and reconfigurability capabilities.

This book is unique in many different ways. Not only does it provide a structured compilation for the research resulting from this extensive program, it also gives an excellent overall insight into various strategic directions in which the entire field of reconfigurable technologies and systems may be heading. In addition, in the four separate parts of the book, the coverage of each of the subjects is comprehensive. Aspects relating to architectures, design methodologies, design tools and a large number of applications are all included. No other book currently in print has such an extensive and overall coverage as this. As such, it is equally suitable for supporting graduate level courses and for practical engineers working in the field of reconfigurable hardware and particularly in FPGAs. We are delighted to see the work and experience of such a large group of researchers and engineers being shared with the reconfigurable community at large.

Patrick Lysaght, Xilinx Corporation
Peter Cheung, Imperial College London
August 2009

Preface

While the idea of creating computing systems with flexible hardware dates back to the 1960s, it was the emergence of the SRAM-based field-programmable gate array (FPGA) in the 1980s that boosted *Reconfigurable Computing* as a research and engineering field. Since then, reconfigurable computing has become a vibrant area with an increasingly growing research community and exciting commercial ventures.

Reconfigurable computing devices are able to adapt their hardware to application demands and serve broad and relevant application domains from embedded to high-performance computing, including automotive, aerospace and defense, telecommunication and networking, medical and biometric computing. Especially in the embedded computing domain with its many and often conflicting objectives reconfigurable computing systems offer new trade-offs. Embedded systems are fueled by microelectronics where one of the currently biggest challenges is the trade-off between flexibility and cost. Reconfigurable devices, especially FPGAs, fill this gap since they provide flexibility at both design-time and run-time. Consequently, in the last years we have seen declining ASIC design starts but continuously increasing FPGA design starts. Continuing advances in the miniaturization of microelectronic components have made it possible to integrate systems with multiple processors on a single chip at the size of a fingernail (system-on-chip (SoC)). Often, the production volumes for SoCs are rather low jeopardizing their economic benefits. On the other hand, modern FPGAs are basically complex SoCs integrating embedded processors, signal processing capabilities, multi-gigabit transceivers and a broad portfolio of IP cores. Thus FPGAs are positioned to become a real alternative to ASICs and ASPPs.

This book is the first ever to focus on the emerging field of *Dynamically Reconfigurable Computing Systems*. While programmable logic and design-time configurability are well elaborated and covered in various books, this book presents a unique overview over the state of the art and recent results for dynamic and run-time reconfigurable computing systems. This book targets graduate students and practitioners alike. Over the last years, many educational institutions began to offer courses and seminars on different aspects of reconfigurable computing systems. We recommend this book as reading material for the advanced graduate level and entrance into own research on dynamically reconfigurable systems. Reconfigurable hardware is not only of utmost importance for large manufacturers and vendors of microelectronic devices and systems, but also a very attractive technology for smaller and medium-sized companies. Hence, this book addresses also researchers and engineers actively working in the field and updates them on the newest developments and trends in run-time reconfigurability.

The book is organized into four parts that present recent efforts and breakthroughs in architectures, design methods and tools, and applications for dynamically reconfigurable computing systems:

Architectures: Three chapters on architectures discuss different dynamically reconfigurable platforms, including multigrained and application-specific architectures as well as an FPGA-based computing system supporting efficient partial reconfiguration.

Design Methods and Tools—Modeling, Evaluation and Compilation: The first part on design methods and tools features four chapters focusing on modeling and evaluation aspects for dynamically reconfigurable hardware, on creating compilers for reconfigurable devices, and on supporting dynamic reconfiguration through object-oriented programming.

Design Methods and Tools—Optimization and Runtime Systems: The second part on design methods and tools comprises six chapters that are devoted to resource allocation in dynamically reconfigurable systems, split into challenging optimization problems that need to be solved during compilation time, e.g., temporal partitioning, and online resource allocation which is provided by a novel breed of reconfigurable hardware runtime systems.

Applications: The last part of the book presents seven chapters with applications of dynamically reconfigurable hardware technology to relevant and demanding domains, including mobile communications, network processors, automotive vision, and geometric algebra.

This book presents the results of a six-years research initiative on dynamically reconfigurable computing systems, initiated and coordinated by Jürgen Teich and funded by the German Research Foundation (DFG) within the Priority Programme (Schwerpunktprogramm) 1148 from 2003 to 2009. To make dynamic reconfigurable computing become a reality this joint research initiative bundled multiple projects and, at times, involved up to 50 researchers working in the topic.

Equivalently, this book summarizes more than 100 person years of research work and more than 20 PhD students have already submitted and defended their theses based on research performed in this initiative. The material presented in this book thus summarizes the golden fruits, major achievements and biggest milestones of this joint research initiative.

We are very grateful to the German Research Foundation for funding and continuously supporting this initiative. We would also like to thank all the researchers contributing to the programme and to this book, the many national and international reviewers as well as the industrial companies that have been steadily supporting our efforts. Additionally, we would like to acknowledge the assistance of Josef Angermeier, Martina Jahn, Enno Lübbers, and Felix Reimann in supporting the editorial process. Last but not least, we thank Springer for giving us the opportunity to publish our results with them.

We hope you enjoy reading this book!

Marco Platzner, Jürgen Teich, Norbert Wehn

Contents

Contributors

Ali Ahmadinia Hardware/Software Co-Design, Department of Computer Science, University of Erlangen-Nuremberg, Erlangen, Germany, A.Ahmadinia@ed.ac.uk

Carsten Albrecht Institute of Computer Engineering, University of Lübeck, Ratzeburger Allee 160, 23538 Lübeck, Germany, albrecht@iti.uni-luebeck.de

Matthias Alles Microelectronic Systems Design Research Group, University of Kaiserslautern, 67663 Kaiserslautern, Germany, alles@eit.uni-kl.de

Josef Angermeier Hardware/Software Co-Design, Department of Computer Science, University of Erlangen-Nuremberg, Erlangen, Germany, angermeier@informatik.uni-erlangen.de

Jürgen Becker Institut für Technik der Informationsverarbeitung (ITIV), University of Karlsruhe, Karlsruhe, Germany, Becker@itiv.uni-karlsruhe.de

Christophe Bobda Department of Computer Science, University of Potsdam, Potsdam, Germany, bobda@cs.uni-potsdam.de

Lars Braun Institut für Technik der Informationsverarbeitung (ITIV), University of Karlsruhe, Karlsruhe, Germany, Braun@itiv.uni-karlsruhe.de

Christian Brehm Microelectronic Systems Design Research Group, University of Kaiserslautern, 67663 Kaiserslautern, Germany, brehm@eit.uni-kl.de

Christopher Claus Institute for Integrated Systems, Technische Universität München, Munich, Germany, Christopher.Claus@tum.de

Thomas Coenen Chair of Electrical Engineering and Computer Systems, RWTH Aachen University, Schinkelstr. 2, 52062 Aachen, Germany, coenen@eecs.rwth-aachen.de

Sven Eisenhardt Department of Computer Engineering, Wilhelm-Schickard-Institute for Computer Science, University of Tübingen, Tübingen, Germany, eisenhardt@informatik.uni-tuebingen.de

Sándor P. Fekete Department of Computer Science, Braunschweig University of Technology, Braunschweig, Germany, s.fekete@tu-bs.de

Jürgen Foag Rohde & Schwarz, Radiocommunications Systems Division, Mühldorfstraße 15, 81674 Munich, Germany, juergen.foag@rohde-schwarz.com

Manfred Glesner Institute of Microelectronic Systems, Technische Universität Darmstadt, Darmstadt, Germany, glesner@mes.tu-darmstadt.de

Philipp Graf FZI Forschungszentrum Informatik an der Universität Karlsruhe, Karlsruhe, Germany, graf@fzi.de

Kim Grüttner OFFIS Institute for Information Technology, Oldenburg, Germany, gruettner@offis.de

Philipp A. Hartmann OFFIS Institute for Information Technology, Oldenburg, Germany, hartmann@offis.de

Christian Haubelt Hardware/Software Co-Design, Department of Computer Science, University of Erlangen-Nuremberg, Erlangen, Germany, haubelt@cs.fau.de

Andreas Herkersdorf Institute for Integrated Systems, Technische Universität München, Munich, Germany, herkersdorf@tum.de

Andreas Herrholz OFFIS Institute for Information Technology, Oldenburg, Germany, herrholz@offis.de

Heiko Hinkelmann Institute of Microelectronic Systems, Technische Universität Darmstadt, Darmstadt, Germany, hinkelmann@mes.tu-darmstadt.de

Michael Hübner Institut für Technik der Informationsverarbeitung (ITIV), University of Karlsruhe, Karlsruhe, Germany, Huebner@itiv.uni-karlsruhe.de

Sorin A. Huss Integrated Circuits and Systems Lab, Computer Science Dept., Technische Universität Darmstadt, Darmstadt, Germany, huss@iss.tu-darmstadt.de

Tom Kamphans Department of Computer Science, Braunschweig University of Technology, Braunschweig, Germany, tom@kamphans.de

Götz Kappen Chair of Electrical Engineering and Computer Systems, RWTH Aachen University, Schinkelstr. 2, 52062 Aachen, Germany, kappen@eecs.rwth-aachen.de

Roland Kasper Institute of Mobile Systems, Otto-von-Guericke University Magdeburg, Magdeburg, Germany, roland.kasper@ovgu.de

Andreas Koch Embedded Systems and Applications Group, Dept. of Computer Science, Technische Universität Darmstadt, Darmstadt Germany, koch@esa.cs.tu-darmstadt.de

Dirk Koch Hardware/Software Co-Design, Department of Computer Science, University of Erlangen-Nuremberg, Erlangen, Germany, koch@cs.fau.de

Roman Koch Institute of Computer Engineering, University of Lübeck, Ratzeburger Allee 160, 23538 Lübeck, Germany, koch@iti.uni-luebeck.de

Tommy Kuhn Department of Computer Engineering, Wilhelm-Schickard-Institute for Computer Science, University of Tübingen, Tübingen, Germany, kuhn@informatik.uni-tuebingen.de

Sebastian Lange Parallel Computing and Complex Systems, Department of Computer Science, University of Leipzig, Leipzig, Germany, langes@informatik.uni-leipzig.de

Enno Lübbers Computer Engineering Group, Department of Computer Science, University of Paderborn, Paderborn, Germany, enno.luebbers@upb.de

Felix Madlener Integrated Circuits and Systems Lab, Computer Science Dept., Technische Universität Darmstadt, Darmstadt, Germany, madlener@iss.tu-darmstadt.de

Erik Maehle Institute of Computer Engineering, University of Lübeck, Ratzeburger Allee 160, 23538 Lübeck, Germany, maehle@iti.uni-luebeck.de

Mateusz Majer Hardware/Software Co-Design, Department of Computer Science, University of Erlangen-Nuremberg, Erlangen, Germany

Michael Meitinger Institute for Integrated Systems, Technische Universität München, Munich, Germany, michael.meitinger@tum.de

Renate Merker Technische Universität Dresden, Dresden, Germany, renate.merker@tu-dresden.de

Martin Middendorf Parallel Computing and Complex Systems, Department of Computer Science, University of Leipzig, Leipzig, Germany, middendorf@informatik.uni-leipzig.de

Norma Montealegre Heinz Nixdorf Institute, Paderborn, Germany, norma@uni-paderborn.de

K.D. Müller-Glaser Institut für Technik der Informationsverarbeitung (ITIV), University of Karlsruhe, Karlsruhe, Germany, Mueller-Glaser@itiv.uni-karlsruhe.de

Wolfgang Nebel Carl von Ossietzky University, Oldenburg, Germany, wolfgang.nebel@uni-oldenburg.de

Bernd Neumann Chair of Electrical Engineering and Computer Systems, RWTH Aachen University, Schinkelstr. 2, 52062 Aachen, Germany, neumann@eecs.rwth-aachen.de

Tobias G. Noll Chair of Electrical Engineering and Computer Systems, RWTH Aachen University, Schinkelstr. 2, 52062 Aachen, Germany, tgn@eecs.rwth-aachen.de

Rainer Ohlendorf Institute for Integrated Systems, Technische Universität München, Munich, Germany, rainer.ohlendorf@tum.de

Julio Oliveira Filho Department of Computer Engineering, Wilhelm-Schickard-Institute for Computer Science, University of Tübingen, Tübingen, Germany, oliveira@informatik.uni-tuebingen.de

Frank Oppenheimer OFFIS Institute for Information Technology, Oldenburg, Germany, oppenheimer@offis.de

Thilo Pionteck Institute of Computer Engineering, University of Lübeck, Ratzeburger Allee 160, 23538 Lübeck, Germany, pionteck@iti.uni-luebeck.de

Marco Platzner Computer Engineering Group, Department of Computer Science, University of Paderborn, Paderborn, Germany, platzner@upb.de

Franz J. Rammig Heinz Nixdorf Institute, Paderborn, Germany, franz@uni-paderborn.de

Felix Reimann Hardware/Software Co-Design, Department of Computer Science, University of Erlangen-Nuremberg, Erlangen, Germany, felix.reimann@cs.fau.de

Wolfgang Rosenstiel Department of Computer Engineering, Wilhelm-Schickard-Institute for Computer Science, University of Tübingen, Tübingen, Germany, rosenstiel@informatik.uni-tuebingen.de

Markus Rullmann Technische Universität Dresden, Dresden, Germany, markus.rullmann@gmx.de

Andreas Schallenberg Carl von Ossietzky University, Oldenburg, Germany, andreas.schallenberg@uni-oldenburg.de

Jochen Schleifer Chair of Electrical Engineering and Computer Systems, RWTH Aachen University, Schinkelstr. 2, 52062 Aachen, Germany, schleifer@eecs.rwth-aachen.de

Tobias Schwalb Institut für Technik der Informationsverarbeitung (ITIV), University of Karlsruhe, Karlsruhe, Germany, Schwalb@itiv.uni-karlsruhe.de

Nils Schweer Department of Computer Science, Braunschweig University of Technology, Braunschweig, Germany, n.schweer@tu-bs.de

Thomas Schweizer Department of Computer Engineering, Wilhelm-Schickard-Institute for Computer Science, University of Tübingen, Tübingen, Germany, tschweiz@informatik.uni-tuebingen.de

Walter Stechele Institute for Integrated Systems, Technische Universität München, Munich, Germany, Walter.Stechele@tum.de

Thilo Streichert Daimler AG, Group Research & Advanced Engineering, Sindelfingen, Germany, thilo.streichert@daimler.com

Marc Stöttinger Integrated Circuits and Systems Lab, Computer Science Dept., Technische Universität Darmstadt, Darmstadt, Germany, stoettinger@iss.tu-darmstadt.de

Jürgen Teich Hardware/Software Co-Design, Department of Computer Science, University of Erlangen-Nuremberg, Erlangen, Germany, teich@informatik.uni-erlangen.de

Christopher Tessars Department of Computer Science, Braunschweig University of Technology, Braunschweig, Germany,

Alexander Thomas ITIV, Universität Karlsruhe, Karlsruhe, Germany, thomas@itiv.uka.de

Steffen Toscher Institute of Mobile Systems, Otto-von-Guericke University Magdeburg, Magdeburg, Germany,

Michael Ullmann Institut für Technik der Informationsverarbeitung (ITIV), University of Karlsruhe, Karlsruhe, Germany, Ullmann@itiv.uni-karlsruhe.de

Jan C. van der Veen Department of Computer Science, Braunschweig University of Technology, Braunschweig, Germany, j.van-der-veen@tu-bs.de

Timo Vogt Microelectronic Systems Design Research Group, University of Kaiserslautern, 67663 Kaiserslautern, Germany, vogt@eit.uni-kl.de

Thorsten von Sydow Chair of Electrical Engineering and Computer Systems, RWTH Aachen University, Schinkelstr. 2, 52062 Aachen, Germany, sydow@eecs.rwth-aachen.de

Norbert Wehn Microelectronic Systems Design Research Group, University of Kaiserslautern, 67663 Kaiserslautern, Germany, wehn@eit.uni-kl.de

Thomas Wild Institute for Integrated Systems, Technische Universität München, Munich, Germany, thomas.wild@tum.de

Peter Zipf Institute of Microelectronic Systems, Technische Universität Darmstadt, Darmstadt, Germany, zipf@mes.tu-darmstadt.de

Part I
Architectures

Chapter 1
Development and Synthesis of Adaptive Multi-grained Reconfigurable Hardware Architecture for Dynamic Function Patterns

Alexander Thomas and Jürgen Becker

Abstract Since the 1980s many reconfigurable architectures have been developed and explored by the industrial and scientific community. However as long as Moore's Law has its validity each technology step forces the developers to rethink their approaches and even more to develop new ideas to be able to keep up with current technological challenges. The main goal of this project was the exploration of existing architectures and development of new ideas and concepts to increase the functional density and flexibility resulting in the new approach, the HoneyComb architecture.

1.1 Introduction

Nowadays, state-of-the-art architectures are very impressive and powerful. Current CPUs with multiple cores on a die are able to reach over 200 GFlops by using single precision arithmetic [12]. In the field of GPUs (graphics processing unit) we already crossed the 1TFlops barrier [10]. The integer performance is equally increased which is more interesting for this project. However, the real throughput of architectures compared to theoretical performance is highly application dependent. Nevertheless, the question arises how this performance could be reached. What are the trade-offs and where are the limits? Are there any alternatives?

The IT industry benefits a lot from Moore's Law [9]. It predicts every 18 month the transistor number is doubled. Counting on this certainty the developer can build even more sophisticated systems and trust in the fact that his technology limit is increased after every 18 month. But what has changed in the last 20 years? Did we get new architectural approaches? No, we still have von-Neumann and Harvard concepts working in our computers and mobile devices. Many dedicated units (ASICs)

Alexander Thomas · Jürgen Becker
ITIV, Universität Karlsruhe, Karlsruhe, Germany, e-mails: `thomas@itiv.uka.de`, `becker@kit.edu`

M. Platzner et al. (eds.), *Dynamically Reconfigurable Systems*,
DOI 10.1007/978-90-481-3485-4_1, © Springer Science+Business Media B.V. 2010

are used to increase the performance and extend the battery life for mobile devices but still there are no really new approaches to find.

Until the year 2004 extra transistors have been used to optimize the execution flow of the processors. Techniques like branch prediction or out-of-order-execution have been developed. All this with the intention to increase the performance but with the prize of many additional gates. Another factor was the maximum frequency which has been raised until technological limits stopped this development and forced the companies to rethink their strategies. Nowadays, multi core solutions dominate the market. Since energy is more expensive the customer appreciates low power solutions and forces the companies to keep the frequencies lower until the next technology step is reached.

However, the overhead within the CPUs gathered in the last 30 years is still there and prevents efficient calculation performance. The best performance compared relatively to area/gate count/power consumption is still given by dedicated ASICs. But the resulting flexibility is very low. An alternative here is given by reconfigurable architectures which promise almost ASIC-like performance and low power consumption with the flexibility of a CPU.

Several architectural concepts are available for reconfigurable systems. One concept is the extension of the von-Neumann or Harvard approaches by introducing modifiable instruction sets. At runtime the programmer could adapt some of the instructions to increase the execution efficiency for a given application. However the memory bottlenecks related to CPUs is also kept in this approach. This limitation can be eliminated by another approach: array-based dynamic reconfigurable architectures. This architecture type consists of a large amount of processing elements which are arranged in a two-dimensional field. The big number of processing units allows very high parallelism and the resulting architecture offers power efficient processing by reducing the voltage and frequency requirements. The problem is however the programmability. Since array-based architectures lack almost completely direct memory access (DMA), the application programmer has to rethink his techniques.

Array-based architectures are divided into two main classes: fine-grained and coarse-grained approaches. Fine-grained architectures like FPGAs [2, 18] are bit logic based and very flexible. Nevertheless, the programming is quite difficult and requires some experience in hardware design. Furthermore, the hardware overhead for bit level reconfigurability is very high what can be seen in the resulting power consumption and design size limitations. The hardware overhead here is defined by the number of required registers or SRAM bits per function.

Coarse-grained array-based architectures are not well explored in comparison to FPGAs. So, mostly scientific approaches are available in this field [4, 7, 11]. Only a few architectural exceptions are available for commercial use [19]. Here, the processing elements are made up of arithmetic logic units (ALUs). The configuration granularity of the communication network is a data vector which reduces the reconfiguration overhead tremendously.

This contribution describes the work of the last six years resulting in a new array-based dynamically reconfigurable architecture, the HoneyComb architecture. The main goal of this project was the combination of advantages of both array-based reconfigurable classes in one architecture: small reconfiguration overhead and flexibility given by bit logic. Furthermore, a set of new features had to be developed to increase the flexibility of the resulting approach, which are described in the following sections. Besides the hardware implementation of the architecture, a set of software tools have been developed to ease the programmability and debugging.

1.2 HoneyComb Architecture

1.2.1 Architectural Considerations

In the first step in the development of a new architecture the determination of a set of known problems is necessary. This has been done e.g. by evaluation of existing architectures that were available to the authors like Xilinx FPGAs, XPP. The following main limitations have been determined which require particular attention:

1. predetermined data types (both)
2. static configurations (coarse-grained approaches)
3. long reconfiguration time (fine-grained approaches)
4. precompiled configuration shapes (both)
5. limited flexibility regarding array fragmentation (both)
6. limited reach ability within the array (both)
7. limited I/O functions (coarse-grained approaches)
8. no power saving support (both)
9. no adaptation regarding application requirements (coarse-grained approaches)

The first point poses some kind of a conflict regarding the array-based architecture classification. Nevertheless, the combination of fine-grained and coarse-grained data types sounds very interesting. The XPP architecture already contains a fine-grained event network for transportation of control data. But the XPP-developers stopped just there and didn't develop the idea any further. This development had to be continued by integration of a more flexible control network with support for a complete set of bit functions and a way to cross the data type domains. By doing so the second point seems to be already solved. It represents the fact that within a configuration all functional units get precompiled operation assignments and are kept static within this configuration. This operation stays the same as long as the configuration is active. But in combination with our flexible control network fine-grained signals can be used to influence the functional units directly. If a list of possible operations is given the fine-grained signals can be simply used to select the next operation and its input and output registers. Dynamically preselected flags (carry, sign, etc.) can be sent back to the fine-grained domain and influence the programmed control logic (FSM, transitional logic). This multi-context technique is quite similar to the way the CPUs are functioning [6].

The third point targets the reconfiguration time of reconfigurable architectures. Usually, all fine-grained architectures (FPGAs) share the same dilemma: Bit level reconfiguration leads to high amount of reconfiguration data. The more reconfiguration data is required to describe an application the longer the reconfiguration will take. It is quite obvious that high reconfiguration time limits the reusability of flexible hardware even more if real time applications have to be executed. High reconfiguration time limits the application of configuration sequencing [3]. Furthermore, most architectures use one single dedicated configuration port to transfer the configuration data to the array. This has to do with the fact that mostly dedicated networks are used for reconfiguration purposes. More ports would lead to higher costs which is considered a waste of resources in this case. Well, to solve both problems from above a shared network for data and configurations has been suggested. This would lead to much higher available bandwidth during reconfiguration and additional resources spent for reconfiguration can be used for application data transfer. Both data types would benefit from this unification and additional resources can be used for both transfer types to speed up the reconfiguration process and to gain additional flexibility for the application routing.

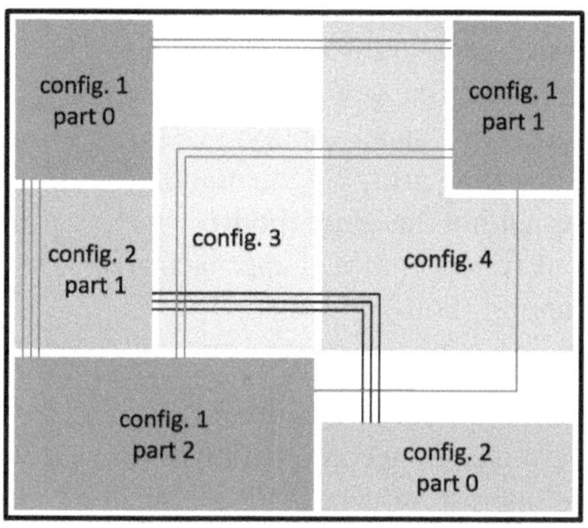

Fig. 1.1 Array fragmentation conflicts: Overlapping communication lines and predefined configuration shapes. Remaining area cannot be used because of not available precompiled configurations.

The next two points target the runtime flexibility on the array level. All known architectural approaches have the same problem regarding configuration placement and routing: the configurations are precompiled and predetermine every detail considering configuration shape and routing. For example reshaping of a configuration structure at runtime is just impossible in case of FPGAs. This would lead inevitable to timing problems and cripple the configuration. A required step would be the re-synthesis of the application which is surely not a runtime feature. In case of coarse-grained architectures reshaping is much simpler since no timing aspects have to be considered. But still recompilation is necessary as well. Point five refers to another limitation for both architecture types. If partially overlapping of configurations is required, maybe because of array fragmentation see Fig. 1.1, regarded applications have to be aware of each other. Otherwise communication lines could be disrupted and the applications crippled. This problem would require precompile considerations, unless the hardware offers support. Our suggestion is online-adaptive hardware-routing. A feature which could establish communication streams for desired output and input ports at cell level during runtime. This feature would lead to flexibility on many levels and solve both problems at once.

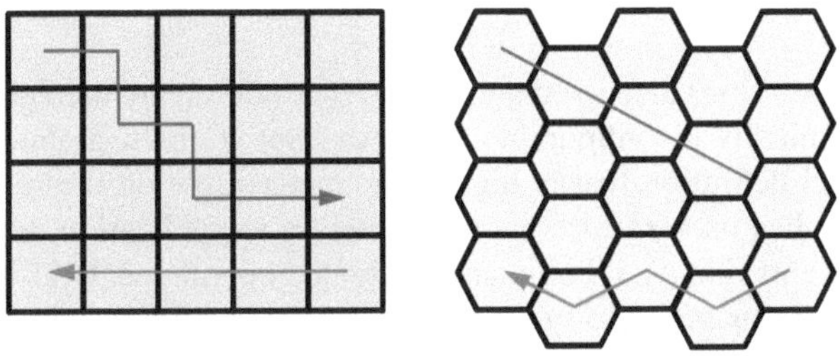

Fig. 1.2 (Color online) Hexagonal shapes reduce resource requirements in diagonal directions: Blue route requires 33% less hops. Vertical and horizontal routings keep their reach ability.

Problem six is obvious and quite simple to solve but yet not faced by prior approaches: rectangular cell structures lead to an overhead for the routing network. Sure, horizontal and vertical routings are no problem. But if an aberration to the ideal routing directions is required additional resources will be wasted. The worst case direction, diagonal routing, leads to increased resource requirements compared to cells with more than four neighbors [14], see Fig. 1.2. Estimations showed that cell shapes with six or eight neighbors are quite promising and reduce the resource requirements by 40% at best. A fact which led to the decision of the six neighbor shaped cells (hexagonal) and gave the new architecture the name HoneyComb.

Point seven refers to the necessity of an adequate I/O interface. In case of FPGAs usually enough resources are available to implement required interface protocols and access patterns. However, in case of coarse-grained architectures using array resources for I/O issues is a waste of resources and should be implemented within a dedicated unit. This units would require less area and energy and allow sophisticated and fast protocols.

Nevertheless the reconfigurable architectures are supposed to be energy efficient. Current approaches lack additional power saving techniques to disable idle cells for example. Nowadays VLSI technologies, such as from TSMC Inc. [17], offer a few ways to reach that goal. Some interesting techniques are: clock gating, voltage gating and voltage scaling. For the HoneyComb architecture decision has been made to stick to clock gating at cell and routing level. Other techniques require special synthesis and layout considerations and will lead to increased area requirements in the backend.

Point nine regards a particular problem of coarse-grained reconfigurable architectures: the application dependency. The more flexibility an architecture offers the more area is required to integrate it on a chip. Even though reconfigurable architectures are meant to be flexible it is still a good idea and in most cases reliable to limit this kind of flexibility at least to a set of supported applications. Therefore an appropriate application-tailored method for reduction and configuration of the RTL model is required.

1.2.2 HoneyComb Overview

Based on considerations discussed above the HoneyComb architecture has been designed: a dynamically reconfigurable runtime-adaptive multi-grained architecture. A VHDL-model definition that is highly parameterizable on the register-transfer-level (RTL) implies its hardware-based adaptation to application demands. Application requirements like ALU count and integrated operations, LUT count and size, available connections between functional units, etc. can be satisfied. Furthermore, the new architecture introduces a set of new features with intention to improve the flexibility and usability of array-based architectures. The most important features are the online adaptive routing technique, multi-grained routing support, multi-grained data paths, multi-context data paths and programmable I/O interfaces. All data transports within the array are fully synchronized and are handled by a synchronous handshake protocol to keep data consistency, see Fig. 1.3.

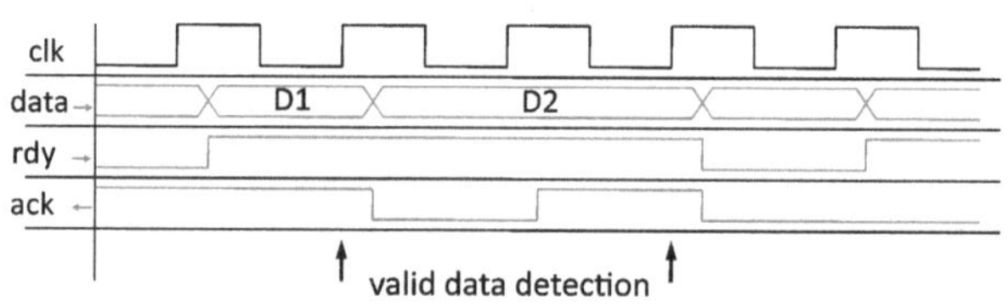

Fig. 1.3 Synchronous handshake protocol: Valid data detection if rdy and ack signals are asserted.

The complete array is made up of three different cell types. For data calculations the data path cells (DPHC) are introduced. These cells are developed to face the challenges of control-based applications in array-based systems. This is accomplished by integrating multi-grained and multi-context features so that the final data path efficiency is raised considerably. Another cell type is the memory cell (MEMHC), which offers memory space inside the HoneyComb array, so that the available memory bandwidth can be locally increased and optimized. Therefore the memory cell offers configurable memory organizations such as random access memory (RAM), first-in-first-out (FIFO) or last-in-first-out (LIFO). The third and last cell type integrates the input/output interfaces (IOHC). The I/O interfaces have been improved by implementing optimized μControllers to provide programmable interface behavior and to enable so-called configuration-sequencing.

Each cell type has an identical composition of basic blocks consisting of a routing unit (RU) and a functional module (FU) [16]. By characterizing the functional module in the RTL specification, the user can decide which cell type he wishes to implement. To ease the array parameterization a MS Excel based application has been developed, which includes the consistency checking and configuration file generation for the VHDL model as well as the software tools, which will be introduced later. The following sections describe the online adaptive routing and the three cell types in detail.

1.2.3 Communication Network and Online Adaptive Routing Technique

Basically, the global communication network of the HoneyComb array consists of all available routing units and the buffered multi-grained interconnect links in between. Therefore, each routing unit uses three different link types to reach the six possible neighbors according to the hexagonal cell shape. Additionally, all routing units within a cell are connected to the cell specific functional modules. Since one of the main design decisions was to support point-to-point connections, the main goal in this topology is the establishment of communication streams between the outgoing ports of the given functional module and the input ports of the target functional module. The reason for this decision is the advantage of the guaranteed bandwidth for the communication, which could not be reached by using packet-based communication.

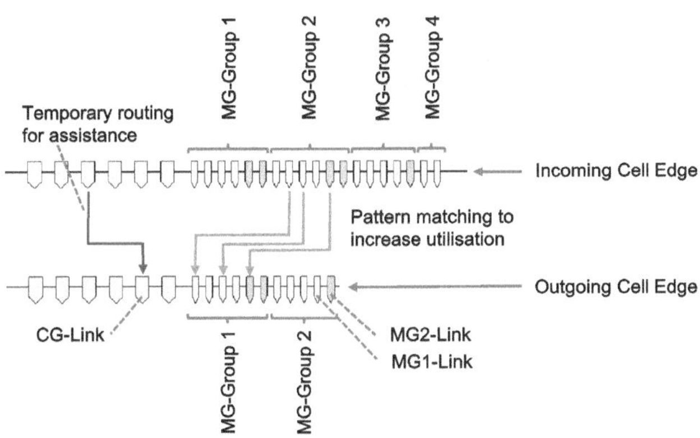

Fig. 1.4 Multi-grained routing technique: Pattern matching increases utilization/required CG routing for current matching pattern transport and source/destination port specification.

The communication network implies three different link granularities, see Fig. 1.4. The first and main link type is the coarse-grained link. In the current version its vector width is 32 bits. This link transports application data as well as configuration data.

The other two multi-grained types (MG1 and MG2) transport fine- and multi-grained application data by composing the desired vector width through a composition of these two link types. Therefore these link types have RTL parameterizable vector widths between $n = 1$ to N for MG1 links and $m = 1$ to M for MG2 links. Preferably, these links should be configured in such a way any vector width between 1 to K bits can be composed. Additionally, a predefined number of given MG1 and MG2 links are grouped in so called multi-grained groups, which finally define the maximum size of multi-grained links. For example, a given MG1 size of $n = 1$ bit and MG2 size of $m = 2$ bits and a group size of 4 links each would allow a routable multi-grained vector of 12 bits $=$ 4 links $\times$ 1 bit $+$ 4 links $\times$ 2 bits. If the given MG vector is partially used, it is still possible to use the remaining multi-grained sub-links to transport multi-grained data. However, this is only possible if the required vector width does not exceed the width of all remaining multi-grained links. To increase the flexibility of this concept, each cell edge can contain several multi-grained groups. The exact number can be configured at RTL by adjusting the given model parameters.

The main advantage of this communication topology is the ability of the hardware to establish a communication stream during runtime. The architecture is able to handle this autonomously, since each routing unit contains the logic and information about its position and the coordinate system of the array. By knowing the destination for a given routing process, each routing unit is capable to calculate the forward direction or to establish a connection to its own functional module if the destination is already reached. The information whether a routing process is about to start or not, is encapsulated in predefined routing instructions and can be generated for coarse-grained streams as well as for multi-grained links. To influence the quality of the resulting streams, several parameters are defined: optimum path bit, fast path counters or configuration mode bit. The optimum path bit forces the hardware to find the shortest path between two cells, which can be easily calculated by considering the current position and the destination. The fast path counter specifies the distance of the used buffers (in hops) between source and destination cell.

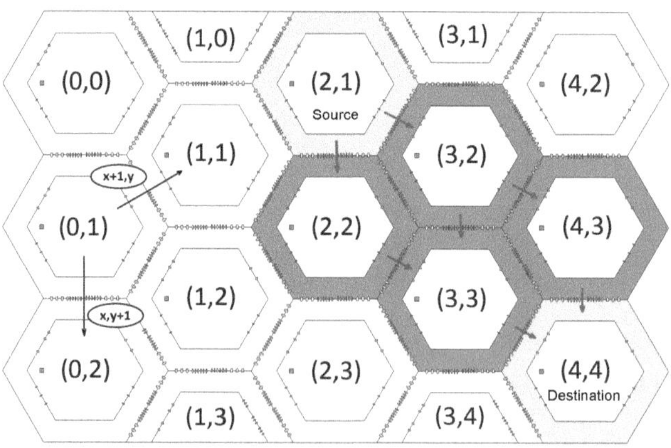

Fig. 1.5 (Color online) HoneyComb coordinate system: black darts indicate main axis of coordinates. Red darts show possible routes for the shortest paths between source and destination cells (yellow).

Basically, the routing technique uses the depth-first-search approach and is based on the given coordinate system, see Fig. 1.5. Thereby, each cell calculates the next direction based on its own position, the destination positions and considering local link utilization for each direction. If the routing process ends up in a dead end, a backtracking algorithm is used to return to the previous cell. This cell blocks the already rejected direction, recalculates the routing and tries another neighbor. If there is no path between source and destination available, the source cell generates an error which can be determined by the host system. The routing algorithm requires three cycles to calculate and establish the routing for each cell hop. In case of backtracking usage an additional clock cycle is required. Once an established stream is not needed anymore an end packet instruction (EPI) is used to terminate the stream and release all used resources.

Coarse-grained streams are used for three different purposes: coarse-grained data stream routing, multi-grained data stream routing and configuration stream routing. The first two purposes are quite obvious and are used for stream establishment and the following application data transfer. In case of the multi-grained data stream routing the coarse-grained stream is only used during the routing process to transport the routing instructions. Once the multi-grained stream has been established the coarse-grained stream will be deleted automatically and the used resource freed.

The purpose of the configuration stream routing is the transport of the configuration data from I/O cells to destination cells. The configuration data can be configuration code for the destination functional module or another routing instructions for the destination routing unit. Since the functional module code and routing instructions for the destination cell are following the configuration stream in the configuration process their destination cell is only specified by the routing instruction for the configuration stream. If the user wishes to move the position of the destination cell he has simply to adjust the coordinates within the routing instruction. This task can be carried out by the runtime system. Therefore the runtime system has to check all available routing instructions within a configuration and adapt the new positions. Any routing conflicts will be solved by the architecture at runtime. Using the online routing technique in hardware reduces the configuration size as well as the reconfiguration time tremendously.

To analyze the runtime progress a set of error messages has been defined. Each routing unit sends error messages to a global status controller so the host system will be noticed by an interrupt if any errors occur. The same controller can be used to deactivate specific cells in case a functional malfunction has been monitored. So, a kind of fault tolerance can be realized as well.

1.2.4 Datapath Cells (DPHC)

The specification of the functional module as a data path module results in a data path cell (DPHC), see Fig. 1.6. The data path structure is held quite complex in this architecture by implementing multiple look-up-tables (LUTs) and arithmetic logic units (ALUs) [15]. This way, the data path consists of two data type domains. The ALUs are part of the coarse-grained domain, whereas the LUTs are used within the fine-grained domain. The fine-grained signals are gained through the conversion of the incoming multi-grained streams, which are separated to single bit signals. Both domains include adequate registers to be able to store generated data, whereas each register is separately configurable to behave like a FIFO. Required depth value is configurable at RTL.

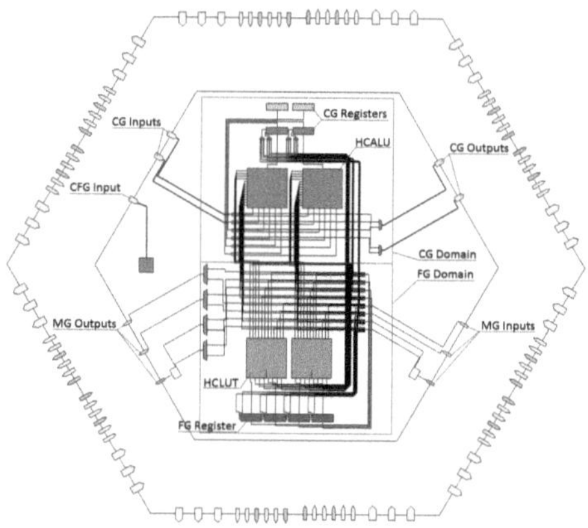

Fig. 1.6 Multi-grained multi-context Datapath HoneyComb Cell (DPHC).

The integrated ALU can be configured to include a context list, which contains alternative configuration parameters, and can be switched at runtime. All ALUs can also generate status flags, indicating the result's characteristics (zero flag, overflow flag, etc.), which could be sent to available LUTs. The addressing of the integrated context lists is done by a set of signals from the fine-grained domain. On the other hand, the combination of the LUTs and the fine-grained registers can be used to program a finite state machine (FSM) with input signals from ALUs, LUTs, registers or other external signals. The resulting structure is able to react on given application's situations and dynamically changes the behavior of the configuration. For more complex applications, control data exchange between cells is possible by using the multi-grained communication network. This feature allows the implementation of control-driven applications by using application-level control mechanisms distributed over several cells.

The complexity of the data path is highly adaptable at RTL by choosing the desired parameters of the VHDL specification. Doing so, the designer can specify the number of implemented ALUs and/or LUTs. An adequate number of registers for each domain can be adjusted as well. The interconnect between all these units is also parameterizable. So, the complete parameter set enfolds about 60,000 single parameters. To master this complexity a new super-configuration method has been developed, see Sect. 1.3.5.

The configurable arithmetic operations within the ALUs are preferably limited to 32 bit integer or fixed-point operations. But it is also possible to use 32 bit single precision floating point operations.

1.2.5 Memory Cells (MEMHC)

Memory cells are composed of memory modules and a coarse-grained register set. The purpose of those modules is to offer memory space within the array, without the necessity to access external system memory. By using the local memory the configurations obtain the complete available memory bandwidth and the application can be processed at full performance. The available memory modes are RAM, FIFO and LIFO. The Fig. 1.7 depicts the structure of the complete MEMHC module.

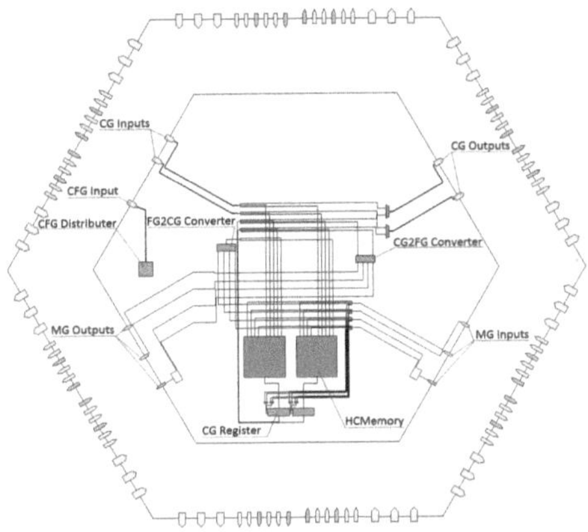

Fig. 1.7 Multi-grained Memory HoneyComb Cell (MEMHC) with flexible memory modules.

Each memory module in the MEMHC is separately usable. If larger memories are required, it is possible to connect two or more modules to a larger logical module. So, applications can use larger memories than offered by a single memory module. Primarily, the memory is capable to store coarse-grained data. By leaving some bits unused, the conversion of the incoming multi-grained vectors to coarse-grained data allows the storage of this kind of data as well. Therefore fine-grained to coarse-grained converters (FG2CG) and coarse-grained to fine-grained converter (CG2FG) are available.

1.2.6 Input/Output Cells (IOHC)

The interfacing of the HoneyComb array to the system is realized by the input/output cells. The core component of these cells is a μController, which is meant to handle interface issues, see Fig. 1.8. The μController is connected to the array by using dual-clocked FIFOs, which are used to separate the system clock domain from the inner-array clock domain. It is possible to slow down the array frequency if the additional speed is not required.

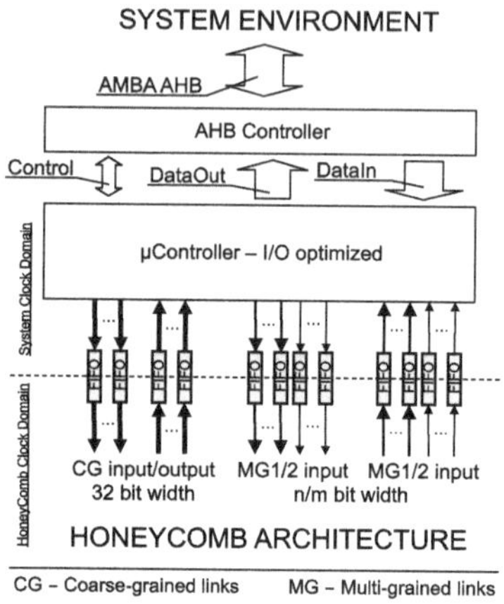

Fig. 1.8 Programmable Input/Output Cell (IOHC) optimized for block transfers with multi-context capabilities.

The data types of the FIFOs are divided into two groups and are related to the global communication type within the array. On one hand there are coarse-grained input/output FIFOs. On the other hand multi-grained input/output FIFOs are used to transport MG1/MG2 data vectors (see Sect. 1.2.3). From a system point of view the programmer has the option to use the packed or unpacked modes, which indicates the way the multi-grained data is stored in or has to be written to the system memory. It is quite obvious that the packed mode can increase the system memory utilization tremendously. Both data types can be evaluated by be IOHC internally instead of being written to the system memory and influence further processing behavior of the μController.

The host system interface can be realized by using the AMBA AHB backbone system. This version is interesting especially if HoneyComb is used as part of a System-on-a-Chip (SoC). In this case each IOHC implies master and slave interfaces to have a direct memory access (DMA) to the system memory as well as direct communication to other bus masters.

Is the HoneyComb architecture used on a single chip solution proprietary protocols should be used. For the final prototype of the HoneyComb architecture the internal interface of the IOHC to the system will be used. This interface uses similar hand shake protocol as described in Sect. 1.2.2. By additional buffering of the I/O signal the external/internal timing requirements can be easily met.

1.2.7 Power Saving Techniques

The dual clocked FIFOs of the IOHC allow the adjustment of the clock frequency. The decrease or increase of the array or I/O frequencies on demand could result in enormous power consumption decrease. If the highest speed mode is not necessary, the frequency can be adjusted to the lowest requirements level and save energy. Current synthesis results indicate the maximum frequency for the IOHC of 400 MHz, the array could run at about 200 Mhz. There is enough potential to slow down the architecture to reduce the power consumption if the maximum performance is not required.

In addition to frequency adjustments clock gating techniques are used the achieve further power savings. Therefore each cell is capable to disable clock networks on two levels. The first level regards the routing units. If no routing activity is determined the routing unit will be simply "clock gated". If the neighbor cell decides to route to this cell the clock will be activated and kept active until all activity is ceased.

The second clock gating level regards the functional modules and is not gated if incoming or outgoing data streams to or from the functional module are active. If the functional module is inactive the second level gate is enabled and the clock network is disabled.

1.3 Tool Support

Beyond the hardware design further work had to be spent into software tools development. First of all an adequate programming interface was required. Therefore, two programming languages have been defined. The first language is assembler based on the low-level architecture code (HoneyComb Assembler). The second language is more abstract and allows using of if-else-structures or for-loops for example (HoneyComb Language). Based on the defined language definitions the HoneyComb Assembler translator and the HoneyComb Language compiler have been developed.

Nevertheless the programming interface is very helpful further tools are required to make debugging possible. Online debugging on array-based architectures would be too expensive in terms of hardware resources. Since every gate is counting to keep the architecture small the design decision has been met to abandon online debugging. Instead the HoneyComb Viewer tool has been designed to visualize internal activities and allow offline debugging.

The HoneyComb architecture is a highly parameterizable model which is coded in VHDL. Since synthesis tools are not bug free and the parameterization causes some synthesis problems as well as slows down the simulation yet another tool (Hierarchy Generator) has been developed to break down the model according to the given parameter set.

The following sub sections describe the languages and the tools for the HoneyComb architecture shortly to give an idea of the functional range.

1.3.1 HoneyComb Assembler and the Hierarchical Programming Model

The HoneyComb assembler reflects closely the structure of the hierarchical HoneyComb architecture. The basic structure of the assembler will be explained here basically. Similar to the HoneyComb structure three completely different hierarchical sections or layers of the language are specified.

The first part (the transport layer) covers the IOHC programming. Data transfer optimized functions like block transfers with various access patterns etc. are defined here. Basic arithmetic and logical operations, program flow instructions, interrupt control and access to input/output FIFOs allow the programming of powerful control applications to influence the configuration sequencing in a user-defined way. The sequence of the configurations in the array are fully controlled by the IOHC. This is called transport layer.

The second part (the communication layer) concerns the routing network. Its main function is the establishment of configuration and data routing streams for coarse-grained as well as multi-grained data types. Therefore just four instructions with proper parameters are defined to control the routing network functions. Since the routing influences the placement on the array any changes to the configuration positions or shapes have to be made by changing this part of the assembler application.

The third part (the configuration layer) of the hierarchical assembler actually includes two parts which cover the programming of the functional modules of the data path cells (DPHC) and memory cells (MEMHC). The character of this code is very structural since low level modules like multiplexers, ALUs, LUTs, etc. have to be specified. Since the data path cells support multi-context functions not only structural considerations have to be made but also sequence of operations over time must be taken into account.

Based on the hierarchical definitions of the assembler the hierarchical programming model is shaped, see Fig. 1.9. The programmer on the assembler level must

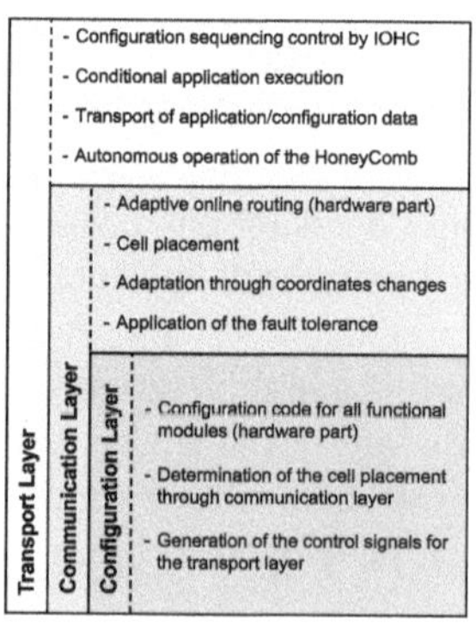

Fig. 1.9 HoneyComb programming model indicating dependencies of the hierarchical architecture structure.

consider the hierarchical influences of the programming model to be able to program the architecture.

The assembler translator is part of a tool conglomeration integrated into a MS Excel application which also includes Super Configuration Generator, Configuration Manager, Configuration Editor and an extended version of the Assembler Translator. The extended version of the Assembler translator is meant to compensate for minor changes to the architecture to keep the application compability.

1.3.2 HoneyComb Language (HCL) and Compiler

The definition of the HoneyComb Language has been inspired by VHDL. Similar to VHDL, HCL contains process definitions, parallel statements and input/output definitions. The process concept has been enhanced by introducing four different process types to satisfy the special needs: functional process, memory process, sub-configuration process and main process. The first two processes are connected to the DPHC and MEMHC and include definitions to target these cell types. The functional descriptions in these processes are highly parallel and cover the configuration layer.

The sub-configuration process is meant to define high-level functions by combining several cell processes or other sub-configuration definitions to larger logical units, including internal communication description and external port definitions. Doing so, it will become easier to handle the configuration sequencing just by addressing these logical names and ports.

The main processes describe the I/O behavior of the array. So, generated code from a main process is directly executed on the IOHCs. The purpose of the main processes is the transport of configurations for data paths and memory cells as well as transport of application data. It is also possible to program conditional transports to allow configuration sequencing (transport layer).

The purpose of HCL is the best possible programming of the architecture by using all available features. By using this language the user gets a way to formulate functional descriptions instead of using the also available native assembler language. Even if high level language support will be developed in the future work the HCL will be used as intermediate representation.

The HCL Compiler is actually a stand-alone application. It translates the HCL descriptions into HoneyComb Assembler descriptions. But still there are some limitations which require more attention in the future. For a fully functional application the compiler requires two operation modes. The first mode is working on a ideal HoneyComb architecture and maps the HCL descriptions in the best possible way considering the HCL application demands. The result is the best performance one can achieve for the given HCL application.

The second operation mode considers a given HoneyComb RTL-configuration and tries to map the HCL application onto it. The mapping success is not always guaranteed. If the mapping was a success the maximum reachable performance of the HCL application may be lower than planned by the programmer which is highly

RTL-configuration dependant. However, this operation mode is not implemented yet and need some attention in the future.

1.3.3 Debugging Tool—HoneyComb Viewer

As already mentioned a visualization tool, the HoneyComb Viewer (HCViewer), for debugging purposes has been developed. Based on the Qt-Framework and the OpenGL API a quite fast and flexible application has been created. An abstract class hierarchy of the HoneyComb architecture has been designed. Many classes have view capabilities and once instantiated they visualize predefined modules of the architecture. Modules like signal lines, ALUs, LUTs and registers with their own representations can use tooltip functions which allow the retrieval of specific debugging information.

To start the debugging work with the HCViewer at first the user has to load the RTL specific configuration file and configure the HCViewer environment. Next, the simulation data which has to be generated through simulation of the HoneyComb VHDL model is loading. Once the simulation data is processed the user can step through the simulation results and in best case verify the correct application execution. For HCViewer example see Fig. 1.12.

Since the complete HoneyComb architecture is coded in the abstract classes it is possible to extend the HCViewer to a fully functional simulator. Therefore the absent functions within the abstract classes have to be added.

1.3.4 Super-Configuration Generator, Configuration Editor and Configuration Manager

Since an ideal architecture approach is far too expensive it is mandatory to restrict the application flexibility to a set of given applications. In most real applications it is already the case that the amount of applications is limited and should be changed only rudimentarily. To extract the HoneyComb application requirements out of the given applications the Super-Configuration Generator has been developed. With this tool already verified and compiled applications can be analyzed and the least common denominator for the RTL specification extracted. The resulting three cell specification templates (DPHC, MEMHC, IOHC) include every component, every feature and every operation to make it possible to build a complete HoneyComb array and support the original applications.

Once the templates have been generated the Configuration Editor and Configuration Manager, see Fig. 1.10, can be used to make some user-defined modifications and create the final HoneyComb array. Final array configuration files for the VHDL model and HCViewer application can be generated and used for further exploration.

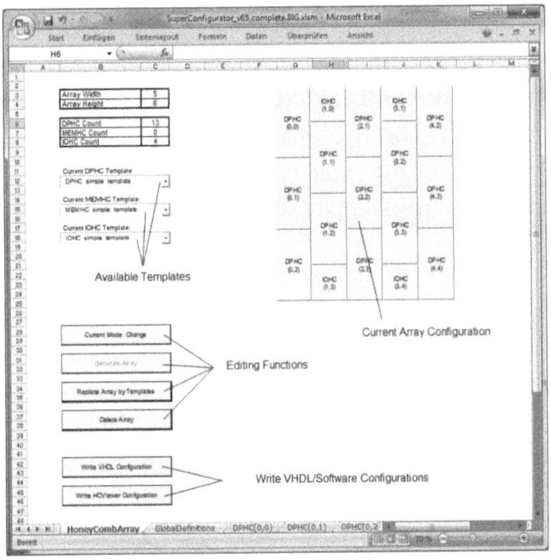

Fig. 1.10 HoneyComb RTL Configuration Manager including Configuration Editor, SuperConfiguration Generator and HC Assembler translator based on MS Excel application.

1.3.5 Hierarchy Generator

As already described above the VHDL model of the HoneyComb architecture is highly parameterizable. The level of the parameterization even impacts the simulation performance and cannot be tolerated for long simulation runs. The lack of support for null-arrays by the ModelSim simulator [8] prevents completely the compilation in this environment, so ActiveHDL simulator from Aldec Inc [1] needed to be used instead. Furthermore the work with synthesis tools also caused a few problems. Due to known bugs Synopsys synthesis tools [13] fail to analyze certain parameterized VHDL expressions which are still static and abort the synthesis process.

To solve these problems another tool, the Hierarchy Generator, has been developed. The main objective of this tools is to analyze the given VHDL model, break down every constant expression including loop unrolling, etc. and finally generate the static VHDL hierarchy.

Besides the fact that after the static hierarchy generation the synthesis tools worked quite well the simulation performance took a leap as expected. The simulation performance accelerated up to a factor ten which is quite impressive. Nevertheless, this acceleration is strong VHDL model dependant and cannot be reached with an average parameterized design.

1.3.6 Application and Synthesis Results

To prove the concept of this work four applications were selected: AES encryption algorithm, 1024-point FFT, recursive iMDCT and Wavelet algorithms. Based on these applications the complete tool chain (Fig. 1.11) has been worked through and the presented results were generated. For better comparison each application was analyzed separately and at the end the complete array with the support for all four applications was generated. The major design decision to stick to homogeneous array structure does not deliver the most cost-effective application dependant results. However, the resulting array allows to reshape and move configurations at maximum flexibility.

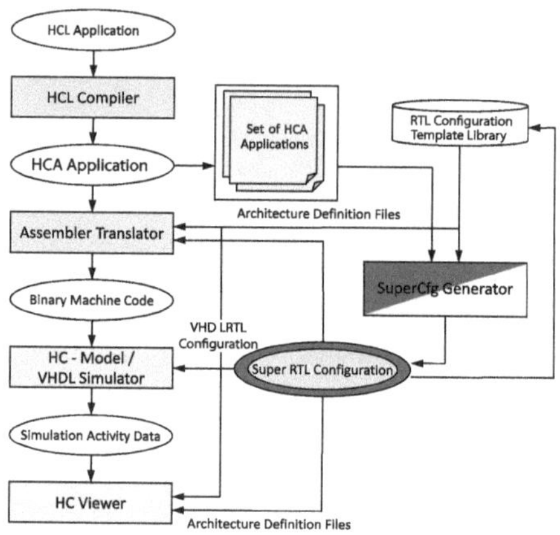

Fig. 1.11 The complete design flow for the applications as well as RTL configurations.

A detailed discussion of the applications cannot be done here so only a short description will be given. The AES algorithm is a block-based encryption algorithm which is working on 4×4 byte blocks. The mapping of this algorithm (Fig. 1.12) required every internal cell of the final prototype. Since the AES algorithm is working iteratively it is not important what the key length for the encryption is (possible values: 128, 192, 256). Once the loop is implemented it is simply a matter of control how often the loop is going to be executed. The loop is mapped as a deep pipeline structure. In addition to the calculation cells (DPHCs) in the pipeline both memory cells (MEMHCs) have been used not only to store the expanded key and further necessary look-up tables but also to extend the pipeline depth by offering FIFO functionality. This way the efficiency of the application could be optimized because more data (over 500 blocks) can be loaded into the pipeline at once so the loading/storing phase overhead can be minimized in comparison to complete calculation time.

The FFT mapping implementation is quite typical to a hardware approach. A butterfly is used for calculations, additional memory is required to store the intermediate results and control logic generates the read and write addresses. The complete application requires 4 DPHCs and 1 MEMHC, whereas the butterfly operation is mapped to one single DPHC and requires 2 cycles for one operation.

The Wavelet algorithm works on whole images to filter the high and low frequencies of the color dispersion. E.g. for the JPEG2000 standard this algorithm has

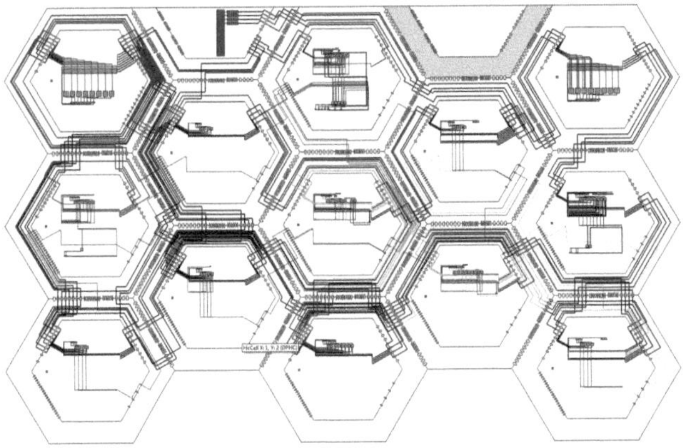

Fig. 1.12 AES256 application running on the final RTL structure of the HoneyComb prototype.

to be performed in 2D, horizontal and vertical. Nevertheless, the mapping on the HoneyComb array took in fact only two cells (DPHCs) to achieve the complete functionality. Because of tight timing constraints between these two cells additional FIFOs have been used to improve the performance.

Table 1.1 Application results regarding resource requirements, configuration time and maximum performance evaluated by using the ASIC prototype configuration limited to 11 DPHCs, 2 MEMHCs, 2IOHCs @ 100MHz.

Application	DPHCs	MEMHCs	Configuration time	Performance
AES256	11	2	6.85 μs	25.6 MB/sec
iMDCT 1 × finger	3	1	24.06 μs	47.6 blocks/sec
iMDCT 7 × fingers	11	2	25.60 μs	333.46 blocks/sec
FFT1024	4	1	7.65 μs	10850 blocks/sec
Wavelet	6	1	3.15 μs	0.6 cycles/pixel

The recursive iMDCT has been implemented according to Nikolajevic/Fettweis suggestion [iMDCT]. Therefore three data path cells are required: two cells for control and multiplexing logic and one cell for the iMDCT finger implementation. One additional memory cell is required for the cosine and spectral values. The performance evaluation was done for 1024 spectral values transformed back to 2048 sample values per block.

All four applications were optimized for integer or fixed-point calculations. The results were verified versus the outputs of the reference applications, which are programmed in C/C++. The final performance and synthesis results are given in Tables 1.1 and 1.2. The ASIC RTL configuration includes all four application resources and represents the prototype configuration. The dynamic power could not be determined for the prototype since the exact application scheduling is not known.

Table 1.2 Synthesis results: application dependent and for the final ASIC prototype.

Application	DPHC area (μm^2)	MEMHC area (μm^2)	IOHC area (μm^2)	Leakage power (mW)	Dynamic power (mW)
AES256	362636	1226638	623680	5.59	146.73
iMDCT 1 × finger	461697	1290972	812529	7.11	66.235
iMDCT 7 × fingers				–	180,19
FFT1024	472477	948950	787658	7.26	75,02
Wavelet	250042	1246197	728551	4.20	87,84
ASIC Prototype	652285	1299802	868313	10.76	–

By exploiting these four applications the decision for the RTL configuration of the prototyping chip was made. The final array configuration will contain 11 DPHCs, 2 MEMHC and 2 IOHCs. The TSMC 90 nm technology within the mini@sic program from Europractice [5] will be used. The maximum possible area of 4×4 mm^2 in this program offers a good opportunity to fabricate a prototype at low price. Current area results of about 11.5 mm^2 offer enough potential for generalization of the array and final layout optimizations.

1.4 Future Work

The most interesting topic for future work would be the extension of the application design flow to high level programming languages like C/C++. Until today all applications for the HoneyComb architecture have been developed in HoneyComb Language which required additional development efforts. A new application flow entry point from C/C++ would give an access to already available high level applications and reduce the application development efforts tremendously.

In this context the development of the HoneyComb language compiler to a full functional compiler is necessary. As already described the compiler is currently limited to compilations for ideal HoneyComb configurations. To ensure the reuse of already built HoneyComb prototypes with new applications the extension of this compiler to support the predefined RTL configurations is necessary.

The man power invested in this project is over six man years by now which covers all the developments described above. However by far not every detail has been analyzed and still much optimization potential has to be explored. In particular new technology features like voltage scaling and voltage gating need additional consideration in a cell based architecture.

Furthermore, the HoneyComb design offers a lot of potential for future optimizations as well. Starting with routing units which are currently multiplexer based and could be extended to general fat trees (GFTs) which are expected to result in a significant area reduction. Other optimization points are the operations within the ALUs which are currently not optimized and are surely worth a look.

For a practical application of this architecture an appropriate runtime environment is required. In cooperation with the ALADYN project (Chap. 12) and based on HoneyComb requirements many ideas have been discussed and developed. Since the implementation of the runtime environment could not been done within this project it remains an important challenge for the future work.

1.5 Conclusion

In the last six years of this project a new innovative highly adaptive dynamic reconfigurable architecture, the HoneyComb architecture, has been developed. New features like adaptive online routing, multi-context data paths, a multi-grained communication network/data paths and programmable I/O interfaces have been integrated and explored. The ability to adapt the RTL configuration to application requirements offers additional potential for area and power savings.

Acknowledgements This work was supported by DFG grant BE 2134/3, project *AMURHA*, as part of the Priority Programme 1148, *Reconfigurable Computing Systems*.

References

1. Activehdl, aldec, inc., http://www.aldec.com/
2. Altera corp., http://www.altera.com/
3. Becker, J., Vorbach, M.: Coarse-grain reconfigurable xpp devices for adaptive high-end mobile video-processing. In: SOC Conference, 2004. Proceedings. IEEE International, pp. 165–166 (2004). doi:10.1109/SOCC.2004.1362392
4. Becker, J., Pionteck, T., Glesner, M.: An application-tailored dynamically reconfigurable hardware architecture for digital baseband processing. In: Integrated Circuits and Systems Design, 2000. Proceedings. 13th Symposium on, pp. 341–346 (2000). doi:10.1109/SBCCI.2000.876052
5. Europractice ic service, http://www.europractice-ic.com/
6. Hennessy, J.L., Patterson, D.A.: Computer architecture. A quantitative approach. 4th edn. Morgan Kaufmann, San Mateo (2006)
7. Kress, R.: A fast reconfigurable alu for xputers. Ph. D. dissertation. Kaiserslautern University, (1996)
8. Modelsim, mentor graphics corp., http://www.mentor.com/
9. Moore, G.E.: Cramming more components onto integrated circuits, reprinted from electronics, vol. 38, nr. 8, April 19, 1965, pp. 114 ff. Solid-State Circuits Newsl., IEEE **20**(3), 33–35 (2006). doi:10.1109/N-SSC.2006.4785860
10. nvidia corp., http://www.nvidia.com/
11. Oppold, T., Schweizer, J.F.O.S.E.W.R.: Crc—concepts and evaluation of processor-like reconfigurable architectures. In: It—Information Technology, doi:10.1524/itit, 49. 3. 157, vol. 49 (2007)
12. Pham, D., Asano, S., Bolliger, M., Day, M., Hofstee, H., Johns, C., Kahle, J., Kameyama, A., Keaty, J., Masubuchi, Y., Riley, M., Shippy, D., Stasiak, D., Suzuoki, M., Wang, M., Warnock, J., Weitzel, S., Wendel, D., Yamazaki, T., Yazawa, K.: The design and implementation of a first-generation cell processor. In: Solid-State Circuits Conference, 2005. Di-

gest of Technical Papers. ISSCC. 2005 IEEE International, vol. 1, pp. 184–1851 (2005). doi:10.1109/ISSCC.2005.1493930
13. Synopsis, inc., http://www.synopsys.com/
14. Thomas, A., Becker, J.: Aufbau- und strukturkonzepte einer adaptiven multigranularen rekonfigurierbaren hardwarearchitektur. In: 17th International Conference on Architecture of Computing Systems (ARCS 2004) (2001)
15. Thomas, A., Becker, J.: Multi-grained reconfigurable datapath structures for online-adaptive reconfigurable hardware architectures. In: VLSI, 2005. Proceedings. IEEE Computer Society Annual Symposium on, pp. 118–123 (2005). doi:10.1109/ISVLSI.2005.51
16. Thomas, A., Becker, J.: New adaptive multi-grained hardware architecture for processing of dynamic function patterns (neue adaptive multi-granulare hardwarearchitektur). In: It—Information Technology, doi:10.1524/itit, 49. 3. 157, vol. 49 (2007)
17. Tsmc ltd., http://www.tsmc.com/
18. Xilinx inc., http://www.xilinx.com/
19. Xpp-iii processor overview (white paper), pact xpp technologies, http://www.pactxpp.com/

Chapter 2
Reconfigurable Components for Application-Specific Processor Architectures

Tobias G. Noll, Thorsten von Sydow, Bernd Neumann, Jochen Schleifer, Thomas Coenen, and Götz Kappen

Abstract Embedded Field Programmable Gate Arrays feature very attractive properties for their use as building blocks of future heterogeneous Systems-on-Chip. In this Chapter strategies are described how to achieve those reconfigurable System-on-Chip components with high energy and area efficiencies. One of the basic ideas is to leverage a-priori knowledge about the structures to be implemented on those arrays to optimize them for classes of applications. This leads to a parameterized target architecture which allows for dedicated adaptation to certain instances on a System-on-Chip. A design flow suitable in a research environment is described. For applications of digital arithmetic, e.g. in digital signal processing, it will be shown how this approach works. As an especially attractive application the use of those components as coprocessors to Application Specific Instruction Processors can be identified. Suitable architectures of such combinations and the resulting features are investigated.

2.1 Introduction

Today's implementation styles of digital computing and logic range from software programming of highly flexible general purpose processors (GPPs) over reconfiguration of field programmable gate arrays (FPGAs) down to the completely inflexible mapping on physically optimized dedicated circuits. A comparison of these implementation styles concerning the two top figures of merit energy efficiency and area efficiency yields in huge differences being as large as several orders of magnitude

Tobias G. Noll · Thorsten von Sydow · Bernd Neumann · Jochen Schleifer · Thomas Coenen · Götz Kappen
Chair of Electrical Engineering and Computer Systems, RWTH Aachen University, Schinkelstr. 2, 52062 Aachen, Germany, e-mails: `tgn@eecs.rwth-aachen.de`, `sydow@eecs.rwth-aachen.de`, `neumann@eecs.rwth-aachen.de`, `schleifer@eecs.rwth-aachen.de`, `coenen@eecs.rwth-aachen.de`, `kappen@eecs.rwth-aachen.de`

M. Platzner et al. (eds.), *Dynamically Reconfigurable Systems*,
DOI 10.1007/978-90-481-3485-4_2, © Springer Science+Business Media B.V. 2010

[26]. Figure 2.1 sketches the results (actually in mW/MOPS = nJ/operations and MOPS/mm^2, respectively) from a long term study performed by our group on the implementation of many applications applying all these different styles [4]. The diagram quantitatively proves which is called the "energy vs. flexibility conflict" or what Bob Brodersen from UCB Berkeley meant by his famous quote "flexibility has a price tag" [5].

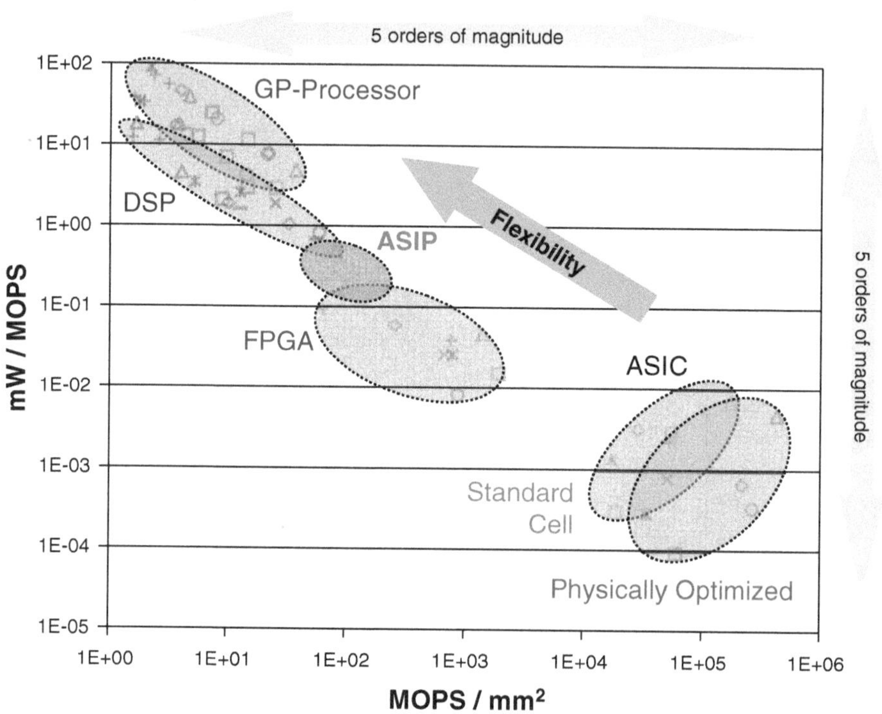

Fig. 2.1 Energy and area efficiency of today's implementation styles for digital computing and logic (all entries properly scaled to a 130-nm CMOS technology).

As can be seen from Fig. 2.1 these alternatives span more than five orders of magnitude in energy efficiency, what is crucial in practically speaking every application today, and in area efficiency, what is crucial as silicon is and will be "not for free". Apart from that, the most important observation from this comparison is that reconfigurable FPGAs feature an attractive compromise between flexibility and these efficiencies (see also [9, 7]). Another important conclusion from Fig. 2.1 is that for an energy and/or area critical System-on-Chip(SoC) each building block should be implemented in that style what is just allowing for the minimal degree of flexibility, resulting in the need for so-called heterogeneous SoC architectures. Consequently, one of the most challenging issues in today's SoC design is to predict or estimate the "right degree" of required flexibility during the specification phase.

A common false conclusion from that comparison is that reconfigurable implementations consume factor of tens or even hundreds of the power and area of dedicated ASIC block implementations without taking flexibility into consideration. But obviously, when it is possible to "freeze" different versions of structures during the specification phase of a SoC tens or even hundreds of them can be implemented as dedicated ASIC blocks on the area occupied by an equivalent reconfigurable block. The potential powering down of all unused blocks allows for a huge increase in energy efficiency. So, in those cases one should not even think about to use more flexible building blocks. Naturally except from multiplexing between these ASIC blocks (i.e. "frozen" block specifications) the flexibility of such an implementation is practically speaking zero.

From the considerations made so far it immediately follows that it would be attractive to have reconfigurable building blocks, so-called embedded FPGAs (eFPGAs), available on-SoC. Moreover that, as the use of building blocks on a SoC is restricted to a very limited class of application, the question arises whether it is possible to trade that a-priori knowledge about the structures to be mapped on those eFPGAs to a further improvement of their energy and/or area efficiency. As this is very similar to the initial idea leading to application specific instruction set processors (ASIPs) a consequent application of eFPGAs is their use as coprocessors to ASIPs leading to so-called reconfigurable ASIPs (rASIPs) and allowing for in-field reconfiguration of the ASIP's instruction set architecture (ISA) [31]. No matter how, as stand-alone building block of a SoC or as coprocessor to an ASIP, since arithmetic structures inherently feature special advantageous properties like iterativity and inherent high locality, eFPGAs appear to be especially attractive as an implementation platform for number crunching arithmetic e.g. in digital signal processing [22].

This chapter reports on efforts in and results of our research towards parameterized eFPGA architectures allowing for adaptation and optimization to different application domains. It will be shown that application class specific eFPGAs indeed close the efficiency gap between conventional FPGA implementations and dedicated ASIC blocks as ASIPs do for the gap between software programmable processors and FPGAs (Fig. 2.1). One of the main challenges results from the fact that a very wide variety of aspects on application, architecture, circuit, and layout level but as well as design and especially mapping aspects have to be considered. The chapter is organized as follows: Sect. 2.2 sketches the structure and features of a parameterized eFPGA architecture. The physical implementation of application class specific eFPGAs is described in Sect. 2.3. Section 2.4 is devoted to the Mapping and Configuration of structures of digital arithmetic onto this architecture. Application examples and results are presented in Sects. 2.5 and 2.6. Finally a short conclusion will be drawn.

2.2 Parameterized eFPGA Target Architecture

In general reconfigurable building blocks are to be distinguished into coarse grain and fine grain architectures (see e.g. [17]). While coarse grain reconfigurable architectures like [28, 21, 16, 30] consist of an array of (very) small and simple programmable components, and therefore in principle suffer from the same disadvantages, fine grain reconfigurable architectures, like commercial FPGAs [18], are built from an array of look-up-table (LUT)-based logic elements (LEs) [3, 10]. Consequently eFPGAs belong to the class of fine grain reconfigurable architectures.

The basic idea of application class specific eFPGAs is to optimize or at least to increase their energy and area efficiency for the structures appearing frequently in this class without sacrifice of flexibility [23, 33, 24]. I.e. any other structure can be mapped onto it, too, but that could result in a worse utilization and efficiency. An obvious initial idea towards application class specific eFPGAs is to optimize the functionality of the LEs according to the most frequent operations at logic level in the given class. In [17] research on optimizing LE core logic for arithmetic applications (e.g. Viterbi decoders) in digital signal processing is described. But having a look on the typical power breakdown of today's commercial FPGAs featuring "conventional" architectures (Fig. 2.2) reveals that the major part of their power dissipation is burned for communication (interconnect between the LEs) and in the clock system (clock generation, distribution, and registers). One reason for this is that conventional FPGAs are optimized for a very wide variety of structures to be mapped on it, even very irregular (e.g. control oriented) structures with very long interconnects (requiring many so-called segmented global routing resources).

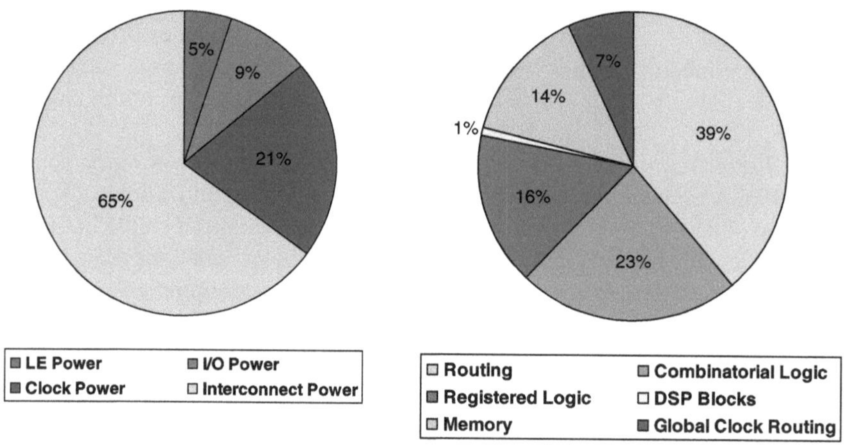

Fig. 2.2 Typical power breakdowns of commercial FPGAs (a: XC4003, [15] and b: Altera [1]).

Of course reducing the area of LEs by customizing it for an application class reduces the interconnect overhead, too, but the total gain to be expected is rather limited. So, in order to attain significant improvements in efficiency the LE and the communication structure have to be optimized according to the typical needs of the given class. In the following that is explained for the important and exemplary class of arithmetic structures as being frequently used in digital signal processing and other applications of digital computing. Similar optimizations can be done e.g. for the application class of protocol processing etc. The most important characteristics of digital arithmetic structures are, that their data or signal flow graphs inherently:

- contain only a limited set of simple bit level basic operations, e.g. gated full adders, comparators, shifters, etc.,
- are iterative in time and functionality, leading to highly regular 2D bit-/function-slice structures, and
- feature a localized connectivity, i.e. in comparison to random or control oriented logic mostly nearest neighbour interconnects, very few medium length interconnects, and rather few global "broadcast" interconnects (control lines etc.).

Without going into detail of conventional FPGA architectures here (refer to [3, 10]) we start from a hierarchical architecture with two-dimensional LE clusters (Fig. 2.3). Obviously this architecture fits well to typical data and signal flow graphs of digital arithmetic and, as will be seen, allows for exploiting the regularity features of it. The dimensions of the 2D LE cluster are part of the (many) parameters allowing for careful adaptation/optimization of this architecture to the requirements of an actual instance of the eFPGA on a SoC. The same is true for the size of the LUTs inside the LEs. As the most frequent basic building block of digital arithmetic is a full adder with some Boolean pre- and/or post-processing a LUT with four inputs (LUT-4) and one or two outputs seems to be a good starting point.

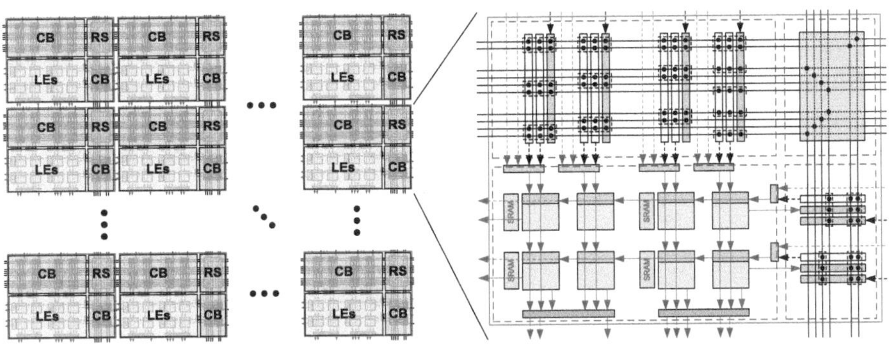

Fig. 2.3 eFPGA target architecture with two-dimensional LE arrays, local interconnect resources, and shared configuration storage components; array dimensions, numbers of interconnect (i.e. bus widths), cross bar switch population, etc. are subject to parameterization.

Like in conventional FPGA architectures the LE arrays are connected to global interconnect lines by so-called connection boxes (CBs) and the communication on this global interconnect lines is controlled by so-called routing switches (RSs), too. As has been already mentioned, with regard to optimizing efficiency, special attention has to be paid to this communication resources.

Figure 2.4 sketches the schematic of a routing switches. The horizontal and vertical bus widths, the available switching point numbers, their positions as well as their connectivity (i.e. the possible ways and directions of interconnect) are parameterized in order to allow for adaptation/optimization to the actual application class. In [23] we have shown that it is not necessary to provide a fully populated routing switch to achieve a good amount of flexibility. In addition, the segmentation of the interconnect can be adjusted by assigning each routing track a certain segment length (corresponding to the number of routing switches that are bypassed before the line connects to the next routing switch).

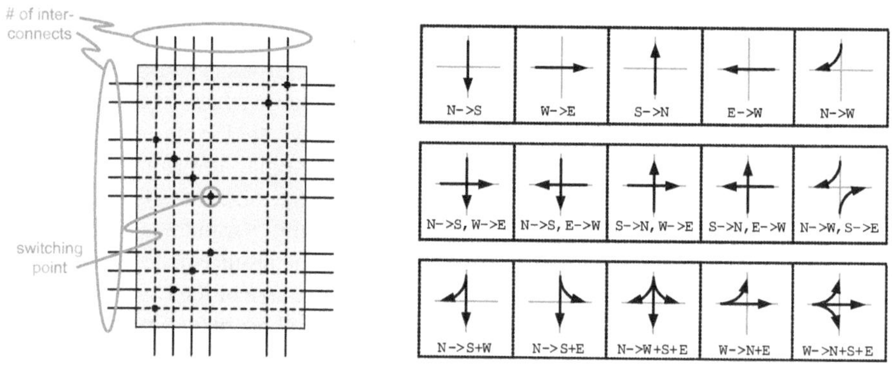

Fig. 2.4 Details of the parameterized routing switches and table of connectivity variants for a single switching point.

Connection box connectivity between the LEs and the global interconnects is specified by parameters, too. As shown in Fig. 2.5 three types of routing channel connectivity are supported: not connected, fully connected, and so-called periodic connectivity. Fully connected tracks offer full population of the connection box, i.e. each track can connect to each according broadcast line of a cluster. However, they

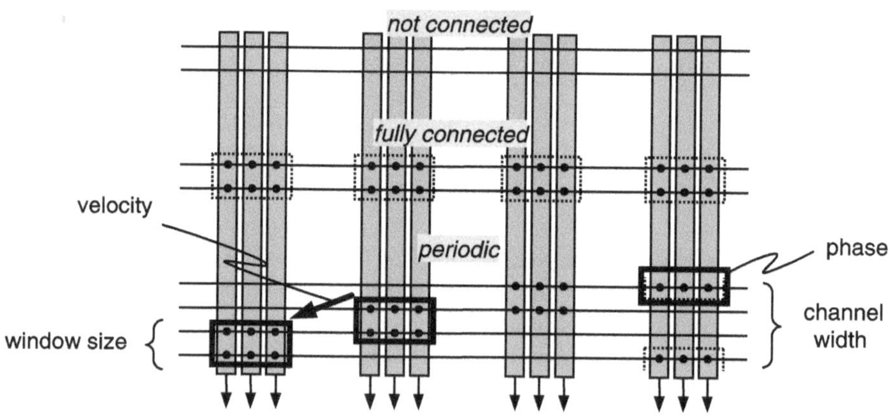

Fig. 2.5 Examples for the connectivity in a horizontal connection box with exemplary periodic interconnect patterns.

have the highest implementation costs. Periodic tracks use a special connection type best suited for arithmetic datapaths, where signals on a bus are typically ordered by the weight of their bits. Accordingly, periodic routing channels have a window of connection points that "slides" across the tracks with a given "velocity". The connection box defined in the architecture template can be composed of any mix of tracks with different channel widths and sliding window specifications.

According to the typical characteristics of data or signal flow graphs of digital arithmetic enumerated above, additional broadcast interconnects over the 2D LE arrays (red lines in Figs. 2.3 and 2.6 and local interconnect multiplexers in so-called dedicated routing boxes (DRBs) are supplied. As shown in Fig. 2.6 these DRBs connect the LE inputs and outputs to the terminals of nearest neighbour LEs and to broadcast lines via area and energy saving multiplexers. By these additional resources the requirements to the global communication resources are dramatically relaxed. Again, also the number of broadcast lines as well as the connectivity of the DRBs are subject to the parameterization.

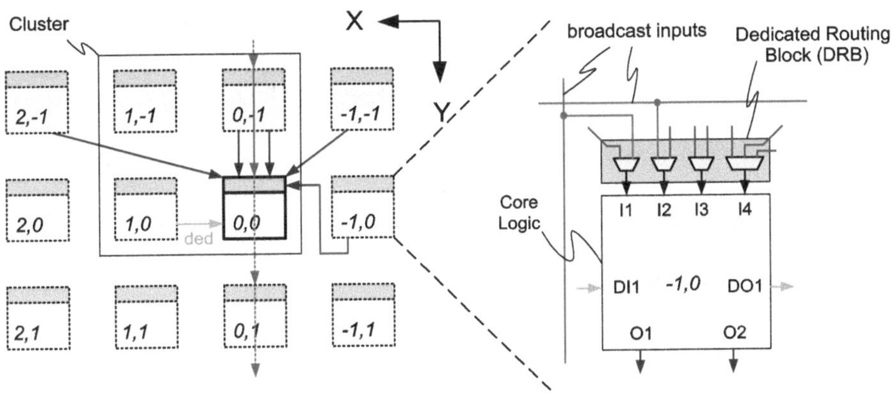

Fig. 2.6 Dedicated routing block connect the LE in- and outputs to nearest neighbour LEs and to broadcast lines.

An especially area consuming part of the eFPGA architecture is the configuration storage built up from SRAM like memory cells. The configuration bits stored in these cells control the functionality of the LEs, the connectivity of the routing switches and connection boxes, etc. In order to significantly improve the area efficiency the a-priori knowledge about regularity can be exploited to reduce the number of configuration storage cells: Thereby one storage location is shared by groups of LEs in a row of the 2D LE arrays and by groups of switching points in the routing switches as well as in the connection boxes. The size of these groups is another parameter of the target architecture. Dynamic reconfiguration of the eFPGA can be simply enabled by doubling the configuration storage cells and using them according to a multiplexed shadow register concept; the according parameter is the number of parallel configurations.

To conclude Table 2.1 shows the most important parameters of the target architectures being subject to the adaptation/optimization to the actual application class of the eFPGA.

Table 2.1 Most important parameters of the target eFPGA architecture

Building block	Parameters	Number of parameters
LE	Number of in-/outputs Functionality Latches/registers	7
Cluster	Number and alignment of LEs Number of in-/outputs per LE and direction Connectivity of connection box	6
Interconnect	Number of routing channels Segmentation length Connectivity of routing switch	12
Configuration	Number of parallel configurations Number of shared configurations	3
Others	Granularity of supply and clock domains	2

The set of all these architecture parameters spans the highly multi-dimensional design space allowing for very dedicated tuning the eFPGA to the requirements in an actual instance, e.g. as a reconfigurable SoC block.

2.3 Physical Implementation of Application Class Specific eFPGAs

All building block components of the eFPGA target architecture were optimized at logic and circuit level (especially towards high speed, small area, and low power dissipation). As an example Fig. 2.7 shows the core logic details of a LE containing three LUT-2, some Boolean post-processing logic, as well as multiplexers.

The required atomic cells required for the LEs, RSs, CBs, and DRBs were layed out in sub- and deep sub-micron CMOS technologies as so-called leaf cells fitting well to our datapath generator (DPG) based automatic layout generation approach [34]. After the parameters for a certain instance of the eFPGA are optimized and fixed the layout can be generated automatically in any shape (quadratic, rectangular, L-shaped, etc.). Additionally all hard macro views, like a VHDL model, required in today's design flows, are generated in the same step. A more detailed discussion corresponding to the design flow depicted in Fig. 2.8 can be found in [25].

Figure 2.9 shows exemplarily the layout of an eFPGA macro featuring 192 LEs, containing about 250,000 transistors and occupying about 0.26 mm^2 silicon area in a 90-nm CMOS technology.

Without going into too much detail here it must be emphasized that in implementing those eFPGAs at deep sub-micron technology physical level the whole palette

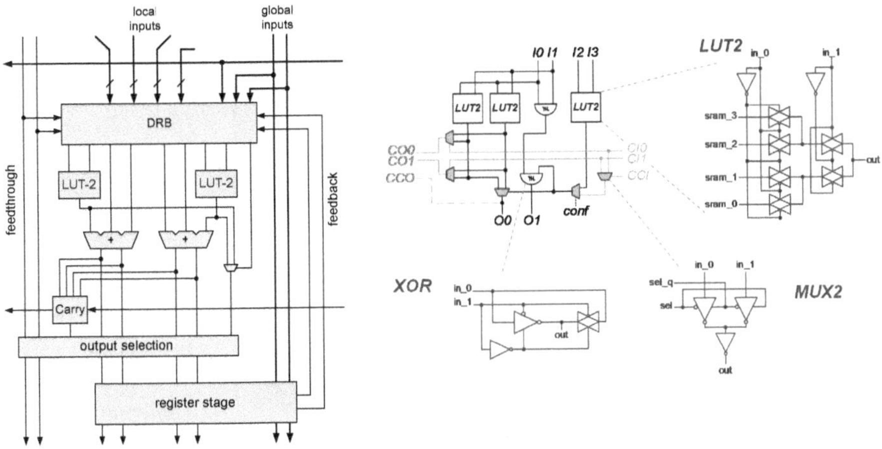

Fig. 2.7 Core Logic and circuit details of a LE containing three LUT-2 some additional glue logic and multiplexers.

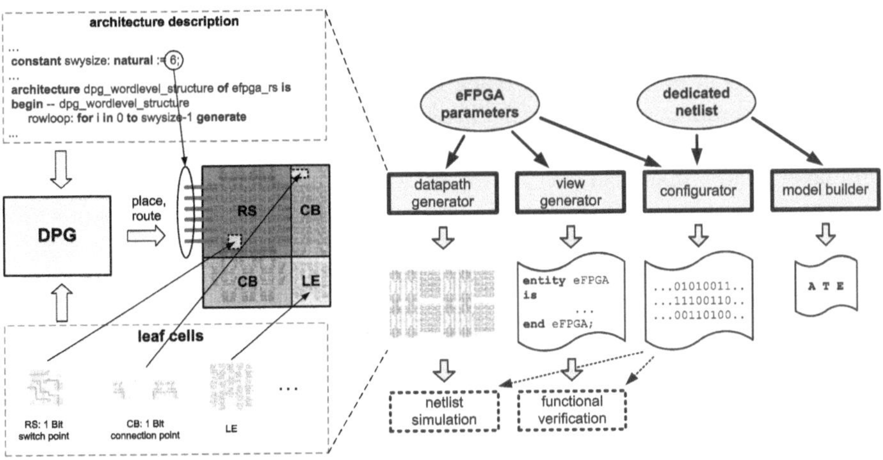

Fig. 2.8 eFPGA design flow including physical implementation, configuration, and model building.

of today's best practice low power strategies used in physically optimized ("full custom") ASIC design can be applied. These strategies range from clock gating or even powering down of non-used sub-blocks, automated device sizing down to automated so-called random modulation (optimal device threshold voltage selection). Moreover that, further even more attractive low power approaches like energy saving clock systems with inter-phase charge balancing and reverse back biasing based leakage current reduction allow for very low power features without compromising the initial goals of the application domain specific eFPGA concept.

According to the design flow depicted in Fig. 2.8 accurate performance and power evaluation of a specific structure to be mapped on the eFPGA can be per-

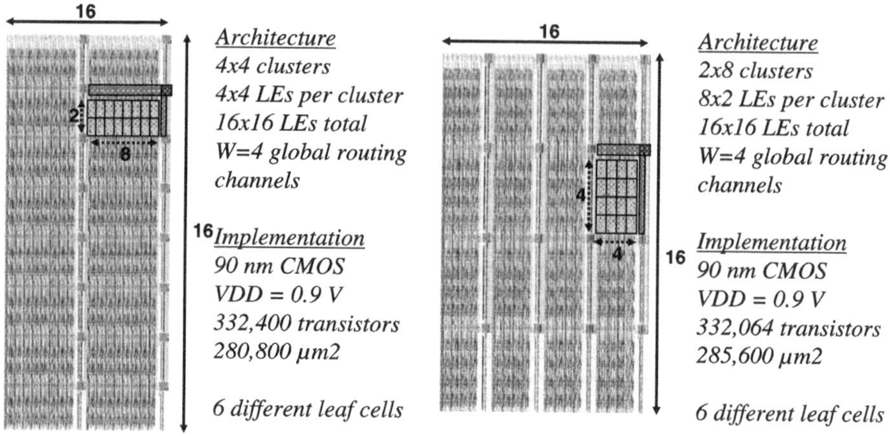

Fig. 2.9 Exemplary layouts of two eFPGA macros.

formed by circuit level netlist simulation. Clearly this is a very time consuming process. In order to accelerate the parameter optimization process in the highly multi-dimensional eFPGA design space it is advantageous to start from analytic cost models to estimate the required silicon area, power dissipation, and delays according to a specific structure mapped to the eFPGA. After RC parasitics extraction from layout and proper characterization of the leaf cells at circuit level ATE models were elaborated allowing for quite accurate performance and cost estimation. From the leaf cell characteristics and the configuration information these models are generated and evaluated in a model builder; shown in Fig. 2.8 on the right.

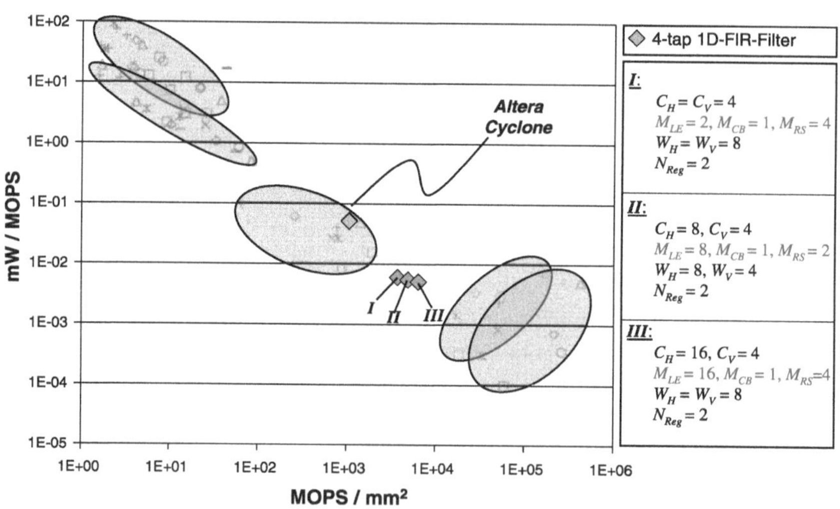

Fig. 2.10 Performance of filter implementation on eFPGA and commercial FPGA.

Figure 2.10 shows the results of a model based parameter evaluation for a 4-tap FIR filter in the design space of Fig. 2.1 in comparison to a commercial FPGA. In [22] it has been shown that the model is quite accurate compared to detailed simulations of layout-extracted netlists.

2.4 Mapping and Configuration

A very challenging task is the mapping of a data or signal flow graph onto the available core logic and routing resources as well as the transformation of the resulting netlist into the configuration bit stream. Mappers for today's commercial FPGAs are highly complex software packages. Most research conducted in the field of (e)FPGA-architectures is based on the VPR design flow [2] which can only be used to handle inflexible, rather old-fashioned standard island-style FPGA architectures and supports only simple LUT-based LEs and a very limited choice of routing switch resources.

Actually for the use with parameterized eFPGA architectures a portable mapper adaptable to a given set of parameters would be required. In order to keep goals realistic we decided to head for a flow quite similar to the design process applied in the DPG based design of dedicated high-performance and low-power arithmetic ASIC blocks: The allocation of the operators in the data or signal flow graph to the LEs is performed manually allowing for maximal exploitation of locality and thereby ensuring lowest possible power dissipation. For an experienced designer this task appears to be rather easy and for most cases of arithmetic structures the allocation is quite obvious.

As a first step in the allocation process Fig. 2.11 illustrates the configuration of the LE core logic shown in Fig. 2.7 for a simple carry-select adder stage.

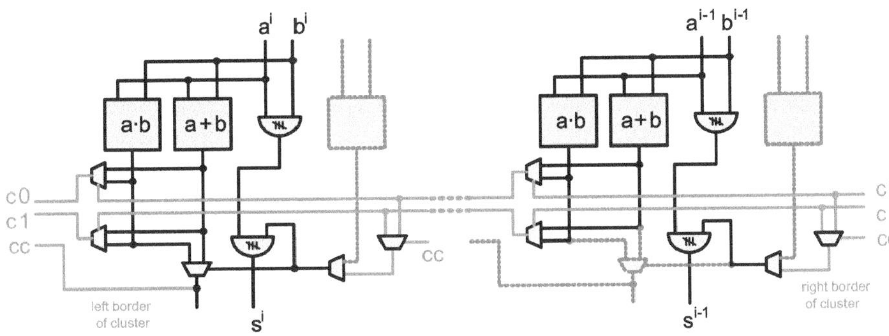

Fig. 2.11 Configuration of the LE core logic shown in Fig. 2.7 for a carry-select adder.

For frequent basic operations a library of those configurations was elaborated (see also configuration table in Fig. 2.14). The allocation maps the basic operations of the structure to be implemented onto such configured LEs. That is exemplarily illustrated in Fig. 2.12 for an inner product accumulation of 4-dimensional vectors (quadMAC operation).

$$S = (X1 \times Y1) + (X2 \times Y2) + (X3 \times Y3) + (X4 \times Y4) + Z$$

The result of that allocation process is a connected netlist of the LEs described in a VHDL model.

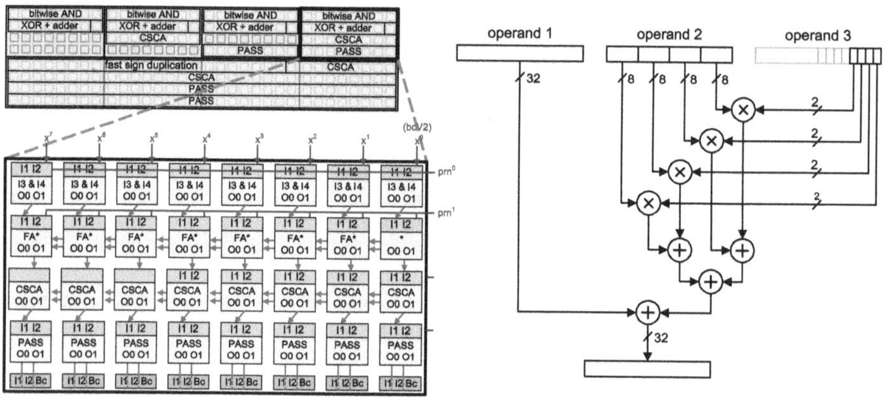

Fig. 2.12 Allocation of the basic operations for a QuadMAC operation.

What remains, is to map the according interconnect between the allocated and functionally fixed LEs to the available routing resources of the eFPGA. In our first experiments described in Sects. 2.5 and 2.6 we performed that interconnect mapping manually, but this is a rather cumbersome and very time consuming process. So, we decided to use the same CAD tools as being used in the DPG based design flow for dedicated ASIC macros: commercial routing tools.

To use these tools the eFPGA routing problem is mapped to an equivalent VLSI routing problem. As the LEs including the nearest neighbour connections are already mapped, they are modelled as black boxes with a given number of input and output ports. Thus, the only components that have to be modelled are connection boxes and routing switches. From these components a model of the complete eFPGA is created using a script. The script includes the most relevant architectural parameters described in Sect. 2.2. The connections that have to be established between the ports of the different LE clusters are described in a netlist. Afterwards a commercial standard VLSI routing tool is employed to find a routing solution using this netlist and the eFPGA model. The configuration bit stream is then generated by analysis of the routing solution.

The components are modelled in a way to allow easy extraction of the eFPGA configuration bits after routing. Bus lines are modelled as metal wires created by the routing tool. Each bus and broadcast direction is assigned to a specific metal layer. Thus, configurable connections in switching points and connection boxes are represented by vias. To restrict the routing tool from creating busses not present in the architecture, the metal layers are masked with keepout areas.

Figure 2.13 shows the model of a switching point configured to connect its right and bottom ports (c). The via created in the equivalent layout represents the established connection. Thus the corresponding configuration bit of the actual switching point is set. Connection boxes are modelled similar to switching points. As metal layers only represent busses, extraction of the routing configuration is reduced to analysis of the via positions is the routing solution, provided by the VLSI routing tool.

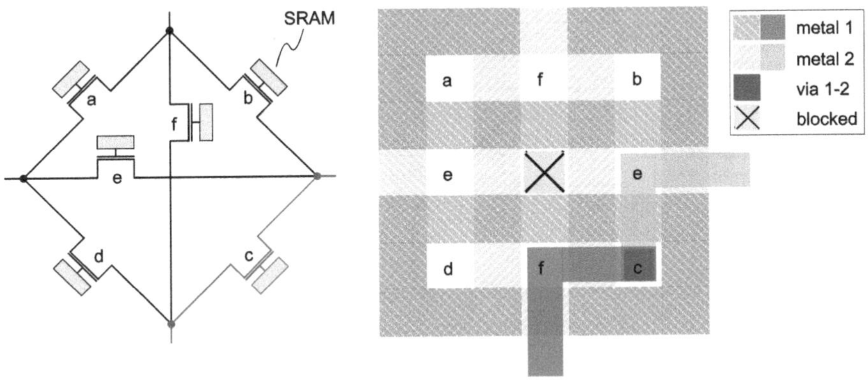

Fig. 2.13 Configuration of a switching point.

The result of application mapping and communication routing described above is a connected "mapped netlist" of the LEs, DRBs, RSs and CBs including their actual configuration. The last step is the transformation of this mapped netlist into a configuration bit stream. A "configurator" creates the complete configuration bit stream based on the mapped netlist, the architecture specifications (parameters) and basic configuration tables for basic eFPGA elements like switching points within an RS. It also uses the information from the layout generator to determine the actual position of all blocks to be configured in the macro. The configuration bit stream is composed of elementary configuration table entries that must be provided for the basic eFPGA elements like RSs or LEs. The elementary tables can be created with small effort, as only few bits are required to configure these basic elements. The bit stream is then concatenated according to the position of the elements in the overall macro. This configuration flow is exemplarily illustrated for a routing switch in Fig. 2.14.

2.5 Examples of (Stand Alone) eFPGAs as SoC Building Blocks

In order to benchmark application class specific eFPGAs against conventional general purpose FPGAs frequent tasks of digital signal processing were implemented

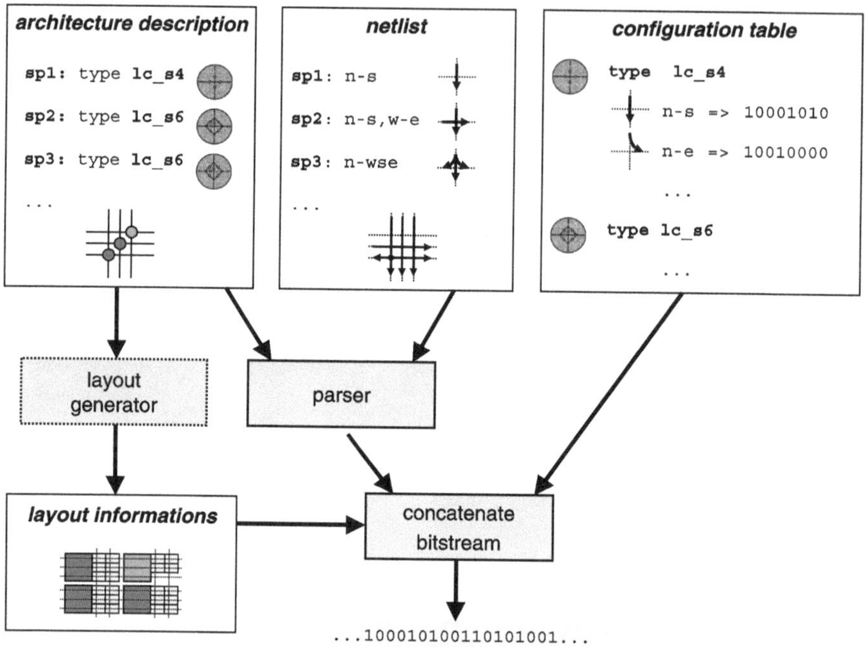

Fig. 2.14 Generation of configuration bit stream.

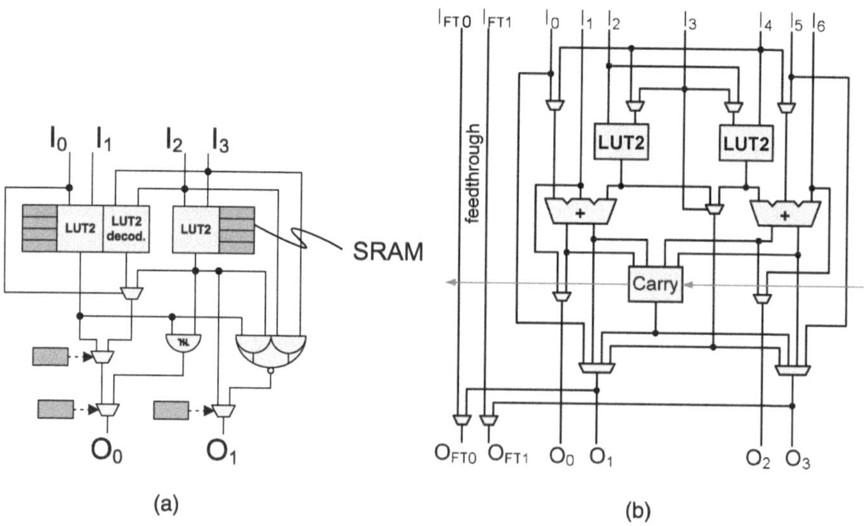

Fig. 2.15 Core logic of the LEs for eFPGA architectures I and II.

on different eFPGA architectures as well as on commercial FPGAs. Two exemplary differently parameterized eFPGA architectures are considered here: Architecture I consists of LEs featuring core logic with three LUT-2 and some additional Boolean logic as depicted in Fig. 2.15 and makes use of configuration storage sharing (Fig. 2.16). Architecture II consists of LEs with more complex (dual) core logic containing two LUT-2, two full adders, additional carry logic, additional multiplexers (Fig. 2.15(b)) and does not make use of configuration storage SRAM sharing for the sake of more flexibility to irregular logic. Additional broadcast line feed through stages allow for building up virtual clusters (Fig. 2.15 (b); for more details refer to [22]).

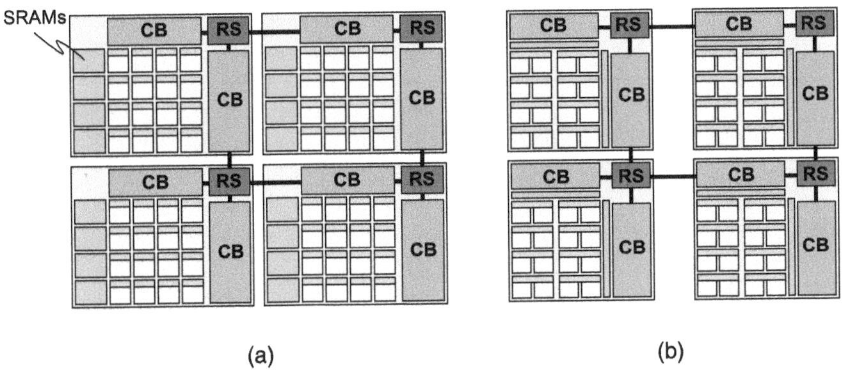

Fig. 2.16 Simplified schematics of eFPGA architectures I and II.

Figure 2.17 shows the results for the implementation of an 8×8 multiplier, a 4-bit MAC unit, a 4-bit butterfly unit, and a 4-tap FIR filter using architecture II in comparison to an conventional FPGA architecture. As can be seen an increase of about one order of magnitude in energy and area efficiency is achievable from application class specific FPGAs.

2.6 Examples of eFPGAs as Coprocessors to Standard RISC Processor Kernels

As arithmetic oriented eFPGAs based on the architecture are attractive SoC components in terms of efficiency and flexibility they were applied for hybrid architectures based on Software-programmable processors coupled with eFPGA accelerators [31]. The main focus was on the coupling mechanisms between the processor core and the eFPGA as these have crucial impact on the overall performance and cost (see also Sect. 2.1). For the evaluation of coupling mechanisms standard processor architectures rather than dedicated ASIPs were used in a first approach to take advantage of the existing tool support [32, 24]. After fundamental coupling concepts were derived refined studies based on an ASIP described using the ar-

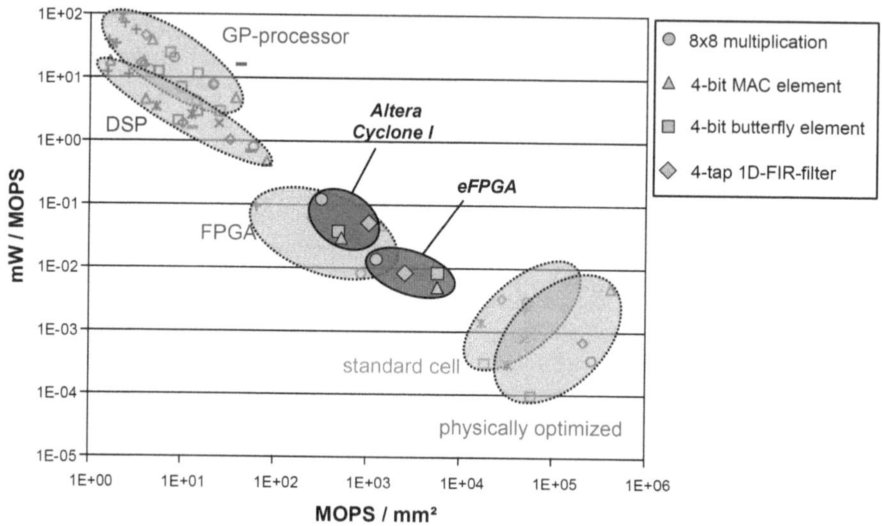

Fig. 2.17 Benchmarking of architecture I for frequent DSP tasks.

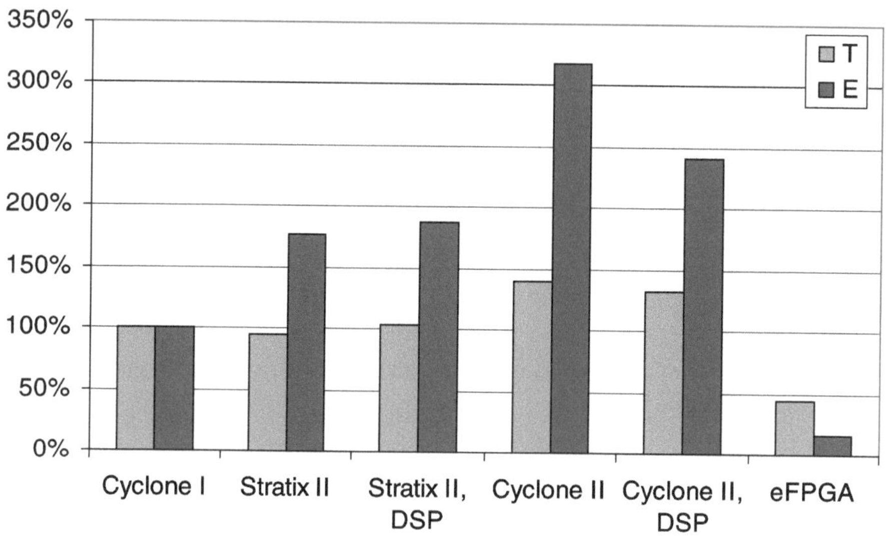

Fig. 2.18 Comparison of execution time (T) and energy consumption (E) of a 4-tap 1D-FIR-filter.

chitecture description language LISA [12] were conducted. In contrast to general purpose FPGAs with embedded processors like e.g. Xilinx Virtex 4, the reconfigurable part of the architecture on the one hand is highly optimized for arithmetic oriented applications. On the other hand the ASIP architecture itself and the coupling between ASIP and eFPGA were tailored in detail to the specific requirements arisen from considered application domains. In Sect. 2.6.2.2 fundamental concepts related to novel ASIP-eFPGA architectures will be illustrated.

2.6.1 General-Purpose Processors Coupled with eFPGAs

For the first approach, two different general purpose processor architectures namely a processor based on a MIPS IV instruction set [19] and an ARM940T [29] were used [32, 24, 31]. These were coupled with an eFPGA accelerator based on the introduced eFPGA architecture II. The coupling is based on multiplexer based logic where the processor has access to dedicated eFPGA resources and the eFPGA communicates with the ASIP via the central register file of the ASIP. In contrast to the ASIP-eFPGA architectures explained in the following subsection the synchronisation scheme between processor and eFPGA was very rudimentary. For these hybrid architectures energy, area and timing information have been acquired by application of appropriate tools. A more detailed discussion can be found in [31, 32, 24]. As a proof of concept a DES encryption and a median-filter have been mapped to this hybrid architecture.

2.6.2 ASIPs Coupled with eFPGA-Based Accelerators

After determining that hybrid processor-eFPGA-architectures take an interesting place within the design space, with the architectures shortly sketched in the previous subsection, refined studies have been realized, in which coupling mechanism, ASIP and eFPGA architecture were designed in detail. Thereby the chosen approach allows for an in-depth design matching of these building blocks.

2.6.2.1 ASIP

In contrast to general purpose processors the utilization of an ASIP yields the opportunity to tailor the instruction set to a given application domain. Furthermore the required enhancements of the processor architecture and the instruction set architecture (e.g. eFPGA custom instructions) necessary to realize an optimum coupling can be accomplished. The ASIP template which was the starting point for all architecture and instruction set modifications is based on DLX-like processor architecture [11]. This processor is based on a Harvard concept with a RISC-like instruction set and a five-stage processor pipeline including enhanced mechanisms like e.g. operand forwarding (see Fig. 2.19). There are a couple of different design tools available to realize an ASIP architecture. In this work a LISA-based design flow has been applied [12, 8]. LISA is a language to realize the processor architecture and the instruction set architecture within one textual description on a high C-like level. Based on such a cycle-accurate LISA-description it is possible to build all required software development tools (Compiler, Assembler, Linker, Simulator etc.) automatically. Furthermore a synthesizable VHDL description of the ASIP architecture is generated. This VHDL description was used to realize a standard-cell based macro of the ASIP. Hence detailed cost and performance values could be acquired.

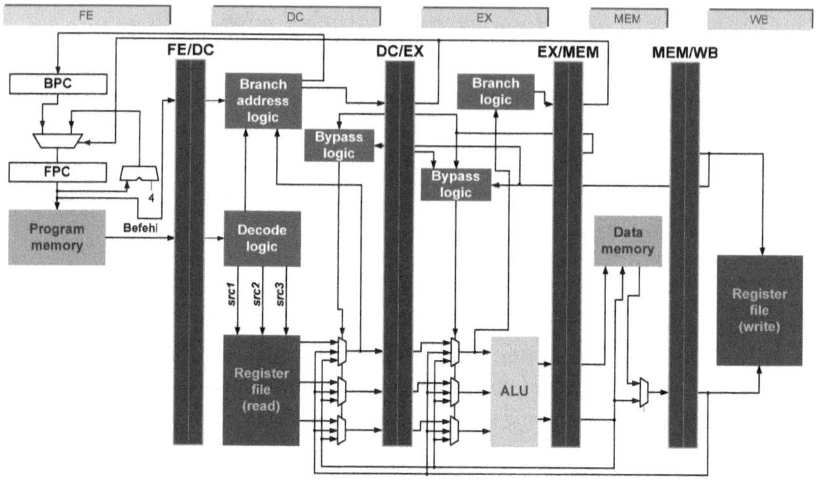

Fig. 2.19 ASIP-template (LT_RISC).

2.6.2.2 ASIP-eFPGA Coupling Mechanisms

Depending on the characteristics of an application the coupling mechanism between processor and an eFPGA-based accelerator must be realized carefully. In general it is useful to map control oriented tasks to the programmable ASIP and parallelizable data paths based on basic arithmetic operations to the eFPGA. There are a couple of influencing factors, like e.g. complexity of the eFPGA operators releasing the ASIP, data requirements of such an operator, etc., which determine the design constraints of the coupling mechanism.

The approach applied in this work allows for a balanced layout of the whole ASIP-eFPGA-architecture particularly with regard to the coupling architecture. There are mainly three different possibilities to realize the coupling architecture:

The control structures are integrated within the control structures of the ASIP pipeline. As both influence each other a fine grain dedicated realization is possible. Thereby the number of building blocks is reduced as an explicit coupling block does not have to be provided. All interfaces are included within the LISA-description of the ASIP. As the coupling structures are integrated within the LISA-description they will be synthesized according to the standard-cell design flow applied here. After the production phase the coupling mechanism is fixed and can not be modified at all. The second variant to implement the coupling is to provide a reconfigurable component beside the arithmetic oriented eFPGA or utilize the eFPGA itself. The usage of reconfigurable component is attractive in terms of flexibility as coupling mechanisms can be adapted and modified even after the production phase just before runtime. Otherwise the cost in terms of energy, timing and area are significantly higher than the dedicated solution described before particularly with regard to arithmetic oriented eFPGAs as they are worse for mapping irregular control structures. The third variant, an attractive alternative, is the combination of both variants and design a heterogeneous solution including dedicated and reconfigurable components thus regarding efficiency as well as flexibility aspects.

In the following the basic coupling concept proposed in this contribution considering the runtime behavior will be introduced. For the eFPGA all data received at the interfaces to the outside world can not be differentiated relating to the meaning of the data. It can not be distinguished if received data words are data samples or control data words. The meaning of the data words is defined by the configuration of the eFPGA. This flexibility yields the opportunity to determine the behavior of the eFPGA just before runtime. To guarantee a high amount of flexibility it is necessary to provide eFPGA interfaces on the ASIP side which are connected to the global resources like e.g. the register file, state registers etc. of the ASIP.

In Fig. 2.20 a simplified block diagram of the fundamental heterogeneous coupling concept for hybrid ASIP-eFPGA architectures is depicted. Starting point for the following considerations is one instruction from the instruction set architecture of the ASIP which is addressed by the program counter (PC) and loaded from the program memory. This VLIW-like instruction consists of two atomic parts which initiate an operation and control flow on the ASIP as well as on the eFPGA. The atomic eFPGA instruction is just a couple of bits which are interpreted by simple dedicated structures and structures configured on the eFPGA. In the simplest case the atomic instruction is just one bit which identifies if an eFPGA operation has to be initiated or not. If an eFPGA operation is activated by the current instruction word a global register on the ASIP is set indentifying the state of an activated eFPGA operation. Further sample or control data words are stored in the register file or a corresponding data memory section which is controlled by a dedicated memory controller. Depending on the operation the part of the register file allocated for eFPGA can be locked for ASIP accesses during the activation of the eFPGA operation. The delay time of an eFPGA operation is deterministic. Therefore a watchdog timer is integrated to initiate the unsetting of the activation flag. After this the data and register file sections allocated by the eFPGA are unlocked and can be utilized for following ASIP or eFPGA operations. This concept avoids the necessity to stall the ASIP pipeline. Data dependencies have to be resolved during programming the ASIP-eFPGA or during compile-time. Thereby resources which have to be exclusively allocated by the eFPGA are determined before runtime. By adding the corresponding resources the described concept has been applied to environments where more than one eFPGA operation can take place in the same time. Altogether this concept provides as much concurrency of ASIP and eFPGA as possible and many options for designing the required coupling mechanism. It is possible to apply a complex eFPGA operator accessing a data memory section as well as a one-cycle operation utilizing the ASIP register file.

Different coupling mechanisms shortly outlined in the following are based on the introduced basic concept and have been implemented to perform an efficiency and performance evaluation. A common implementation of the basic concept was not possible due to restrictions of the ASIP development tools. More details for each implemented coupling mechanism are provided in [31]. The loosely-coupled (LC) mechanism, which was implemented, is provided for more complex eFPGA

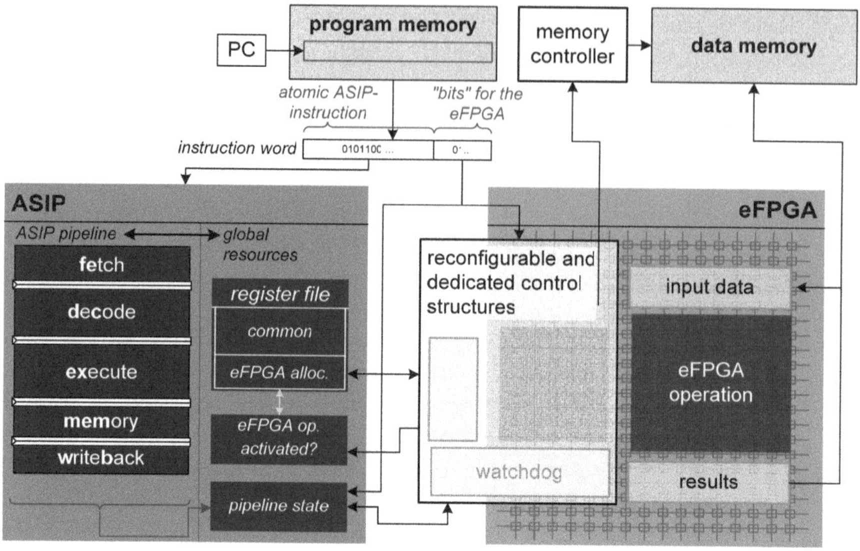

Fig. 2.20 Fundamental coupling concept for hybrid ASIP-eFPGA architectures.

operators, that consume relatively more runtime on the eFPGA. Data words are communicated via the ASIP register file. The synchronization is realized by a specific watchdog based interrupt mechanism. The loosely-coupled mechanism with access to a data memory section (LC-MEM) is based on the same principles like the LC mechanism. Instead of using the register file for data communication a dedicated memory section is allocated. The tightly-coupled (TC) scheme is used for less complex eFPGA operations consuming just a few computation cycles. For data communication the ASIP register file is applied. As the coupling structures are designed rudimentary the ASIP pipeline is synchronized by a stall during an eFPGA operation.

2.6.2.3 Performance and Cost Evaluation

In the context of modeling the cost and performance of the overall ASIP-eFPGA architecture the partial models for the individual building blocks have been integrated to a comprehensive model. The ASIP has been realized as standard-cell macro in a 90 nm technology based on the VHDL-model generated from the LISA-description. The physical cost and performance have been acquired by using application dependent switching factors on the terminal interfaces of the ASIP. The switching factors have been determined by a cycle-accurate simulation of an overall functional VHDL-model. The overall functional VHDL-model can be constructed by integration of the functional VHDL-description of the ASIP which also includes a behavioral model of program and data memory as well as the automatically generated VHDL-models related to the eFPGA. For memory related costs and performance the tool CACTI has been applied [6]. The physical cost for the eFPGA accelerator has been obtained from the eFPGA design flow introduced in Sect. 2.3. The overall cycle-accurate functional model yields detailed runtime profiles for each mapped application on one hand. On the other hand the overall efficiency in terms of area and energy efficiency have been determined by summing up the individual contributions of each building block.

2.6.2.4 Examples of eFPGAs as Coprocessors to Application Specific Instruction Processors (rASIPs)

This subsection presents a software defined radio (SDR) approach for a Global Navigation Satellite System (GNSS) receiver as another application example which is mapped to the ASIP-eFPGA architecture. If realized in software the critical component of the GNSS receiver is the correlator channel, because of the massive parallelism in this block. Generally, the correlator channel is used to generate a local signal and synchronize it with the incoming satellite signal [13, 27, 20]. If local and incoming signal are synchronized in frequency and phase of code and carrier, the receiver can extract two types of data required to estimate the user position. First, the navigation data broadcasted at a frequency of 50–250 Hz which includes all information to calculate the position of the transmitting satellite. Second, the flight time of the satellite signal and thus the distance between satellite and receiver. The flight time is calculated based on the delay which has to be introduced to achieve synchronization between local and incoming signal. A standard correlator channel, analog to digital conversion (ADC) and correlator control are shown in Fig. 2.21. In a first stage the incoming signal is multiplied with the in-phase and quadrature samples (I, Q) of the carrier frequency estimate which is generated using a digitally controlled oscillator (DCO). In the second stage the baseband signal is multiplied with a satellite dependent pseudo random noise (PRN) code (P) and the difference of an early and late version (EML) of the same PRN code. Finally, the samples are accumulated for each path and the result are transferred to the correlator control which adjusts PRN code generator and DCO to keep incoming and local signal synchronized.

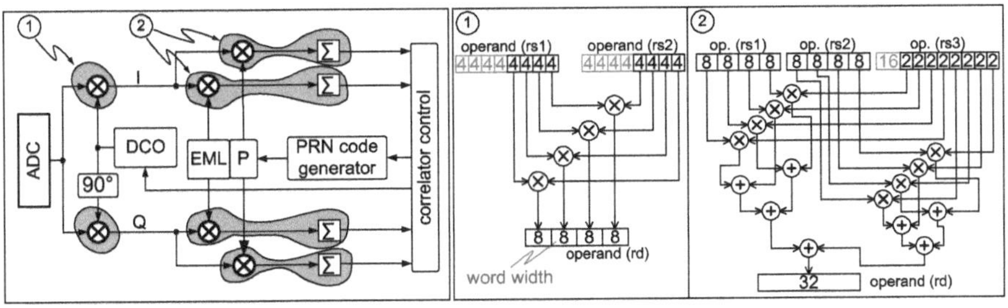

Fig. 2.21 Exemplary eFPGA operator (right) to realize a part of the GPS-Receiver (left).

In this application example two main ideas are implemented. First, the ASIP exploits the parallelism in the incoming data to achieve real-time requirements of the correlation in software. Therefore, two eFPGA custom instructions (CI) for baseband mixing and code correlation representing the eFPGA operators have been introduced in the ASIP's instruction set. They are marked in Fig. 2.21 as 1 and 2. Second, the implementation using an ASIP-eFPGA architecture allows the trade-off between sensitivity and the number of channels which can be processed in real-time. This trade-off is achieved by using a high-precision mode (HP) with an 0.5 dB increased signal-to-noise (SNR) compared to the low-precision (LP) mode. Therefore, the LP mode implies a digitized satellite and local carrier signal of 2 bit and a 4 bit signal at the output. In the HP mode the input signal and local carrier are 4 bit and half the number of samples can be processed in real-time. The mode can be switched at run-time setting an ASIP general purpose register [14].

Table 2.2 Area, energy and delay comparison for different architectures

	LT_RISC	LT_RISC with multiplier	ASIP-eFPGA		ASIP-eFPGA with multiplier	
			LP	HP	LP	HP
Area [mm^2]	0.633	0.655	0.803	0.914	0.829	0.939
Delay [μs]	9191	2545	237	413	184	322
Energy [μJ]	125.8	50.1	4.77	8.12	4.04	7.66

Table 2.2 summarizes the results derived using the design methodology and the power estimation flow presented in the preceding sections. The results for the LT_RISC and the ASIP-eFPGA architecture are derived for a clock frequency of 220 MHz and a 90 nm CMOS technology. Software correlation results are derived for one PRN code phase and one fixed carrier frequency. It can be seen that the power consumption and chip area is clearly dominated by the program and data memory with 8 kB each. For the LT_RISC with multiplier the memory area portion is 84% of the overall macro area. This amount decreases for an ASIP-eFPGA with multiplier in HP-mode to 59%, while the eFPGA area contributes with 26%. Figure 2.22 classifies the derived results in the efficiency diagram. It turns out that the implementations of the software correlation using the LT_RISC are less efficient than the realization on a state-of-the-art DSP (Texas Instruments, TMS320C642). Nevertheless, enhancing the ASIP with an arithmetic oriented eFPGA yield results which fall right between the DSP and the FPGA cluster and increase the efficiency about half an order of magnitude, while preserving the flexibility of programmable solution.

2.7 Conclusion

In this chapter it has been shown how to tailor embedded FPGAs to application classes in order to achieve improved energy and area efficiencies. A parameterized

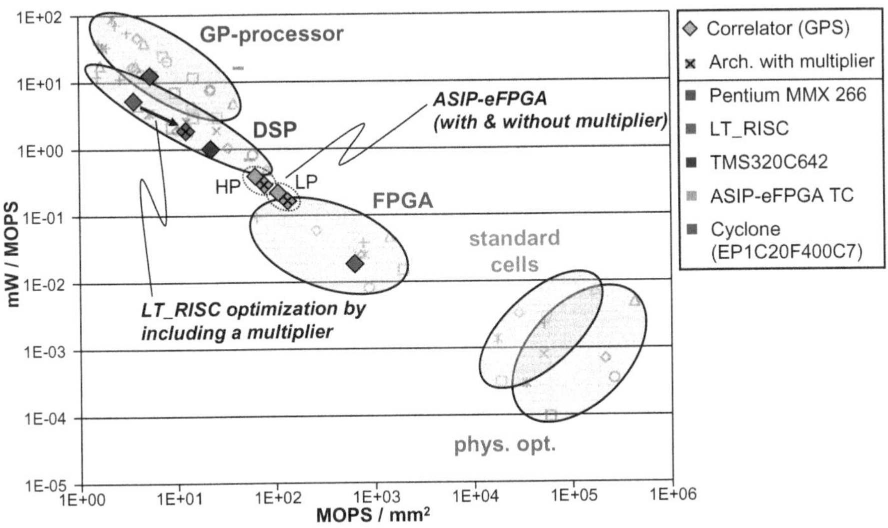

Fig. 2.22 Performance of a eFPGA based GNSS receiver.

target architecture featuring a high degree of flexibility and allowing for dedicated optimization has been described. Applying a design flow suited to perform research in that domain it was shown for basic but frequently used structures that the efficiency improvement potential is as high as one order of magnitude. Apart from using those eFPGAs as stand-alone building blocks of a SoC they appear to be very well suited to implement coprocessors to software programmable kernels. On exemplary application vehicles this strategy has been worked out for the case of eFPGA/ASIP coupling and the feasibility and attractiveness of this concept has been shown. Further work has to be performed on the challenging task of implementing a suitable mapper supporting this approach in practical industrial design.

Acknowledgements This work was supported by DFG grants No 281/9-1 and No 281/9-2, project *Reconfigurable Components for ASIP Architectures*, as part of the Priority Programme 1148, *Reconfigurable Computing Systems*.

References

1. Altera: Altera quartus ii 5.1 handbook, vol. 2. (Dec. 2005). Altera. http://www.altera.com/literature/lit-index.html
2. Betz, V., Rose, J.: VPR: A new packing, placement and routing tool for FPGA research. In: Proc. 7th International Workshop on Field-Programmable Logic and Applications (FPL '97), pp. 213–222 (1997)
3. Betz, V., Rose, J., Marquardt, A.: Architecture and cad for deep-submicron fpgas. In: Kluwer International Series in Engineering and Computer Science. Springer, Berlin (1999)

4. Blume, H., Feldkämper, H.T., Noll, T.G.: Model-based exploration of the design space for heterogeneous systems on chip. J. VLSI Signal Process. Syst. **40**, 19–34 (2005)
5. Brodersen, R.: Future system-on-a-chip design issues. In: Short Course on System-on-a-Chip Design, ISSCC 2003, San Francisco (Feb. 2003)
6. Company, H.P.D.: Cacti 5.0. http://www.hpl.hp.com
7. Compton, K., Hauck, S.: Reconfigurable computing: a survey of systems and software. In: ACM Computer Survey, vol. 34, pp. 171–210 (2002)
8. CoWare Corporation: Processor designer reference manual (2007). http://www.coware.com
9. DeHon, A.: The density advantage of configurable computing. In: Computer, vol. 33, pp. 41–49 (2000)
10. George, V., Rabaey, J.M.: Low energy fpgas—architecture and design. In: Kluwer International Series in Engineering and Computer Science. Springer, Berlin (2001)
11. Hennessy, J., Hennessy, J.L., Goldberg, D., Patterson, D.A.: Computer Architecture: A Quantitative Approach, 1st edn. Morgan Kaufmann, San Mateo (1996)
12. Hoffmann, A., Schliebusch, O., Nohl, A., Braun, G., Wahlen, O., Meyr, H.: A methodology for the design of application specific instruction set processors (ASIP) using the machine description language lisa. In: Proc. IEEE/ACM 2001 (2001)
13. Kaplan, E.D., Hegarty, C.J.: Understanding gps: Principles and Applications, 2nd edn. Artech House, Boston (2006)
14. Kappen, G., Pieper, V., Kurz, L., Noll, T.G.: Implementation and analysis of an sdr processor for gnss software correlators. In: Proceedings of the 21th International Technical Meeting of the Satellite Division of the Institute of Navigation ION GNSS, Savannah, GA, pp. 2258–2267 (2008)
15. Kusse, E., Rabaey, J.: Low-energy embedded FPGA structures. In: IEEE Symp. on Low Power Electronics and Design, pp. 155–160 (1998)
16. Lange, H., Schröder, H.: Evaluation strategies for coarse grained reconfigurable architectures. In: Proceedings International Conference on Field Programmable Logic and Applications, pp. 586–589 (2005)
17. Leijten-Nowak, K.: Template-based embedded reconfigurable computing. PhD thesis, University of Eindhoven (2004)
18. Lewis, D., Ahmed, E., Baeckler, G., Betz, V.: The stratix ii logic and routing architecture. In: Proceedings of the 2005 ACM/SIGDA 13th International Symposium on Field-Programmable Gate Arrays, Monterey, California, USA (2005)
19. MIPS Technologies Inc.: MIPS R10000 microprocessor user's manual, version 2.0 (1997)
20. Misra, P., Enge, P.: Global positioning system: Signals, measurements, and performance, 2nd edn. Ganga-Jamuna Press, Lincoln (2006)
21. Miyamori, T., Olukotun, K.: Remarc: reconfigurable multimedia array coprocessor. In: Proceedings 6th International Symposium on Field Programmable Gate Arrays. ACM Press, Monterey (1998)
22. Neumann, B.: Modellierung und analyse arithmetikorientierter eFPGA-architekturen. PhD thesis, Chair of Electrical Engineering and Computer Systems, RWTH Aachen University (2008)
23. Neumann, B., von Sydow, T., Blume, H., Noll, T.G.: Design and quantitative analysis of parametrisable eFPGA-architectures for arithmetic. In: Advances in Radio Science-Kleinheubacher Berichte, Miltenberg (2006)
24. Neumann, B., von Sydow, T., Blume, H., Noll, T.G.: Application domain specific embedded fpgas for flexible isa-extension of asips. J. VLSI Signal Process. (2008)
25. Neumann, B., von Sydow, T., Blume, H., Noll, T.G.: Design flow for embedded fpgas based on a flexible architecture template. In: Proceedings Conference on Design, Automation and Test in Europe, DATE, pp. 125–131. European Design and Automation Association, München (2008)
26. Noll, T.G.: Application specific eFPGAs for soc platforms. In: International Symposium on IEEE VLSI Design, Automation and Test, VLSI-TSA, p. 28 (2005)
27. Parkinson, B., Spilker, J.J.: Global Positioning System: Theory and Applications, vol. I. American Institute of Aeronautics and Astronautics, Washington (1996)

28. Pionteck, T., Staake, T., Stiefmeier, T., Kabulepa, L.D., Glesner, M.: Design of a reconfigurable aes encryption/decryption engine for mobile terminals. In: Proceedings International Symposium on Circuits and Systems, ISCAS (2004)
29. Samsung: Arm940t (sam443). http://www.samsung.com
30. Technologies, P.X.: Smexpp smart media processor. http://www.pactxpp.com
31. von Sydow, T.: Modellbildung und analyse heterogener ASIP-eFPGA-architekturen. PhD thesis, Chair of Electrical Engineering and Computer Systems, RWTH Aachen University (2008)
32. von Sydow, T., Korb, M., Neumann, B., Blume, H., Noll, T.G.: Modelling and quantitative analysis of coupling mechanisms of programmable processor cores and arithmetic oriented eFPGA-macros. In: Proc. ReConFig'06, pp. 252–261 (September 2006)
33. von Sydow, T., Neumann, B., Blume, H., Noll, T.G.: Quantitative analysis of embedded FPGA architectures for arithmetic. In: Proceedings International Conference on Application Specific Systems, Architectures and Processors Conference 2006 (ASAP '06), pp. 125–131 (September 2006)
34. Weiss, O., Gansen, M., Noll, T.G.: A flexible datapath generator for physical oriented design. In: Proc. ESSCIRC 2001, pp. 408–411 (2001)

Chapter 3
Erlangen Slot Machine: An FPGA-Based Dynamically Reconfigurable Computing Platform

Josef Angermeier, Christophe Bobda, Mateusz Majer, and Jürgen Teich

Abstract Dynamically partially reconfigurable architectures combine high performance and flexibility. They offer a novel possibility to dynamically load and execute hardware modules, previously only known for software modules. In order to realize these promises, the following dilemmas had to be solved: the too often limited memory of reconfigurable architectures for many data-intensive applications, the restricted communication possibilities for partial hardware modules, the unflexible tool flow for partial module design, and the IO-pin dilemma, that the placement of hardware modules, with requirements for input and output signals to the periphery, was predetermined to a single position. These were physical restrictions and technical problems limiting the scope or applicability of dynamically partially reconfigurable architectures. This led us to the development of a new FPGA-based reconfigurable computer called *Erlangen Slot Machine*, a platform for interdisciplinary research on dynamically reconfigurable systems. It leverages many architectural constraints of existing platforms and allows a user to partially reconfigure hardware modules arranged in so-called *slots*. The uniqueness of this computer stems from a) a new slot-oriented hardware architecture, b) a set of novel inter-module communication techniques, and c) concepts for dynamic and partial reconfiguration management.

Josef Angermeier · Mateusz Majer · Jürgen Teich
Hardware/Software Co-Design, Department of Computer Science, University of Erlangen-Nuremberg, Erlangen, Germany,
e-mails: `angermeier@informatik.uni-erlangen.de`, `teich@informatik.uni-erlangen.de`

Christophe Bobda
Department of Computer Science, University of Potsdam, Potsdam, Germany,
e-mail: `bobda@cs.uni-potsdam.de`

M. Platzner et al. (eds.), *Dynamically Reconfigurable Systems*,
DOI 10.1007/978-90-481-3485-4_3,

3.1 Introduction

No FPGA-based platform on the market today provides a solution to the problems of design automation for dynamically reconfigurable hardware modules and their efficient and flexible relocation. The purpose of the *Erlangen Slot Machine (ESM)* [16, 17, 8, 6, 7] is to overcome many of the deficiencies of existing FPGA-based reconfigurable computers by providing:

- A new flexible FPGA-based reconfigurable platform that supports relocatable hardware modules arranged in so-called *slots*.
- Tool support for the development of run-time reconfigurable computation and communication modules using new inter-module communication paradigms.
- A powerful reconfiguration manager which enables various preprocessing stages for fast bitstream manipulation. We call the preprocessing stages *plugins*. For example, a relocation plugin can be selectively activated before a bitstream is uploaded to the FPGA. As reconfiguration times in the range of seconds [15] are not sufficient for applications that require a fast reaction to external events, the ESM provides a hardware reconfiguration manager which is the foundation for reconfiguration times in the range of milliseconds.

3.2 Drawbacks of Existing Dynamically Reconfigurable Systems

The growing capacities provided by FPGAs as well as their partial reconfiguration capabilities have made them the ultimate choice for many time-constrained and computationally challenging applications. Xilinx FPGAs [23] are the only devices on the market with large capacity and the ability to support partial reconfiguration. The Virtex series offers enough logic for efficiently implementing applications with high demand of resources, e. g., arising in video, audio and signal processing as well as in other fields like automotive applications.

There are, however, many problems open concerning module relocation: One particular problem is, for example, that, in order to connect a module to other modules and/or pins, signals are often required to pass through other modules. We call those signals used by a given module and crossing other modules *feed-through signals*. Using feed-through lines to access resources has, however, two negative consequences, as illustrated in Fig. 3.1:

- Difficulty of design automation: Each module must be implemented with all possible feed-through channels needed by other modules. Because we only know at run-time which module needs to feed through a signal, many channels reserved for a possible feed-through become redundant.
- Relocation of modules: Modules accessing external pins are no more relocatable, because they are synthesized for fixed locations where a direct signal line to these pins is established.

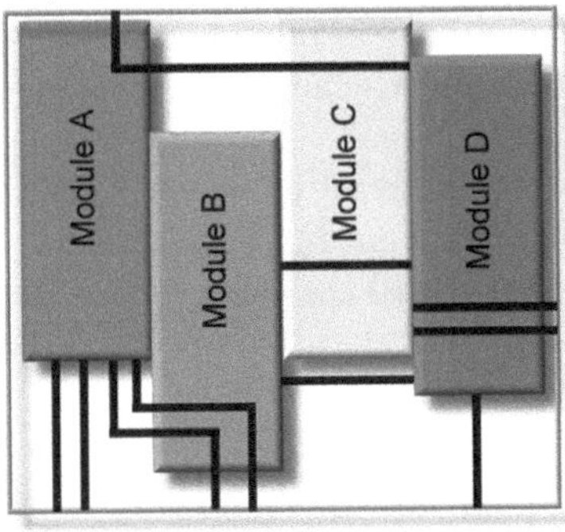

Fig. 3.1 The feed-through line problem with relocatable modules. Placing a new module B into slot two requires that the new module provides all feed-through lines needed by slot one and three. This fact disables any module relocation and makes it impossible to place modules with different feed-through requirements into the other slots.

Many FPGA-based reconfigurable platforms such as [9, 22, 18, 4, 19, 20] offer various interfaces for audio, video capturing and rendering and for communication. However, each interface is connected to the FPGA using dedicated pins at fixed locations. Modules with access to a given interface such as a VGA (video graphics adapter, see Fig. 3.2) must be placed in the area of the chip where the FPGA signals are connected, thus making a relocation impossible. Until now, no platforms on the market provide a solution to these problems.

Fig. 3.2 Pin distribution of a VGA module on the RC200 platform. It can be seen that the VGA Module occupies pins on the bottom and right FPGA borders. In consequence, only a narrow part on the left hand side is available for dynamic module reconfiguration.

Before we present in detail the architecture of the *Erlangen Slot Machine* and some required factors for innovation, we summarize the most the important problems limiting the use of partial and dynamic reconfiguration on current existing FPGA-based reconfigurable computers:

1. *Limitation of partial reconfiguration support on actual FPGAs*: Very few FPGAs allowing partial reconfiguration exist on the market. These few FPGAs,

like the Virtex series by Xilinx [23], impose nonetheless some restrictions on the least amount of resources that can be reconfigured at a time, for example, a column-wise reconfiguration.

2. *I/O-pin dilemma*: Existing platforms include I/O peripherals like video, RAMs, audio, ADC (analog to digital converter) and DAC (digital to analog converter) connected at fixed pins of the FPGA device. This has the consequence that a module cannot be relocated because of these pin constraints or making a relocation very difficult. Another problem related to pins is that the pins belonging to a given logical group like video, and audio interfaces are often not situated closely to each other. On many platforms, they are spread around the device. A module accessing a device will have to feed many signal lines through the FPGA to many different components. This situation is illustrated in Fig. 3.2: Two modules (one of which is a VGA module) are shown. The VGA module uses a large number of pins at the bottom part of the device and also on the right hand side. Implementing a module without feed-through lines is only possible on the two first columns on the left hand side. The effort needed for implementing a reconfigurable module on more than two columns together with the VGA module is very high. This situation is not only present on Celoxica boards [9]. XESS boards [22], Nallatech boards [18], and Alpha boards [4] face the same limitations. On the XF-Board [19, 20] from ETH Zurich, the peripherals are connected to one side of the device. Each module accesses I/Os through an operating system (OS) layer implemented on the left and right part of the device. Many other existing platforms like the RAPTOR board [14], Celoxica RC1000 and RC2000 [9] are PCI systems that require a workstation for operation. The use in stand-alone systems as needed in many embedded systems is not possible.
3. *Inter-module communication dilemma*: Modules placed at run-time on the FPGA device typically need to exchange data among each other. Such a request for communication is dynamic due to run-time module placement. Dynamically routing signal lines on the hardware is a very cumbersome task. For efficiency reasons, new communications paradigms must be investigated to support such dynamic connection requests such as *packet-based DyNoCs* [3] or principles of *self-circuit routing*.
4. *Local memory dilemma*: Modules requiring larger amounts of local memory cannot be implemented since a module can only occupy the memory inside its physical slot boundary. Storing data in off-chip memories is therefore the only solution. However, existing FPGA-based platforms often have only one or two external memory banks and their pin connections are spread loosely over the borders of the FPGA.

With these limitations in mind, we designed a new and FPGA-based reconfigurable computer called the *Erlangen Slot Machine* (ESM). Its architecture circumvents many of the above problems and was built to ease the research on implementing dynamically reconfigurable hardware modules including their run-time management and inter-module communication.

3.3 The Erlangen Slot Machine

The main idea of the Erlangen Slot Machine (ESM) architecture is to ease the application development as well as the research in the area of partially reconfigurable hardware. The advantage of the ESM platform is its unique slot-based architecture which allows the slots to be used independently of each other by delivering peripheral data through a separate crossbar switch as shown in Fig. 3.3. The ESM architecture is based on the flexible decoupling of the FPGA I/O-pins from a direct connection to a crossbar interface chip. This flexibility allows to place application modules at run-time in any available slot independently. Thereby, run-time placement is not constraint any more by physical I/O-pin locations as the I/O-pin routing is done automatically in the crossbar, and the I/O pin dilemma is thus solved in hardware.

3.3.1 Architecture Overview

The ESM platform (see Fig. 3.3) is centered around an FPGA serving as the main reconfigurable engine and an FPGA realizing the crossbar switch. They were separated into two physical boards (see Fig. 3.10) called *BabyBoard* and *MotherBoard* and are implemented using a Xilinx Virtex-II 6000 and a Xilinx Spartan-II 600 FPGA, respectively. Figure 3.3 shows the slot-based architecture of the ESM consisting of the Virtex-II FPGA, local SRAM memories, configuration memory and a reconfiguration manager. The top pins in the north of the FPGA connect to local SRAM banks. These SRAM banks serve to solve the problem of restricted intra-

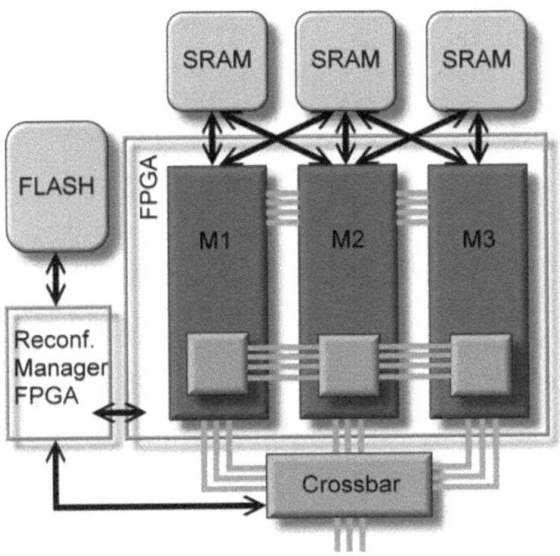

Fig. 3.3 ESM Architecture overview. The architecture of the BabyBoard is refined in Fig. 3.4. The MotherBoard is shown in Fig. 3.5.

module memory in case of, e.g., video applications. The bottom pins in the south connect to the crossbar switch. Therefore, a module can be placed in principle in any free slot and have its own peripheral I/O-links together with dedicated local external memory. Each slot of up to 6 slots can access each one local SRAM bank.

3.3.2 The BabyBoard

3.3.2.1 Computation and Reconfigurable Engine

The reconfigurable engine of the ESM computer is a printed circuit board that features a Xilinx Virtex II-6000 FPGA, several SRAMs and a reconfiguration manager implemented as well on an FPGA. Due to the restriction[1] in the reconfiguration of Virtex-II FPGAs, we adapted our architecture to match the following properties:

- *Solving the I/O-pin dilemma*: On-line placement of modules on a reconfigurable device, in this case the FPGA, is done by downloading a partial bitstream that implements the module on the FPGA. Relocation, i.e., placing a module into a location different from the one for which it was synthesized. Relocation can be done only if all the resources are available and structured in the same way in the designated placement area at compile-time (e.g., slot 0) and at run-time (e.g., slot 3) location. This includes also the I/O-pins used by the module. We solved the I/O-pin dilemma on the ESM by avoiding fixed connections of peripherals to the FPGA. As shown in Fig. 3.4, all the bottom pins from the FPGA are connected to an interface controller realizing a crossbar and implemented itself using a Xilinx Spartan-II FPGA. At run-time, the crossbar configuration is also adapted, to connect FPGA pins to peripherals automatically based on the slot position of a placed module. Notably, this I/O-pin rerouting scheme is achieved without reconfiguration of the crossbar FPGA. This makes it possible to establish any connection from one module to peripherals dynamically.
- *Solving the memory dilemma*: Memory is very important in applications like video streaming in which a given module often must exclusively access a picture frame at a time for computation. However, as we mentioned earlier, the capacity of the available BlockRAMs in FPGAs is limited. External SRAM memory has therefore been added to allow storage of large amounts of data by each module. To allow a module to exclusively access its external memory bank, 6 SRAM banks are located at the north border of the FPGA. In this way, a module will connect to peripherals from the south, while the north will be used for temporally storing computation data. According to the 6 memory banks which can be connected on the top, the device is divided into a set of elementary slots called *micro-slots* A to V (see Fig. 3.4). In order to use an SRAM bank in the north, a module must have at least a width of three micro-slots (creating slots S1 to S6). The *Erlangen Slot Machine* owes its name from this arrangement of reconfig-

[1] The reconfiguration can be done only in chunks of full CLB columns.

urable slots. This modular organization of the device simplifies the relocation, primary condition for a viable partially reconfigurable computing system. Each module moved from one slot to another will encounter equal resources. The architecture of the BabyBoard is illustrated in Fig. 3.4 in more details.

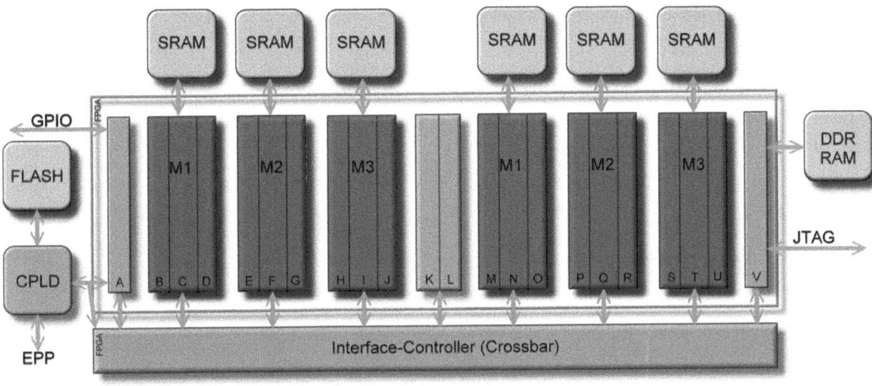

Fig. 3.4 Architecture of the ESM BabyBoard. Slots A to V denote micro-slots that provide the module and reconfiguration granularity. Three consecutive micro-slots define a macro-slot. Each macro-slot (S1 to S6) can access one full external SRAM bank. In terms of slice count, a micro-slot occupies 768 slices (4 CLB columns) on the FPGA. Slots A, K, L and V are special micro-slots as slots A and V interface external pins and slot K, L contain BlockRAM.

3.3.2.2 The Reconfiguration Manager

Apart from the main FPGA, the BabyBoard also contains the configuration circuitry. This consists of a CPLD, a configuration FPGA (a small Spartan II FPGA) implementing the reconfiguration manager (Sect. 3.5) and a Flash memory, see also Fig. 3.4.

- The CPLD is used to download the Spartan-II configuration from the Flash upon power-up. It also contains board initialization routines for the on-board PLL and the Flash.
- The reconfiguration management is implemented on the Spartan-II FPGA. This device contains a circuit to perform module relocation while loading a new partial module bitstream. Its architecture and functionality will be described in details in Sect. 3.5.
- The Flash provides a capacity of 64 MBytes, thus enabling the storage of up to 32 full configurations or of a few hundred partial module bitstreams.

3.3.2.3 Memory

Six SRAM banks of size 2 MB each are vertically attached to the board to the north of the device, thus providing external memory to six macro-slots (denoted as S1 to S6 in Fig. 3.4) for temporal data storage. The SRAMs can be also used for shared memory communication between neighbor modules, e.g., for streaming applications. They are connected to the FPGA in such a way that the reconfiguration of a given module will not affect the access to other modules.

3.3.2.4 Debug Lines

Debugging capabilities are offered through general purpose I/O provided at regular distances between the micro-slots. A JTAG port provides debug capabilities for the main FPGA, the CPLD and the Spartan-II.

3.3.3 The MotherBoard

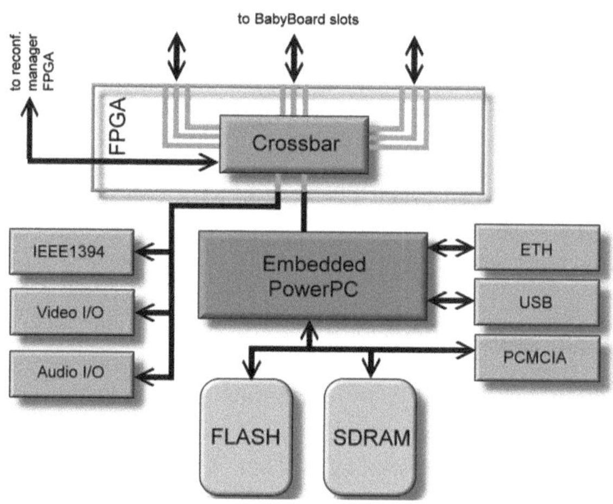

Fig. 3.5 Architecture of the ESM MotherBoard. The PowerPC serves as the main controller of the ESM computer and is running Linux. Its memory bus is connected directly to the crossbar for memory-mapped communication with the reconfiguration manager on the BabyBoard.

The MotherBoard provides programmable links from the FPGA to all peripherals for multimedia and communication such as IEEE1394, USB, Ethernet, PCMCIA, Video and Audio-I/Os. The physical connections are established at run-time through a programmable crossbar implemented statically on a Spartan-II chip on the MotherBoard. Video capture and rendering interfaces as well as high speed communication links also exist on the MotherBoard on which the BabyBoard is mounted through four connectors (see Fig. 3.5). A PowerPC processor (MPC875) is the core of the MotherBoard. It is used to control the complete ESM through a software programming interface. In particular, it manages the flow of data on the MotherBoard as well as provides the interfaces to the external world, e.g., Ethernet and USB. Upon start-up, one can log-in into the ESM like a full Linux-based computer system. The PowerPC of the ESM is used for application development and for testing and initiating the dynamic hardware reconfigurations. It provides all required operating system functions for module management.

3.4 Inter-module Communication

One of the central limiting factors for the wide use of partial dynamic reconfiguration not yet addressed is the problem of inter-module communication. Each module that is placed on one or more slots on the device must be able to communicate with other modules. The ESM provides and supports four main paradigms for communication among different modules (see Fig. 3.6): The first one is a direct communication using bus-macros between adjacently placed modules (see Fig. 3.6(a)). Secondly, shared memory communication using SRAMs or BlockRAMs is possible (see Fig. 3.6(b)). However, only adjacent modules can use these two communication modes. For modules placed in non-adjacent slots, we provide a dynamic signal switching communication architecture called *Reconfigurable Multiple Bus* (RMB) [2, 1] (see Fig. 3.6(c)). Finally, the communication between two different modules can also be realized through the external crossbar (see Fig. 3.6(d)). These four communication modes will be explained more closely in the following.

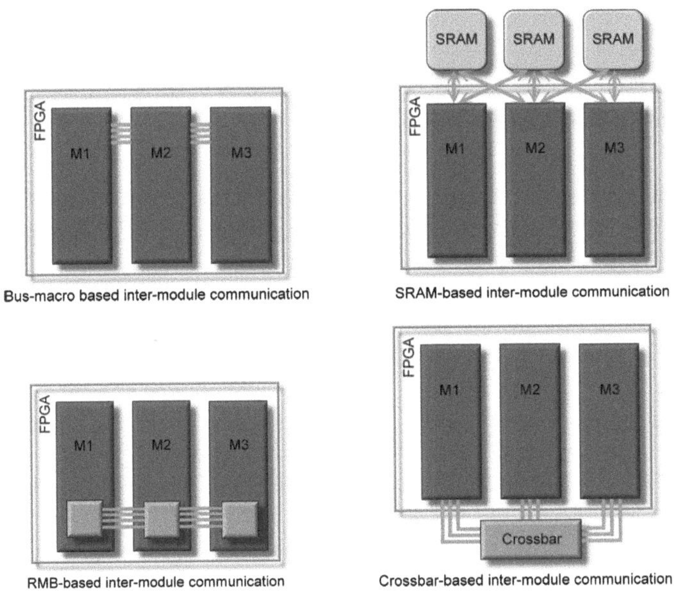

Fig. 3.6 Inter-module communication possibilities on the ESM: (a) bus-macro, (b) shared memory, (c) reconfigurable multiple bus (RMB), (d) external crossbar.

3.4.1 Communication Between Adjacent Modules

On the ESM, bus-macros are used to realize a direct communication between adjacently placed modules, providing fixed communication channels that help to keep the signal integrity upon reconfiguration. Because only four 1 bit signals can be passed for each bus-macro, the number of bus-macros needed for connecting a set of n signals between two placed modules is $n/4$.

3.4.2 Communication via Shared Memory

Communication between two neighboring modules can be done in two different ways using shared memory: First, dual-ported BlockRAMs can be used for implementing communication among two neighbor modules working in two different clock domains. The sender writes on one side, while the receiver reads the data on the other side. The second possibility uses external RAM. This is particular useful in applications in which each module must process a large amount of data and then sends the processed data to the next module, as it is the case in video streaming. On the ESM, each SRAM bank can be accessed by the module placed below as well as those neighbors placed right and left. A controller is used to manage the SRAM access. Depending on the application, the user may set the priority of accessing the SRAM for the three modules. In Sect. 3.6, we will present a video streaming case study that uses this way of communication.

3.4.3 Communication via RMB

In its first mentioning, the Reconfigurable Multiple Bus (RMB) architecture [12, 21, 2] consists of a set of processing elements or modules, each possessing an access to a set of switched bus connections to other processing elements. The switches are controlled by connection requests between individual modules. On the ESM, an RMB is realized by a one-dimensional arrangement of switches between N slots (see Fig. 3.7). In our FPGA implementation, the horizontal arrangement of parallel switched bus line segments allows for the communication among modules placed in the individual slots. The request for a new connection is done in a wormhole fashion, where the sender (a module in slot S_k) sends a request for communication to its neighbor (slot S_{k+1}) in the direction of the receiver. Slot S_{k+1} sends the request to slot S_{k+2}, etc., until the receiver receives the request and returns an acknowledgment. The acknowledgment is then sent back in the same way to the sender. Each module that receives an acknowledgment sets its switch to connect two line segments. Upon receiving the acknowledgment, the sender can start the communication. The wired and latency-free connection (circuit routing) is then active until an explicit release signal is issued by the sender module. The concept of an RMB was first presented in [21] and extended later in [12] with a compaction mechanism for quickly finding a free segment. However, it has been never implemented in real hardware.

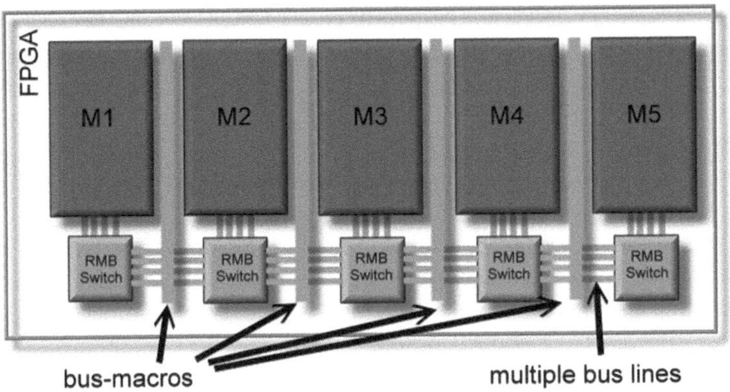

Fig. 3.7 FPGA implementation of an RMB (Reconfigurable Multiple Bus) architecture providing circuit-switched bus connections between individual modules placed in slots $\{S1, \ldots, S5\}$.

In our implementation [2] of the RMB on Xilinx Virtex FPGAs, we separated the RMB switches from the modules. In this way, we provide a uniform interface to designers for connecting modules to the multiple line switches. The implementation of the RMB structure on an FPGA Virtex II 6000 with four processors and four parallel 16 bit lines reveals an area overhead of 4% with a frequency of 120 MHz on the controller [1]. In [1], we have summarized area and data speed numbers in terms of (a) different numbers of modules, (b) different numbers of parallel bus segments, and (c) bitwidths of each bus segment. As shown on Fig. 3.7, bus-macros are used at the boundary of modules and controllers (switches) to insure a correct operation upon reconfiguration.

We were able to show that a module reconfiguration can take place columnwise at the same time other modules are communicating on the chip without any signal interference. This is possible by storing the states of the RMB switches in regions of BlockRAM that are physically unaffected by partial reconfiguration.

3.4.4 Communication via the Crossbar

Another possibility of establishing communication among modules is to use the crossbar. Because all the modules are connected to the crossbar via the pins at the south of the FPGA, the communication among two modules can be set in the crossbar as well.

3.5 Reconfiguration Manager

Each dynamically reconfigurable computer requires an operating system for the initialization of executable application modules and their run-time supervision. The main tasks of such an operating system in case of the ESM are (a) scheduling of application modules, (b) management of free slots including slot segmentation and partitioning, (c) loading, unloading and relocation of application modules into slots, (d) configuration of peripheral devices, (e) configuration of the crossbar, and (f) bitstream management.

Initial experiments provided evidence that the most-time critical operations must be executed in hardware in order to keep the reconfiguration time of a minimum. The loading, unloading and relocation of modules may be considered to be the most time-critical tasks which have therefore been implemented in a dedicated hardware reconfiguration manager. All other system tasks can implemented in C and executed on the PowerPC embedded processor (see Fig. 3.3) belonging to the ESM MotherBoard. These two parts of the operating system are linked via a simple communication bus as shown in Fig. 3.3. This hardware/software interface is realized through a set of elementary reconfiguration instructions passed from the PowerPC to the reconfiguration manager on the BabyBoard using Memory-Mapped I/O. The minimal set of basic instructions that the reconfiguration manager implements are summarized as follows:

- LOAD (load bitstreams to their pre-compiled position)
- UNLOAD (unload bitstreams to deactivate a running module)
- RELOCATE_AND_LOAD (relocate bitstreams to a different slot position before loading)

In an implementation, the reconfiguration manager was built in hardware and located in a Spartan-II 400 FPGA which is connected to the main FPGA via the SelectMAP interface.

During normal operation, the bitstream data will be loaded from the flash memory located on the BabyBoard (see Fig. 3.4). However, bitstreams must be first downloaded and stored into the flash memory. Here, two methods are supported. The first method uses a parallel port interface implemented directly in the reconfiguration manager to download the configuration data from a host PC to the flash memory. The second method uses the Ethernet port of the PowerPC processor on the MotherBoard to download bitstreams from a remote host. In order to support these and also many other reconfiguration scenarios, we developed a very flexible, plugin-based reconfiguration manager architecture that will be described in the following.

3.5.1 Flexible Plugin Architecture

Figure 3.8 depicts the flexible architecture of the reconfiguration manager hardware. It consists of (a) a MicroBlaze microcontroller [23], and (b) a data crossbar switch is located between the plugins. The crossbar plugin shown in this figure connects the reconfiguration manager control implemented in software on the MicroBlaze to the ESM MotherBoard in order to establish the communication link to the PowerPC shown in Fig. 3.3.

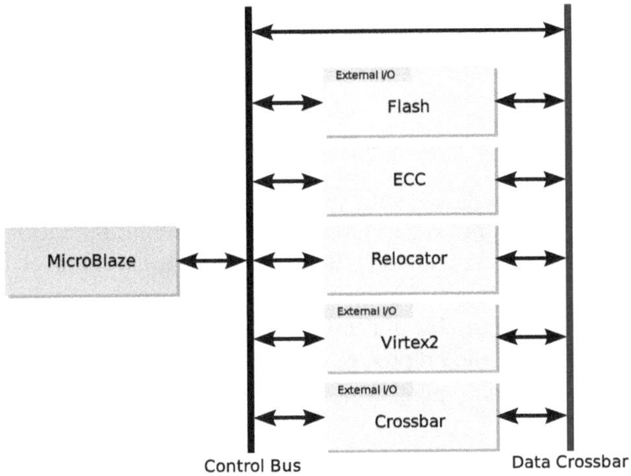

Fig. 3.8 Flexible architecture of the ESM reconfiguration manager with plugins such as Flash, ECC, module relocator and other possible plugins.

All plugin modules are connected to two communication interfaces: The control bus connects plugins to the MicroBlaze for initialization and control. The data crossbar connects to the data input and output ports of each plugin and its connection setup is also controlled by the MicroBlaze which is programmed in assembly language.

In order to upload a hardware module from flash to the FPGA, the following sequence of steps has to be performed:

- A command is sent from the PowerPC to the MicroBlaze to upload a bitstream to the FPGA without the use of any other plugins.
- The MicroBlaze connects the output of the flash plugin to the Virtex-II plugin input through a write into the configuration register of the data crossbar.
- Next, the MicroBlaze initializes the flash plugin with the start address and length of the bitstream.
- The MicroBlaze then enables the SelectMAP interface in the Virtex-II plugin.
- Then, the MicroBlaze enables the flash plugin to start reading the bitstream.
- The flash plugin sends the bitstream to the Virtex-II plugin byte by byte as long as its ready signal is true (if not, the flash plugin has to wait).
- While the flash and the Virtex-II plugin are running in parallel, the MicroBlaze checks periodically if any of the plugins has finished its operation.
- Only if after finishing one command, the MicroBlaze can execute a new command, and, for example, reinitializes the plugins and the data crossbar.

If one load command has been executed and another load follows, then the procedure starts from second step, because the data crossbar has already been set. The addition of plugins to the reconfiguration manager is simple. Any new module must have a fixed control bus interface and a fixed data crossbar interface. With these standard interfaces, the plugin can be directly controlled through the MicroBlaze assembly program. The data crossbar uses a parametrized HDL description which can be configured at design-time to the number of actually instantiated plugins.

3.5.2 Reconfiguration Scenarios

Depending on the operating system requirements, different operations need to be performed on each bitstream in general. Before a bitstream is uploaded to the FPGA, it can pass through any number of additional plugins. The order in which a bitstream passes the plugins is configurable at run-time through the setup of the data crossbar switch. This allows a flexible preprocessing of the bitstream prior to being loaded. Only the number of available plugins in the reconfiguration manager has to be determined at design-time.

Based on the introduced reconfiguration manager architecture depicted in Fig. 3.8, several reconfiguration scenarios are possible. Some of these are depicted in Fig. 3.9. In the first scenario, only a basic upload of a bitstream is performed. Therefore, the data flows from the flash plugin output directly through the data crossbar to the Virtex-II plugin input. If an error-correction (ECC) is desired, then the flash output data can be sent to the ECC plugin before going to the Virtex-II plugin. This case is shown in Fig. 3.9(b). In the third scenario, the bitstream is read from the flash, error-corrected and relocated before being sent to the Virtex-II plugin for upload (see Fig. 3.9(c)). Here, the crossbar is configured by the MicroBlaze processor in such a way that the output of each plugin is connected to the input of its neighboring plugin. The fourth scenario depicted in Fig. 3.9(d) shows how the bitstream data is delivered by the PowerPC through the MotherBoard crossbar. The bitstream is subsequently error-corrected and relocated prior to its upload.

The plugins that are currently implemented for the reconfiguration manager are: ECC plugin, decompression plugin and relocator plugin which can translate a bitstream on the fly to any slot location on the FPGA by directly manipulating the address offsets in the bitstream at load-time.

3.5.3 Implementation Results

The reconfiguration manager was implemented including the MicroBlaze microcontroller, parallel port interface plugin, flash memory interface plugin, Virtex-II SelectMAP plugin, an OPB (open peripheral bus) interface implementing the con-

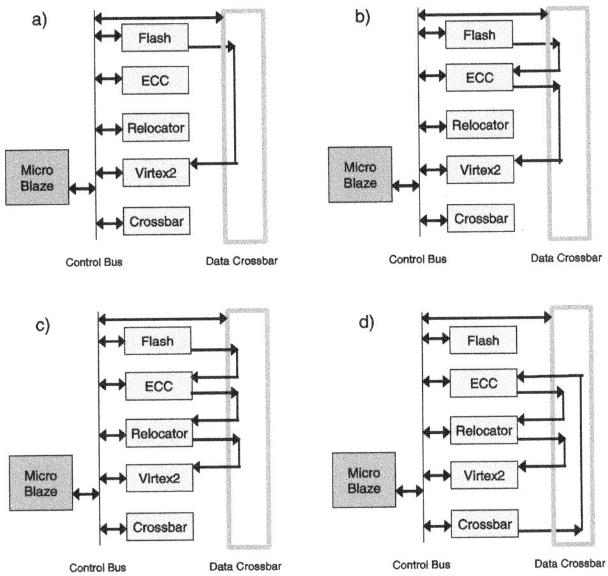

Fig. 3.9 Four different reconfiguration scenarios supported by the ESM reconfiguration manager.

trol bus and the data crossbar. The control bus is a 32 bit OPB bus, while the data crossbar is an 8 bit full duplex crossbar.

The flash plugin interface is able to sustain a data rate of 10 Mbyte/s in a conservative and tested timing setup. As the SelectMAP interface can upload bitstreams at a rate of 50 MByte/s, an additional decompression plugin would accelerate the reconfiguration time when used on compressed bitstreams.

The final board implementation of the BabyBoard and MotherBoard is shown in Fig. 3.10. The reconfiguration manager has been implemented in the Spartan-II 400 FPGA which is located close to the 64 MByte flash device and the main Virtex-II 6000 FPGA. Technical data sheets as well as software, primer applications, and user information is available at http://www.r-space.de.

The separation of BabyBoard and MotherBoard was made in order to customize the ESM architecture to other application domains such as automotive. In order to do so, a new MotherBoard could be designed to have different peripherals such as CAN, LIN, FlexRay controllers, and A/D and D/A converters.

3.6 Case Study: Video and Audio Streaming

Video streaming can be defined as the process of performing computations on video data streams. Many video algorithms process the data stream picture-by-picture. Usually, a picture frame is transmitted pixel-by-pixel and therefore can be processed on a pixel-by-pixel basis. However, since a lot of algorithms require the neighbor-

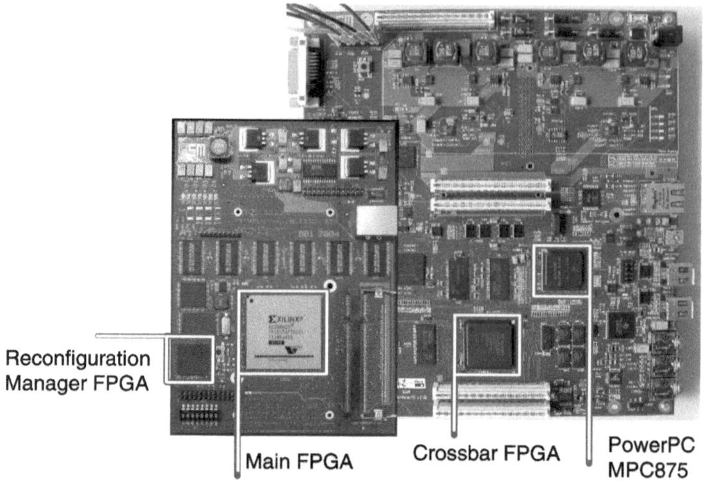

Fig. 3.10 Implementation of the ESM BabyBoard (left) and MotherBoard (right). Technical data sheets as well as software and application codes are available at http://www.r-space.de.

hood of a pixel, e.g., for filtering, often at least one complete frame must be stored and processed before the next one can be accessed. Capturing the neighborhood of a pixel is often done using a sliding window [6] the size of which varies according to the size of a neighbor region. A given set of buffers (FIFO) is used to update the window. The number of FIFOs varies according to the size of the window. In each step, a pixel is read from the memory and placed in the lower left cell of the window. Up to the upper right pixel which is disposed, i.e., output, all the pixels in the right part of the window are placed at the queue of the FIFO one level higher.

In the field of video compression, the processing is usually done in a block-by-block basis, different from the sliding window concept. However, the overall structure is almost the same.

As shown in Fig. 3.11, the architecture of a video streaming system is usually built on a modular basis. The first module buffers with the image captured from an image source. This can be a camera or a network module which collects the picture data through a network channel, or any other source. The frames are alternately written to the SRAM banks RAM1 and RAM2 by the capture module. The second module collects the picture from RAM1 or RAM2 if this RAM module is not in use by the first module, builds the sliding windows and passes it to the third module which processes the pixel and saves it in its own memory or directly passes it to the next module. This architecture presents a pipelined computation in which the computational blocks are the modules that process the image frames. RAMs are used to temporally store image frames between two modules, thus allowing a frame to stream from RAM to RAM and the processed pictures to the output. An adaptive video streaming system is characterized by its ability to optimize the computation performed on the video stream according to changing environmental conditions. In most cases, only one module on the computation chain must be changed while the

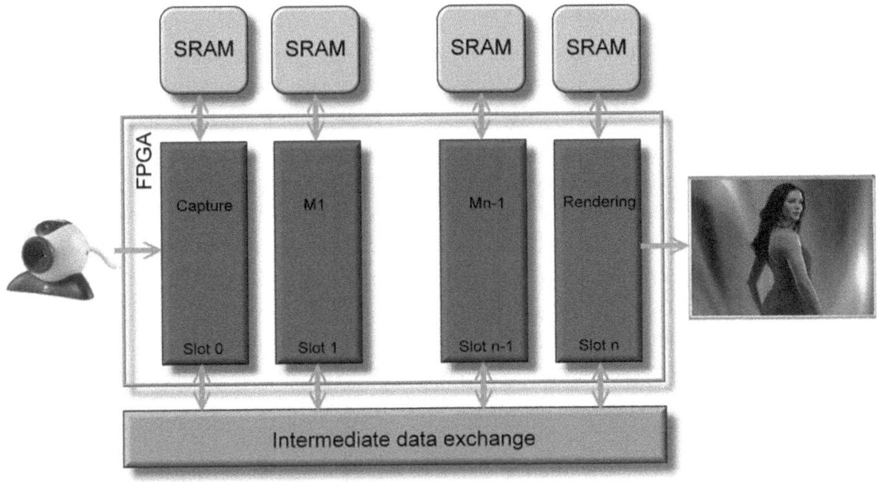

Fig. 3.11 A modular architecture for video streaming as implemented onto the slot-based structure of the ESM.

other keep running. The video capture module, for example, can be changed if we want to optimize the conversion of pixels to match the current brightness or the current landscape. It is also possible to change the video source from a camera to a new one with different characteristics. In an adaptive system, the functionality of a module on the computation path should be changed very fast without affecting the rest of the system. This can be done by providing some parameters to the module to instruct it to switch from one algorithm to the next one. However, the structures of the basic algorithms are not always the same. A Sobel filter [13], for example, cannot be changed into a Laplace filter by just changing the parameters. This is also true for a Median-operator which cannot be replaced by a Gauss-operator by just changing parameters. Network and camera require two different algorithms for capturing the pixels. In many cases, the complete module should be replaced by a module of the same size, but different in its structure while the rest of the system shall keep running.

Our ESM architecture fulfills the prerequisites for a modular pipelined and adaptive system for video streaming. In the system architecture presented before, we divided the device into slots, which each of them can implement a given module. RAMs are provided to the north of the device while the southern pins can be used by modules to communicate with the rest of the environment.

3.7 Usage of the ESM in Different Fields

Built in order to make partial hardware reconfiguration become a reality, the ESM platform has shown its benefits as a general interdisciplinary platform already in several quite different application fields and other research projects as well:

- *Reconfigurable Networks (ReCoNets)*: In the ReCoNets project (see Chap. 11), reconfigurable nodes are connected together to form a network of reconfigurable computers. Novel procedures for self-repair and intelligent partitioning were developed to achieve a higher level of fault tolerance. In order to guarantee short repair times in case of node defects, the placement of tasks is optimized and replicated nodes are created. The ESM platform has been integrated and used in this network. According to Chap. 11, applications taken from automotive networking have been shown to provide sophisticated implementations for hardware and software tasks that may migrate within the network.
- *Reconfigurable Operating Systems (ReCoNos)*: The group of Prof. Platzner developed new aspects of operating systems for reconfigurable hardware based on the ESM platform (see Chap. 13). Hereby, it was shown for the first time that operating system resources could be shared between software programs and reconfigurable hardware modules, e.g. for synchronization.
- *Partial Module Visualisation*: The group of Prof. Becker is known for their research on dynamic 2D routing and placement (see Chap. 12). The ESM platform provided here an ideal experimentation platform due to its large FPGA without integrated processors and the unfragmented resources. The external PowerPC was applied for on-line reconfiguration of the routing calculations. Furthermore, a visualizer of reconfigurable modules was developed and demonstrated at FPL 2008:
- *Reconfigurable Videoengines (AutoVision)*: The ESM was also applied to develop a reconfigurable driver assistance system. The group of Prof. Stechele (see Chap. 18) working on reconfigurable video engines which adopt to the current driving situation in order to increase driving compfort and prevent car accidents. The ESM platform was applied because of its flexibility, and the sufficient available memory. Results of joint work have been published in [10, 5]. Notably, partially reconfigurable video engines applied to automotive applications were demonstrated at the CeBit 2008, see Fig. 3.12.
- *Partitioning Strategies*: The group of Prof. Merker applied the ESM for the implementation of parallel algorithms, because (1) the FPGA provided sufficient resources for the implementation, (2) local SRAM allowed the implementation of tasks, which needed a lot of local storage, and (3) the communication structures of the ESM offered new opportunities for the exchange of data between tasks. Furthermore, the ESM was used to develop new partitioning strategies.
- *Task Preemption*: Despite the possibility to execute several hardware tasks in parallel on an FPGA, partial reconfiguration runs typically sequential. There exists only one reconfiguration port which is used exclusively during the reconfiguration of a task (module) on all available platform. Single processor scheduling algorithms for task reconfiguration with preemption have been evaluated in a real time application implemented on the Erlangen Slot Machine. Besides allowing reconfigurable connections of peripherals to pins of the FPGA, the Virtex-II FPGA of the ESM allows to host applications requiring quite a large number of slots. This has been used to study and develop preemption in the reconfiguration phase, see [11].

Fig. 3.12 The group of Prof. Stechele and Prof. Teich at the CeBIT 2008 demonstrating the partially reconfigurable video modules on the Erlangen Slot Machine.

- *Security of ECC implementations*: The Erlangen Slot Machine was finally also used in the project of securing ECC implementations against differential power analysis by Prof. Huss, see Chap. 19.

3.8 Conclusions

We have presented a new dynamically reconfigurable computer architecture called Erlangen Slot Machine (ESM) that was built for reasons that many brilliant ideas for reconfigurable computers and for dynamic resource management cannot be efficiently and directly transfered using currently available technology, mainly because of I/O-pins, memory, and inter-module communication constraints. Although new generations of FPGAs do provide new reconfiguration opportunities such as smaller entities of reconfiguration, the ESM is a unique stand-alone reconfigurable computer that has shown to bridge this gap by providing (a) new architectural concepts to avoid the above physical problems and restrictions, (b) new inter-module communication concepts, as well as (c) an intelligent module reconfiguration management.

The development of the ESM board was a huge task, and would have been impossible without a tight joint work on different fields, therefore we feel deeply indepted to the people mentioned in Table 3.1 and would like to thank them for their hard work.

Table 3.1 List of people involved in the development of the ESM computer.

Person	Task
Mateusz Majer	Project coordinator, firmware, board design, video demonstrator
Josef Angermeier	Operating system, firmware, ESM tutorial, taillight engine demonstrator
Bruno Kleinert	Reconfiguration manager driver
Thomas Stark	Crossbar driver
Ulrich Batzer	Taillight engine demonstrator
Matthias Kovatsch	Taillight engine demonstrator
Amouri Abdulazim	Main FPGA visualizer
Diana Göhringer	RMB communication
Andre Linarth	Motherboard PCB
Ding Li	Crossbar design
Peter Asemann	Programmer's interface
Christian Freiberger	Reconfiguration manager
Jan Grembler	Video demonstrator
Felix Reimann	RMB communication
Christoph Lauer	Module relocator

Acknowledgements Supported by DFG grant TE 163/14-2, project *ReCoNodes*, as part of the Priority Programme 1148, *Reconfigurable Computing Systems*. We would also like to acknowledge the DFG for providing monetary support to build 20 prototypes of the ESM boards that are used today in the above projects but also on a lending base at some universities world-wide. Thanks also to Patrick Lysaght at Xilinx.

References

1. Ahmadinia, A., Ding, J., Bobda, C., Teich, J.: Design and implementation of reconfigurable multiple bus on chip (RMBoC). Technical Report 02-2004, University of Erlangen-Nuremberg, Department of CS 12, Hardware-Software-Co-Design (2004)
2. Ahmadinia, A., Bobda, C., Ding, J., Majer, M., Teich, J., Fekete, S., van der Veen, J.: A practical approach for circuit routing on dynamic reconfigurable devices. In: Proceedings of the 16th IEEE International Workshop on Rapid System Prototyping, Montreal, Canada, pp. 84–90 (2005)
3. Ahmadinia, A., Bobda, C., Majer, M., Teich, J., Fekete, S., van der Veen, J.: DyNoC: A dynamic infrastructure for communication in dynamically reconfigurable devices. In: Proceedings of the International Conference on Field-Programmable Logic and Applications, Tampere, Finland, pp. 153–158 (2005)
4. Alpha Data Ltd.: ADM-XRC-II Xilinx Virtex-II PMC (2002). URL http://www.alpha-data.com/adm-xrc-ii.html
5. Angermeier, J., Batzer, U., Majer, M., Teich, J., Claus, C., Stechele, W.: Reconfigurable HW/SW architecture of a real-time driver assistance system. In: Proceedings of the Fourth In-

ternational Workshop on Applied Reconfigurable Computing (ARC). Lecture Notes in Computer Science (LNCS), pp. 149–159. Springer, London (2008)
6. Bobda, C., Ahmadinia, A., Majer, M., Ding, J., Teich, J.: Modular video streaming on a reconfigurable platform. In: Proceedings of the IFIP International Conference on Very Large Scale Integration, Perth, Australia, pp. 103–108 (2005)
7. Bobda, C., Majer, M., Ahmadinia, A., Haller, T., Linarth, A., Teich, J.: Increasing the flexibility in FPGA-based reconfigurable platforms: The Erlangen Slot Machine. In: Proceedings of the IEEE Conference on Field-Programmable Technology, Singapore, Singapore, pp. 37–42 (2005)
8. Bobda, C., Majer, M., Ahmadinia, A., Haller, T., Linarth, A., Teich, J., Fekete, S.P., van der Veen, J.: The Erlangen Slot Machine: A highly flexible FPGA-based reconfigurable platform. In: Proceeding IEEE Symposium on Field-Programmable Custom Computing Machines, pp. 319–320 (2005)
9. Celoxica Ltd.: Rc2000 Development Board (2004). URL http://www.celoxica.com/products/boards/rc2000.asp
10. Claus, C., Stechele, W., Kovatsch, M., Angermeier, J., Teich, J.: A comparison of embedded reconfigurable video-processing architectures. In: Field Programmable Logic and Applications, 2008. FPL 2008. International Conference on, pp. 587–590 (2008). doi:10.1109/FPL.2008.4630015
11. Dittmann, F., Weber, E., Montealegre, N.: Implementation of the reconfiguration port scheduling on the Erlangen Slot Machine. In: FPGA '09: Proceeding of the ACM/SIGDA International Symposium on Field Programmable Gate Arrays, pp. 282. ACM, New York (2009). doi:10.1145/1508128.1508192
12. ElGindy, H.A., Somani, A.K., Schröder, H., Schmeck, H., Spray, A.: RMB—a reconfigurable multiple bus network. In: Proceedings of the Second International Symposium on High-Performance Computer Architecture (HPCA-2), San Jose, California, pp. 108–117 (1996)
13. Gonzalez, R., Woods, R.: Digital Image Processing. Prentice Hall, New York (2002)
14. Kalte, H.M., Porrmann, U.R.: A prototyping platform for dynamically reconfigurable system on chip designs. In: Proceedings of the IEEE Workshop Heterogeneous Reconfigurable Systems on Chip (SoC), Hamburg, Germany (2002)
15. Krasteva, Y., Jimeno, A., Torre, E., Riesgo, T.: Straight method for reallocation of complex cores by dynamic reconfiguration in Virtex II FPGAs. In: Proceedings of the 16th IEEE International Workshop on Rapid System Prototyping, Montreal, Canada, pp. 77–83 (2005)
16. Majer, M., Teich, J., Ahmadinia, A., Bobda, C.: The Erlangen Slot Machine: A dynamically reconfigurable FPGA-based computer. J. VLSI Signal Process. Syst. **47**(1), 15–31 (2007)
17. Majer, M., Teich, J., Bobda, C.: ESM—the Erlangen Slot Machine. http://www.r-space.de
18. Nallatech, Inc.: FPGA Boards
19. Platzner, M., Thiele, L.: XFORCES—executives for reconfigurable embedded systems. http://www.ee.ethz.ch/~platzner
20. Steiger, C., Walder, H., Platzner, M., Thiele, L.: Online scheduling and placement of real-time tasks to partially reconfigurable devices. In: Proceedings of the 24th International Real-Time Systems Symposium, Cancun, Mexico, pp. 224–235 (2003)
21. Vaidyanathan, R., Trahan, J.L.: Dynamic Reconfiguration: Architectures and Algorithms. IEEE Computer Society, Los Alamitos (2003)
22. Xess Corp: FPGA Boards. http://www.xess.com
23. Xilinx, Inc.: MicroBlaze CPU. http://www.xilinx.com

Part II
Design Methods and Tools—Modeling, Evaluation and Compilation

Chapter 4
Models and Algorithms for Hyperreconfigurable Hardware

Sebastian Lange and Martin Middendorf

Abstract A new concepts for run time reconfigurable systems called hyperreconfiguration is investigated in this project. Hyperreconfigurable architectures can dynamically change their reconfiguration potential to adapt to the current needs of a computation and therefore gauge high flexibility with high reconfiguration overhead against small flexibility with reduced overhead. Within this concept we also propose the use of multi-level reconfiguration where higher-level hyperreconfiguration operations define the flexibility of the system for reconfiguration while lower-level reconfiguration operations alter the system's functionality within the limits set by the preceding hyperreconfigurations.

4.1 Introduction

Dynamically reconfigurable architectures or systems can adapt their function and/or structure to suit the changing needs of a computation during run time (e.g., [1]). The advantage of run time reconfiguration is that it allows new algorithmic solutions for many applications. A principle problem of such architectures is that the high flexibility offered by a large number of reconfigurable components leads to a large amount of information required for reconfiguration. This large amount of data transfer makes run time reconfigurations time critical operations. Therefore, new approaches for dynamic reconfiguration have been proposed in the literature in the last years to cope with this problem. Compression methods for the reconfiguration bit stream can reduce the amount of data that has to be transferred to the chip [2]. The compression methods can be enhanced by a reordering of the frames in the configuration data [24] or by considering differential configuration data [5, 23]. Also

Sebastian Lange · Martin Middendorf
Parallel Computing and Complex Systems, Department of Computer Science, University of Leipzig, Leipzig, Germany, e-mails: langes@informatik.uni-leipzig.de, middendorf@informatik.uni-leipzig.de

M. Platzner et al. (eds.), *Dynamically Reconfigurable Systems*,
DOI 10.1007/978-90-481-3485-4_4, © Springer Science+Business Media B.V. 2010

special hardware that allows to address a set of reconfiguration bits that receive the same value has been designed (an example is the wildcard mechanism of the Xilinx XC6200 [3]). Computation of the reconfiguration bits directly on the chip can also prevent long loading time (see [6, 25, 26]). For partial reconfigurable architectures it is advantageous to reuse certain frames (see Chap. 8).

Another concept to make run time reconfiguration faster has been introduced by the authors in [10, 13]. It uses the fact that algorithms or computations typically consist of different phases where during each phase only a fraction of the reconfiguration potential of the underlying architecture is needed. The idea is to make the reconfiguration potential itself reconfigurable, i.e., to adapt the actually available set of reconfigurable resources (and thus the number of reconfiguration bits that are necessary to define the state of the architecture during reconfiguration) to the actual needs during run time. Then the smaller the actual reconfiguration potential of an architecture is the smaller is the amount of reconfiguration information that has to be transferred during reconfiguration and therefore the faster is a corresponding reconfiguration step. Architectures that follow this concept are called *hyperreconfigurable* [10, 13]. Since the reconfiguration takes place on different levels in this architectures they are also *multi-level reconfigurable*. In this chapter we give an introduction and overview on some of our work on hyperreconfigurable architectures. Because of the page limit we can neither give proofs nor algorithms. Also, some topics can not be covered here. These are hyperreconfigurable architectures with caches for (hyper-)contexts (for details see [16]), multi-task hyperreconfigurable architectures (for details see [11, 14]), online strategies for hyperreconfigurable architectures (for details see [19]), and applications to control systems (for details see [9]).

This chapter is organized as follows. In the next Sect. 4.2 the concept of hyperreconfigurable architectures is introduced. Example Architectures and test applications that are used to obtain experimental results are described in Sect. 4.3. Section 4.4 discusses the algorithmic problem to decide when hyperreconfiguration operations should be executed and which hypercontexts should be used. Section 4.5 discusses different levels of granularity for the reconfiguration of multi-level hyperreconfigurable architectures. A comparison of partially reconfigurable architectures to hyperreconfigurable architectures is presented in Sect. 4.6.

4.2 Hyperreconfigurable Machines

In this section we introduce the concept of hyperreconfiguration for reconfigurable architectures with two levels of reconfiguration. This concept is then extended to more than two levels of reconfiguration.

2-level Reconfiguration A *configurable system* is considered here as a parameterizable electronic system for information manipulation with a set $P = \{p_1, \ldots, p_n\}$ of parameters. We assume that all of the parameters are vital for the system's operation. In order to be operational, a configurable system must be given a sequence $\bar{c}$ of n parameter values (the ith value assigned to p_i). Let $\bar{C}$ be the set of all sequences of parameter values that define a possible state of the system, i.e., they are feasible. A system is *run time reconfigurable* when some of the parameter values can be changed multiple times and at run time. In this chapter we assume that all parameters in P are run time reconfigurable. We call a sequence $\bar{c}_1, \ldots, \bar{c}_n$ of parameter values where $\bar{c}_i$ is a possible value for p_i a *context*. The operation that assigns new values to the parameters is called (ordinary) reconfiguration (operation).

The idea of hyperreconfiguration is to parameterize the reconfiguration potential of a system itself by introducing a two-level run time reconfiguration scheme. On the upper level of reconfiguration, the system provides an additional set $P' = \{p'_1, \ldots, p'_n\}$ of parameters that adapts the reconfiguration potential of the system. In other words, the upper level of reconfiguration restricts the set of run time parameters that are available for reconfiguration. On the lower level, only those parameters of the run time reconfigurable system that were made available by the previous upper level reconfiguration can be assigned new values during reconfiguration. Systems that provide this capability are called *hyperreconfigurable systems*. An operation that assigns new values to the parameters in P' is called *hyperreconfiguration (operation)* or *upper level reconfiguration*. We call a corresponding sequence $h = h_1, \ldots, h_n$ of parameter values where h_i is a possible value for p'_i a *hypercontext*. In this chapter we assume that p'_i is a boolean parameter. If its value is '1' then p_i can be assigned a new value during a reconfiguration and otherwise it is not available for reconfiguration (in this case it might have a default value or the old value). Hence, a hypercontext $h = h_1, \ldots, h_n$ defines a subset $\{p_i \mid h_i = 1, i = 1, \ldots, n\}$ of P which can be assigned a new value during reconfiguration. Let $|h| = |\{p_i \mid h_i = 1, i = 1, \ldots, n\}|$. It should be noted that a more general definition for hyperreconfigurable systems has been given in [10, 13].

Concerning an algorithm or a computation that runs on a reconfigurable system we are in this chapter only interested in those aspects of the algorithm that have to do with reconfiguration. The specific computation operations that are executed between reconfiguration operations are not relevant here. Thus, we assume that an algorithm that performs m reconfiguration operations is characterized by a sequence $c = c_1, \ldots, c_m$ of context requirements where a *context requirement* c_i is a sequence $c_i = c_{i1}, \ldots, c_{in}$ of boolean values. Hence, for each reconfiguration operation there exists exactly one corresponding context requirement. If $c_{ij} = 1$ for context requirement c_i then the parameter p_j might be assigned a new value during the corresponding reconfiguration and otherwise it will not be assigned a new value. Hence, a context requirement defines a subset of parameters from P that might be assigned a new value during the corresponding reconfiguration operation. Let C be the set of context requirements that are possible. Often it will not be possible to determine in advance which parameters an algorithm will assign new values in a configuration step. This is typically the case when a context depends on data that are computed at run time. But for the concept of reconfigurable architectures it is enough when an upper bound on the requirements that will actually be needed during run time can be given. For example, it might be possible to know in advance that the routing requirements will be low during a certain phase of the algorithm even when the exact routing scheme is not known in advance. Note, that several methods for resource estimation on reconfigurable architectures have appeared in the literature (e.g., see [4]).

In this chapter we consider only one special type of hyperreconfigurable systems which is called the *switch model* of hyperreconfigurable systems (for other models see, e.g., [13]). Such systems are characterized by the following additional assumptions: (i) every parameter $p_i \in P$ is boolean, (ii) $\bar{C} = \{0,1\}^n$, i.e., every sequence of length n with values in $\{0,1\}$ is a feasible context, (iii) $C = \{0,1\}^n$, i.e., every subset of P can be specified by an algorithm as a context requirement, (iv) a reconfiguration operation is possible only if the actual hypercontext *satisfies* the corresponding context requirement, i.e., for each $i \in [1,n]$ with $c_i = 1$ in the corresponding context requirement the actual hypercontext has $p'_i = 1$.

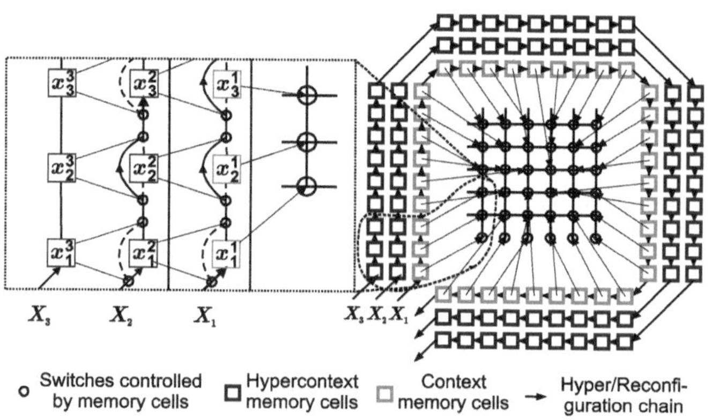

Fig. 4.1 2-level reconfigurable switch box with three reconfiguration chains and memory cells x_i^k, $k \in \{1,2,3,\}$.

To illustrate the concept of hyperreconfiguration consider the example of a switchbox for an ordinary reconfigurable architecture (e.g., an FPGA (Field Programmable Gate Array)) as shown in Fig. 4.1. The switchbox has a set $X = \{x_1,\ldots,x_n\}$ of switches. The state of each switch x_i is defined by the value of parameter p_i which is stored in a corresponding memory cell x_i^1 (e.g., an SRAM cell). During reconfiguration the memory cells are chained sequentially to form a reconfiguration chain which acts similarly to a shift register, shifting in one bit $\bar{c}_i$ of the new context $\bar{c}$ and shifting out one bit of the old context at each time step. This is done until each value $\bar{c}_i$ is stored in the memory cell x_i^1 of its corresponding switch x_i, $i = 1,\ldots,n$. Thus for an ordinary reconfigurable switchbox it is necessary to define the state of every switch, i.e., n values $\bar{c}_i$ have to be transferred into the reconfiguration chain.

For a hyperreconfigurable switchbox it is assumed that a single memory cell x_i^2 (hypercontext memory cell) is added to the architecture for each switch x_i. Each hypercontext memory cell x_i^2 manipulates two additional switches—one in front and one behind the corresponding (context)-memory cell x_i^1. These switches control whether the memory cell x_i^1 is part of the shift register and the reconfiguration bits are sent through the memory cell during reconfiguration, or bypass it and thereby exclude the switch x_i from reconfiguration. Loading a new hypercontext is done analogously to reconfiguration: the (hypercontext) memory cells form a shift register and the bits of the new hypercontext are shifted in. In the example of Fig. 4.1 two of the three shown hypercontext memory cells contain '1' which means the corresponding context SRAM cells are included in the current chain of context memory cells. The other hypercontext memory cell contains '0' which means the corresponding context memory cell is bypassed and therefore not included in the current chain of context SRAM cells. Since the number of bits in the hypercontext is always the same, the time for a hyperreconfiguration step does not change.

To measure the costs (time) for (hyper-)reconfiguration operations two cost measures are used: (i) $init(h_{prev}, h)$ is the cost of performing a hyperreconfiguration that brings the machine from hypercontext h_{prev} into a new hypercontext h, (ii) $cost_h(\bar{c}_{prev}, \bar{c})$ denotes the cost of an ordinary reconfiguration that brings the machine from context $\bar{c}_{prev}$ into context $\bar{c}$ when the machine is in hypercontext h. For the switch box model the time is measured as the number bits that have to be transfered, i.e., the time to load the (hyper-)context bits into the (hyper-) context memory cells (assuming one time unit is the time to load one bit). Thus, $init(h_{prev}, h) = init(h_h) = n$ and $cost_h(\bar{c}_{prev}, \bar{c}) = |h|$ is the number of context memory cells that are included into the reconfiguration chain under hypercontext h.

An implementation of an algorithm $C = c_1, \ldots, c_m$ on a hyperreconfigurable machine is characterized by a partition of C into substrings $S_1, \ldots, S_r$ (i.e., $C = S_1 \ldots S_r$) and hypercontexts $h_1, \ldots, h_r$, $r \geq 1$ such that each context requirement in S_i is satisfied by h_i (because h_i is the hypercontext of the hyperreconfiguration that is executed before the first reconfiguration of S_i). The corresponding total reconfiguration costs are $r \cdot n + (\sum_{i=1}^{r} |h_i| \cdot |S_i|)$ where $|S_i|$ is the length of S_i, i.e., the number of context requirements in S_i. It is assumed that a hyperreconfiguration is always performed before the first reconfiguration step. Figure 4.2 shows an example of an algorithm and its implementation on a hyperreconfigurable machine.

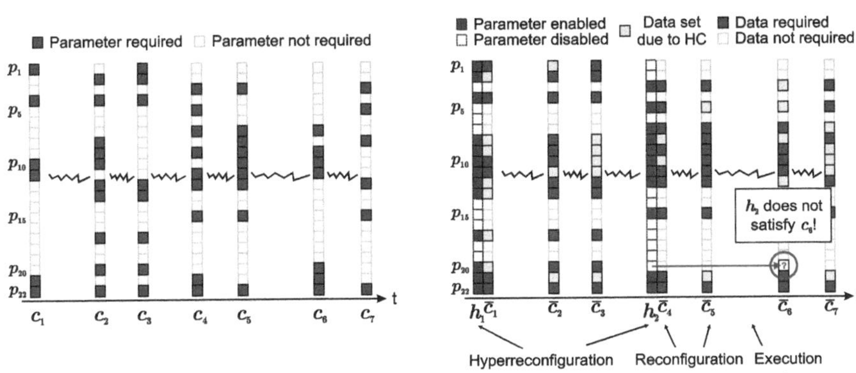

Fig. 4.2 Algorithm as a sequence $c_1, \ldots, c_n$ of context requirements for 20 parameters (switches) and a sequence of contexts and hypercontexts that would correspond to a valid implementation if h_6 had switch p_{20} enabled.

Multi-level Reconfiguration So far we have considered machines with 2 levels of reconfiguration but the concept can be extended to a higher number of reconfiguration levels [18]. An L-level hyperreconfigurable machines (in the switch model), $L \geq 2$, has a set $X^{(k)} = \{x_1^k, \ldots, x_n^k\}$ of memory cells for each level $k \in [1, L]$. Analogously to the 2-level hyperreconfigurable machines, the content of one such memory cell defines the presence of the corresponding memory cell in the reconfiguration chain of its subsequent lower level. Thus, a memory cell x_i^k is part of the k-level reconfiguration chain if and only if the content of memory cell x_i^{k+1} is set to '1'. Likewise, the contents of the memory cells in the 2-level reconfiguration chain define the availability of the corresponding switches for reconfiguration. The configuration chain of the highest level L always contains all n memory cells. In the following we call the contents of the memory cells belonging to the k-level reconfiguration chain the k-level context and denote it by h^k, $k \in [1 : L]$. Furthermore, we refer to the reconfiguration operation changing a k-level context as a *k-level hyperreconfiguration*, $k \in [2 : L]$. Figure 4.2 shows an example of an algorithm and its implementation on a hyperreconfigurable machine.

The amount of reconfiguration data for a k-level hyperreconfiguration is given by the number of memory cells in the corresponding reconfiguration chain, determined by h^{k+1}. A k-level hyperreconfiguration requires also the reconfiguration of all levels i, $2 \leq i < k$. During the execution of algorithm $C = c_1, \ldots, c_m$, the L-level hyperreconfigurable machine performs the sequence of operations $H_1 S_1, \ldots, H_r S_r$. Here H_i stands for a sequence of hyperreconfigurations at level k_i to level 2, $k_i \in [2 : L]$, and S_i stands for a sequence of 1-level reconfigurations, which use only those parts of the system that are made available by the 2-level context as defined by H_i. The costs of a of a sequence $H = h^k h^{k-1} \cdots h^2$ of hyperreconfigurations at levels k to 2 are $init(H) = init(h^k) + init(h^{k-1}) + \cdots + init(h^2) = |h^{k+1}| + |h^k| + \cdots + |h^3|$ (exception: if $k = L$ then $init(h^k) = n$). The total reconfiguration costs of algorithm C are $\sum_{i=1}^{r} init(H_i) + \sum_{i=1}^{r} |h_i^2| \cdot |S_i|$.

Heterogeneous Multi-level Reconfiguration Multi-level reconfigurable architectures allow for very flexible control of the reconfigurable resources since the reconfiguration chains contain a memory cell for each switch at each reconfiguration level. The number of memory cells required to implement the concept of multi-level reconfiguration thus increases linearly with the number of reconfiguration levels. On architectures where a great number of architectural parameters exist, the implementation overhead induced by multi-level reconfiguration can severely diminish the advantage of faster reconfiguration. To address this problem an extension of the concept of multi-level reconfigurable architectures is introduced which allows for different numbers of reconfiguration levels for each switch $x_i \in X$ [15, 17, 18, 20]. Formally, a *heterogeneous L-level hyperreconfigurable machine* has for each switch x_i a distinct number of L_i reconfiguration levels, $1 \leq L_i \leq L$. For each switch there exists a set of memory cells $Y_i = x_i^2, \ldots, x_i^{L_i}$. $X^{(k)}$ is the set of memory cells of the form $\{x_i^k \mid k \in [2 : Y_i], i \in [1 : n]\}$. Any subset of $X^{(k)}$ can form a reconfiguration chain depending on the content of the memory cells on the reconfiguration levels above. But memory cells on the highest level (i.e., memory cells $x_i^{L_i}$, $i \in [1 : n]$) are always part of their reconfiguration chain. Costs and hyperreconfiguration operations are defined analogously as for multi-level hyperreconfigurable architectures.

4.3 Example Architectures and Test Cases

In order to evaluate and validate our concepts fine-granular and coarse-granular hyperreconfigurable model machines have been used.

Fine-granular Hyperreconfigurable Machine The fine-granular machine that has been designed for this evaluation is called Simple Hyper-Reconfigurable Architecture (SHyRA). It can be can be seen as a model of a simple a 1-dimensional FPGA. The SHyRA (see Fig. 4.3) consists of a set of reconfigurable LUTs, a storage for intermediate values is facilitated by a file of single bit registers, which are connected to the LUTs by a multiplexer (MUX) and a demultiplexer (DeMUX). These multiplexers are implemented as full crossbars from all registers to all input ports of each LUT and respectively from each LUT output to all registers for the demultiplexer. The system interfaces to the outside world through the content of the register file. The architectural design should be regarded as an architectural template, which can be parameterized with the number of registers and LUTs to expose and the number of inputs per LUT.

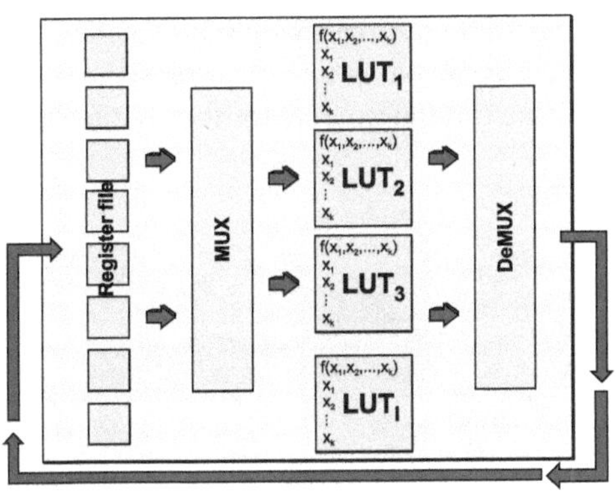

Fig. 4.3 Simple Hyper-Reconfigurable Architecture (SHyRA).

The execution of a computation on SHyRA is cycle-based. Each computational cycle consists of the following two steps:

1. Input stimuli are propagated from the registers through the multiplexer to the corresponding inputs of the LUTs.
2. A new output value is generated by each of the LUTs.
3. The resulting values transferred to their destination register in the register file by using the demultiplexer.

Hyperreconfiguration and reconfiguration operations can be performed before every computational cycle. The structure of SHyRA prohibits the use of LUTs in sequence in one cycle. In order to implement complex logic functionality exceeding the capabilities of single LUTs, multiple cycles are required to evaluate the complete circuit. Due to this restriction, application processes are forced to make extensive use of reconfiguration operations.

In order to facilitate an easier access to SHyRA, a compiler tool was implemented. This compiler translates structural VHDL descriptions into reconfiguration data suitable for SHyRA. The structural VHDL descriptions hereby make use of the Xilinx simulation primitives found in the SimPrim library. The implementation of this compiler is based on previous work described in [8].

The reconfiguration data of the SHyRA consist of three parts: (i) data for configuration of the multiplexer, (ii) data defining the truth tables of the individual memories, (iii) data for configuration of the demultiplexer. The Switch model of hyperreconfigurable architectures fits well to the SHyRA, because for each of the three types of reconfigurable resources every reconfiguration bit can be viewed as a binary switch.

The test cases implemented on the SHyRA consist of an 8 bit ripple carry adder, an LED decoder, and a 4 bit counter circuit. The resulting reconfiguration data describe 79 reconfigurations for the adder circuit, 224 for the LED decoder and 48 reconfigurations for the counter. The sequences of context requirements were extracted as follows: if a bit changes its value between subsequent reconfigurations is included in the corresponding context requirement. In order to determine the context requirement of the first reconfiguration operation, it was assumed that all parameters (reconfiguration data) have been set to zero prior to the first reconfiguration. The resulting sequences of context requirements for the counter is presented in Fig. 4.4.

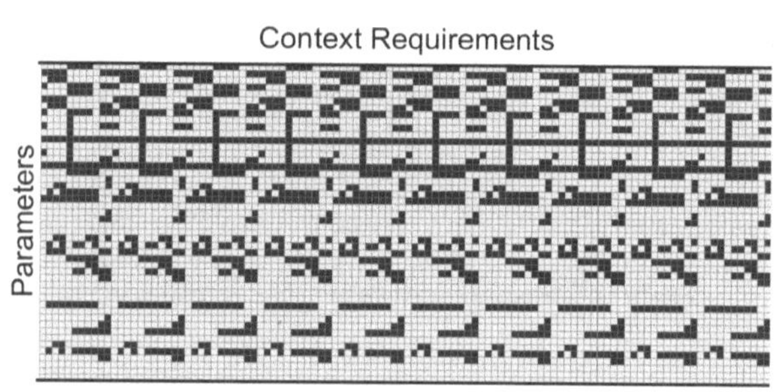

Fig. 4.4 Sequence of context requirements for the counter application.

Coarse-granular Hyperreconfigurable Machine For a coarse-grained example of reconfiguration, a VLIW processor is seen as a hyperreconfigurable machine where the functional units are the reconfigurable resources. A reconfiguration step takes place every clock cycle. Executing an algorithm on a VLIW processor thus closely resembles a rapidly reconfiguring system. Because of the high number of functional units a VLIW code word frequently encodes idle operations for units currently not usable due to a lack of parallelism in the algorithm. The concept of hyperreconfigurability can be employed to limit the effect of these idle operations.

The TMS320C6201 signal processor from Texas Instruments was used for the experiments. It features four types of functional units (L = logic, M = multiply, D = load/store, S = shift) with two units of each type. The functional units are divided into two sets, with each set containing one of the two units of each type. Each set of units uses its own file of registers. The sharing of data between functional units of different sets is implemented through cross-connection circuitry, which allows only one unit of a set to read data from the register file of the other set. The processor is accompanied by a C compiler framework, which produces highly optimized and parallelized code. The compiler flow also allows for an output of the assembly code which includes the assignment of instructions to corresponding functional units. This assignment provides information about the usage of individual functional units and thus their need for reconfiguration.

As test cases two algorithms, a Finite Impulse Response (FIR) filter and a vector summation, were implemented in C. The compiler framework was used to obtain parallelized VLIW code. The resulting code, which essentially represents the reconfiguration data of the processor, was then executed. During execution, the state (idle or busy) of each functional unit was recorded at each clock cycle, yielding an 8-bit vector for every clock cycle. The sequence of these vectors for each execution of the algorithms can be viewed as a sequence of context requirements for a hyperreconfigurable system that functions according to the Switch model. The sequence of 63 (44) context requirements was obtained for an execution of the FIR filter (respectively vector summation) on a data set of 124 (respectively 62) values.

4.4 The Partition into Hypercontexts Problem

An important problem that emerges for a hyperreconfigurable machine and a given algorithm (i.e, a sequence of context requirements) is to define when hyperreconfigurations should be done and how corresponding hypercontexts are defined such that the context requirements of the algorithm are satisfied and the total hyperreconfiguration costs are minimized [12, 13]. Formally we define (the definition for L-level reconfigurable machines can be done analogously).

Partition into Hypercontexts (PHC) problem: Given a hyperreconfigurable machine and a sequence $C = c_1 \cdots c_m$ of context requirements. Find a partition of C into substrings $S_1, \ldots, S_r$ (i.e., $C = S_1 \cdots S_r$) and hypercontexts $h_1, \ldots, h_r$, $r \geq 1$ such that every context requirement in S_i is satisfied by h_i and the total reconfiguration costs are minimal.

Theorem 4.1. *The PHC problem is NP-complete for general* 2*-level reconfigurable machines* [13].

Theorem 4.2. *The PHC-problem can be solved on heterogeneous L-level reconfigurable machines in the switch model in time* $O(m^3 + m^2(m^2 + n)(L - 2) + m^2 n)$ [18, 7].

It should be noted that for $L = 2$ the run time can be improved to $O(m^2 n)$. For a given algorithm it is also interesting to find the optimal number of reconfiguration levels so that the reconfiguration costs become minimal. Formally, we define

Reconfiguration Level (RLP) problem: Given a hyperreconfigurable machine and a sequence $C = c_1 \cdots c_m$ of context requirements. Find a number L of reconfiguration levels so that the solution of the corresponding PHC problem for L-level reconfigurable machines is minimal.

The corresponding problem for heterogeneous machines is to find for each switch the best number of reconfiguration levels, given that L is the maximum number of reconfiguration levels, $L \geq 2$. This problem is called the *Heterogeneous Reconfiguration Level* (H-RLP) problem.

Theorem 4.3. *The RLP problem is solvable for multi-level reconfigurable machines in the switch model in polynomial time* $O(nm^5 + m^6)$ [18].

Theorem 4.4. *The H-RLP problem is NP-hard for heterogeneous multi-level reconfigurable architectures in the switch model even when the maximum number of reconfiguration levels is only* 2 [18].

Since the H-RLP is NP-hard a polynomial time heuristic *Het_Rlp_Heur* has been developed. As a starting point, consider a homogeneous L-level reconfigurable machine with n switches and an algorithm $C = \{c_1, \ldots, c_m\}$. As a first step, the PHC problem is solved which results in a cost minimal assignment of upper level contexts to context requirements. Keeping the context assignment fixed, the change of reconfiguration costs when decreasing the number of reconfiguration levels L_i for a single switch x_i by 1 is determined. Observe that lowering the number of reconfiguration levels has two effects: (i) the overall reconfiguration costs are decreased by the number z_i of L_i-level hyperreconfigurations, (ii) the overall costs are increased by the number of $(L_i - 1)$-level hyperreconfigurations w_i during which the memory cell $x^{L_i - 1}$ is not part of the $(L_i - 1)$-level reconfiguration chain. Therefore, it is only advantageous to decrease the number of reconfiguration levels L_i of switch x_i if $z_i \geq w_i$. After the numbers of levels have thus been adjusted for all switches the PHC problem is solved again. The heuristic strategy repeats these steps until no further reduction of the number of reconfiguration level can be achieved.

4.4.1 Experiments and Results

In the following, we experimentally evaluate the merits of multi-level reconfigurable architectures. The minimum total reconfiguration costs of the five sample applications (described in Sect. 4.3) have been computed for homogeneous and heterogeneous multi-level reconfigurable architectures and compared to standard 1-level reconfigurable machines. Note, that the results for the heterogeneous machines have been obtained by using heuristic *Het_Rlp_Heur* whereas the other results are exact solutions. Furthermore, the number of reconfiguration levels yielding overall minimal reconfiguration costs are determined by solving the RLP problem for homogeneous architectures and for heterogeneous architectures.

The results presented in Table 4.1 show that the introduction of multi-level reconfiguration strongly reduces the reconfiguration costs for all test applications. The cost reduction compared to standard 1-level reconfiguration ranges for the homogeneous case from 21.3% for the vector sum to 87.0% for the LED decoder. Already, the introduction of one additional reconfiguration level leads to a reduction between 21.3% for the vector sum to 80.1% for the LED decoder in the homogeneous case. The optimal number of reconfiguration levels is between 2 and 4 in the homogeneous case. Adding more reconfiguration levels than optimal leads to a linear increase of the reconfiguration costs in all five test applications.

Table 4.1 Minimal total reconfiguration costs obtained for the five test application an architecture using full reconfiguration (1-level) and on homogeneous (Hom.) and heterogeneous (Het.) multi-level reconfigurable architectures with numbers of reconfiguration levels ranging from 2 to 8; costs of the RLP and Het-RLP-Switch solutions are printed in bold.

	8-Bit Adder		Counter		LED Decoder		FIR LMS		Vector Sum	
	Hom.	Het.	Hom.	Het.	Hom.	Het.	Hom.	Het.	Hom.	Het.
1 level	4424		5280		34720		504		496	
2 level	1458	1458	3761	3761	6898	6898	**397**	**397**	281	281
3 level	1254	1244	**3479**	**3442**	5010	4994	405	397	**276**	270
4 level	**1245**	**1184**	3516	3442	**4527**	4485	413	397	280	**268**
5 level	1291	1184	3553	3442	4550	4351	421	397	286	268
6 level	1345	1184	3590	3442	4619	**4347**	429	397	292	268
7 level	1399	1184	3627	3442	4773	4347	437	397	298	268
8 level	1453	1184	3664	3442	4930	4347	445	397	304	268

The introduction of heterogeneous multi-level reconfiguration leads in four of the five tests cases to a small (less than 5%) further reduction of the reconfiguration costs compared to the homogeneous case. This holds also when the maximum number of reconfiguration levels is the same as for the best homogeneous case. It should be noted, that in the latter case (and often also in the former case) an additional advantage of the heterogeneous machines that they need less memory cells. Moreover, the cost reduction for the heterogeneous machine might become larger using an improved heuristic for the RLP problem.

In order to study the impact of individual reconfiguration levels on the total reconfiguration costs the portion of the reconfiguration costs caused by each individual level k, $k \in [1 : L]$ is depicted in Fig. 4.5 for the 8-bit adder. The figure shows clearly that the lowest reconfiguration level generates the largest portion of the total reconfiguration costs (this holds also for the other four test applications). This result is to be expected as the lowest reconfiguration level provides the actual data of the architecture's reconfigurable units. Mostly the higher-level reconfiguration operations are less often performed because each k-level hyperreconfiguration requires a reconfiguration of all lower levels. On the other hand the number of memory cells which are included in the reconfiguration chain can only decrease for lower levels. Therefore, it is possible that a higher level has a larger part of the reconfiguration costs as a lower level. An example is the 8-bit adder in Fig. 4.5 on a homogeneous machine with $L = 15$. In this case, the accumulated costs for the 15-level hyperreconfigurations is 79 whereas for it is only 54 for the 14-level hyperreconfigurations. It can also be seen that the higher levels have smaller reconfiguration costs for the heterogeneous machines compared to the homogeneous machines.

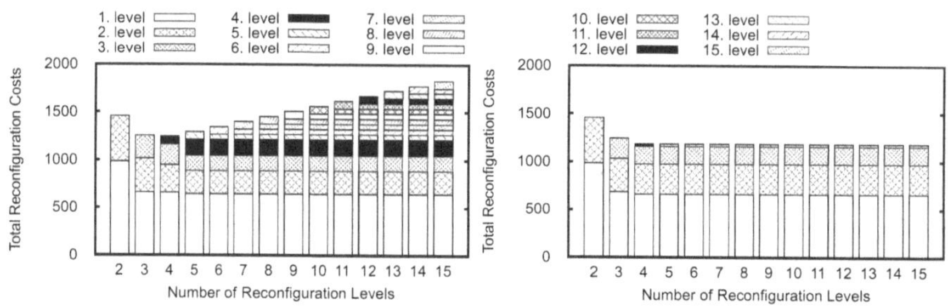

Fig. 4.5 Reconfiguration costs on different levels for the 8-bit adder on homogeneous multi-level reconfigurable machines (left) and heterogeneous multi-level reconfigurable machines (right).

4.5 Diverse Granularity in Multi-level Reconfigurable Systems

Whereas multi-level reconfigurable systems significantly reduce the costs, they require a large number of additional memory cells to store upper level contexts. When looking at the reconfiguration behavior of individual run-time parameters, there typically exist groups of parameters, which are always changed together. For example, all reconfiguration bits of a single LUT or the switches of a switchbox change their values together because they are functionally linked. The availability for reconfiguration of such a group of parameters can therefore be controlled by a single memory cell in the hypercontext. This reduces the number of necessary memory cells. Furthermore, as hypercontexts contain less memory cells, the cost of a hyperreconfiguration operation decreases as well. Because the number of parameters controlled by an individual memory cell can differ between cells, we call this approach diverse granularity.

A possible disadvantage of diverse granularity is that it reduces the flexibility of reconfiguration of the system. The entire group of parameters controlled by a single memory cell must be reconfigured, even if only one parameter requires new data. Hence, the design of the architecture is a trade-off between high flexibility of reconfiguration, which requires many memory cells and a more efficient implementation with less memory cells but also less flexibility. Thus, during the design of a multi-level reconfigurable architecture with diverse granularity, two main questions have to be answered: How many memory cells should be used for each level and which memory cells (or switches) should a memory cell control? This problem is called the Hypercontext Design (HD) problem, formally defined as follows.

Hypercontext Design (HD) problem: Given an algorithm as a sequence of context requirements $S = c_1, \ldots, c_m$ for a set of n switches with memory cells $X^{(1)} = \{x_1^1, \ldots, x_n^1\}$ and an integer L. Define an L-level reconfigurable machine in the switch model with sets $X^{(k)} = \{x_1^k, \ldots, x_{n_k}^k\}$, $k \in [2 : L]$ of memory cells for chosen numbers $n \geq k_2 \geq \cdots \geq k_L \geq 0$, and define a partition π_k of $X^{(k)}$, $k \in [1 : L-1]$ into n_{k+1} sets $X_1^{(k)}, \ldots, X_{n_k+1}^{(k)}$ such that the total reconfiguration cost for the given algorithm are minimized, when executed on the defined architecture. The partition of $X^{(k)}$, $k \in [1 : L-1]$ into $n_k + 1$ subsets defines then an assignment of the jth memory cell on level $k+1$, $j \in [1 : n_{k+1}]$ to all memory cells in the jth subset of the partition. This assignment means that either all cells in the jth subset of the partition are included in a context (when the jth memory cell on level $k+1$ is set to 1) or all are excluded (when the jth memory cell on level $k+1$ is set to 0).

The above definition of the HD problem allows resulting architectures to contain memory cells that control an arbitrary subset of lower-level memory cells. We therefore call this the *set model of granularity*. For many systems, grouping arbitrary sets of memory cells is not desirable (e.g., because the wiring becomes too complex). Therefore we introduce a variant of the HD problem where it is required that all memory cells (or switches) which are controlled by a single memory cell must form a consecutive subchain of their corresponding reconfiguration chain on the level below. The resulting problem that is analogous to the HD problem is called the Interval Hypercontext Design (I-HD) problem.

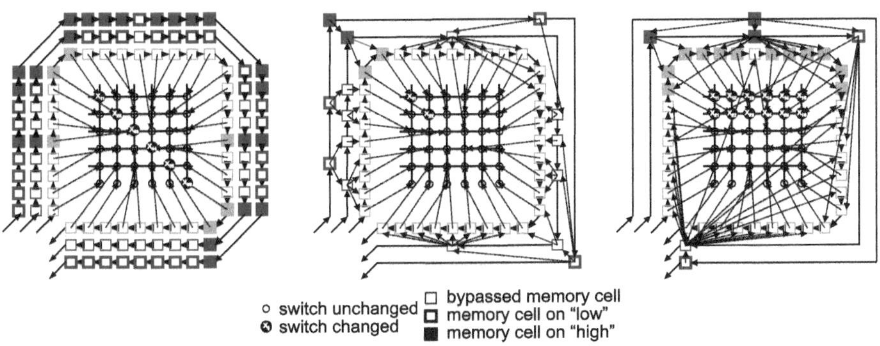

Fig. 4.6 Switch box with different granularity; full model (left), interval model (middle), set model (right).

Theorem 4.5. *The HD problem in the switch model is NP-hard for 2-level reconfigurable architectures* [20].

Theorem 4.6. *The I-HD problem in the switch model is NP-hard for 2-level reconfigurable architectures* [20].

We consider a special case of the I-HD-Switch problem, where suitable hypercontexts and the time steps for all hyperreconfiguration operations have been predetermined for the given sequence of context requirements. This special case is called the *I-HD with Fixed Hypercontext Assignment* (Fixed-I-HD) problem. The corresponding problem for the HD problem is called *HD with Fixed Hypercontext Assignment* (Fixed-HD) problem.

Theorem 4.7. *Given an algorithm as a sequence m context requirements and an L-level reconfigurable architecture, $L \geq 2$ with n switches. An exact solution to the Fixed-I-HD problem in the switch model can be found in polynomial time $O(mn^2 + n^2(n+m)(L-2))$* [7, 20].

Theorem 4.8. *The Fixed-HD-Switch problem is NP-hard* [20].

As simple observation for the Fixed-HD problem is that the memory cells of a given solution can be conceptually permuted for each level. This observation has been used in [18] to develop a polynomial time heuristic for the Fixed-HD problem that uses the exact algorithm for the Fixed-I-HD and heuristically finds a good permutation of the memory cells so that the corresponding Fixed-I-HD is good.

So far it was shown that the two subproblems of I-HD, the optimal partitioning of a sequence of context requirements into hypercontexts and the definition of an optimal architecture, can be solved exactly and efficiently in the switch model. However, this does not contradict the NP-hardness of the I-HD problem. Each optimal solution to either subproblem presumes the previous optimal solution of the other subproblem. The partitioning into hypercontexts assumes an optimal architecture has been defined whereas the Fixed-I-HD-Switch problem requires an optimal partitioning into hypercontexts to generate it. Heuristic strategies for finding a good solution in polynomial time which are based on solving both subproblems exactly have been designed in [18]. A similar strategy has been used to develop heuristics for the HD problem. A difference is that the second subproblem, the Fixed-HD problem, is NP hard and can therefore not be solved optimally within a polynomial time heuristic (if NP≠P). Therefore, the polynomial time heuristic for solving the Fixed-HD problem is used to solve this subproblem. The heuristic has a parameter ξ which gives a weight to the reconfiguration costs corresponding to all levels $\geq$2. For a large values of ξ the benefit of reducing the number of memory cells increases. Thus, it is possible to get architectures with a different number of memory cells for different values of ξ.

The results for the I-HD and the HD problem are shown in Table 4.2 for the five test applications and are compared with 1-level reconfiguration and multi-level reconfiguration in the *full model*, i.e., the number of memory cells on all levels is n. It can be seen that (with one exception) the concept of diverse granularity reduces both the total reconfiguration costs as well as the number of memory cells used. The set model shows better results than the interval model for all test cases. However, this is not surprising, because the set model provides more flexibility in choosing which switches are grouped together. Note also that the cost reduction achieved for the two coarse-grained test samples is smaller than for the fine grained samples. One reason is that often in the fine grained applications switches have a more similar usage pattern.

Table 4.2 Number of switches (#Switches), number of reconfiguration operations (#Reconfigurations), average switch use, (minimal) total reconfiguration costs (Costs) with associated number of reconfiguration levels (in parenthesis) for the five test applications: 8 bit ripple carry adder, counter, LED decoder, FIR filter and vector summation.

	Adder	Counter	LED Decoder	FIR LMS	Vector Sum
#Switches	79	48	224	8	8
#Reconfigurations	56	110	155	63	62
Average switch usage in %	6.3	30.3	2.4	38.7	29.6
1-level reconfiguration					
Costs	4424	5280	34720	504	496
PHC multi-level reconfiguration					
Lowest cost (# rec. levels)	1245 (4)	3479 (3)	4527 (4)	397 (2)	278 (3)
#memory cells	237	96	652	8	16
I-HD (interval model)					
Lowest cost (# rec. levels)	1175 (4)	3351(4)	3968 (5)	386 (3)	268 (3)
#memory cells	136	34	390	7	13
HD (set model)					
Lowest cost (# rec. levels)	1042 (2)	2620 (3)	3774 (3)	390 (2)	269 (3)
#memory cells	16	13	123	9	7

The results of the heuristic for different values of ξ which result in different total numbers n_{cell} of memory cells on reconfiguration levels ≥ 2 are shown in Fig. 4.7 for the adder. Note that the extreme cases for n_{cell} are $(L-1) \times n$, when each level ≥ 2 has n memory cells, and $L-1$ when each level ≥ 2 has only 1 memory cell. The figure shows that the use of a small number of memory for a 2-level-reconfigurable architecture already can reduce the reconfiguration costs strongly. Architectures with more reconfiguration levels that have been constructed by the heuristic need more memory to cells to obtain the same total reconfiguration costs as for lower levels. Note, that it can occur that the minimum reconfiguration costs are lower for a higher number of reconfiguration levels (e.g., in the interval model in the left part of Fig. 4.7).

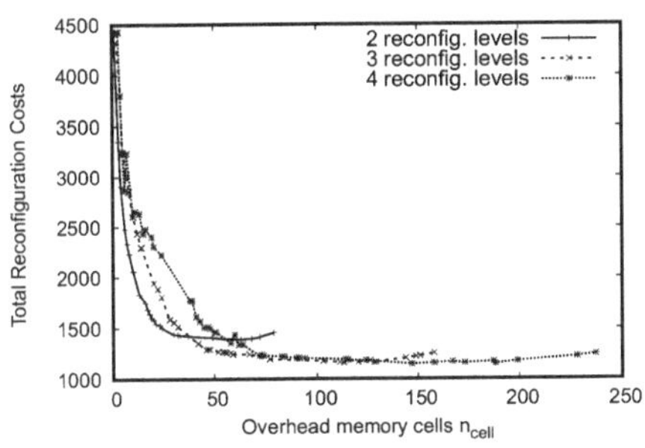

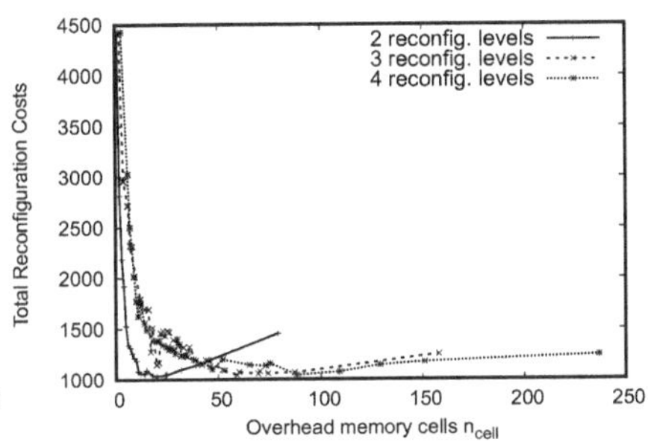

Fig. 4.7 Number n_{cell} of memory cells on all reconfiguration levels ≥ 2 and reconfiguration costs for architectures as found by the heuristic for the adder; interval model (left), set model (right).

Comparing architectures for the full model and the interval model with similar reconfiguration costs it can be seen that n_{cell} is for the adder (counter, LED decoder, FIR LMS, vector sum) by a factor 2.6 (respectively 2.6, 7.2, 2.7, 1.1) larger for interval model than for the full model (when both models have 2-levels of reconfiguration). This factor is 3.6 (respectively 3.4, 4.2, 1.2, 1.9) for 4 levels of reconfiguration. Note, that in all cases the cost difference is less than 1% (with one exception where it is 6%) but the costs of the interval model are never larger than for the full model. Thus, the relative advantage of the interval model depends on the test application and on the number of reconfiguration levels.

We have assumed here that an architecture is designed for a single given algorithm. This might not be realistic for some practical cases. But in that case it is also possible to use our methods for a set of typical algorithms (further discussion [7]).

4.6 Partial Reconfiguration and Hyperreconfiguration

Current partially reconfigurable FPGAs define atomic groups of reconfigurable resources, which can only be reconfigured together: we call them frames similar to the notation used by Xilinx Virtex FPGAs (see also Chap. 3). Each of these frames consists of a number of reconfigurable resources such as logic blocks and interconnect resources. The size of a frame (i.e., the number of reconfiguration bits) is constant for all frames in the architecture. In order to initiate a reconfiguration during runtime, the reconfiguration subsystem of the FPGA is first given the location of the frame at which to start reconfiguring and the number of frames for which new data will be provided. Then, the corresponding data for the specified frames are transmitted. If frames that are not adjacent in the reconfiguration chain must be reconfigured, several separate reconfiguration operations have to be performed. This greatly adds to the reconfiguration costs because a reconfiguration operation requires additional padding data to be transmitted before the data of the first frame to ensure the correct operation of the reconfiguration subsystem. Furthermore, the control data necessary for addressing the sought frames induces a substantial overhead as these addresses grow with the number of addressable frames and have to be specified at the start of each partial reconfiguration operation.

Clearly, complex systems such as FPGAs provide trade-offs with respect to many aspects of their design. Due to the topic of this project we will focus on the reconfiguration overhead and consider partially reconfigurable FPGAs from an abstract point of view to compare it to multi-level reconfigurable architectures (see also [21, 22, 7]).

4.6.1 Frame Model of Partially Reconfigurable Architectures

Partially reconfigurable architectures in the *frame model* comprise of an ordered set of n reconfigurable units $X = \{x_1, \ldots, x_n\}$ where the order is according to the sequence on the reconfiguration chain. A context requirement $c \in C$ is defined as

$c \subset X$, the subset of units that have to be reconfigured during a reconfiguration operation. The architecture partitions the set of units X into a set of k non-overlapping intervals $F = \{f_1, \ldots, f_k\}$, which are the *frames* of the architecture. Given a context requirement $c \in C$ and a frame $f \in F$, all units $x \in f$ have to be reconfigured if $f \cap c \neq \emptyset$ and none of them is reconfigured otherwise. A context $\bar{c}$ describes the reconfiguration data for those frames that receive new reconfiguration data. Partially reconfigurable architectures can be modeled formally as two-level reconfigurable architectures. The lower reconfiguration level comprises the transfer of the actual data for reconfigurable units as defined by the context. The information of the upper reconfiguration level designates which reconfigurable units receive data. In the case of partially reconfigurable architectures, this consists of the initial padding data, the number of frames to send and the address of the first frame. Conceptually, we have thus assigned the transmission of the reconfiguration data to the lower reconfiguration level, whereas the upper reconfiguration level describes the parameters of the reconfiguration operation.

The cost model for this architecture is called *frame cost model* and is described in the following. The cost $\underline{cost}(\bar{c})$ involved with the reconfiguration of a single context $\bar{c}$ (lower-level reconfiguration) is given by the sum of the sizes of the frames that are reconfigured. The cost $\overline{cost}(\bar{c})$ of reconfiguring the upper level context encompasses the cost for initiating new reconfiguration operations, the address of the first data frame and the number of consecutive frames to reconfigure. Here we propose to ignore the costs for initiation because they are not necessary conceptually and therefore might lead to an unfair evaluation of the partially reconfigurable model. A reconfiguration operation is initiated for the first frame of a context that requires new data. Further reconfigurations have to be performed when two frames must be reconfigured in sequence but are not adjacent in the reconfiguration chain. The frame address and the number of consecutive frames are both binary values of length $\log_2|F|$. Formally we have

$$\underline{cost}(\bar{c}) = \sum_{i=1}^{|F|} \begin{cases} |f_i| & f_i \cap \bar{c} \neq \emptyset \\ 0 & \text{otherwise} \end{cases}$$

$$\overline{cost}(\bar{c}) = \sum_{i=1}^{|F|} \begin{cases} |f_i| + 2\lceil \log_2|F| \rceil & (\bar{c} \cap f_i \neq \emptyset) \wedge (\bar{c} \cap f_{i-1} = \emptyset \vee i = 1) \\ 0 & \text{otherwise} \end{cases}$$

The total reconfiguration costs for sequence S are then given by $cost(\bar{c}) = \sum_{\bar{c} \in S} (\overline{cost(\bar{c})} + \underline{cost(\bar{c})})$. Similar to the concept of diverse granularity for multi-level reconfigurable architectures, it is worthwhile to investigate the effect of using heterogeneously sized frames.

Interval Hypercontext Design in the Frame Model (I-HD-Frame) problem: Given an algorithm A as a sequence of context requirements $S = c_1, \ldots, c_m$ for a ordered set $X = \{x_1, \ldots, x_n\}$ of reconfigurable units. Find a partition $\pi = \{f_1, \ldots, f_k\}$ of X with $1 \geq k \geq n$ where each f_i is a subinterval of the reconfiguration chain. The partition is chosen in such a way, that the execution of A on the corresponding partially reconfigurable architecture with heterogeneous frame sizes produces minimal total reconfiguration costs according to the frame cost model.

Theorem 4.9. *The I-HD-Frame problem can be solved in time* $O(\log n \cdot n^4 m)$ [21].

The frame cost model allows for a formal treatment of the reconfiguration approach taken by partially reconfigurable FPGAs. The distinction that is made between upper and lower-level reconfiguration costs highlights the conceptual similarities of partially reconfigurable FPGAs and hyperreconfigurable architectures. A difference is that hyperreconfigurable systems can employ patterns in the reconfiguration usage spanning multiple reconfiguration operations, partially reconfigurable architectures are confined to achieve a reduction of the reconfiguration cost based on the patterns found within a single frame.

4.6.2 Results

Figure 4.8 shows the reconfiguration costs of a 2-level hyperreconfigurable machine with diverse granularity in the interval model and of a partially reconfigurable machines in the frame model for an 8 bit ripple carry adder ((similar results have been obtained for other test applications)). For both types of machines the reconfiguration costs are given for the homogeneous and the heterogeneous version. In addition, the figures shows the reconfiguration costs of a stochastic model which assumes that each reconfigurable unit has a certain probability to be reconfigured. This model has been investigated theoretically in [7] with respect to optimal frame size and minimal reconfiguration costs for the partially reconfigurable machines. The stochastic model uses the same parameters as the adder application, i.e., the same number of reconfigurable units, same number of context requirements, and the same probability for a reconfigurable unit to be reconfigured (for the adder this probability is 0.063).

The results show that hyperreconfigurable systems produce less reconfiguration costs than partially reconfigurable FPGAs for the test applications. This holds for the homogeneous and the heterogeneous architectures. The difference increases with a larger number of frames (i.e., group of switches). The reconfiguration cost is high when only one frame is used, in which case both hyperreconfiguration and partial reconfiguration perform a full reconfiguration of the system. For both types of architectures there exists an optimal number of frames and the cost increases when more frames are used. This increase is strong for partial reconfiguration and small for hyperreconfiguration. This steep increase in the reconfiguration costs of partially reconfigurable architectures is due to the overhead caused by addressing individual frames.

The graphs pertaining to partially reconfigurable architectures contain several points at which there is a sudden increase in the reconfiguration costs. At these points, where the number of frames equal powers of two, the size of a frame address increases by one, accumulating to the aforementioned increase.

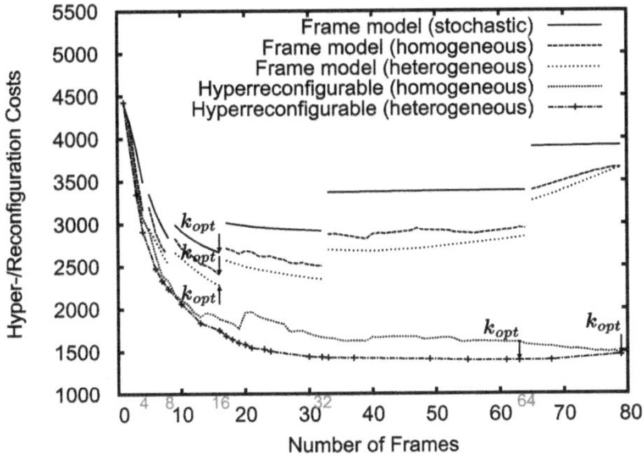

Fig. 4.8 Total reconfiguration costs for the adder on different reconfigurable architectures: frame model (stochastic, homogeneous, heterogeneous) and 2-level reconfigurable machine in the interval model (homogeneous = equal lengths intervals, heterogeneous = different lengths intervals).

It is evident from the figure that the stochastic cost model projects higher reconfiguration costs. The reason is that the assumption that the reconfigurable units have independent probabilities to be reconfigured represents an unfavorable for frame-based partial reconfiguration with homogeneous frame sizes.

4.7 Conclusions

We have given an introduction and overview to hyperreconfigurable systems. Such systems have several layers of reconfiguration operations which are used to reduce the reconfiguration overhead in order to make dynamic reconfiguration faster. Higher-level hyperreconfiguration operations define the flexibility of the system for reconfiguration while lower-level reconfiguration operations alter the system's functionality within the limits set by the preceding hyperreconfigurations.

Acknowledgements This work was financed by the German Research Foundation (DFG) through the project *Models and Algorithms for Hyperreconfigurable Architectures* within the priority program 1148 *Reconfigurable Computing Systems*.

References

1. Compton, K., Hauck, S.: Configurable computing: a survey of systems and software. ACM Comput. Surv. **34**(2), 171–210 (2002)
2. Dandalis, A., Prasanna, V.K.: Configuration compression for FPGA-based embedded systems. In: Proc. ACM Int. Symp. on Field-Programmable Gate Arrays, pp. 173–182 (2001)

3. Hauck, S., Li, Z., Rolim, J.: Configuration compression for the Xilinx XC6200 FPGA. IEEE Trans. Comput.-Aided Des. **8**, 1107–1113 (1999)
4. Kannan, P., Balachandran, S., Bhatia, D.: On metrics for comparing routability estimation methods for FPGAs. In: Proc. 39th Design Autom. Conf., pp. 70–75 (2002)
5. Koch, D., Teich, J.: Platform-independent methodology for partial reconfiguration. In: Proc. 1st Conference on Computing Frontiers, pp. 398–403 (2004)
6. Köster, M., Teich, J.: (Self-)reconfigurable finite state machines: Theory and implementation. In: Proc. Design. Autom. and Test in Europe, pp. 559–566 (2002)
7. Lange, S.: Hyperreconfigurable systems. Formal concepts, architectures, and applications. Dissertation (*in German*), Faculty of Mathematics and Computer Science, University of Leipzig (2008)
8. Lange, S., Kebschull, U.: Virtual hardware byte code as a design platform for reconfigurable embedded systems. In: Proc. DATe 03 (2003)
9. Lange, S., Middendorf, M.: Hyperreconfigurable architectures as flexible control systems. In: ARCS 2004 Workshop "Dynamically Reconfigurable Systems". LNI, vol. P-41, pp. 175–184 (2004)
10. Lange, S., Middendorf, M.: Hyperreconfigurable architectures for fast runtime reconfiguration. In: Proc. 2004 IEEE Symposium on Field-Programmable Custom Computing Machines (FCCM'04), pp. 304–305 (2004)
11. Lange, S., Middendorf, M.: Models and reconfiguration problems for multi task hyperreconfigurable architectures. In: Proc. Rec. Arch. Workshop (RAW 2004), p. 8 (2004)
12. Lange, S., Middendorf, M.: The partition into hypercontexts problem for hyperreconfigurable architectures. In: Proc. of the International Conference on Field Programmable Logic and Applications (FPL 2004). LNCS, vol. 3205, pp. 251–260 (2004)
13. Lange, S., Middendorf, M.: Hyperreconfigurable architectures and the partition into hypercontexts problem. J. Parallel Distrib. Comput. **65**(6), 743–754 (2005)
14. Lange, S., Middendorf, M.: Multi task hyperreconfigurable architectures: Models and reconfiguration problems. Int. J. Embed. Syst. **1**(3/4), 154–164 (2005)
15. Lange, S., Middendorf, M.: On the design of two-level reconfigurable architectures. In: Proc. Int. Conf. on Reconfigurable Computing and FPGAs (ReConFig05), p. 8 (2005)
16. Lange, S., Middendorf, M.: Cache architectures for reconfigurable hardware. In: Proc. Int. Conference on Engineering of Reconfigurable Systems and Algorithms (ERSA'06), p. 8 (2006)
17. Lange, S., Middendorf, M.: Granularity aspects for the design of multi-level reconfigurable architectures. In: Proc. IEEE Int. Conf. on Field Prog. Techn. (ICFPT 06), pp. 9–16 (2006)
18. Lange, S., Middendorf, M.: On multi-level reconfigurable architectures. In: Proc. 13th Reconfigurable Architectures Workshop (RAW 2006), p. 8, IEEE (2006)
19. Lange, S., Middendorf, M.: Online strategies for the reconfiguration of two-level reconfigurable architectures. In: Proc. Workshop on Dynamically Reconfigurable Systems (DRS) (2007)
20. Lange, S., Middendorf, M.: Design aspects of multi-level reconfigurable architectures. J. Signal Process. Syst. Signal Image Video Technol. **51**(1), 23–37 (2008)
21. Lange, S., Middendorf, M.: On the reconfiguration costs of models for partially reconfigurable FPGAs. In: Proc. IV Southern Programmable Logic Conference, pp. 111–118 (2008)
22. Lange, S., Middendorf, M.: Spp1148 booth: Hyperreconfigurable architectures. In: Proc. Int. Conference on Field Programmable Logic and Applications (FPL 2008), p. 353 (2008)
23. Malik, U.: Configuration encoding techniques for fast fpga reconfiguration. PhD thesis, University of New South Wales (2006)
24. Pan, J., Mitra, T., Wong, W.F.: Configuration bitstream compression for dynamically reconfigurable FPGAs. In: Proc. of the 2004 IEEE/ACM International conference on Computer-aided design (ICCAD'04), pp. 766–773 (2004)
25. Sidhu, R., Wadhwa, S., Mai, A., Prasanna, V.: A self-reconfigurable gate array architecture. In: Proc. FPL. LNCS, vol. 1896, pp. 106–120 (2000)
26. Wadhwa, S., Dandalis, A.: Efficient self-reconfigurable implementations using on-chip memory. In: Proc. FPL. LNCS, vol. 1896, pp. 443–448 (2000)

Chapter 5
Evaluation and Design Methods for Processor-Like Reconfigurable Architectures

Sven Eisenhardt, Thomas Schweizer, Julio Oliveira Filho, Tommy Kuhn, and Wolfgang Rosenstiel

Abstract This chapter focuses on the utilization of fast reconfiguration to optimize area, performance, and power. The results are quantified by a synthesizable architecture model. In order to assure good applicability of the research, a C-compiler is co-developed with the architecture. This chapter provides an overview of the optimization techniques and a summary of current evaluation results.

5.1 Introduction

Reconfigurable systems provide the ability to reuse architectural resources over time. Especially for applications which require frequent adoptions during execution it is beneficial to use dynamically reconfigurable fabric. Such an application, for example, is the physical layer of the WiMax [24] standard. This layer defines the scalable OFDMA digital modulation scheme. Scalable means that each channel bandwidth may vary between 1.25 MHz and 20 MHz depending on the demanded traffic. Due to this, the scalable OFDMA requires its underlying hardware architecture to support fast Fourier transforms (FFTs) with a high throughput rate and adaptable to an input vector size varying between 128 and 8192 points. This adaption to the required input vector size can be accomplished by dynamic reconfiguration whereas interruptions of the data stream due to reconfiguration have to be avoided. Therefore reconfiguration time becomes a crucial factor in choosing a suitable target architecture for an application.

Sven Eisenhardt · Thomas Schweizer · Julio Oliveira Filho · Tommy Kuhn · Wolfgang Rosenstiel
Department of Computer Engineering, Wilhelm-Schickard-Institute for Computer Science, University of Tübingen, Tübingen, Germany,
e-mails: eisenhardt@informatik.uni-tuebingen.de, tschweiz@informatik.uni-tuebingen.de, oliveira@informatik.uni-tuebingen.de, kuhn@informatik.uni-tuebingen.de, rosenstiel@informatik.uni-tuebingen.de

M. Platzner et al. (eds.), *Dynamically Reconfigurable Systems*,
DOI 10.1007/978-90-481-3485-4_5,

Applying dynamic reconfiguration to traditional FPGAs is a difficult task. Their fine granularity induces an extensive size for configuration data and exchanging only parts of the circuit usually results in fragmentation.

Newly developed reconfigurable architectures can be reconfigured within one clock cycle. We call such architectures *processor-like reconfigurable architectures*. They usually have a coarser granularity and require less configuration data than traditional FPGAs so that multiple configuration contexts can be stored inside the reconfigurable array and reconfiguration keeps pace with execution.

As illustrated in Fig. 5.1, the ability for fast reconfiguration involves both, benefits and costs. The main benefit of a processor-like reconfigurable architecture is its flexibility. If the application's functionality or its requirements change over time, the corresponding configurations can be loaded with low latency. Processor-like reconfiguration allows to instantiate and to execute within one clock cycle exactly that part of circuit that is needed in this cycle. An additional benefit is a better tradeoff between area requirements and performance.

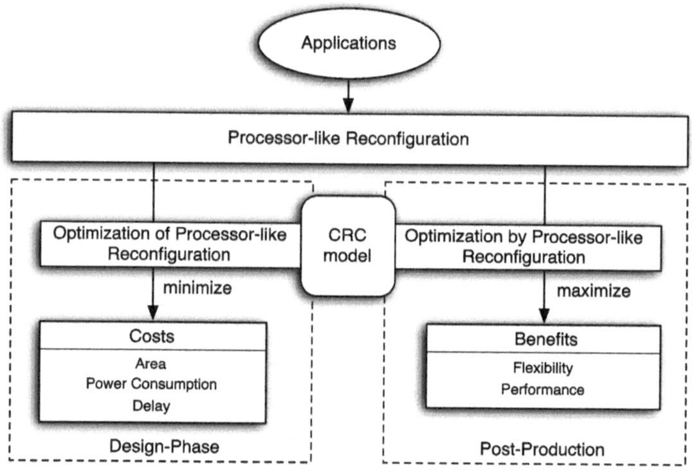

Optimizing benefits and costs of processor-like reconfiguration.

Fig. 5.1 Optimizing benefits and costs of processor-like reconfiguration.

However, these benefits are achieved at cost of additional hardware. But the costs of processor-like reconfiguration are not only to be considered from a financial point of view. We look at costs such as the increased power consumption that drains the battery of a mobile device. Also by the term costs we refer to the additional delay induced by processor-like reconfiguration for loading contexts in each clock cycle.

Costs for processor-like reconfiguration should be minimized during the architecture design phase whereas its benefits can be maximized after production. Architecture optimizations during the design phase can reduce the extra hardware overhead.

And novel compiler techniques must consider reconfiguration during the application mapping phase.

To achieve both goals, minimize costs and maximize benefits, a general model for processor-like reconfigurable architectures is necessary—the *CRC model* (Configurable Reconfigurable Core, CRC). This model can be adapted to the requirements of the compiler and an application domain. Synthesizing instances of the CRC model enables a detailed evaluation, which cannot be accomplished with commercial architectures.

The remainder of this chapter is organized as follows: The CRC model as well as benefits and costs of processor-like reconfiguration are discussed in Sect. 5.2. Techniques to optimize the design of processor-like reconfigurable architectures are demonstrated in Sects. 5.3 and 5.4. Methods to optimize switching between different applications are presented in Sect. 5.5.

5.2 Benefits and Costs of Processor-Like Reconfiguration

The ability to reconfigure within one clock-cycle provides an additional degree of freedom that can be exploited for optimization. However, processor-like reconfiguration comes with additional costs.

As described in the previous section, configuration memory is required to store multiple contexts and a control unit has to select the active context for execution. Configuration memory is expensive in terms of area and reading the configuration memory in each clock cycle increases the delay. Two issues are important to quantify these benefits and costs: The architecture that provides all elements required to support processor-like reconfiguration. And a compiler capable of using this form of reconfiguration. Before we address these two issues the CRC model is presented.

5.2.1 CRC Model

The CRC model was developed to represent a wide range of processor-like reconfigurable architectures. In its most general specification, only a few features are defined. As depicted in Fig. 5.2, it consists of a rectangular array of *processing elements* (PE) that are connected by a reconfigurable interconnect network. Each PE consists of a functional unit (FU) for word-wide arithmetic and logic operations, a register set, and a context memory that defines several configurations for the PE. At the beginning of each clock cycle, a context is selected by a control unit that can vary significantly for the various architectures. Since the interconnect network also varies, it is not further specified in the general CRC model.

Instances of the CRC model are specified by refinement of the general model. The instances are described as a transaction-level model in SystemC, and at the register-transfer level in Verilog. We use the SystemC implementations for system-

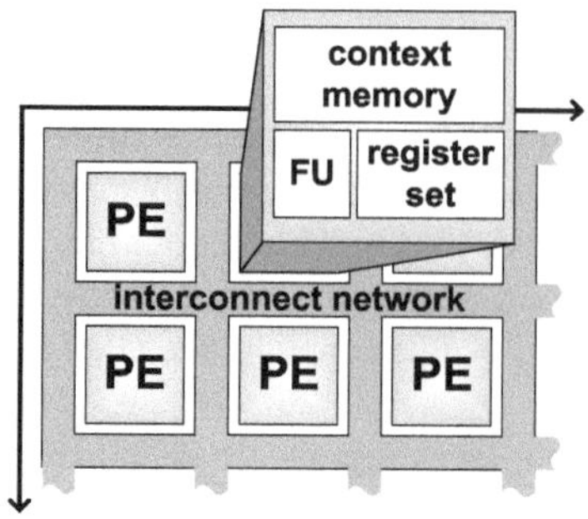

Fig. 5.2 General specification of the CRC model [21].

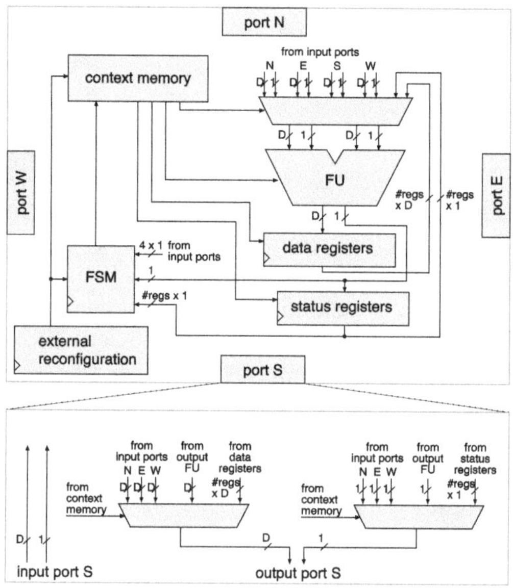

Fig. 5.3 A PE of a possible architecture instance. D denotes the width of the data path that can vary for different implementations [20].

level evaluations. The Verilog implementations are synthesized using commercial tools, and they are simulated and analyzed at the gate-level for detailed evaluations.

Figure 5.3 shows the PE of a possible instance of the CRC model that implements a nearest neighbor interconnect network and includes a control unit that implements an finite state machine (FSM). By doing so, arrays of arbitrary sizes can be created easily. However, we can also create instances of PEs that share one control unit.

The configuration memory is subdivided into context memory and memory in the control unit that stores the state transitions of the FSM. Both memories are configured from outside the reconfigurable fabric when the device is booted or when required by the application which is called external reconfiguration.

The output of the FSM selects an entry in the context memory. The output of the context memory selects the operands for the FU. The operands can come from the ports north, east, south, or west (N, E, S, W) or from internal registers of the PE. Ports can be connected to a neighboring PE or they can be ports of the device if the PE is located at the border of the array. The context memory also determines the operation performed by the FU and destination registers for the data and 1-bit status outputs of the FU if the result is to be stored in a register of the PE. The status output is used for carry signals and for the results of compare operations. To determine the next state, the status output of the FU, the output of the status registers, and the status signals of the four ports are connected to the FSM. The FSM and the register are synchronized by a common clock. If the control unit is shared among a number of PEs, the according lines are connected to the external FSM and the same context is selected for all PEs sharing the FSM.

The lower part of Fig. 5.3 shows the port S module of the PE in detail; the port N, E, and W modules are equivalent. The data and status input is routed directly to the inside of the PE.

The context memory determines what data and status signals are available at the output ports. This is done for data and status signals independently. The source for the output ports can be the output of the FU, the output of any of the registers, and the input port of any of the other ports. The output of one of the other ports is used to implement the nearest neighbor interconnect network. A PE is able to simultaneously execute an operation in the FU and to route data and status signals from one neighbor to another.

The CRC model is very flexible and can be modified by setting various parameters. By resizing configuration memory, specifying the capabilities of the FU, changing the data path width etc. a great variety of model instances can be designed. In order to facilitate the generation of CRC model instances, an *architecture description language* (ADL) was defined. It comprises key features to allow fast modeling and analysis of such architectures, namely: representation of processing element array (ir)regularities, flexible and concise description of interconnection network, and connection to the external environment. Based on the ADL it is possible to automatically generate a SystemC based simulator or a Verilog architecture template. Based on such templates it is possible to synthesize instances which can be used to estimate the required area, power, and performance for sample applications.

5.2.2 Compiler

Similar to the architecture, also the compiler has to support processor-like reconfiguration. If the compiler does not make use of this ability to switch contexts within one clock cycle, the result might be suboptimal in respect of area, performance and power. Therefore we had to develop a compiler that can be adjusted to possible architecture instances of the CRC model. We use applications described in C as input of the compiler. The application mapping process of our compiler resembles that of the high-level synthesis. It comprises two phases: *scheduling* and *binding*.

Scheduling partitions the application's description into control steps such that all operations in a control step execute in the same clock cycle. In our case the scheduling must also consider the partitioning in different contexts, for example, control steps can be merged together into one context to meet performance constraints.

The binding task assigns the operations within each clock cycle to available hardware units. A resource such as functional, storage, or interconnection unit can be shared by different operations or data transfers if they are mutually exclusive. For example, two operations assigned to two different contexts are mutually exclusive since they will never execute in parallel. We consider two different binding subtasks:

operation binding assigns each operation in a context to a processing element.

interconnection binding assigns a data transfer to interconnection resources such as multiplexer, bus, port, etc.

In the remainder we also refer to interconnection binding as routing.

In cooperation with the research group of Prof. Fekete (priority program project ReCoNodes, see Chap. 10), techniques for simultaneous scheduling, placement, and routing are developed on the basis of integer linear programming [5].

5.2.3 Evaluation

To quantify the area trade-off between statically and processor-like reconfigurable architectures, we used one statically reconfigurable architectures and different instances of the CRC model. The statically reconfigurable architecture was generated from the CRC model by removing the control unit and the configuration memory. In this case each PE features only one data and one status register.

We synthesized the PEs using a commercial synthesis tool targeting a 130 nm standard cell technology. For all PEs a target clock speed of 200 MHz was specified. The area of a PE is estimated as the cell area of the resulting gate-level netlist. To obtain area estimations for an architecture instance being composed of several identical PEs, we summed up the area of all PEs being part of the instance.

To specify a performance constraint for the evaluation, we use the initiation interval (*II*), i.e. the number of clock cycles that are available to consume a set of input values.

Table 5.1 shows the results of mapping four example applications to both, statically and processor-like reconfigurable architectures. For each application/*II* pair, the area of a customized IP-core featuring exactly the required number of FUs is provided in the table. The number of contexts provided by the core, and with it the number of states and registers, is rounded up to the next power of two. The interconnect network and register requirements are not considered and we set the number of PEs equal the number of required FUs. The last column in the table compares this IP-core to a customized statically reconfigurable architecture that uses the same *II* but does not reuse the FUs.

Table 5.1 Area results for example applications.

Application	*II*	FUs	Contexts	Area of IP-core with min. # FUs	Area compared to static arch.
rgb2yiq	1	21	1	0.610 mm^2	100.0%
rgb2yiq	3	7	3	0.322 mm^2	52.7%
ellipf	1	26	1	0.756 mm^2	100.0%
ellipf	2	13	2	0.468 mm^2	61.9%
ellipf	4	7	4	0.322 mm^2	42.6%
ellipf	8	4	8	0.280 mm^2	37.0%
rgb2cmyk	1	9	1	0.262 mm^2	100.0%
rgb2cmyk	1	7	3	0.322 mm^2	132.0%
resampling	1	39	1	1.134 mm^2	100.0%
resampling	1	28	2	1.007 mm^2	88.9%

For deep sub-micron process technologies, the interconnect delay contributes significantly to the overall delay of digital circuits. In addition to the metal wire delays, the reconfigurable routing switches contribute to the interconnect delay of reconfigurable architectures.

Processor-like reconfiguration allows it to utilize reconfiguration as a third dimension for routing by redirecting communication through the time domain. By doing so, the connection lengths may be reduced as illustrated in Fig. 5.4 for an 8-point fast Fourier transform (FFT) implementing the Cooley-Tukey algorithm. When moving from $II = 1$ to $II = 2$, the longest connections can be reduced to nearest-neighbor connections by executing the upper and the lower half of the graph in different contexts.

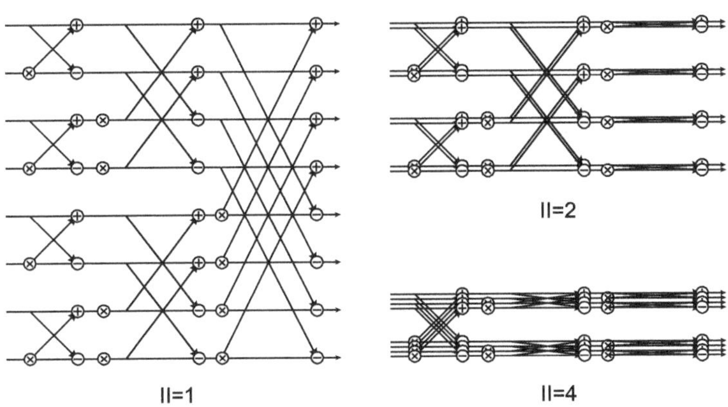

Fig. 5.4 Reduction of an 8-point FFT's interconnect requirements by redirecting communication through the time domain [19].

By doubling the number of contexts again ($II = 4$), the connections in the middle of the graphs can also be reduced to nearest-neighbor connections. Due to the regularity of the Cooley-Tukey algorithm, the same considerations hold for n-point FFTs at different *II*s in general (with n and *II* being powers of two).

Our experiments demonstrated that redirecting communication through the time domain can compensate for the performance overhead imposed by processor-like reconfiguration or even outperform a statically reconfigurable architecture [19].

In the following sections we discuss several approaches to optimize the cost-benefit ratio of processor-like reconfigurable architectures.

5.3 Specialization/Instruction Set Extension

Processor-like reconfiguration implies extra area, performance and power costs, as discussed in Sect. 5.2. The configuration memory increases the architecture area and the static power consumption. Reconfiguration at a cycle-by-cycle basis increases the clock period and dynamic power consumption. One way to decrease these extra costs is to specialize the architecture based on a set of representative applications called an *application domain*. Specialization changes the number of FUs and their instruction set, the size and number of storage resources, and the topology of the interconnection network to better perform some specific tasks required by the target domain. For example, a special interconnection network may provide routing paths only between critical resources; such network requires less configuration bits than a generic one because there are less routing paths available. Specialization leads frequently to less architecture flexibility, which is reflected in a smaller configuration memory area footprint. The final product of specialization is a *domain specific architecture*, which addresses more efficiently the requirements of its target domain at a lower configuration cost.

There are several approaches to design domain specific reconfigurable architectures. The first approach formalizes the problem as a design space exploration (DSE) [17, 16, 3, 14]. As a starting point, a template is used in which architecture resources, such as interconnection topology and datapath width, are easily configured by setting parameters. Specialization consists then in finding the parameterization that optimizes the architecture to the application domain. An extensive DSE example can be seen in, applied to the development of the ADRES architecture. Mei [15, 4] shows that DSE methods expose several design trade-offs, such as the size versus complexity of network interconnection, and distributed versus local register allocation. DSE methods allow to quickly check different architectural options; however, finding a good architecture instance requires an efficient method for pruning the huge search space.

The second approach consists in allocating customized processing units as peripheral components to the array. The work in [10, 11] suggests (1) to share critical functional resources that occupy large area, such as multipliers, and (2) to pipeline performance critical functional units. Bansal et al. [2] evaluates the impact of using several and different functional units within the PEs. They showed that, in coarse grained reconfigurable architectures, better performance may be achieved by increasing the number of FUs in individual PEs compared to architectures where each PE has only one functional unit.

Within our work, specialization means extending the functional units with custom modules. These modules—called *custom instructions*—have been successfully employed in the industry [23] and academia [9] for the specialization of general purpose and VLIW processors. The following example shows how clusters of operations within the applications can be used to specialize the FUs. Figure 5.5 shows the data flow graph (DFG) of a real world application—the trilinear interpolation filter kernel—used in the resampling phase of the ray casting algorithm. This application can be mapped in a CRC instance using a pipeline of processor elements. During the mapping, each operation in the DFG is assigned to one PE in the architecture; that requires a total of 28 PEs organized in a 12 stages pipeline.

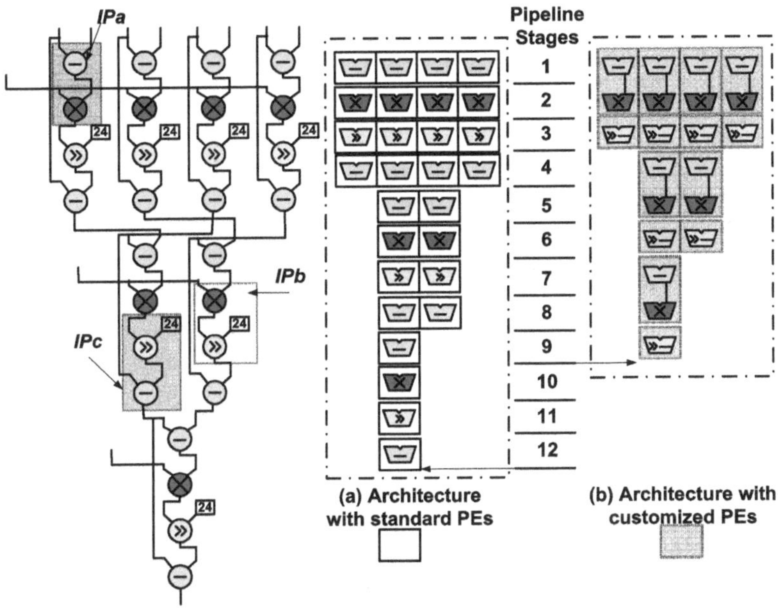

Fig. 5.5 Trilinear Interpolation mapping on architectures with (a) standard and (b) customized PEs [18].

A closer observation of the trilinear interpolation DFG reveals a very regular structure, where some subgraphs corresponding to similar clusters of instructions emerge frequently. Three examples of such clusters (IP_a, IP_b, and IP_c) are stressed in the depicted DFG. We call these subgraphs *instruction patterns*. If the PEs are extended with custom instructions to execute the instruction patterns IP_a and IP_c, the mapping of the trilinear interpolation requires only 14 PEs.

Table 5.2 depicts the costs for area and power of the two architectures after synthesis in a 130 nm technology. The PEs of the specialized architecture requires more area. However, the overall necessary area and consumed dynamic power are 49%

and 48% smaller, respectively. That is explained by the fact that only half the amount of PEs is necessary. In some cases, the whole instruction pattern may be implemented with an execution delay shorter than one clock period. For example, the *shift* ($\gg$) operation in the instruction pattern IP_c can be eliminated. IP_c can then be implemented using a subtractor with one shifted input. For these cases, we also have significant performance improvement. While the throughput is still the same, the number of necessary pipeline stages (latency) drops from 12 to 9.

Table 5.2 Impact of instruction specialization.

	#PEs	PE area (μm^2)	Total area (mm^2)	Dynamic power (mW)
Standard PEs	28	96.42	2.70	177.64
Customized PEs	14	97.14	1.36	91.39

5.3.1 Methodology

In our research, we use instruction patterns to extend the instruction set of PEs. As discussed previously, instruction patterns are clusters of operations that frequently appear in the set of applications representing one target domain. These applications or some of their computing intensive loop kernels are described using a data flow graph (DFG). We formalize the DFGs representing each application as a graph $G(V, E)$, where nodes V correspond to primitive operations or input/output data, and edges E represent data dependencies between them. Each vertex $v \in V$ represents an operation with a corresponding hardware implementation cost $\mathrm{C}(v)$ and an execution delay $\delta(v)$. The DFG allows a systematic extraction of instruction patterns in three steps: operation cluster identification, instruction pattern evaluation, and instruction pattern selection. We detail these phases in the following. The operation cluster identification phase lists all subgraphs in the DFG that present all the following characteristics:

- number of input nodes smaller or exactly equal to a maximal constraint N_{in}
- number of output nodes smaller or exactly equal to a maximal constraint N_{out}
- an accumulated implementation cost smaller or equal to a maximal cost constraint C_{max}
- a critical path delay smaller or equal to a maximal execution delay constraint δ_{max}
- convexity

N_{in}, N_{out}, C_{max}, and δ_{max} guarantee that the listed operation clusters meet the design constraints for new custom instructions. Convexity is an important property to make sure that, when executing one operation, all input values will be available to the FU.

The instruction pattern evaluation phase groups the previously listed operation clusters, such that clusters in the same group have the same subgraph structure.

We say that all the clusters in one group follow or implement the same instruction pattern. The representativity of each instruction pattern is then assessed according to its *recurrence* and possible *conflicts*. Recurrence is the number of clusters in the DFG that follow the instruction pattern. Two distinct instruction patterns are in conflict, if any of their instances in the DFGs have a common vertex. Figure 5.5 helps understanding these definitions. Assuming that the set of extracted instruction patterns is $\Phi = IP_a, IP_b, IP_c$, we can observe the pattern IP_a can be found seven times in the DFG, and thus the recurrence of IP_a is seven. Patterns IP_a and IP_c are not in conflict, because none of their instances share a common operation in the DFG. On the other hand, IP_b is in conflict with all other patterns. It makes no sense to simultaneously select two patterns for implementation if they are in conflict.

The instruction pattern selection phase selects a small number of patterns which, together, maximize the reuse of specialized instructions when mapping the targeted applications (DFGs). The idea is to concentrate the execution within a minimal, or at least a few, number of specialized units, and thus to reduce the number of required PEs. The output of the selection phase is a set of instruction patterns that will be implemented as customized instructions.

5.3.2 Study Case: Multipoint FFT for Scalable OFDMA Based Systems

We present our results using one real world application domain: the multipoint FFT processors used in scalable OFDMA systems.

We use the physical layer of the WiMax [24] standard as a real world use case scenario to introduce our results. This layer defines the scalable OFDMA digital modulation scheme. Scalable means that each channel bandwidth may vary between 1.25 MHz and 20 MHz depending on the demanded traffic. Due to that, the scalable OFDMA requires its underlying hardware architecture to support fast Fourier transforms (FFTs) with a high throughput rate (up to 480 MSamples/s), implemented in a stream oriented way, and with a input vector size varying between 128 and 8196 points. Such set of FFT algorithm constitute here the target domain.

To provide high throughput rates, we mapped the FFT in the pipelined scheme as depicted in Fig. 5.6. We partitioned the algorithm in two parts or *Fields*. *Field A* is a pipelined implementation of the radix-8 FFT algorithm. The purpose of *Field A* is to diminish the storage demand dividing the further FFT processing into 8 $\frac{N}{8}$-point-FFTs. Each one of this smaller FFTs are processed 8 stripes of PEs called *Field B*. These fields realize a radix-2^2SDF algorithm proposed in [8]. Each one of the *Fields B* are attached to one output of the *Field A*. They receive one sample per cycle and process the stream of samples in sequential and pipelined way. Each operation represented in the DFG of Fig. 5.6 correspond to a complex arithmetic operation.

The starting architecture instance was dimensioned to meet the constraints of the WiMax standard. We started with PEs designed to support complex addition and subtraction arithmetic, and storage units dimensioned for complex numbers. To

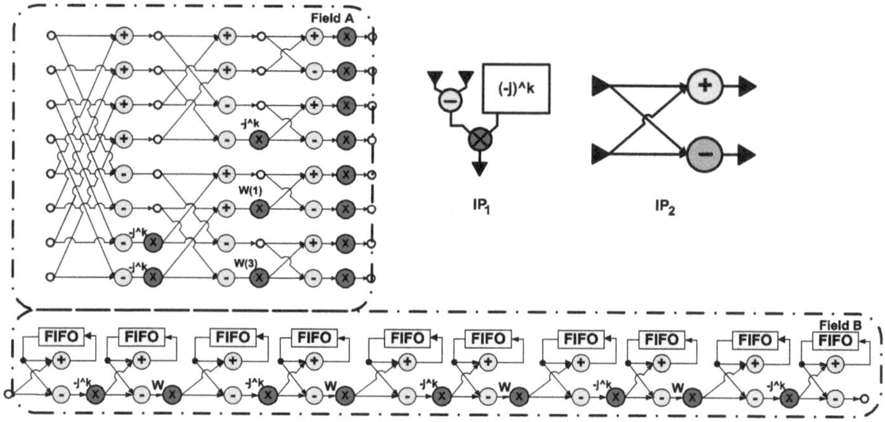

Fig. 5.6 FFT mapping [18].

guarantee a future fair evaluation when comparing with the specialized instances, we dispensed unnecessary units, such as logical and comparison operations. We also embedded multipliers on PEs only when necessary. The *Field A* required then 37 PEs, where only 13 of those comprised multipliers. Each one of the *Fields B* has 29 PEs, where 9 of them have multipliers. *Field B* is also attached to a bank of FIFOs, but we do not consider them when presenting our results because they are a particular requirement for the implementation of the R2^{2}SDF algorithm, which is not altered after we specialize the PEs. Additionally, seven contexts are considered in each field. They provide the necessary flexibility when considering different input vector sizes and for particularities during the mapping of the R2^{2}SDF algorithm.

We identified two promising instruction patterns using the methodology discussed in Sect. 5.3.1. They are depicted as IP_1 and IP_2 in Fig. 5.6. For that, we considered operation clusters with a maximum of three input and two output ports, and containing up to 2 functional units, as a PE hardware cost constraint. The instruction pattern IP_1 offers an alternative implementation which throws out the multiplier because the multiplication term is always one of the complex numbers $(-j)^k, k = 0, 1, 2, 3$. It can be implemented as a subtraction followed with some possible exchange of real and imaginary terms and their signals. The instruction pattern IP_2 corresponds to the so called FFT *butterfly*. The pattern can be found spread along all the DFG and covers a large number of operations. We specialized the architecture in two steps to allow a progressive and detailed analysis. In the first architecture instance we implemented only IP_1 as a customized instruction. In the second and final architecture instance, we include special instructions for both patterns.

We described all three architecture instances in Verilog, using the CRC Model. Where necessary, custom instructions and the corresponding modification on the context memory were added manually to the PEs. All other hardware resources within the PEs, such register banks and routing multiplexers were not modified. Table 5.3 shows the comparison between the three architecture instances considering the overall number of PEs, total silicon area, the number of configuration bits for each context, and the area per operation. Results for area are obtained after synthesis and using a commercial tool.

Table 5.3 Impact of instruction specialization—FFT

	#PEs	Total area (mm^2)	#Configuration bits	Area per operation (μm^2)	Dynamic power (mW)	Leakage power (mW)
Arch0	269	7.09	4104	26.35	750.13	1.44
Arch1(IP_1)	226	5.64 (−20%)	3502	20.96	627.61 (−16%)	1.12 (−22%)
Arch2(IP_1 and IP_2)	133	4.12 (−41%)	2686	15.31	534.42 (−28%)	0.84 (−41%)

The insertion of IP_1 as customized instruction eliminates 43 PEs with multipliers. That alone accounts to a reduction of 20% in area. When the FFT *butterflies* (IP_2) are embedded in the instruction set of the PEs, the number of necessary PEs falls to the half. That is easily explained. Each *butterfly* concentrate in one PE the execution of two basic operations and they cover 100% of the DFGs. The simultaneous adoption of these two instruction patterns as customized instructions leads to an architecture instance with up to 41% less area.

The adoption of instruction patterns as customized instructions not only decrease the number of necessary PEs, but also improves the efficiency of architectural resource usage. We use the number of configuration bits per context to demonstrate that. This metric is used in the literature [15] to assess the complexity of control resources in CGRAs. Our initial architecture spent 4104 configurations bits per context, which corresponds to about 15 bits/operation. The final specialized architecture needs only 2686 bits per context, improving the efficiency to only 10 bits/operation. The same conclusion can be obtained if we observe the silicon area used per executed operation. The initial architecture uses 26 μm^2 per operation. This area usage is improved up to 15 μm^2 per instruction in the final instance.

It is also important to compare the final architecture instance obtained using our method with state-of-art FFT designs proposed in the literature. Table 5.4 depicts this comparison using design silicon area, throughput and power consumption. All designs present similar operating frequency (∼200 MHz).

Table 5.4 Comparison with state of the art FFT designs [18].

	ASIC1 [13]	ASIC2 [12]	Altera FPGAs	Arch2 (IP_1 and IP_2)
Area (mm^2)	1.25	3.09	n.a.	4.12
Throughput (MSamples/s)	160	1000	195	1600
Power consumption (mW/MSample)	0.34	0.175	n.a.	0.33 @ 8196 points 0.26 @ 1024 points 0.16 @ 128 points
Clock frequency (Mhz)	250	250	220	200
Technology	180 ηm	180 ηm	Stratix III	130 ηm
Input vector size	1024	128	256, 1024, 4096	256–8196

The design of Liu et al. [13] proposes an array with few complex elements strongly connected among each other. It was optimized for a low memory fingerprint. They achieve the smallest area usage, but due to their sequential execution they only achieve 160 MSamples/s. Lin et al. [12] uses the same algorithm we use. They propose a high throughput design (up to 1 GSamples/s) for a 128 points FFT. Both designs have their input vector size fixed at design time. The third design is proposed by Altera for FPGAs [1]. They are more flexible and can be customized for calculating 256, 1024 or 4096 points FFTs. They achieve a throughput of around 200 MSamples/s. Unfortunately, no numbers for overall area were available. Our design has a larger silicon area usage due to the high hardware parallelism, but it is still comparable with the realistic designs. Due to the massive hardware parallelism we achieve up to 1.6 GSamples/s and our design is flexible enough to accept all input vectors between 128 and 8196 points.

5.4 Optimizing Power

As discussed in Sect. 5.2 processor-like reconfiguration causes additional power consumption. We propose two architecture improvements to reduce the power consumption: instruction set extensions and dual supply operating voltages.

5.4.1 Optimizing Power by Instruction Set Extensions

The instruction set extension methodology presented in Sect. 5.3 not only decreases the total implementation area, but also improves dynamic and static power consumption. Table 5.3 presents the dynamic and static power consumption for the FFT example. The insertion of customized instructions accounts to a reduction of 16% in dynamic power and 22% in static power due to leakage currents. When the FFT *butterflies* are embedded in the instruction set of the PEs, gains in dynamic and static power are 28% and 41%, respectively. Results for estimated power are obtained after synthesis and using a commercial tool.

5.4.2 Optimizing Power by Dual-V_{DD} Architectures

Dual supply operating voltage allows to exploit the slack time between different operations to reduce the power consumption by executing the faster operations on lower voltage. To realize the proposed approach we extended our CRC model with voltage islands. For that purpose we partitioned a processing element (PE) of this model into different voltage regions. Our approach is similar to the clustered voltage scaling technique used at gate level. However, we apply that idea to the functional unit (FU) of dynamically reconfigurable processor architectures at operation level. In our approach no frequency adaption is necessary. Therefore, we improve power and energy consumption [22].

5.4.2.1 The Dual-V_{DD} Architecture Model

We augmented the CRC model with dual-V_{DD} capabilities as depicted in Fig. 5.8. We used a commercial tool targeting a 90 nm multi-voltage standard cell technology to synthesize and analyze the two different PE types considering the following design aspects.

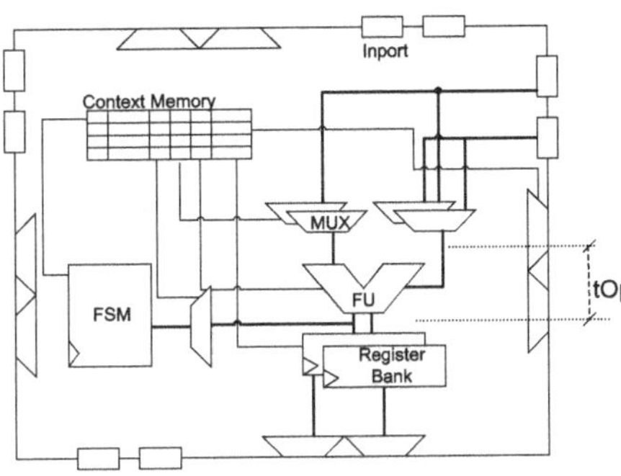

Fig. 5.7 Standard processing element (PE_H).

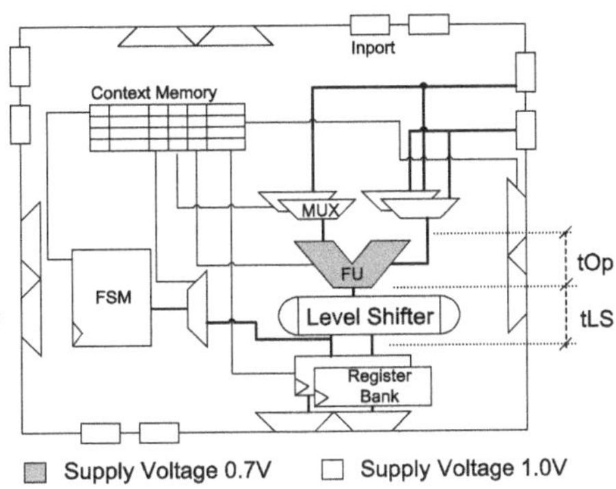

Fig. 5.8 Dual-V_{DD} processing element (PE_L).

The PE_H type, displayed in Fig. 5.7, is much like the original processing element of the CRC model. Its functional unit is designed to operate in the same voltage level (1.0 V) as its neighboring elements within the PE. Therefore, no additional components such as level shifters are necessary. PE_H is meant to execute operations in the critical path or operations that would violate our timing constraints if executed on lower voltage. The PE_H type typically consumes more power, but it is faster.

The PE_L type is a dual-voltage module where all the components are supplied with 1.0 V except the functional unit which is supplied with 0.7 V. We altered the design of the PE to accommodate a low voltage FU without violating the design timing constraints of the PE of PE_H type. First, we redesigned the FU with a new supply voltage. The multiplier module was left out because it did not meet the timing constraints when executed in 0.7 V. Multiplication must then be always executed in the PE_H type. Second, we built in a level shifter at the output of the functional unit. At this point, the signal voltage level must be converted back to 1.0 V in order to appropriately stimulate the register bank and output multiplexers. A level shifter preceding the input ports of the functional unit is not necessary because the higher voltage from the environment is already an adequate input stimulus for the FU modified circuit. As a result of these modifications, PE_L is functionally and structurally (except for the multiplier) equal to the PE_H type. However, their circuit netlists differ from each other because the synthesis tool tries to keep the timing constraints while considering the new power supply conditions.

These two PE types may now be used to compose dual voltage instances of the CRC array. The number and spatial distribution of each PE type is decided at the architecture design time. Such design decisions are highly dependent on the target application and its mapping strategy.

5.4.2.2 Delay and Area

To present our results, we estimate the execution delay as indicated in Figs. 5.7 and 5.8. This delay consists of the time for a given operation to be executed in the FU (tOp). For PEs of PE_L type, it must be considered that the delay also includes the time spent on the voltage level shifter (tLS). Table 5.5 compares the delay of the data path section composed of operation and level shifter delay on a PE of PE_L type with the delay of operations running on the FU of PE_H type. The state-reg column denotes the time from the rising clock edge until the result is available to be stored in one of the registers. The corresponding paths, being composed of the FSM, the context memory, and the operation, are subject to a timing constraint during synthesis.

For both PE types, a timing constraint of 3.75 ns was specified for synthesis. One can see that this timing constraint can be met for all operations executed on the PE of PE_H type. As indicated in Table 5.5, the multiply operation is the critical component and only this operation yields a violation of the timing constraint on the PE of PE_L type. It is obvious that the multiplication is the slowest operation executed on the PE of PE_H type. This means that we can execute the other operations on lower voltage, because the timing constraint is not violated by the additional delay due to level shifters and increased delay on lower voltage. All in all, 33 level shifters (32 at the data output, 1 at the flag output) are inserted at the output of the FU in a 32-bit PE. This leads to an area increase of 4.5%.

5.4.2.3 Power Estimation

The mapping of applications onto dynamically reconfigurable processor architectures can be done in different ways. To validate our approach and to obtain the power estimations, we mapped the luminance calculation ($xy = (c1 * xr + c2 * xg + c3 * xb + c4) \gg c5$) of the RGB to YIQ conversion from the Embedded Microprocessor Benchmark Consortium (www.eembc.org) Consumer Benchmark under two different mapping strategies. Applying the multi-context pipelined execution power consumption of a PE of PE_L type compared to a PE of PE_H type is reduced up to 13.3% in this example. Our experiments show that for a chained execution, the power consumption of a FU plus level shifter power reduces up to 39% compared to a FU on high voltage level and thereby the power consumption

Table 5.5 Comparison of delay of operations performed on a Dual-V_{DD} processing element and a standard processing element [22].

Op.	PE_L FU (ns)	PE_L LS (ns)	PE_L FU + LS (ns)	PE_L state-reg (ns)	PE_H FU (ns)	PE_H state-reg (ns)
*	–	–	–	–	2.67	3.61
+	2.25	0.31	2.56	3.44	0.99	2.27
–	2.23	0.29	2.52	3.44	0.94	2.26
$\ll$	2.09	0.31	2.40	3.48	1.36	2.49
==	1.74	0.20	1.94	2.81	0.74	1.90
!=	1.66	0.20	1.86	2.73	0.58	1.86
>	1.71	0.20	1.91	2.78	0.75	1.91
>=	1.56	0.20	1.76	2.64	0.68	1.84
<	1.33	0.20	1.53	2.63	0.53	1.84
<=	1.33	0.20	1.53	2.63	0.53	1.84
and_d	0.78	0.31	1.09	2.19	0.26	1.52
or_d	0.77	0.31	1.08	2.19	0.26	1.52
xor_d	0.78	0.31	1.09	2.19	0.26	1.56
not_d	0.34	0.33	0.67	1.67	0.26	1.39
and_s	0.58	0.20	0.78	1.64	0.57	1.39
or_s	0.98	0.20	1.18	1.97	0.57	1.39
xor_s	1.30	0.20	1.50	2.38	0.59	1.41
not_s	0.31	0.20	0.51	1.38	0.25	1.07

of a PE of PE_L type reduces up to 22.1% compared to a PE of PE_H type. The total power consumption of the dual-V_{DD} architecture instance decreases by 10.5% compared to a single voltage architecture instance performing a chained execution of the luminance calculation.

5.5 Optimizing External Reconfiguration

The previous sections focused on processor-like reconfiguration which is performed by switching between contexts in each clock cycle. This kind of reconfiguration is only possible if these contexts have been previously loaded into the configuration memory within the reconfigurable array. Since the configuration memory is limited and the actual applications are not known at design time, it is likely that an application's resource demand exceeds the resources provided by the target architecture. This requires external reconfiguration, i.e., writing into the context memory from outside the reconfigurable array. Usually external reconfiguration is performed when the device is powered up and initialized. In order to allow temporal partitioning of an application, however, external reconfiguration has to be performed during execution and therefore becomes a time critical process. Without optimizations external reconfiguration causes a suspension of processing. The higher the frequency of external reconfiguration the more crucial is the optimization of reconfiguration time.

External reconfiguration is typical for single-context devices like FPGAs. In this book several approaches are demonstrated for the optimization of dynamic reconfiguration of FPGAs. Since FPGAs provide reconfiguration on bit-level, the configuration size for an FPGA is considerably larger than for coarse grained reconfigurable architectures, a fact which eventually results in longer reconfiguration times. The configuration overhead can be reduced if the reconfiguration times are masked through configuration prefetching. Configuration prefetching can be performed on single-context architectures by partitioning the fabric into reconfigurable areas of which some areas are being executed while others can be reconfigured. This, however, has the drawback that not all computation resources can be used at the same time. For the processor-like reconfigurable architectures the availability of multiple contexts allows a different approach to configuration prefetching.

Processor-like reconfigurable architectures are not intended to execute entire applications but to accelerate data- and computation intensive parts of streaming applications. In the remainder of this section such application parts are referred to as kernels. Naturally such kernels have to be computed in a sequence that depending on the application is previously known or unknown. Therefore the reconfiguration of one kernel to a successive kernel is considered.

5.5.1 Multi-Context Configuration Prefetching

Depending on the kernel currently being executed on the processor-like reconfigurable architecture, a number of contexts remain inactive during execution. The corresponding entries of the configuration memory can be changed without interfering with the execution of active contexts. It is therefore not necessary to partition the fabric into reconfigurable areas, since inactive contexts can be reconfigured while other contexts are being executed. In this way, the functional units of all PEs remain available for processing. To ensure that enough contexts remain inactive during processing in order to allow prefetching, the number of contexts available for one kernel can be constrained easily through according compiler settings.

The two graphs depicted in Fig. 5.9 illustrate the principle of multi-context configuration prefetching for a reconfiguration between two kernels K and K'. The graphs show the state of the 32 contexts of a sample architecture over time whereas the light shade illustrates execution and black shade signals reconfiguration activity.

In the upper graph the execution is suspended until kernel K' has been configured. In the lower graph the inactive contexts are used for prefetching. After kernel K' has been configured into a subset of the inactive contexts it can be activated within one clock cycle by switching contexts. Thereby the reconfiguration time is masked and suspension is avoided.

Even though the effective reconfiguration time is eliminated in the example, the output data stream of the reconfigurable device is still interrupted for the latency of kernel K'. The interruption can be further reduced by stage-wise reconfiguration if the kernels are executed in a pipeline.

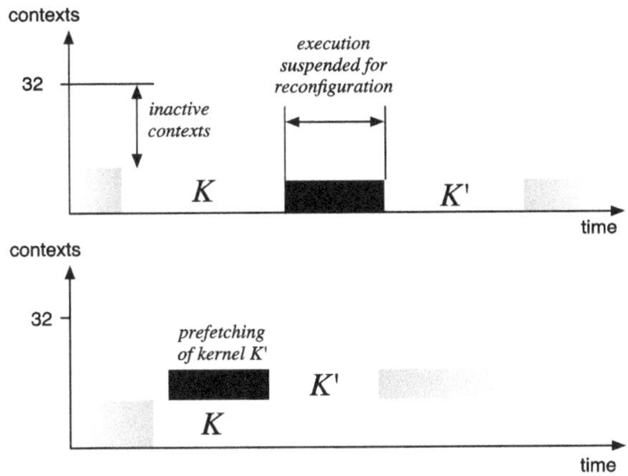

Fig. 5.9 External reconfiguration performed without (upper graph) and with configuration prefetching (lower graph) [6].

5.5.2 Speculative Configuration Prefetching

The prefetching method presented in the previous section requires knowledge about the execution sequence of kernels. However, it is possible that the selection of the subsequent kernel depends on user interaction or on the result of the kernel being currently executed. If the subsequent kernel remains unknown, one possible approach would be to configure a random kernel speculatively and then hope that actually it will be selected. To increase the probability of the right kernel being prefetched, for each kernel that is executed the remaining inactive contexts could be configured with a set of candidate kernels. We refer to this method as speculative configuration prefetching.

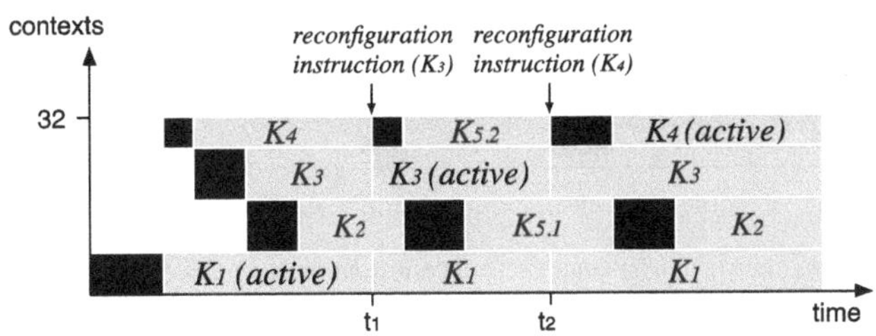

Fig. 5.10 Example for speculative prefetching of configurations.

Figure 5.10 illustrates a sample execution with applied speculative configuration prefetching. In this graph the gray shaded areas identify for each context the kernel that has been configured. The kernels labeled as *active* are currently being executed whereas the black shaded contexts are being reconfigured. In the beginning kernel K_1 is being executed. This execution time is used to prefetch a set of kernels

$\{K_2, K_3, K_4\}$. At time t_1 the subsequent kernel K_3 is disclosed. Since this kernel has already been speculatively prefetched the contexts can be switched with no time overhead. During execution of K_3 kernel K_1 remains in the contexts but kernel K_5 is being prefetched. At time t_2 kernel K_4 is required by the system. This time, however, the needed kernel was not prefetched. Therefore it has to be reconfigured and execution is suspended until the according contexts have been configured.

The longer the reconfiguration time of a kernel the greater is the profit of prefetching this kernel if it is actually selected. We have developed an approach that optimizes the expected profit of prefetching. The kernel selection is based on the kernel characteristics and also takes into account a previous knowledge of the developer about kernel selection probabilities.

5.5.3 Experimental Results

The presented methods were evaluated using the fast fourier transform (FFT) kernels which were discussed in Sect. 5.3.2. One characteristic of scalable OFDMA is the frequent change of kernels with different vector input sizes. Therefore the experiments considered changing FFT kernels.

Each kernel mapping was based on an implementation of the Cooley-Tukey algorithm. Table 5.6 shows the characteristics of the different kernels: the number of samples, the number of required contexts, and the configuration time for loading the kernel into the configuration memory.

Table 5.6 Characteristics of the different FFT mappings.

FFT Size	Contexts	Config. time (clock cycles)
8	1	1440
16	1	1440
32	2	2880
64	4	5760
128	8	11520
256	16	23040
512	32	46080

The first experiments were based on known kernel sequences. For each possible kernel combination a cycle accurate reconfiguration schedule was generated based on the kernel characteristics. During runtime, depending on the upcoming kernel the according schedule was used as basis for a stage-wise configuration prefetching. The interruption of the output data stream during kernel reconfiguration highly depended on the two kernels that were reconfigured and varied between 0 and 954 clock cycle. In average, however, the interruption was reduced by 48% [6].

In the second experiment, the subsequent kernel was selected randomly. Two simulation runs of one second were performed based on the same random kernel sequence, once without and another run with speculative configuration prefetching.

For each simulation the ration of reconfiguration time to processing time was determined. These two ratios were then compared to evaluate the effectiveness of speculative prefetching. As a result the reconfiguration time overhead could be reduced by 38% in average [7].

5.6 Conclusion

In this chapter we have evaluated the benefits and costs of processor-like reconfiguration. To enable this research, we have created the CRC model as a general model for processor-like reconfigurable architectures. Furthermore a compiler was developed that takes advantage of fast reconfiguration and that can be adapted to different model instances. In this way the CRC model facilitates the exploration of architectural alternatives paired with novel compiler techniques. Based on a practical example we demonstrated how to take advantage of processor-like reconfiguration.

We also would like to highlight the industrial usability of such architectures. Recently, first commercial products that include processor-like reconfigurable devices were introduced to the market. These processors already exploit some of the optimization potential that was evaluated in this chapter.

Acknowledgements This work was supported by DFG grant RO 1030/13, as part of the Priority Programme 1148, *Reconfigurable Computing Systems*.

References

1. Altera Corporation: FFT MegaCore function user guide—v6.1. Online: http://www.altera.com (2004)
2. Bansal, N., Gupta, S., Dutt, N., Nicolau, A.: Analysis of the performance of coarse-grain reconfigurable architectures with different processing element configurations. In: Proc. of the WASP (2003)
3. Bossuet, L., Gogniat, G., Philippe, J.L.: Generic design space exploration for reconfigurable architectures. In: Proc. of the IPDPS, pp. 163–171 (2005)
4. Bouwens, F.J., Berekovic, M., Kanstein, A., Gaydadjiev, G.N.: Architectural exploration of the ADRES coarse grained reconfigurable array. In: Proc. of the ARC, pp. 1–13 (2007)
5. Brenner, J.A., v. d. Veen, J.C., Fekete, S.P., Filho, J.O., Rosenstiel, W.: Optimal simultaneous scheduling, binding and routing for processor-like reconfigurable architectures. In: Proceedings of the 17. International Conference on Field Programmable Logic and Applications, Spanien (2006)
6. Eisenhardt, S., Oppold, T., Schweizer, T., Rosenstiel, W.: Optimizing partial reconfiguration of multi-context architectures. In: IEEE ReConFig, Cancun, Mexico (2008)
7. Eisenhardt, S., Oliveira, J., Kuhn, T., Rosenstiel, W.: Speculative configuration prefetching for multi-context architectures. In: Workshop on Synthesis and System Integration of Mixed Information Technologies (SASIMI), Okinawa, Japan (2009)
8. He, S., Torkelson, M.: Designing pipeline FFT processor for OFDM (de)modulation. In: Proc. of the ISSSE, pp. 257–262 (1998)

9. Jain, D., Kumar, A., Pozzi, L., Ienne, P.: Automatically customising vliw architectures with coarse grained application specific functional units. In: Proc. of the SCOPES, pp. 17–32 (2004)
10. Kim, Y., Kiemb, M., Choi, K.: Efficient design space exploration for domain-specific optimization of coarse-grained reconfigurable architecture. In: SoC Design Conf. (2005)
11. Kim, Y., Kiemb, M., Park, C., Jung, J., Choi, K.: Resource sharing and pipelining in coarse-grained reconfigurable architecture for domain-specific optimization. In: Proc. of the DATE, pp. 12–17 (2005)
12. Lin, Y.W., Liu, H.Y., Lee, C.Y.: A 1-GS/s FFT/IFFT processor for UWB applications. IEEE J. Solid-State Circuits **40**(8), 1726–1735 (2005)
13. Liu, Z., Song, Y., Ikenaga, T., Goto, S.: A VLSI Array Processing Oriented Fast Fourier Transform Algorithm and Hardware Implementation. IEICE Trans. Fundam. **E88-A**(12), 3523–3530 (2005)
14. Mei, B., Vernalde, S., Verkest, D., Lauwereins, R.: Design methodology for a tightly coupled VLIW reconfigurable matrix architecture: a case study. In: Proc. of the DATE, pp. 1224–1229 (2004)
15. Mei, B., Lambrechts, A., et al.: Architecture exploration for a reconfigurable architecture template. IEEE Des. Test **22**(2), 90–101 (2005)
16. Miramond, B., Delosme, J.: Design space exploration for dynamically reconfigurable architectures. In: Proc. of the DATE, pp. 366–371 (2005)
17. Oliveira, J., Schweizer, T., Oppold, T., Kuhn, T., Rosenstiel, W.: Tuning coarse-grained reconfigurable architectures towards an application domain. In: International Conference on Reconfigurable Computing and FPGAs(ReConfig) (2006)
18. Oliveira, J., Kuhn, T., Rosenstiel, W.: Evaluating the impact of customized instruction set on coarse grained reconfigurable arrays. In: Proceedings of the International Conference on Field-Programmable Technology (ICFPT), Taiwan (2008)
19. Oppold, T.S., Eisenhardt, W.R.: Optimization of area and performance by processor-like reconfiguration. In: International Parallel and Distributed Processing Symposium (IPDPS)—Reconfigurable Architectures Workshop (RAW), Long Beach, USA (2007)
20. Oppold, T., Schweizer, T., Kuhn, T., Rosenstiel, W., Kanus, U., Straßer, W.: Evaluation of ray casting on processor-like reconfigurable architectures. In: International Conference on Field Programmable Logic and Applications (FPL), Tampere, Finland (2005)
21. Oppold, T., Schweizer, T., Oliveira, J.F., Eisenhardt, S., Rosenstiel, W.: Crc—concepts and evaluation of processor-like reconfigurable architectures. It—Information Technology, doi:10.1524/itit, 49. 3. 157 **49**(3) (2007)
22. Schweizer, T., Oppold, T., Filho, J.O., Eisenhardt, S., Blocher, K., Rosenstiel, W.: Exploiting slack time in dynamically reconfigurable processor architectures. In: International Conference on Field Programmable Technology (ICFPT), Kitakyushu, Japan (2007)
23. Tensilica, Inc.: Xtensa configurable processors. Online: http://www.tensilica.com (2008)
24. WiMax Forum: Mobile WiMax. Online: http://www.wimaxforum.org/ (2006)

Chapter 6
Adaptive Computing Systems and Their Design Tools

Andreas Koch

Abstract While reconfigurable adaptive computing has many proven advantages over conventional processors, in practice, it is often limited to niche applications. This situation, which we aim to resolve with our research, is often linked to the lack of programming languages for adaptive computers that are familiar to software developers. We present a compile flow capable of translating general-purpose C programs to hybrid hardware/software applications for execution on an adaptive computer and give an overview of the required advances in compiler technology as well as in computer architecture and operating system design.

6.1 Introduction

As demonstrated numerous times, reconfigurable computing can have significant advantages over conventional processors for a wide range of applications [28]. Despite these advantages, however, it is only rarely employed outside of academic settings.

One of the key reasons for this discrepancy is the difficulty of actually programming a reconfigurable computer. Most commonly, this is done by designing a compute architecture for the algorithm from scratch, which is then described in a hardware design language such as VHDL or Verilog.

While this approach can result in very high-performance implementations, it requires programmers to be experienced in computer architecture, digital logic design and hardware design languages. Only very few software developers actually have these skills. Thus, the power of reconfigurable computing remains unavailable to most potential users.

Andreas Koch
Embedded Systems and Applications Group, Dept. of Computer Science, Technische Universität Darmstadt, Darmstadt Germany, e-mail: `koch@esa.cs.tu-darmstadt.de`

M. Platzner et al. (eds.), *Dynamically Reconfigurable Systems*,
DOI 10.1007/978-90-481-3485-4_6,

In recent years, many attempts have been made to close this gap and lift the abstraction level of reconfigurable computer programming to that of conventional software-programmable processors.

To this end, we have been working on Comrade, a compiler for automatically translating general-purpose ANSI C programs for computers containing both a conventional and a reconfigurable processor. As we will describe below, the choice of C with its pointers and possibly irregular control flow has significant effects on both the compile flow as well as the target computer architecture and its operating system.

Our DFG-funded project "Adaptive Computing Systems and their Design Tools" was initiated prior to the DFG Priority Program 1148, but has been associated with the Program right from the start due to the thematic closeness. Since the schedules of our on-going project and the concluding Priority Program are thus out-of-phase, this report will concentrate on the major results achieved during the era of the SPP. Section 6.7 will give some perspective on the issues we are addressing in our current research.

6.2 Execution Model

In contrast to traditional research on High-Level Synthesis (HLS) [8] our target architecture is assumed to always contain a conventional software-programmable processor (SPP) in addition to a reconfigurable processor. While the compute-intensive parts of a program can be implemented in a spatially distributed fashion for high-performance, other parts of the program that are either unsuitable (e.g., I/O using `printf` or similar functions) or that are only used rarely and would not justify the permanent allocation of computing area (e.g., error handling) are left in software on the SPP.

This target architecture also avoids a basic problem of High-Level Synthesis, which aimed to translate the *entire* program into hardware: If the input program contained a construct (e.g., function calls, irregular control flow, dynamic memory allocation, etc.) that the specific HLS algorithm could not handle, the translation was aborted completely. While we also intend to translate our input language to the widest practical degree, our flow can always fall back to the SPP to execute program parts that the flow cannot process yet due to implementation limitations, or that would exceed the capacity of the reconfigurable device. This allows incremental development of the compiler, with increasing parts of the profitable computations of a program being moved for acceleration to the reconfigurable device.

We call such an architecture an *adaptive computer system* (ACS). To be more precise, we differentiate between the underlying reconfigurable device (RD), which can be an either an FPGA or a coarse-grained reconfigurable array (CGRA), and the reconfigurable compute unit(s) (RCU) that can be mapped to it.

When combining multiple processing elements (such as the SPP and RCU of an ACS), the manner of their interaction must be specified. This is done by the *execution model* (discussed in greater detail in [22]).

Different ACS compilers employ different models, which differ mainly in the granularity of the SPP/RCU partitioning.

The ASH system [1] can switch between the SPP and a hardware accelerator (note: not an RCU, ASH targets ASICs) only at procedure boundaries. This rather coarse granularity leads to an *entire* procedure being ineligible for hardware acceleration even if only a single non-compilable construct is present. ASH exceeds classic High-Level Synthesis by allowing software functions to be called from the hardware accelerator, however.

Other systems, such as GarpCC [2], Nimble [27] and our own Comrade, allow a finer-grained partitioning *within* of functions. For the following discussion, consider the sample program shown in Fig. 6.1.

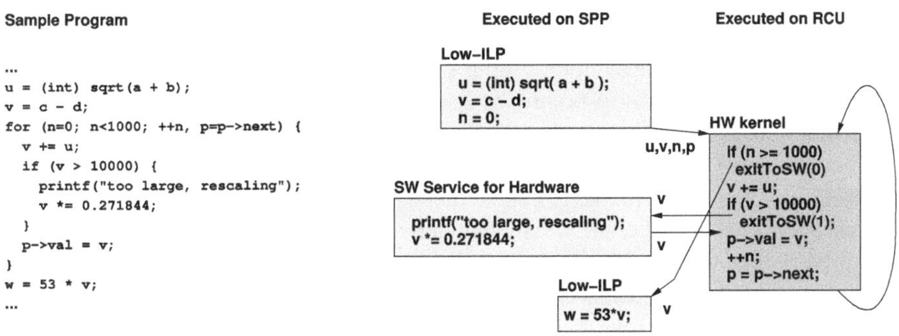

Fig. 6.1 Example in the Comrade execution model [22].

The program contains a typical hardware kernel (a loop) which is surrounded by code with a low degree of instruction level parallelism (ILP) statements. For this example, we assume that the functions sqrt and printf are not efficiently compilable to an RCU. Thus, they, as well as the rest of the low-ILP code, should be left on the SPP.

All three of these finer-partitioning compilers can successfully perform this operation. However, they differ in their handling of the RCU-unsuitable code within the loop kernel. Using dynamic profiling, they might discover that the condition $v > 10000$ occurs only rarely, and thus moving the loop to the RCU is profitable despite the infrequent switches to the SPP (for the printf and the floating-point multiplication). Such a switch requires exchanging the live variables between the two processors and can have a significant overhead. After a switch to the SPP, GarpCC and Nimble then execute the entire remainder of the current loop iteration in software (they generate both hardware and software versions of each kernel). Only when re-entering the loop for another iteration is the decision made whether to continue on the SPP or switch back to the RCU.

The model we use in Comrade (shown in Fig. 6.1) is even more fine-grained: We can now switch between individual statements, moving just the printf and the float multiplication to the SPP as a so-called *software service*. After completion of the service, execution switches back to the RCU immediately. Note that, with the finer granularity, *fewer* live variables need to be transferred between the two processors. In this fashion, a single hardware kernel can access multiple software services, each with just the code for the requested function.

The exchange of live variables has far-reaching architectural consequences when *pointers* are to be supported (as we have to do for C in Comrade). In this case, address arithmetic performed on the SPP and RCU must be compatible, and both processors must share a coherent view of memory (reads and writes can be performed by both sides). This will be discussed in greater detail in the next section.

6.3 ACS Architecture

The Comrade model of execution requires a number of capabilities from its underlying computing platform to be practical. As explained above, our model requires low-latency SPP-RCU communication for the exchange of the live variables. Note that this does not have to be a high-bandwidth link, since generally only very few variables have to be exchanged (due to the small scope of our software services). Pointer addresses must be freely exchangeable across the SPP-RCU boundary, and both processors must have high-throughput access to a shared main memory, possibly in the presence of virtual memory (e.g., when running the ACS under a full-scale Linux OS). For security reasons, the RCU must of course respect the access permissions imposed by memory protection. If these apply at the process level (the general case for Unix variants), the code of the SPP software *within* the hybrid SPP/RCU process should also be inaccessible to a possibly rogue RCU. Furthermore, the RCU must be prevented from interfering with OS scheduling decisions, e.g., by denying other processes (or the OS itself) access to main memory. Finally, the software part of a hybrid SPP/RCU process should execute at full speed, without slow-down due to the RCU.

A much more detailed discussion of these aspects can be found in [21–23].

6.3.1 Reconfigurable System-on-Chip Architecture

To achieve these goals and actually make the Comrade execution model feasible on real hardware, we implemented suitable ACS architectures as reconfigurable systems-on-chip (rSoC), at first using the Xilinx Virtex II Pro-based ML310 platform. The II Pro FPGAs embed SPPs (300 MHz PowerPC 405 CPUs) into a reconfigurable logic array suitable as RCU. Since the interfaces, both between SPP and RCU as well as the rest of the system (memory, I/O, etc.) are mainly realized using reconfigurable logic, they can be changed to fit the requirements we formulated above.

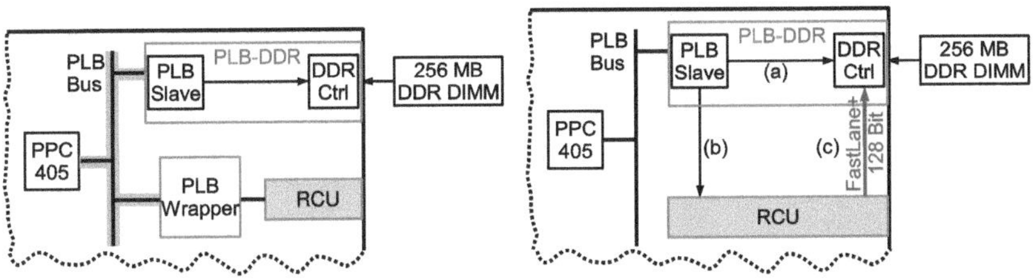

(a) via PLB [23] **(b)** via FastLane+ [24]

Fig. 6.2 RCU integration techniques.

The default system-on-chip architecture supported by Xilinx EDK design tools is shown in Fig. 6.2(a). Note that both SPP-RCU communication as well as memory accesses from the RCU have to pass through the single Processor Local Bus (PLB). While the multi-purpose use of such standard busses allows the flexible composition of SoC architectures, the performance suffers: The standard approach achieves a transfer rate of just 213 MB/s (of the theoretically possible 1600 MB/s) for RCU memory accesses while the SPP executes a (mostly idle) Linux.

To improve RCU-to-memory bandwidth, we developed the FastLane+ architecture (Fig. 6.2(b)): Now, the RCU is attached directly to main memory using a dedicated 128b wide bus. This increases the bandwidth available to the RCU significantly (now to 1424 MB/s, again, while Linux is idling on the SPP), fulfilling one our requirements stated above. But in addition, FastLane+ ensures that the SPP has override priority when accessing memory and can even terminate RCU-initiated transfers. This guarantees that the OS (and programs chosen by the OS scheduler) are never starved by lack of memory bandwidth (a critical aspect for OS operations such as interrupt handling or time-critical disk or network I/O).

Since the RCU, acting as a slave, can be accessed from the SPP via the PLB, SPP-RCU transfers (such as the live variable exchange required when starting the RCU, or performing a software service on the SPP) can also be performed quite quickly (just 20 ns/40 ns per 32b variable read/written).

6.3.2 Operating System Integration

Improving the data transfer between SPP, RCU and memory is necessary to implement our execution model in a real system, but does not suffice on its own. To also operate securely under a protected virtual memory scheme while exchanging pointers, running software applications at full speed (unhindered by the presence of an RCU), and allowing low-latency RCU-SPP switches (for performing software services), the purely hardware architectural measures described above are inadequate: We now need support from the operating system. To demonstrate the feasibility of our approach even when running a full-scale OS, we chose Linux (which continues to grow market share in the embedded systems area) over a less demanding, but functionally more limited real-time kernel.

6.3.2.1 RCU-SPP Signalling

Normally, the RCU raises an interrupt to get the attention of the SPP (e.g., to provide a software service). Even in a Linux version patched for low-latency responses, the processing of such an interrupt takes 62 μs on the ML310 platform. The normal path (labeled as such in Fig. 6.3) sketches how an interrupt passes through numerous layers in the Linux kernel, before it finally reaches the handler in the software main thread of the hybrid HW/SW user program.

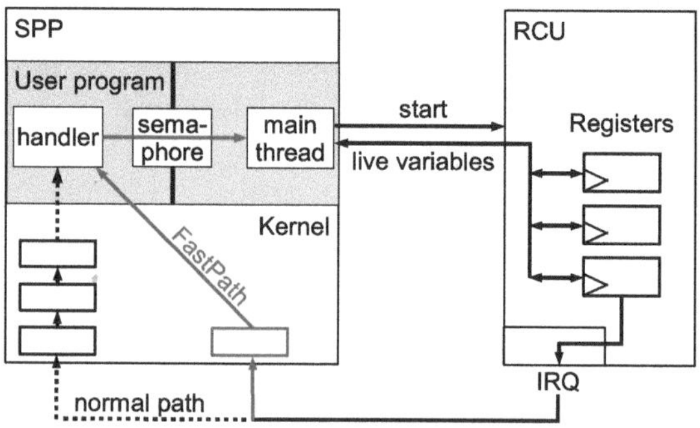

Fig. 6.3 RCU-SPP Signalling [23].

Our FastPath approach takes a number of measures to improve the interrupt response latency. First, a dedicated interrupt vector on the PowerPC 405 is assigned the RCU, allowing the bypassing of the kernel interrupt processing layers. Then, instead of handling the interrupt in a device driver and using mechanisms such as file descriptors or similar to forward it to the user program, FastPath can jump directly to a previously registered callback function which executes in context of the user program (has access to global variables and is subject to access limitations). The callback is executed in a separate thread in parallel to the existing threads of the user program. As usual, execution can be synchronized using a semaphore. But in FastPath, this semaphore is realized using a dedicated special SPP register (instead of an in-memory data structure that would also cause bus traffic). By proceeding in this fashion, we can reduce software service latency down to the 9.6...2.7 μs, an improvement of 6.5× to 23× over the original implementation. The FastPath latency on the comparatively slow 300 MHz PowerPC 405 outperforms even special sub-kernel facilities (such as RTAI and LXRT) running on multi-GHz desktop CPUs [26].

The accelerated RCU-SPP signalling has a tremendous effect on the practical partitioning granularity for SPP-RCU execution (shown in Fig. 6.4). To give an example interpretation: Assuming the RCU executes an algorithm $HW_{accel} =$

10 times faster than the SPP, requires an average $t_{overhead} = 6.15$ μs for the RCU-SPP signalling, and disregarding the time for the software service itself (which would also need to be executed when running only on the SPP), we can accept a switch to a software service every $t_{SSW} = 55.4$ μs and still achieve an effective acceleration of 9× over the SPP. With FastPath, the quick SPP-RCU switches required by our execution model are demonstrated to be achievable on real hardware/software platform.

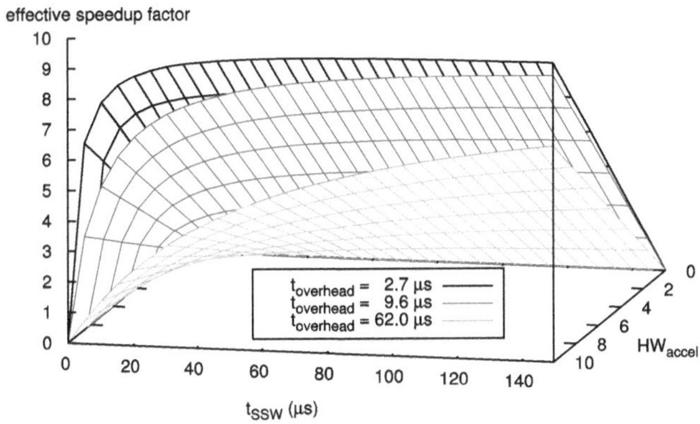

Fig. 6.4 Effective speed-up vs. RCU-SPP signalling delay [23].

6.3.2.2 Shared Virtual Memory

Fast SPP/RCU accesses to a shared physical memory (as described in Sect. 6.3.1) can be implemented purely in hardware. But for safety and security reasons [31], protected virtual memory is becoming more common even in traditional embedded OS such as LynxOS, VxWorks, and of course embedded Linux. To support our model of execution even when running the ACS under such an OS, RCU memory addressing and accessing must be coordinate with the SPP-side memory management unit (MMU). We have implemented and evaluated two significantly different solutions to this problem.

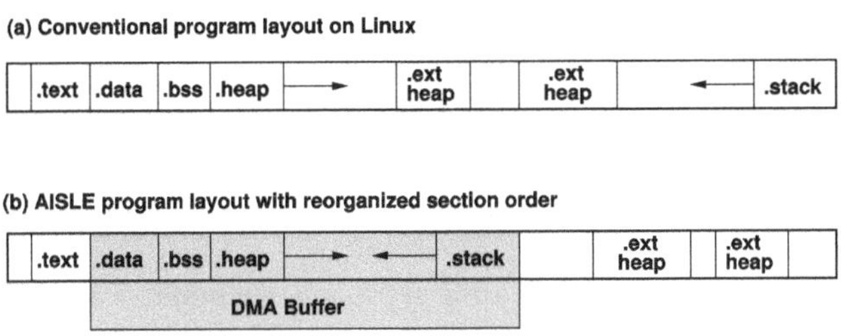

Fig. 6.5 Conventional and AISLE Program Layouts [22].

Figure 6.5 shows the simpler of the two approaches, the Accelerator-Integrating Shared Layout for Executables (AISLE) [22]. In this model, we map all data regions (uninitialized, initialized, heap, and stack) into a so-called DMA buffer by modifying the ELF program loader [34]. A DMA buffer is a contiguous address range accessible to the RCU that is locked into physical memory (not subject to being paged to and from disk). This also ensures that the same virtual address will always map to the same physical address. Thus, virtual $\leftrightarrow$ physical address translation consists of just adding/subtracting constant offsets. Analogously, addresses generated by the RCU can easily be constrained by a simple bounds check to always lie within the buffer, protecting other processes from a potentially rogue RCU. AISLE also protects the executable code (the so-called .text segment) of the user program itself from the RCU by keeping it *outside* the buffer. In this fashion, both SPP and RCU operate on the same virtual addresses and can transparently process data stored in the shared memory (no inefficient copying between SPP- and RCU-accessible memories is required). Reference [22] explains AISLE in greater detail, also covering topics such as cache coherency between SPP and RCU, and size management of the DMA buffer.

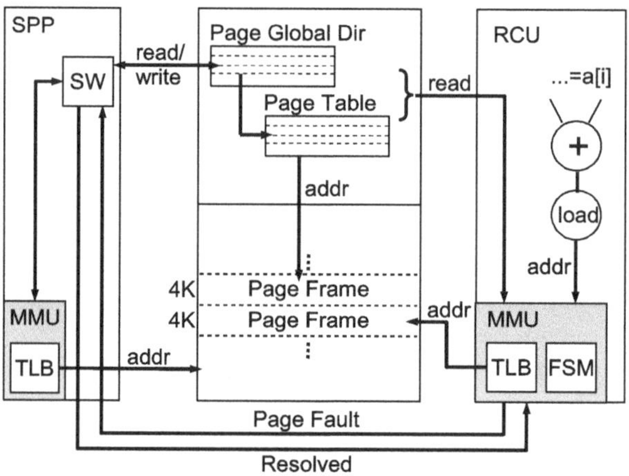

Fig. 6.6 SPP-RCU shared virtual addressing with PHASE/V [23].

While AISLE provides the desired functionality in a very efficient fashion (as will be shown below), it does have a number of limitations: Once the DMA buffer is allocated, it can be resized only with considerable overhead. In many cases, this will lead to the maximal size buffer being allocated to the RCU. Since, by its very nature, the DMA buffer does not participate in paging, it will always occupy its full size in physical memory.

As a second, potentially more flexible approach, we implemented the Processor-Hardware Accelerator Shared Environment with Virtual addressing (PHASE/V, shown in Fig. 6.6) [23]. Here, the RCU fully participates in all virtual memory operations, including demand paging, variable virtual-physical mappings, and discontiguous physical memory ranges. With demand paging now supported, physical memory is only allocated when needed and released when required otherwise (in contrast to the statically pre-allocated DMA buffer of AISLE). Analogously to the SPP, the RCU uses a translation-lookaside buffer (TLB) to cache virtual-physical translations (taking only a single clock cycle for a translation), but is fully capable of performing a walk of the page mapping tables itself to determine as-yet unknown mappings on a TLB miss. If no physical page is found, the RCU uses the FastPath signalling scheme to request the SPP to handle the page fault (e.g., load the missing page from disk). In the reverse direction, the OS on the SPP informs the RCU when it alters the mappings (e.g., when a memory page is swapped out to disk), causing the RCU to synchronize its TLB with the one in the SPP.

6.3.3 Evaluation

To stress-test the performance of the different schemes, we use the traversal of a linked-list, randomly distributed in memory [23]. AISLE and PHASE/V have similar performance, until the number of virtual pages in the working set exceeds the RCU-TLB capacities (which occurs between 16 K and 32 K list elements). Then, PHASE/V begins to slow down (e.g., taking 2.3× the time of AISLE at 128 K list elements). However, even in this extreme case, the RCU remains 23% faster than a pure-software version running of the SPP. Thus, the often quoted maxim that reconfigurable computing is mainly useful for stream processing has been invalidated by our research: RCUs can also be faster than SPPs even for highly irregular pointer-chasing applications. In such scenarios, performance will be limited by the memory bandwidth, which we can fully exploit with FastLane+ on the RCU.

More detailed evaluations are presented in [22, 23]. On of the experiments also demonstrates that an active RCU only marginally slows the pure software processes scheduled by the OS to execute on the SPP. In summary, by taking the appropriate architecture and OS measures, we have now created an environment which enables the practical use of the compute model of Sect. 6.2.

6.4 Hardware/Software Co-compilation Flow

Now that we have defined and implemented a suitable reconfigurable target architecture, we can examine the programming tools we developed to make the technology accessible to software developers.

6.4.1 Overview

The heart of the flow is the Comrade compiler. As usual for modern compilers, it is organized in a multi-pass manner. First, common compiler operations are performed (lexing, parsing, machine independent optimization). At this step, the intermediate representation (IR) of the compiler front- and middle-end is exported from the system (as a C program) and subjected to dynamic profiling. As a result, actual execution frequencies can now be back annotated into the IR to guide further processing.

Based on the profiling data, program is partitioned for SPP and RCU execution (see Sect. 6.4.2 for more details). The software part destined for the SPP is enriched with interface code (to exchange the live variables at SPP/RCU execution boundaries and start the RCU), exported as a C program and fed into a conventional software C compiler.

The part to be executed on the RCU needs to be processed further by Comrade. It is first transformed into a control flow graph (CFG) in Static Single Assignment form, which forms the base of Comrades hardware-centric intermediate representation, the Control Memory Data Flow Graph (CMDFG, see Sect. 6.4.3). Hardware-specific optimizations (e.g., parallelizing memory accesses) are then performed by transforming the CMDFG. The hardware for the RCU is finally generated from the optimized CMDFG as separate data path and dynamically scheduled controller (see Sect. 6.4.4) in the form of RTL Verilog netlists. These are then synthesized together with the fixed rSoC architecture of the ACS (Sect. 6.3.1), and handed to the FPGA vendor tools for implementation (map, place, route, bitstream generation).

6.4.2 Profile-Based Inlining and Partitioning

Dynamic profiling data is gathered early in the compile process after some machine independent optimizations (e.g., goto removal) have already been performed. This profile is first used to inline only the most heavily used functions. In this way, we avoid the code size explosion (specifically, the associated area requirements on the RCU) that can occur when inlining indiscriminately.

For partitioning, multiple versions of the loops making up the potential RCU kernels are created. This is done in an inside-out fashion, starting with just the innermost loop(s) and then widening the scope (possibly merging formerly separate loops into one kernel candidate) until the RCU area is exceeded. Figure 6.7 shows an example of this for two loops: The CFG shown in (a) is expanded into a CFG that has the entire loop nest executing on the SPP (left), the outer loop on the SPP and the inner on the RCU (middle), and the entire nest executing on the RCU (right).

We then examine each of these candidate RCU kernels by building valid *paths* of operations that can be natively executed on the RCU. These will generally bypass external I/O and all functions which were not inlined (either due to adverse profiling data or lack of source code). Such operations will be marked as potential software services. *Valid* means for path that it will pass *through* the candidate region (enter and leave it). Path construction first constructs a valid path and then expands it, adding rejoining sub-paths in order of decreasing execution frequency. This is done until the entire candidate is covered by RCU-suitable paths and software services, or the RCU area is exhausted. More details on these steps can be found in [11, 12]. As

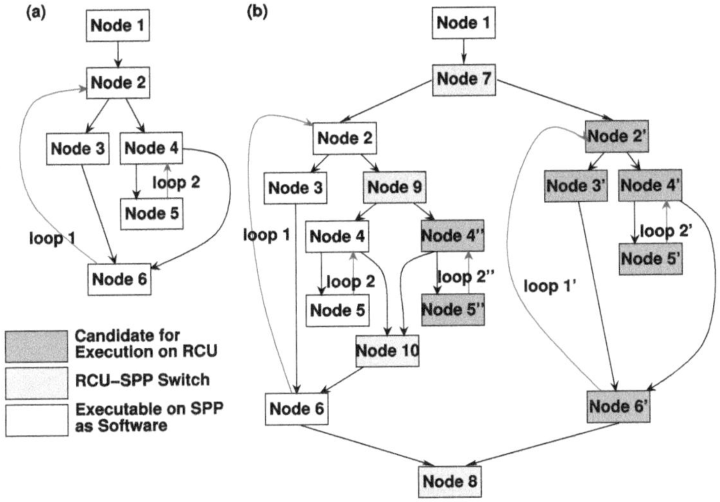

Fig. 6.7 Generating alternate SPP-RCU partitionings [13].

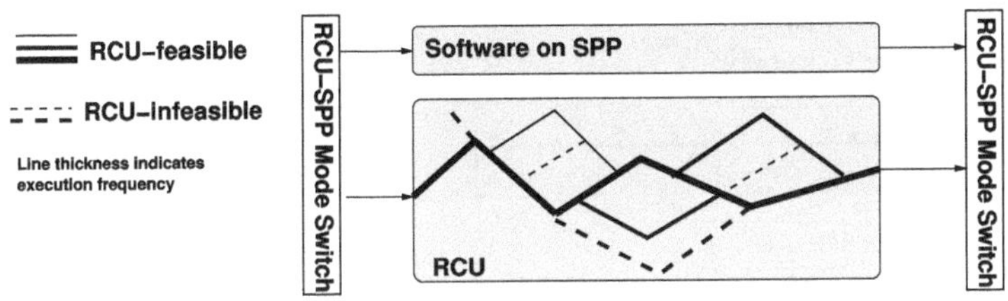

Fig. 6.8 RCU path construction [13].

examples, 80% of the instructions of the Versatility wavelet image [30] compressor and the ADPCM audio coder can potentially be moved to on the RCU in this fashion.

6.4.3 CMDFG Intermediate Representation

We experimented with a number of intermediate representations for hardware/software co-compilation. This included various SSA forms, such as plain SSA [4], array SSA [16], and our own initial attempt for a data-flow controlled form heavily emphasizing parallel execution [14] (DFC-SSA, which later proved too difficult to schedule). None of these proved sufficiently expressive.

To fill this need, we developed Control Memory Data Flow Graphs (CMDFG), described in greater detail in [6, 7]. The CMDFG expresses control flow (extracted from the intermediate SSA-CFG created during compilation), inter-operator data flow (as a classical data flow graph) and memory dependencies (such as those computed by alias analysis). It is somewhat similar to the Program Dependence Web

[3] and Program Dependence Graph [5], which are also low-level fine-grained intermediate representations. However, they do not express memory dependencies at all and rely on constructs such as γ and β functions or data-flow switches to guide computation. The CMDFG is closer to the actual hardware: It employs distributed multiplexers for this purpose, with control edges running from a condition to the multiplexer input nodes to determining the selected data.

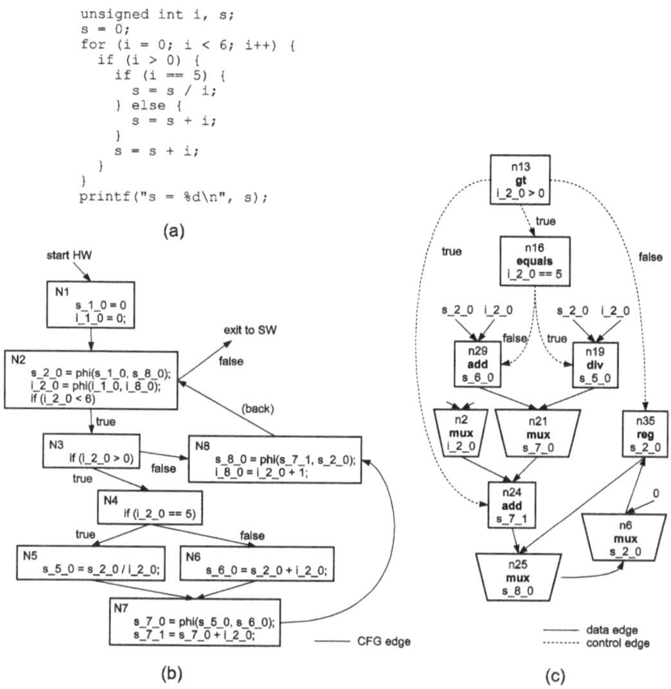

Fig. 6.9 (a) Sample program, (b) SSA-CFG, (c) CMDFG [6].

Figure 6.9 shows a very simple example of such a CMDFG (for simplicity, without memory edges). For the source code shown in (a), the SSA-CFG is shown in (b). The body of the loop is represented by the CMDFG shown in (c). For clarity, this omits the increment of the index variable and the test of the loop condition. Note that the true condition of n13 controls both the nested if statement as well as the assignment to s at the same level. In contrast to many previous hardware compilers, Comrade can easily handle even complex nested control structures in loops (including other loops!) using the CMDFG.

Figure 6.10 shows a simple example how memory edges explicitly express memory dependencies. In the CDFG without memory edges (b), the load and store of the sample program (a) could execute in arbitrary order and potentially violate the write-after-read requirement. The CMDFG (c) enforces correct execution, with the aid of two additional constructs: A Memory Forwarder (MF) executes if allowed

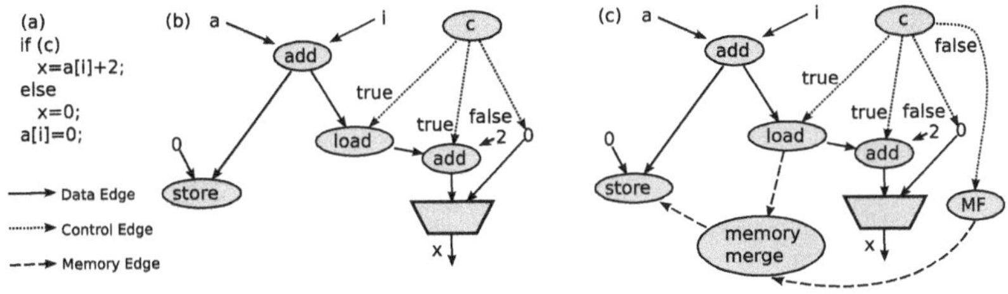

Fig. 6.10 (a) Sample program, (b) CDFG, (c) CMDFG.

by the incoming control edge, and allows execution of its successor memory nodes. The memory merge node allows execution of its successor memory nodes if at least one of its direct memory predecessors was executed. In this fashion, the write a[i]=0 will only execute if c was true, and the load in x=a[i]+2 has already completed, *or* c was false, and the MF node executed, leading to the execution of the memory merge node, which in turn allows the store node to execute.

By selectively modifying the memory dependence graph, we can easily express parallel memory accesses once we have proven their independence, e.g., by alias analysis methods. The memory aspects of the CMDFG are discussed in greater detail in [7].

6.4.4 CoCoMa Controller Model

Another core result of our research was creation of new dynamic scheduling semantics for the CMDFG. Beyond the capabilities of traditional dependency graphs (e.g., control data flow graphs), which have a node execute and produce a result as soon as all of its operands are available (and a possibly incoming control edge is also valid), we can now express the *deletion* of results. This can extend to stopping a calculation already in progress once it has been determined that the result will not be needed. Since this relation is transitive, an entire chain of unneeded calculations can be stopped, and the operators be made available for the next set of operands.

These semantics are defined by the Comrade Controller Micro-architecture (CoCoMa), which also describes their mapping into hardware. At the abstract level, we now a structure similar to a Petri net with two kinds of tokens: *Activate Tokens* (AT) indicate the presence of data at the source of an edge, Cancel Tokens (CT) erase an AT (and its associated data item) when they meet. ATs generally move in the direction of data flow, while CTs move in the opposite direction.

A simple example for the inner if of the Fig. 6.9(a) is shown in Fig. 6.11 for three successive clock cycles (assuming the addition operator has a latency of one cycle, and the division takes longer). In cycle 0, the values for s_2_0 and i_2_0 have been computed (both of this happens in block N2 of the SSA-CFG). Thus, ATs are present on all outgoing edges of the associated multiplexers n6 and n2. Both the addition and division operators in CMDFG nodes n29 and n19 now start to compute (since

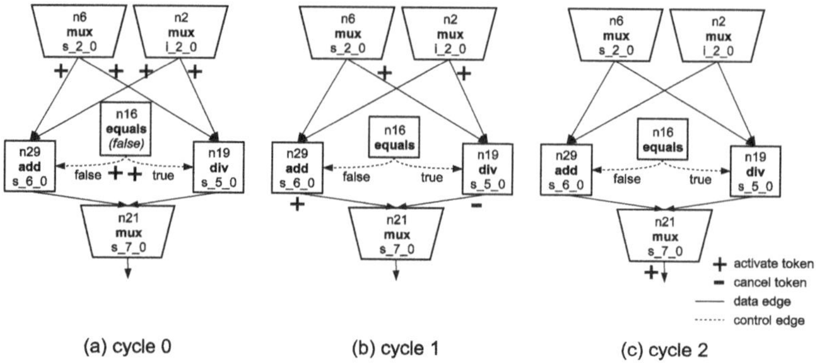

Fig. 6.11 Interaction of Activate and Cancel Tokens [6].

all of their operands are available). But the condition i == 5 is assumed to have evaluated to *false* in this example. Thus, only the operator associated with this state, namely the addition, is allowed to complete (an AT travels along the control edge labeled *false* to node n29). On completion, the operator consumes its operands (and their associated ATs) on the incoming data edges from n6 and n2, and produces the result and an AT at its output in cycle 1. The other branch of the condition i == 5 is handled as follows: The division node at the destination of the control edge also receives an AT in cycle 0. However, since the condition state *false* does not match the *true* requirement on the control edge, the AT is turned into a CT at the output of the unneeded division operator in cycle 1. Note that the division has been started in cycle 0 (the same as the addition), but is now canceled in cycle 1. In cycle 2, the addition result and its AT travel *downward* through n21 for use in further computation. The CT at the division operator has now traveled upwards and erases the incoming operands (and their ATs) which were not consumed by the terminated division. Thus, the entire CMDFG subgraph is now available for the next loop iteration. In static scheduling, this would only happen after the division could be completed (even though its result would not be needed).

From this basic scheme, more complex behaviors can be derived. For example, the CMDFG memory dependencies (described in Sect. 6.4.3) are also modelled as ATs travelling along memory edges. Also, more complex transition rules are required when describing deeply nested structures, especially loops. The best formulation of these is still the subject of active research.

Despite its complexity, CoCoMa scheduling, which described in greater detail in [6], can lead to significant speedups, even when compared with other more advanced dynamic schemes such as lenient execution [1]. It does have additional cost when compiled into hardware, though: Each data register now requires an additional two bits to hold an AT and CT each. Furthermore, the token transition rules need to be implemented as logic networks, also requiring RCU area. As shown in [7], the CoCoMa overhead for realistic applications is in the range of 600–1700 slices, which should be considered acceptable given current RCU device sizes.

6.5 Infrastructure

Beyond the execution model, an appropriate practically realizable ACS architecture and the core compile flow, we developed a number of adjunct technologies.

6.5.1 Parametrized Module Library

Many steps of the Comrade compile flow need target hardware-specific data. This ranges from area/delay estimates for specific operations to information about their physical and logical interfaces down to (possibly pre-placed) structural netlists. All of these aspects are encapsulated in the Generic Library for Adaptive Computer Environments (GLACE), which we introduced in [29].

GLACE provides all of the basic operators (arithmetic, logic, memory) as well as the system interfaces (I/O registers, SPP-RCU signalling) we need for C-to-hardware compilation in a parametrized fashion [19]. For example, we can retrieve from GLACE data about a 27×18 bit unsigned multiplier. The underlying data is organized and accessed by the Flexible API for Module-based Environments [18], which defines both passive (design data model) and active components (query/reply scheme for interacting with the module library). The data model is organized into different *views*, each representing a subset of aspects relevant for a specific stage of the compile flow. Examples include *behavior* (which lists operator semantics), *synthesis* (giving are/delay and interface data), *topology* (holding placement information), and *place* (which encompasses pre-placed EDIF netlists).

GLACE data, accessed via FLAME, is used in Comrade e.g., for the hardware/software partitioning step. It guides both the expansion of nested kernels from pure software to pure hardware as well as the construction of RCU-executable paths through the candidate kernels. It also provides the pre-placed operator netlists for the datapath, which are then combined with the CoCoMa controller and the system interface into a single RCU configuration.

6.5.2 Physical Design Aspects

We have always also considered the physical design aspects of reconfigurable computing. This began with Structured Design Implementation methodology [17], which described how to assemble regular datapaths from pre-placed and pre-routed parametrized modules. We have carried forward this intent into the *topology* and *placed* views of FLAME.

As an example, Fig. 6.12 shows the regular layout of an 8 bit unsigned multiplier on a Xilinx Virtex FPGA and the associated FLAME *topology* view. This describes the extent of the layout (separating regular and irregular parts) as well as the locations of external ports (which are organized here at a pitch of two bits

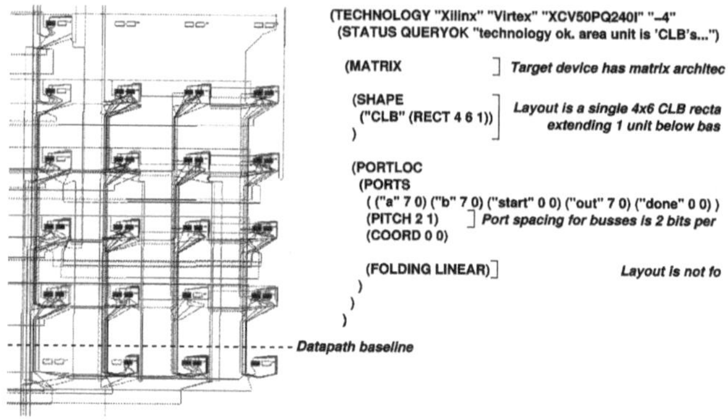

Fig. 6.12 Pre-placed 8-bit multiplier and FLAME description [29].

per CLB height). The regular parts of a module have a consistent horizontal data flow, consisting of regularly spaced bits. Above and below this regular area, irregular components (such as module-local controllers and overflow computation) can be arranged. A purely-module based approach is inefficient, however, when composing more complex logic expressions from individual modules: Each operator would be mapped to its own separate module (which would at least be one LUT column long), even though multiple operators could conceivably be mapped together into a single LUT column (as long as the number of LUT inputs is not exceeded). To solve this problem, we have developed a universal generator for logic modules that accepts arbitrarily complex logic expressions and performs inter-operator logic optimization and mapping to generate a single regularly pre-placed module for the entire expression [35, 20].

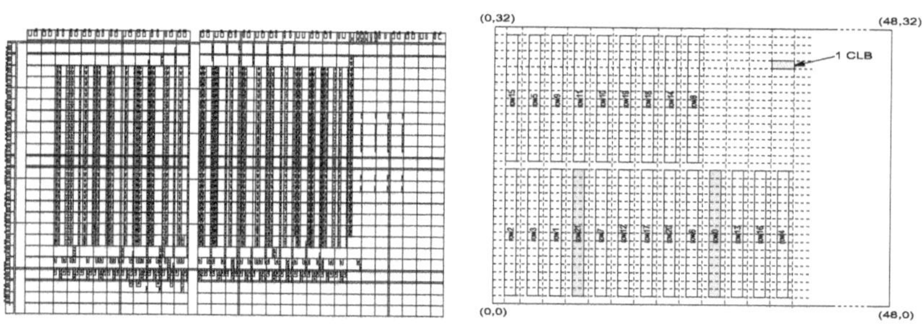

(a) Purely horizontal **(b)** Clustered horizontal sub-datapaths

Fig. 6.13 Regular placement of Wavelet datapath [33].

For high performance, the regular layout style also should be preserved when assembling an RCU datapath from multiple modules. This requires custom tools, since the vendor CAD tools are not specialized for datapath layout. An easy way to achieve regular inter-module placement is a linear arrangement of modules, this is an approach that has already been used in earlier projects [27]. An example of such a linear layout for a kernel of a Wavelet image compression algorithm [30] is shown in Fig. 6.13(a).

However, this technique obviously does not scale with increasing datapath complexity: At some point, the linear distances become excessive and the performance decreases (despite the regular linear layout). To this end, we have developed a combined approach ClaP [33] that partitions the datapath into clusters of linear sub-datapaths (shown in Fig. 6.13(b) for the same circuit). Within each cluster, a strictly horizontal data flow is preserved, but less tightly connected modules can be placed in different clusters, allowing the exploitation of the vertical dimension. The core of ClaP is a simulated annealing algorithm with different moves (inter-cluster, intra-cluster, move entire clusters, . . .). For the given example, this vertical stacking of horizontally arranged clusters leads to a delay reduction of ca. 20% over the purely linear solution.

6.5.3 *Reconfiguration Scheduling*

The kernels currently extracted by Comrade from real C applications (without exploiting alias or loop iteration space analysis) have a maximum size of a few hundred operators and often take up just 3000. . . 6000 cells (4-LUT + flip-flop) of RCU area. Considering that even medium-sized commodity FPGAs have more than 17,000 cells, a study of how to use all of this space has much potential.

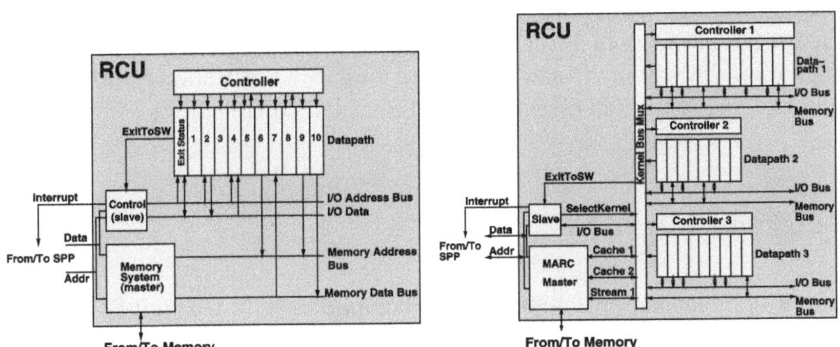

(a) One kernel per configuration **(b)** Multiple kernels per configuration

Fig. 6.14 (a) Initial and (b) new on-chip architecture.

Figure 6.14(a) shows the initial RCU architecture generated by Comrade. It consists of interfaces to the rest of the system (a slave to exchange live variables with the SPP and a master to perform independent memory accesses) and the controller/datapath for one kernel.

Orthogonally to trying to extract more complex kernels from the C programs (see Sect. 6.7), we can employ the unallocated RCU area to hold the controllers/datapaths for *different* kernels on the RCU. In this manner, we avoid potentially very slow reconfigurations and simply *switch* between kernels that have been merged into a single configuration. Such an architecture is shown in Fig. 6.14(b). The SPP can now initiate an appropriate switching of the kernel bus multiplexer by performing a slave mode write to the RCU. Thus, a switch between different kernels requires just a single clock cycle instead of the many milliseconds of a reconfiguration.

While Comrade does not yet exploit this scheme, we have already developed the algorithms for computing how the kernels should be merged into configurations so that the number of reconfigurations is minimized over the entire program execution [15]. For rapid calculation of an estimated solution (independent of the detailed execution trace of the program), we use a heuristic that constructively builds clusters of configurations following the nested loop structure of the program. To evaluate the quality of the heuristic and to compute the optimal solution when algorithm run-time is less critical, an alternate exact approach using dynamic programming was also designed. Since the exact approach evaluates the complete trace of kernel execution order during a program run, its own run-time can be lengthier for more complex programs. In general, the heuristic computed results close to the optimal solution for our experiments, but required less than 1/10 s execution time. The optimal algorithm generally had run-times between 1/10 s and 100 s.

The results of configuration merging are extremely promising. From the Wavelet image compression application Versatility [30], we can extract eight kernels. When using separate configurations for these kernels, the execution of the hybrid hardware/software application will require 5381 reconfigurations. After performing configuration merging, the heuristic will reduce that to five reconfigurations, while the optimal solution can even get it down to just four reconfigurations. Given that reconfiguration times in many cases dominate the entire application run-time, configuration merging will be an essential part of a refined compile flow.

6.6 Lessons Learned

The sheer implementation complexity of a compile flow from a high-level language down to hardware is tremendous. Even for pure software compilers, the significant infrastructure requirements to get a flow working at all (not innovating yet) are considered serious impediments to current research by notable compiler experts [10]. The efforts to bring up a software compiler are dwarfed by the *additional* hardware architecture and design tasks necessary for a *hardware/software* co-compiler such as Comrade. The compiler part alone of Comrade currently consists of more than 300,000 lines of C++ and Java code, excluding the Stanford SUIF2 compiler framework [32] which is used as the front-end. Also not included are the system interfaces (SPP, memory, including caching and streaming) that were formulated in Verilog and VHDL and the operating system modifications (see Sect. 6.3). Our research continues to be significantly hampered by our unfortunate choice of SUIF2 as the base for the compiler. While SUIF2 was being hailed as a future-proof successor of the earlier SUIF1 system, which was successfully used in the Nimble project [27], SUIF2 development quickly faltered and even today the system does not have all of the capabilities of its less modular predecessor. In retrospect, we should have chosen the much more capable Open64 system [9], which was already available at the project's start, but was discarded to its convoluted implementation and steep learning curve Today, more alternatives are available: Open64 continues to be developed and is very powerful especially in the area of parallelizing loop transformations. While the more recent LLVM [25] is not yet as advanced as Open64 in that area, it offers a cleaner internal structure and a well-integrated compile flow encompassing many scalar optimizations and even just-in-time compilation. Especially the later would allow for interesting research in just-in-time hardware generation.

The choice of a suitable intermediate representation for the RCU part of the compile flow also turned out to be pivotal for the project. Existing representations, both for pure software (e.g., SSA-CFG) or pure hardware (e.g., DFG and CDFG) lacked the expressive power to handle the complex RCUs capable of accessing memory in their own, which were our aim with Comrade. While the CMDFG has fulfilled that role admirably (we have yet to discover practical limitations), it took much experimentation (e.g., with DFC-SSA and Array-SSA forms) to develop it. For similar research, we highly recommended to spend significant effort developing an IR suitable for *all* phases of the project before beginning to work on individual passes.

Analogously, the CMDFG execution semantics and their realization in the dynamic CoCoMa controller have also advanced the project significantly. In contrast to pure DSP or scientific computing C code, the general purpose C code we are aiming to compile is not amenable to efficient static scheduling: In much practical code, simply too many data, control and (possibly hidden) memory dependencies exist. Modern out-of-order SPPs handle such temporally distributed programs by discovering these dependencies at run-time and re-ordering execution around them. The CMDFG model allows a similar approach for our spatially distributed computations: The CoCoMa dynamic scheduling resolves dependencies at run-time (but see Sect. 6.7).

As described in the preceding sections, our core focus is still the compiler and the ACS architectures supporting its execution model. However, when compiling more complex programs potentially containing many RCU kernels (something we have not done yet), it is obvious that the reconfiguration overhead on current reconfigurable devices (generally FPGAs without configuration caches etc.) will become excessive and most likely negate any possible speed-up actually achievable by RCU execution. At this stage, our currently under-utilized work on reconfiguration scheduling will become crucial for high performance.

Over the course of the project, we have worked predominantly with Xilinx FPGAs. Starting with the XC4085XL in 1998, we progressed through the Virtex XCV1000 (our long-time work horse), Virtex II Pro and now Virtex 5 FX FPGAs. Up to and including the Virtex device, we managed to keep our physical design tools (floorplanning, pre-placed module generation) synchronized with the latest FPGA features. However, with FPGA complexity increasing (more and more special purpose blocks becoming available), and the quality of conventional logic synthesis and design implementation tools finally improving (e.g., good support for physical synthesis), we believe that it is no longer practical for academic research to cover the entire flow from high-level input language down to layout generation within a single project. We have thus downsized our own efforts in the physical design area (which achieved speed-ups in the 20...30% range) to focus on the compiler's optimization and hardware-generation stages (where we often see speed-ups of 2× or more). In practice, this means that the compiler will no longer generate pre-placed floorplanned EDIF netlists for the RCU (datapath, controller, system interface), but instead stop at a RTL description that is exported to a commercial logic synthesis system for actual mapping.

6.7 Future Work

Now that the compile flow and the associated ACS architecture are fully operational, we can let the initial results of front-to-back compilation experiments guide our future research (some of which has already started).

First, it becomes obvious that we should move larger parts of an input program to the RCU in order to profit from acceleration. While our fine-grained execution model does support this already, the scope of the current SPP/RCU partitioner turns out to be to restrictive. Instead of the current profile-based inlining/profile-based path construction, we are currently implementing full-scale path profiling to discover critical paths flowing through multiple functions [24] without the need for inlining. These whole-program paths can then be used as the basis for RCU kernels.

Second, we need to improve the degree of ILP in the generated hardware. While our current CMDFG-to-CoCoMa translation does support speculation within loop iterations, we do not yet speculate across iteration boundaries. To resolve this, we can introduce *token queues* that allow the computation of predecessor nodes (e.g., the loop index calculation) to re-start in a new iteration even though not all of their successors have consumed their tokens (and the associated data items) yet. Different branches of the CMDFG can thus execute different loop iterations. Other techniques which will be investigated are loop transformations for increased parallelism (using the different cost models of spatial computation, e.g., availability of many registers compared to SPP) and speculative memory accesses (both read and write). The latter capability does require active support from the memory system. Thus, we will continue our research in ACS architecture issues.

6.8 Conclusions

With the Comrade compiler, our DFG project demonstrates the feasibility of compiling from general-purpose ANSI C into hybrid applications, executing both in software on a conventional processor and hardware-accelerated kernels on a reconfigurable compute unit. The flow draws on our advances in compiler construction, high-level synthesis, computer architecture and physical design automation to achieve this goal. Now that the system is functional and can be evaluated on real hardware fully supporting its execution model, we can proceed to actually improve the quality of results. We have already discovered a number of research areas, ripe with potential for significant improvements, that will be tackled in the next stage of work.

Acknowledgements This work was supported by DFG grant Ko 1721/1-2, project *Adaptive Rechensysteme und ihre Entwurfswerkzeuge*, associated with the Priority Programme 1148, *Reconfigurable Computing Systems*.

References

1. Budiu, M.: Spatial computation. PhD thesis, Carnegie Mellon University, Computer Science Department. Technical report CMU-CS-03-217 (2003)
2. Callahan, T.J., Hauser, J.R., Wawrzynek, J.: The Garp architecture and C compiler. Computer **33**(4), 62–69 (2000). doi:10.1109/2.839323
3. Campbell, P.L., Krishna, K., Ballance, R.A.: Refining and defining the program dependence web. Cs93-6, University of New Mexico, Albuquerque (1993)
4. Cytron, R., Ferrante, J., Rosen, B.K., Wegman, M.N., Zadeck, F.K.: Efficiently computing static single assignment form and the control dependence graph. ACM Trans. Program. Lang. Syst. **13**(4), 451–490 (1991). doi:10.1145/115372.115320
5. Ferrante, J., Ottenstein, K.J., Warren, J.D.: The program dependence graph and its use in optimization. ACM Trans. Program. Lang. Syst. **9**(3), 319–349 (1987)
6. Gädke, H., Koch, A.: Accelerating speculative execution in high-level synthesis with cancel tokens. In: Proc. Intl. Workshop on Applied Reconfigurable Computing (ARC) (2008)
7. Gädke, H., Stock, F., Koch, A.: Memory access parallelization in high-level language compilation for reconfigurable adaptive computers. In: Proc. Intl. Conf. on Field Programmable Logic and Applications (FPL) (2008)
8. Gajski, D.D., Ramachandran, L.: Introduction to high-level synthesis. IEEE Des. Test **11**(4), 44–54 (1994). doi:10.1109/54.329454
9. Group, O.S.: Open64—the open research compiler. http://www.open64.net/
10. Hall, M., Padua, D., Pingali, K.: Compiler research: the next 50 years. Commun. ACM **52**(2), 60–67 (2009). doi:10.1145/1461928.1461946
11. Kasprzyk, N.: COMRADE—Ein Hochsprachen-Compiler für Adaptive Computersysteme. PhD thesis, Technische Universität Braunschweig (Germany) (2005)
12. Kasprzyk, N., Koch, A.: Verbesserte hardware-software partitionierung für adaptive computer. In: Proc. Conference on Architecture of Computing Systems (ARCS) (2004)
13. Kasprzyk, N., Koch, A.: High-level-language compilation for reconfigurable computers. In: Proc. Intl. Conf. on Reconfigurable Communication-centric SoCs (ReCoSoC) (2005)
14. Kasprzyk, N., Koch, A., Golze, U., Rock, M.: An improved intermediate representation for datapath generation. In: Proc. International Conference on Engineering of Reconfigurable Systems and Algorithms (ERSA) (2003)
15. Kasprzyk, N., van der Veen, J., Koch, A.: Configuration merging for adaptive computer applications. In: Proc. Intl. Conf. On Field-Programmable Logic (FPL) (2005)

16. Knobe, K., Sarkar, V.: Array SSA form and its use in parallelization. In: Proceedings of the 25th ACM SIGPLAN-SIGACT Symposium on Principles of Programming Languages, pp. 107–120 (1998). http://doi.acm.org/10.1145/268946.268956
17. Koch, A.: Regular datapaths on field-programmable gate arrays. PhD thesis, Tech. Univ. Braunschweig (Germany) (1997)
18. Koch, A.: FLAME: A flexible API for module-based environments (EIS TR 2004-01). Technical report, Tech. Univ. Braunschweig, Dept. of Integrated Circuit Design (E.I.S.) (2004)
19. Koch, A.: FLAME library specification (EIS TR 2004-02). Technical report, Tech. Univ. Braunschweig (2004)
20. Kunz, J.: Eine placer-modul-erweiterung für den "universal generator for logic circuits on fpgas". Master's thesis, Tech. Univ. Darmstadt (2007)
21. Lange, H., Koch, A.: Design and system level evaluation of a high performance memory system for reconfigurable SoCi platforms. In: Proc. HiPEAC Workshop on Reconfigurable Computing (2007)
22. Lange, H., Koch, A.: An execution model for hardware/software compilation and its system-level realization. In: Proc. Intl. Conf. on Field Programmable Logic and Applications (FPL) (2007)
23. Lange, H., Koch, A.: Low-latency high-bandwidth hw/sw communication in a virtual memory environment. In: Proc. Intl. Conf. on Field Programmable Logic and Applications (FPL) (2008)
24. Larus, J.R.: Whole program paths. SIGPLAN Not. **34**(5), 259–269 (1999)
25. Lattner, C.: Llvm: An infrastructure for multi-stage optimization. Master's thesis, University of Illinois at Urbana-Champaign (USA) (2002). http://www.llvm.org/
26. Laurich, P.: A comparison of hard real-time linux alternatives (2004). http://www.linuxdevices.com/articles/AT3479098230.html
27. Li, Y., Callahan, T., Darnell, E., Harr, R., Kurkure, U., Stockwood, J.: Hardware-software co-design of embedded reconfigurable architectures. In: DAC '00: Proceedings of the 37th Conference on Design Automation, pp. 507–512. ACM, New York (2000)
28. Lysaght, P., Rosenstiel, W. (eds.): New algorithms, architectures and applications for reconfigurable computing. New Algorithms, Architectures and Applications for Reconfigurable Computing (2005)
29. Neumann, T., Koch, A.: A generic library for adaptive computing environments. In: International Conference on Field Programmable Logic and Applications (FPL) (2001)
30. Ponnuswamy, K.P., Kumar, S., Pires, L., Ponnuswamy, S., Nanavati, C., Golusky, J., Vojta, M., Wadi, S.: A benchmark suite for evaluating configurable computing systems—status, reflections, and future directions. In: in Proc. ACM/SIGDA Int. Symposium on Field Programmable Gate Arrays (FPGA'00), pp. 126–134 (2000)
31. Rose, G.: Using the microprocessor MMU for software protection in real-time systems. Technical report, LynuxWorks, Inc. (2009)
32. Stanford: The SUIF2 compiler system. http://suif.stanford.edu/suif/suif2/
33. Thorns, F.: ClaPi—clustering and placement. Master's thesis, Tech. Univ. Braunschweig (Germany) (2002)
34. Tool interface standard (TIS) executable and linking format (ELF) specification version 1.2. TIS Committee (1995)
35. Wewetzer, C.: A universal generator for logic circuits on FPGAs. Master's thesis, Tech. Univ. Braunschweig (2005)

Chapter 7
PolyDyn—Object-Oriented Modelling and Synthesis Targeting Dynamically Reconfigurable FPGAs

Andreas Schallenberg, Wolfgang Nebel, Andreas Herrholz, Philipp A. Hartmann, Kim Grüttner, and Frank Oppenheimer

Abstract Dynamic Partial Reconfiguration (DPR) is a promising technology ready for use, enabling the design of more flexible and efficient systems. However, existing design flows for DPR are either low-level and complex or lack support for automatic synthesis. In this chapter, we present a SystemC^{TM}-based modelling and synthesis flow using the OSSS+R framework for reconfigurable systems. Our approach addresses reconfiguration already on application level enabling early exploration and analysis of the effects of DPR. Moreover it also allows quick implementation of such systems using our automatic synthesis flow.

7.1 Introduction

Dynamic Partial Reconfiguration (DPR) is the ability of FPGAs, to change some parts of their programming while the remaining (static) parts keep operating. With current hardware description languages like Verilog, VHDL, or SystemC [10], DPR can only be expressed at a very low implementation level. There is no support for explicitly expressing the change of design components at runtime in these HDLs. More importantly, it is not possible to explore the impact of reconfigurable subsystems on the performance and behaviour of the system as a whole in early design phases.

Andreas Schallenberg · Wolfgang Nebel
Carl von Ossietzky University, Oldenburg, Germany,
e-mails: andreas.schallenberg@uni-oldenburg.de,
wolfgang.nebel@uni-oldenburg.de

Andreas Herrholz · Philipp A. Hartmann · Kim Grüttner · Frank Oppenheimer
OFFIS Institute for Information Technology, Oldenburg, Germany,
e-mails: herrholz@offis.de, hartmann@offis.de, gruettner@offis.de,
oppenheimer@offis.de

Note: Parts of this book chapter have been previously published in [20, 21].

M. Platzner et al. (eds.), *Dynamically Reconfigurable Systems*,
DOI 10.1007/978-90-481-3485-4_7,

Additionally, using DPR manually requires a lot additional design effort, since it affects both static and dynamic parts of the design. This time-consuming and error-prone work makes DPR prohibitive for practical use in real products. There has been quite some research to overcome this. Unfortunately, the proposed solutions required a significant change in modelling the static part of the design. Today's technologies for DPR affect the possible system architectures in several areas:

Timing

- The actual replacement of system components during run-time usually results in delays, during which the affected area is *not usable*.
- Static components need to be aware of this possible temporal unavailability of the dynamic components.
- The effect of the reconfiguration delays on a system's latencies, throughput etc. needs to be considered.
- Without proper insight into the required and resulting timing of the application, the performance of dynamically reconfigurable systems might be sacrificed.

Communication

- To ensure correct behaviour, communication across the border of static and dynamic components has to be designed carefully.
- The dynamic parts need to abort communication to the static part before any reconfiguration in a controlled way. After reconfiguration it needs to establish communication again.

Behaviour

- Depending on the application, it might be necessary to preserve and restore the internal state of a dynamically reconfigured component.
- The static parts need to know how to modify the dynamic parts.

These issues make the exploration and design of dynamically reconfigurable applications difficult and time consuming. To reduce this design burden, the specific aspects of DPR should be visible in the abstract application model. If DPR is found to be beneficial, the proposed solution should be implemented without the need to recode the model. If DPR is found to be not beneficial, there has to be a quick path to a static solution without DPR. Whatever choice is made, the tool flow must allow use of standard synthesis and simulation tools, e.g. SystemC and VHDL tools.

Consequently, there is a need for an efficient design flow providing an appropriate level of abstraction to capture the essential properties of an application as well as enabling the integration of DPR. OSSS+R is a SystemC based design methodology enabling algorithmic specification in C/C++, functional simulation and automated synthesis. Our extension to the set of available modelling primitives and simulation abilities is done in terms of a SystemC domain-specific library, available under the LGPL license. Simulation can be done with any IEEE 1666–2005 standard conforming simulator. The designer identifies potential candidates for dynamic reconfiguration, marks them and observes the effects by simulation. The model can be directly fed into the *Fossy* synthesis tool [7], generating VHDL. Feeding resulting files into an FPGA synthesis tool quickly yields bitfiles and initial, approximate configuration times. A back-annotation of these times into the abstract model allows performance evaluations.

In this chapter we present the main modelling [19, 20] and automatic synthesis [21] capabilities of OSSS+R. They are illustrated by an adaptive car audio system modelling example. The synthesis flow is presented, based on a reduced system covering reconfigurable crypto algorithms. The overall design flow starts from a pure C++ application level model which is refined to a high level OSSS+R model and finally synthesised to a register transfer level model. We further present results of the final implementation of the reconfigurable crypto algorithms on a Xilinx ML-401 development board using the Xilinx *Early Access Partial Reconfiguration* design flow [24].

7.2 Related Work

There are other frameworks which allow both modelling and synthesising of dynamic reconfigurable systems.

One example is Pebble [13], a low level HDL providing specific statements for reconfiguration. One is a mux/demux encapsulation of logic variants. The control inputs of these muxes are used as reconfiguration conditions. The logic variants are to be exchanged during reconfiguration. Additionally, a `RECONFIGURE_IF` statement allows an alternative specification. The specification does not cover reconfiguration times. A compiled model can be simulated using the Rebecca simulator. The authors demonstrated synthesis for an Xilinx 6200 FPGA.

JHDL [3] is a structural hardware description language based on Java. It provides reconfigurable elements, called `PRSocket`, which can receive a `Reconfigure(int)` call, requesting a specified implementation. Depending on the argument, new circuit nodes are created. JHDL can be simulated and synthesised. The system clock needs to be stopped during reconfiguration, which makes modelling of reconfiguration times impossible.

T.K. Lee et al. [12] used RT C to describe a reconfigurable system. The reconfigurable elements are tasks, which are grouped in structs, arrays or unions. The grouping determines the replacement, e.g. members of a union are mutually exclusive. These groupings allow influence on control complexity, area demands and design performance. An example model was transformed into Handel-C and RT-Pebble, with tool assistance and some manual work. It was then implemented on a Celoxica RC1000-PP board.

The DCS toolset [14] accepts specially crafted VHDL as its input, containing all implementations of the configurable components. It also needs auxiliary scheduling and timing information. The toolset then generates a simulation model for debugging. Since the design is already given in VHDL, FPGA vendor tools are used to implement the design [18].

There are other approaches, based on SystemC (like OSSS+R), which allow simulation but do not have tool-assisted synthesis.

SyCERS [1] is a framework allowing modelling of run-time reconfiguration, intended to explore design alternatives. The functionality to be replaced is represented by functions inside modules. These functions are called from the body of `SC_THREAD`s and `SC_METHOD`s. By using function pointers to change the function at simulation time the dynamic behaviour is achieved. Simulation is done by mapping to the Caronte architecture, containing a microprocessor to access the reconfigurable modules.

In [2] a modelling framework using dynamic thread spawning is used. The framework implements an additional layer to the SystemC kernel providing required features like dynamic ports. Using this layer, threads can be replaced at runtime, expressing dynamic behaviour.

The ReChannel [16] library allows modelling of run-time reconfiguration at different levels of abstraction. Though ReChannel guides the designer during iterative refinement to lower levels of abstraction it does not provide automatic synthesis.

In these approaches, the management of the dynamic resources (which thread may use which resources at what instant) needs to be specified manually. An exception to this is used in the ADRIATIC project [15]. In this approach, mutual exclusive modules are to be described as bus slaves. Then multiple slaves are grouped and wrapped in a dynamic reconfigurable fabric (DRCF) which switches among them and acts as a physical bus slave itself. The bus master requests a specific logical slave by its bus address which is then utilised by the DRCF to enable the requested logical bus slave. The DRCF introduction is done at RT level.

The given list of SystemC-based approaches does not include those requiring a modified simulation kernel. Further SystemC-based approaches can be found in [4, 11, 23].

7.3 Methodology

Our group has previously investigated concepts for object oriented hardware specification, simulation and synthesis [8, 17]. This lead to the OSSS language [5]. We also introduced a basic concept to handle reconfigurable architectures [19]. We do not require the designer to explicitly express a demand for reconfiguration. Instead, he uses certain containers and relies on an automatically generated infrastructure for administrative tasks. We regard reconfigurability as an orthogonal aspect to the functional specification of the circuit.

Exploiting reconfiguration is a decision of architecture. Consequently, when we describe parts targeted to be reconfigurable we stick as close as possible to well-known concepts from VHDL or Verilog. We address the following aspects of the design process:

1. The designer should focus on the desired functionality and be able to separately address the aspect of reconfiguration.

2. Interaction between processes competing for limited and shared resources (reconfigurable areas and reconfiguration ports) should be handled automatically.
3. Design space exploration shall be supported by quick changes between dynamic and static implementation.
4. Modifying configuration parameters should be easy.
5. The design specification shall be as close to a non-reconfigurable specification as possible.
6. The specification should be simulated and synthesised "as-is", that is, without any pre-processing.

The designer implicitly constructs the necessary infrastructure and explicitly parameterises only the necessary parts. This is comparable to a runtime system and a software operating system removing design burdens like process scheduling and memory management from the programmer. The designer describes the dynamic behaviour of the application first. Technological aspects can be added to the model later in a separate step.

7.3.1 General Concept

The OSSS+R configuration specification concept bases on polymorphism. There is a fundamental difference between polymorphism known from software languages like Java or C++ and the polymorphism which is used in OSSS. Software uses pointers as basis for its concept of polymorphism. Pointers in C++ have a formal type. The actual type of the object being pointed to has to be of a class which is derived, or identical to the formal type of that pointer.

The hardware description language OSSS does not allow pointers. Pointer synthesis is a difficult task [22] that tends to lead to very inefficient designs. However, OSSS has polymorphic containers which behave different depending on their actual type. They can be copied, their methods can be invoked and they can be used in signals just like normal objects. OSSS+R provides the same concept in another construct, the `osss_recon` container. The container is instantiated with a specializing formal class type (like a software pointer's type). This type has to be derived from the predefined class `osss_object`. It is the base type for all actual types of the container's content. The container must be instantiated as a member of a SystemC module. Therefore, it has a well-defined location in the structure of the design. An `osss_recon` container cannot be used for local variables, signals, ports or method parameters. However, a container's content can be copied from or to local variables, signals etc.

An `osss_recon` container represents an area within a reconfigurable device. Fig. 7.1 illustrates a relation between DPR-FPGAs and reconfigurable containers. Such reconfigurable container provide a formally specified interface to every possible actual content. The relationship between a reconfigurable area on a DPR-FPGA device and its static environment on the same device is similar: The content may

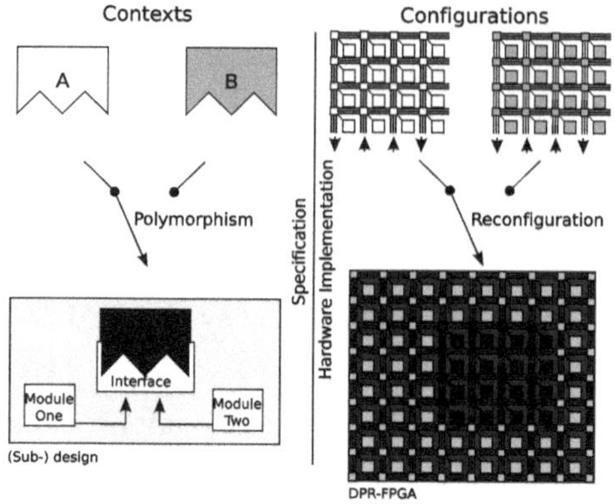

Fig. 7.1 Analogy between polymorphism and reconfiguration [20].

change but in order to be integrated into the static neighbourhood it requires a static interface to the outside.

Exchanging an object in an `osss_recon` container with another one (possibly of the same type) corresponds to a change in the flip-flop states, which we call an *object switch*. If the class types are different it also corresponds to a change of the programmed logic on the DPR-FPGA in that area (reconfiguration). We call this a *class switch*.

For such an area within the FPGA device we call the combination of logic configuration (information given in a partial configuration bitstream) plus the sum of the flip-flop states a *context*. This corresponds to the implementation of the actual class and the member attribute values.

We allow accesses to a reconfigurable container by a context's name or anonymously. Anonymous accesses always manipulate the context that is currently physically present in the reconfigurable area. We abbreviate the formal type used for this anonymous access with *ACFT*. The ACFT is given by the specialization class for the reconfigurable container.

In contrast, if there is a unique identification for a context we call it a *named* or *persistent context* which also includes an identifier in the OSSS+R model. We describe those contexts using the `osss_contexts` container. While a named context is configured on a reconfigurable area it is both accessible using the named context identifier or referring to it as an anonymous context using the `osss_recon` container. We abbreviate the formal type of named contexts with *NCFT*. These *named contexts* can be used as polymorphic containers themselves, just like the reconfigurable containers. The actual type of the named context's content (*NCAT*) just needs to be identical or derived from the formal type (NCFT). The lifetime of a named context is unlimited since it is always declared as a member of a module. Each named context is associated statically with exactly one reconfigurable container.

Table 7.1 Relations between pure software modelling, DPR hardware, and OSSS+R.

Software (C++)	Hardware	OSSS+R
class T { };	Partial bitstream	class T: public osss_object { };
Attribute values	Flip-flop states	Attribute values
Polymorphic pointer T * x;	Reconfigurable area	Reconfigurable container osss_recon⟨T⟩x;
Object being pointed to T a, b;	–	Context osss_context⟨T⟩a,b;
Address of object	–	Identifier
No fixed binding	–	Binding a to x a(x);
Modifying content of address x x = T();	Reconfiguration (writing partial bitstream)	Assignment to anonymous context x = T();
Modifying address x = &a; *x = T();	State preservation and reconfiguration	Assignment to named context a = T();

Table 7.1 gives an overview of the relations between pure software modelling, DPR hardware, and OSSS+R. It shows some code fragments in C++ and to corresponding expression in OSSS+R.

7.3.2 Lifetime and Conflict Management

There may be multiple named contexts for each reconfigurable container and multiple reconfigurable containers on a device. But there is one single configuration controller per device. Accessing a named context is simply done by using the identifier in an assignment or a method call. This automatically instructs the configuration infrastructure to make sure the correct context is available in the reconfigurable area and therefore possibly demand a reconfiguration, if a class switch is to be performed during the context switch. Since there may be multiple `osss_recons` demanding a reconfiguration at a time an arbitration mechanism is needed to solve the accesses to the configuration resource (e.g. a Flash or EPROM providing configuration streams).

There also may be concurrent access requests to the same `osss_recon` (by `SC_CTHREADs`). Handling this requires a second arbitration mechanism. Both are hidden in the configuration infrastructure and can (but do not need to) be customised using user-defined arbitration algorithms. The invocation of the arbitration and its specification is completely transparent to the designer.

Another task performed by the configuration infrastructure is the handling of the attribute values. Whenever a context is being swapped out of the reconfigurable area the flip-flop states have to be saved. Otherwise the context could not be completely restored later on. This is also handled automatically by the configuration infrastructure and not visible to the designer. For synthesis, methods will be added which allow reading and writing of attributes. In OSSS+R accesses to named contexts and reconfigurable containers cannot be interrupted. Additionally, contexts are only exchanged while no access (e.g. a method call) is active on the reconfigurable container. Therefore we are able to determine points in control flow which represent idle states in the object. For saving and restoring objects, that is, to make swapping of objects functionally invisible to its users, it is sufficient to know the objects' attributes at these idle states.

Instead of directly expressing a demand for reconfiguration, the source code may contain demands for certain contexts. This leaves the following tasks up to the automatically generated infrastructure:

- Book keeping (which context is in which reconfigurable container)
- Detecting conflicts (is the desired context available, checked by its ID)
- Solving conflicts (swapping of context states, possibly including reconfiguration steps)
- Delaying accesses (halting processes until conflicts are resolved).

Before an access is performed, the client process automatically requests permission from the access controller. The client process specifies a desired context. The access controller keeps track of the current context which is present in the reconfigurable area. When it detects a mismatch between the active and desired context an object switch is initiated. If the active context is a persistent one, the current attributes are saved in an external storage. Next, class switches are detected. The access controller tracks the actual classes of all contexts and the current class. If the requested context has a different class than the actual one, a class switch is necessary. The access controller instructs the configuration controller with this task. After completion, the access controller checks if the desired context is a named one, so attributes have to be restored. Finally the client process is granted access rights.

When using the anonymous container only (no named contexts), the named context handling logic is not generated for that `osss_recon` container.

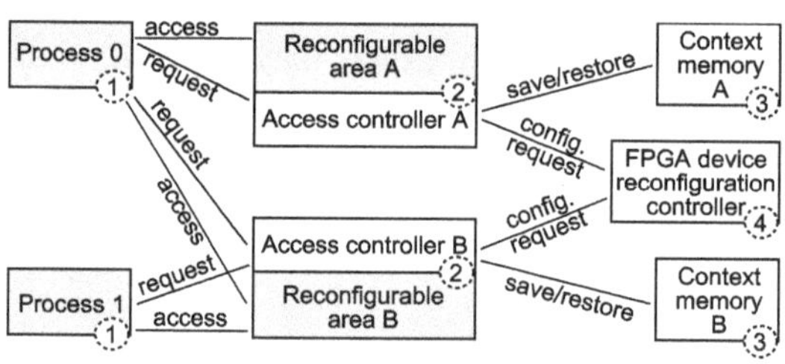

Fig. 7.2 Abstract architecture [20].

Fig. 7.2 shows a generic structure of a design consisting of two user processes (1), two reconfigurable areas (2) (with access arbiters attached), two context attribute storages (one for each reconfigurable area, (3)) and a configuration controller taking care of access conflicts to the device's reconfiguration port (4). The white boxes are specified implicitly in the model when certain keywords, e.g. `osss_context`, are used. They represent the configuration infrastructure. The designer need not care about their instantiation, use or interaction. He just has to set some architectural parameters (e.g. configuration times or the selection of arbitration algorithms) when the timing and scheduling is to be specified. Simulating access and configuration controllers requires adding further timing information to the model. Saving and restoring the classes' attributes from or to a memory requires time which is dependent on the size of the attributes. Also the configuration times depend on the class type and specific hardware device type. This information is expressed using `OSSS_DECLARE_TIME` statements. Initially the designer has to guess values since they are unknown before bitstream generation. After synthesis they may be replaced by more precise ones.

7.4 Derived Interface Classes

The described approach allows to model systems containing reconfigurable areas. All accesses to these areas are expressed as normal object use (method calls and assignments). However, the need for a static interface at the signal level requires the use of a common interface class at the logical level. Each `osss_recon` is associated with exactly one formal interface class (ACFT) which defines the available methods. Every named context for that reconfigurable container needed to have the same formal interface class (NCFT = ACFT). This even applies if the named contexts actually contain instances of derived classes (NCAT derived from NCFT).

The restriction NCFT = ACFT means forcing the designer to use the very same class for the declaration of the `osss_recon` container and its associated `osss_context` containers. It is possible to instantiate objects of derived types (NCAT derived from NCFT) in a named context container but it is not possible to invoke methods specified in the derived class using that container. If the NCFT could be a class derived from a ACFT, additional methods would become usable via the named context.

This means not having a fixed interface at the level of program logic anymore but adapting it to the actual type. For synthesis we still need a fixed signal level interface. A static analysis has to find a unified interface that enables access to all possible contents. Since named contexts are statically bound to reconfigurable containers, no dynamic creation of contexts is possible. Both ACFT and NCFTs are statically known, so we can walk through all NCFT classes and the ACFT in order to calculate a virtual but static unified interface class. Previously this class had to be explicitly present in the model and the designer had to create it manually. With our proposed methodology this has been automated. This simplifies modular design, since it allows adding new classes to the class tree incrementally and using them in named contexts.

The reconfigurable instance acts as a self-adapting server component. It always provides the correct interface implementation as requested by the user.

The static binding of named contexts to reconfigurable containers implies that no context migration from one reconfigurable container to another is possible. This is not a severe limitation since the designer can explicitly copy the context's content to ordinary objects or other contexts. In other words, migration of contexts has to be expressed explicitly by the designer.

7.5 Modelling Example: Car Audio System

We present a modelling example as illustrated in Fig. 7.3 to demonstrate the OSSS+R approach. It covers a car audio system consisting of two audio sources (a music player module and a navigation speech module) requesting services from a reconfigurable processing area and a mixer module which combines the results. A security module rarely uses the same reconfigurable area to perform cryptographic operations (unlock the system if the correct key is present).

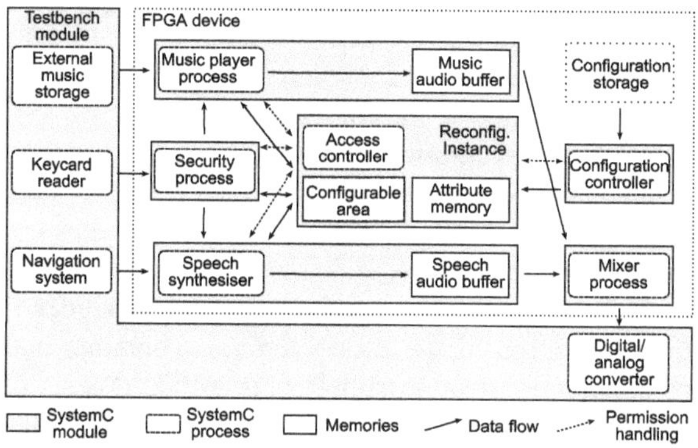

Fig. 7.3 Block diagram of the car audio system modelling example [20].

Three modules (music playback, navigation speech and security) perform their tasks partly on the reconfigurable area. The music player is able to decode different compressed audio formats (MP3, Vorbis and AAC). The decoding must be fast enough to prevent an audio-buffer underrun if the reconfigurable area is used temporarily by another module. The navigation module has sporadic need for decoding speech when a text is to be processed. The text is also provided as compressed audio samples. The implementation of the player is re-used for the navigation system. The mixer module has to fetch audio data from these two sources and to produce an output signal. It does not need access to the reconfigurable area. The security module performs cryptographic operations on the reconfigurable area. It is allowed to interrupt the output stream when a new keycard is inserted and a different access key needs to be checked. From the modelling point of view each of these modules requires an individual named context. These contexts are instances of classes from the same class tree (see Fig. 7.4).

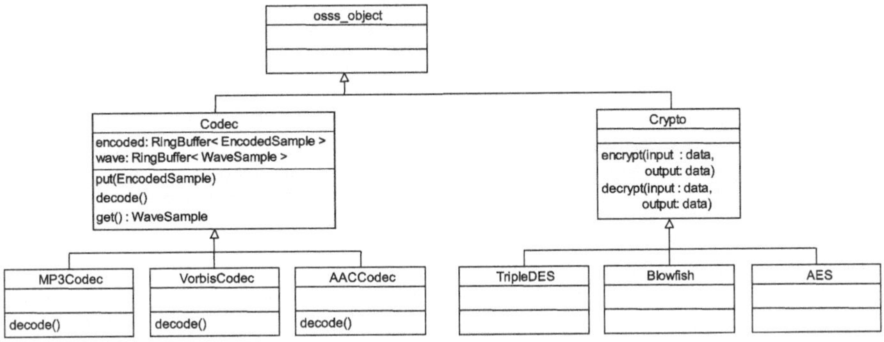

Fig. 7.4 Class tree of the car audio system modelling example.

The music player uses a named context of type *Codec*. It receives data from the music storage (located in the testbench) via a polymorphic signal of the formal type *EncodedSample*. The samples have actual types like e.g. *MP3Sample*. Their parent class provides a getDecoder() method which is overloaded and implemented by the derived classes and returns an appropriate codec. The instantiation of the codec is done by assigning the returned decoder directly to the named context. The decoder itself is one of the classes *MP3Codec*, *AACCodec* or *VorbisCodec*. This way the music player can be designed with just knowing the classes *Codec* and *EncodedSample*. The codecs can be individually added or removed to the model without changing the player module. This also applies to the navigation module since it shares its implementation with the music player.

7.5.1 Coding Style: From C++ Polymorphism to OSSS+R

In a SystemC design flow the design entry may be a C/C++ description of the application's core algorithms. This is to be refined into a hardware description, using SystemC modelling elements. SystemC presents itself as a library, not a language, so one may also use all features of C/C++. While this allows faster simulation and easier modelling, the drawback is, that such a model might use features which are (typically) not synthesisable. Manual recoding would be required to obtain a synthesisable model written in Verilog or VHDL for example. A designer may be tempted to avoid all non-synthesisable features in the first place, however this would sacrifice advantages like more abstract modelling.

OSSS+R encourages the designer to use C++ features, since classes, objects and inheritance are synthesisable. For more complex components, like concurrently used objects, OSSS+R provides synthesisable containers to reduce the dilemma described before. A group of objects where each member is accessed rarely overlapped with other members of the same group is a good candidate for reconfiguration. Additionally, polymorphic pointers in the C++ model are a hint to dynamic objects which also make good candidates.

In the initial C++ model of the car audio system example some objects were implemented using polymorphism. The different implementations of the abstract Codec interface were accessed through a polymorphic pointer. This way, the referenced codec could easily be switched from one implementation e.g. Vorbis to MP3 without having to change its interface.

Assuming the codec objects are used mutually exclusive and switches are rare events. To let the codecs share the same physical reconfigurable area the polymorphic pointer is replaced by an OSSS+R reconfigurable object. The reconfigurable object uses the same generator base class as the polymorphic pointer and provides the same C++ syntax for accessing the object.

The C++ model contains a polymorphic pointer `wg` which is initialised using one of the available generator classes.

```
Codec * my_codec;
my_codec = new VorbisCodec();
// for all samples to be decoded
my_codec->put(enc_sample);
my_codec->decode();
decoded_sample = my_codec->get();
// end for
// switch to MP3
my_codec = new MP3Codec();
// for all samples to be decoded
my_codec->put(enc_sample);
my_codec->decode();
decoded_sample = my_codec->get();
// end for
// ...
```

For OSSS+R, this pointer is moved inside the SystemC module `DecoderModule` and transformed into a reconfigurable object:

```
SC_MODULE( DecoderModule ) {
  // ...
  osss_recon< >           my_reconfigurable_area;
  osss_context< Codec > my_codec;

  SC_CTOR( DecoderModule ) {
    my_reconfigurable_area.reset_port( reset );
    my_reconfigurable_area.clock_port( clock );
    // binding context my_codec to reconfigurable area
    my_codec(my_reconfigurable_area);

    SC_CTHREAD( work, clock.pos() );
    reset_signal_is(reset, true);
    // process work can access context my_codec
    uses( my_codec );
  }
};
```

Within the module constructor, the reconfigurable object is bound to the process `work`, enabling `work` to access the object. Similarly, the object could be bound to even more processes. As you remember, the reconfigurable object automatically provides a built-in scheduler, serialising all incoming requests.

As shown in the implementation of the process `work`, the new variable is used almost like the original C++ pointer, the only difference being the assignment to an object, where `new` is omitted.

```
void CodecModule::work() {
  my_codec = VorbisCodec();
  // forall samples to be decoded
  my_codec->put(enc_sample);
  my_codec->decode();
  WaveSample decoded_sample = my_codec->get();
  // end for
  // switch to MP3
  my_codec = MP3Codec();
  // for all samples to be decoded
  my_codec->put(enc_sample);
  my_codec->decode();
  decoded_sample = my_codec->get();
  // end for
  // ...
}
```

The resulting implementation performs a reconfiguration, whenever the runtime class of `my_codec` changes, possibly caused by an assignment. However, if the run-time class matches the previous one, only the object's attributes are modified.

7.5.1.1 Devices and Timing

To reflect reconfiguration and context switch times during simulation with proper timing, OSSS+R supports timing annotations provided by the designer. Timing annotations are defined as part of the target platform definition. They are given for a combination of platform and class type, e.g. *Virtex 4* and *VobisCodec*. Designers may specify the time needed for a reconfiguration and the time needed to store the state of a class instance:

```
OSSS_DECLARE_TIME(            // Timing:
   virtex4,                   // Platform
   VorbisCodec,               // Class name
   sc_time(   3, SC_MS),      // Context save/restore
   sc_time(1100, SC_MS));     // Reconfiguration time
```

Initially during the modelling phase, the specified times are rough estimates by the designer. Later on, when the final implementations of the configurations are available, the exact reconfiguration times can be obtained through the size of the partial bitstreams and the performance of the chosen reconfiguration controller. These timings can then be back-annotated to the initial model, providing the exact timing behaviour within the application model. If the model shows some unexpected or unwanted behaviour due to this reconfiguration times, these issues can be traced back to the OSSS+R model. This is much more convenient than debugging RT level code.

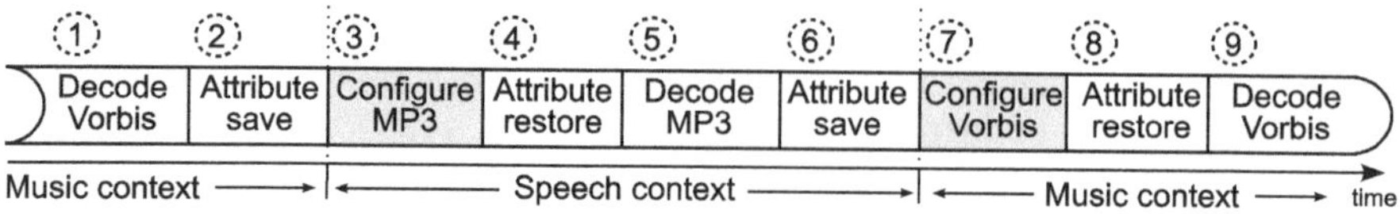

Fig. 7.5 Timeline for car audio system scenario [20].

Table 7.2 Guessed phases (as shown in Fig. 7.5) durations.

Process	Duration	Phases
Vorbis attribute save/restore	3.0 ms	2, 8
Vorbis configuration	1100.0 ms	7
MP3 attribute save/restore	2.0 ms	4, 6
MP3 configuration	1000.0 ms	3
MP3 playback	2129.8 ms	5

7.5.2 Simulation

The designer now may want to use such a model e.g. to estimate the sizes of buffers. This can be important to make sure that the music does not stop playing while the navigation system uses a different codec to decode some text messages.

Fig. 7.5 shows an example scenario. The music player is decoding a Vorbis stream (1). It proceeds until the music buffer is filled or the stream is interrupted, then it releases the reconfigurable container. Since we have chosen a fair access scheduler (Round Robin) for the reconfigurable container the navigation system will be granted access. The context used by the navigation module differs, therefore the attributes for the music context have to be saved (2). Now a class switch is required (Vorbis to MP3) and the area is configured with a different configuration (3). After that, the attribute state from the last usage of the speech decoding context is restored (4). The spoken text is a MP3 stream and has a decoded length of five seconds play time. After decoding (5) the reconfigurable container is released again. The music player may proceed, so the navigation context is being stored (6), the Vorbis codec is configured (7) and the music context's attributes are restored (8). Then decoding continues (9). Note that the designer explicitly states (1), (5) and (9) only.

One design decision would be to determine the size of the music buffer. It should be large enough to contain enough samples to bridge the time during which the music decoding is interrupted. It must provide enough data through the phases (2) to (8). Table 7.2 contains the phase durations as declared in additional statements in the SystemC elaboration phase (except for the MP3 playback time).

A 128 kbit/s MP3 stream consists of frames of 27 ms length [6, 9] (926 frames for a five second piece of speech). Each frame takes 2.3 ms to decode (assuming a decoder running at 24 MHz) which results in a total processing time of 2,129.8 ms. The phases (2) to (8) should last 4,237.8 ms. The first decoded frame from phase (9) is available 2.3 ms later. The music buffer has to be large enough to contain 4,240.1 ms of audio data (186,989 samples at 44.1 kHz sample rate).

The simulation allows to include the influence of additional aspects on the required buffer size since it also covers overhead caused by data exchange between modules or handling of permissions. Such overhead cycles are difficult to calculate manually. In our example the testbench request speech decoding at $t = 0$. The music player releases the reconfigurable container at that time. At $t = 1{,}002.418$ ms the first sample of decoded speech is available and by $t = 3{,}320.975$ ms all speech samples are processed. At $t = 4{,}251.839$ ms a music buffer underrun occurs since the next music frame is available at $t = 4{,}570.111$ ms. In order to cover the overhead the buffer has to be 13 frames (at least 330,011 ms) larger than the spreadsheet calculation indicated. These extra 13 frames are necessary due to overhead cycles.

7.6 Synthesising OSSS+R

Since the audio codecs in the modelling example in Sect. 7.5 do not contain real functionality, we present the synthesis flow of only a (slightly simplified) crypto subsystem. Three different cryptographic algorithms have been converted from existing implementations in C or VHDL to corresponding user-classes (see Fig. 7.4): Triple-DES, Blowfish, and AES.

While Triple-DES and Blowfish are rather symmetrical algorithms, the implementation of AES encryption and decryption is quite different. Therefore, two separate classes for encryption and decryption (AES, AES^{-1}) have been created, to reduce the size of the reconfigurable area on the FPGA.

An additional simplification is the replacement of the named context `osss_context<Crypto>` by direct accesses to an equivalent `osss_recon<Crypto>`. This was needed, since the current implementation of the OSSS+R synthesis tool *Fossy* does not support named contexts, yet. However, the design is easy to understand and well-suited to explain problems and solutions of DPR and to illustrate the overall OSSS+R design flow.

7.6.1 *From OSSS+R to RT Level*

Figure 7.6 presents the flow from an OSSS+R model to a final FPGA implementation. Initially, the OSSS+R model is simulated to validate its behaviour using the OSSS+R simulation library. The model is then automatically synthesised to register transfer level (RTL) using the *Fossy* tool. First, OSSS+R specific language elements are replaced with equivalents composed of SystemC components. Then synthesis of the resulting SystemC model to RTL is performed, including class tree synthesis, implicit to explicit FSM transformation etc. The output can be either SystemC or VHDL. The generated VHDL may then be further processed by FPGA vendor tools, e.g. the Xilinx ISE tool suite. Once bitstreams are obtained, the reconfiguration times can be calculated and back-annotated into the original OSSS+R design.

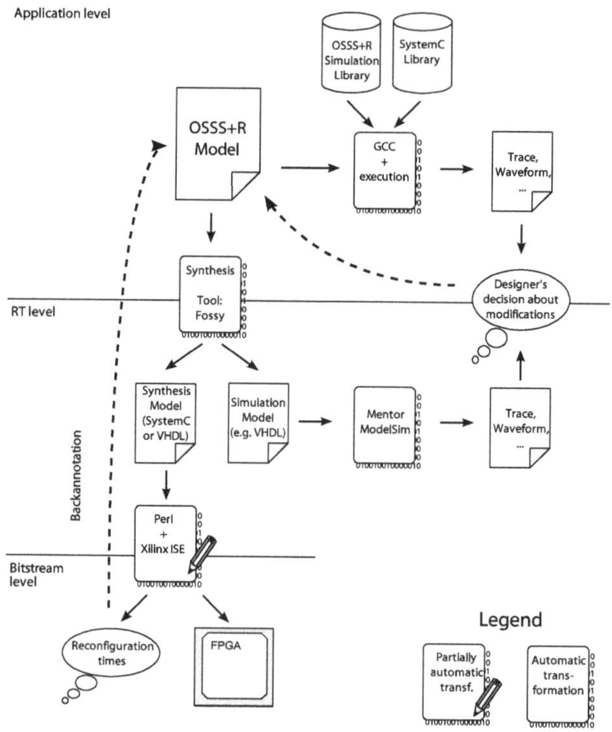

Fig. 7.6 Flow from OSSS+R (C++) model to FPGA implementation [21].

In Fig. 7.7(a) block-diagram of the generated RTL architecture of the crypto subsystem is shown. The square, gray boxes are generated by *Fossy* representing infrastructure components which are needed to implement the dynamic partial reconfiguration. While these components are automatically provided and instantiated as simulation models by the simulation library, they are not synthesisable as such and have to be replaced by synthesisable equivalents.

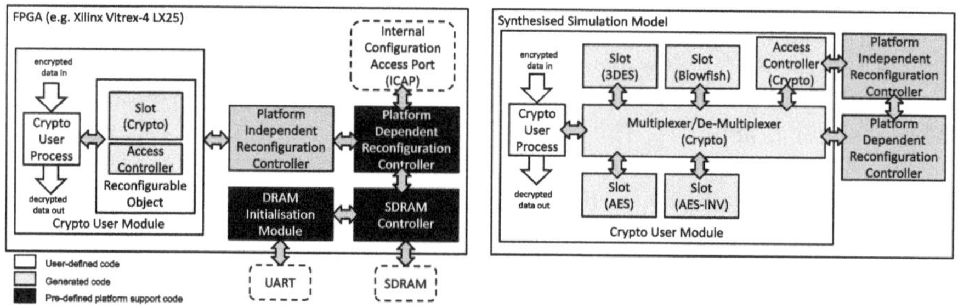

(a) Synthesis block diagram

(b) RT level simulation model block diagram

Fig. 7.7 Synthesis structure of Adaptive Crypto example.

Recon-Object For each reconfigurable object a corresponding reconfigurable area, called *slot* is generated. This slot may take any of the classes that have been mapped to the reconfigurable object. Each of the classes is generated as a stand-alone module, communicating with other modules by a signal-based protocol. Thanks to their polymorphic nature, all of the classes within one reconfigurable object can share the same physical interface although the signal interpretation varies during runtime. After implementation, each of the classes will be represented by its own partial bitstream.

Each slot is managed by an accompanying component controlling the access to the slot, detecting needs for reconfigurations and initiating reconfiguration requests. Operations on the reconfigurable object, e.g. method calls, inside the crypto user module, are replaced by a signal-based protocol to the access controller and the slot. Each request to a reconfigurable object is first directed to the access controller, to schedule it with other pending requests. If the access is granted and the requested configuration is activated the process directly communicates with slot.

Reconfiguration Controller If an access controller detects the need to perform a reconfiguration, a request for reconfiguration is sent to the **p**latform **i**ndependent part of the **r**econfiguration **c**ontroller (PIRC). The PIRC is automatically generated by *Fossy*. In our example, only one access controller is requesting services, so the PIRC does not need to be equipped with a scheduler to resolve conflicts.[1] Additionally, the PIRC translates requested class types and location information to bitstream numbers. The translated requests are serviced by the **p**latform **d**ependent **r**econfiguration **c**ontroller (PDRC) part. A PDRC is implemented manually once for a given platform, e.g. an FPGA prototyping board, and can be re-used for multiple applications. Platform dependent blocks are shown in black in Fig. 7.7(a).

Method Calls The user processes contain accesses (method calls and assignments) to reconfigurable objects and their contexts. These accesses are replaced by a signal level protocol between user processes and access controllers (for permission handling and reconfiguration) and user processes and slots (for method calls and assignments). In a reconfigurable system, a single user process may communicate with a set of different slot implementations, each having their individual interface signal interpretation. Due to the strong type system in the original model, it is guaranteed that the user process always uses the correct signal interpretation. After the replacement of all OSSS+R specific elements with synthesisable SystemC equivalents, the RTL model is generated as SystemC or VHDL.

RTL Simulation Model Typically, a designer wants to check the result of any automatic transformation by at least simulating its result. However as standard HDLs do not support the expression of DPR, the result of the OSSS+R synthesis cannot be simulated as such. As a solution to this, *Fossy* can generate an RTL simulation model, which can be simulated with any standard HDL simulator and is shown in Fig. 7.7(b).

---

[1] If more than one *Recon Object* is bound to the same device, an (potentially user-defined) arbiter is automatically synthesised to serialise reconfiguration requests.

In this model, all possible configurations of a slot are instantiated in parallel and connected to a multiplexer structure. For the simulation model, a pseudo PDRC is generated which controls the select inputs of the multiplexers. For each reconfiguration request, instead of writing bitstreams to an FPGA configuration port, the PDRC mimics the behaviour by waiting for as long as the configuration would take in the real system. The waiting time is taken from the timing specifications which have been given by the designer in the original model (see Sect. 7.5.1.1). After a first implementation these values may also be replaced with the reconfiguration times of the final partial bitstreams. The pseudo PDRC provides the same interface as the original, so despite the multiplexer structure, the rest of the model is identical to the synthesis model. This way the application can be simulated with standard HDL simulators and will show the same behaviour as the reconfigurable design.

From RTL to Bitstreams If the simulation is successfully validated using the RT level simulation model, the RT level synthesis model can be transformed to gate level and bitstreams using FPGA vendor tools. The synthesis of OSSS+R has been developed to be platform independent. However, to support the DPR features of a target platform, the model usually has to be tailored to a vendor specific tool framework. Typically, this includes creating a specific top level, some pinout description files, a floor-planning file etc. We have implemented this vendor specific adaption for the *Early Access Partial Reconfiguration Flow* (EAPR) [24] from Xilinx.

7.7 Evaluation

Using the EAPR flow, we have successfully implemented the generated RTL model of the crypto example application on an ML401 development board from Xilinx. The PDRC has been designed manually, using the Virtex4 ICAP directly with a maximum bandwidth of roughly 600 MBits/sec. Table 7.3(a) shows the size of the partial bitstreams and their resulting reconfiguration times. It should be noted, that in this example the actual algorithms have been injected manually as HDL IP blocks in a automatically generated empty slot.

To get a picture of the overhead introduced by the reconfiguration infrastructure Table 7.3(b) shows the usage of FPGA resources for PIRC, PDRC, and access controller. Compared to the total resources of the FPGA the overhead is rather small. While the overhead for PDRC and PIRC is more or less constant, the resource usage for access controllers would increase with the number of slots and the use of a scheduler.

Fossy generates an implementation for the *empty* base class `Crypto` as well. The area cost for this slot, which has the synthesized method interface, but an empty data-path inside the methods, shows the overhead for implementing method-based communication. In the given example, the interface is quite wide (128 data bits in each direction), therefore the cost is already quite high. However, if the methods actually contain computations the registered inputs would be required anyhow.

Table 7.3 Results of evaluation.

(a) Bitstream sizes and configuration times

	Slices	Size [bytes]	Configuration time [μs]
Triple DES	1,830	203,872	254.8
Blowfish	1,083	175,812	219.8
AES	1,265	159,388	199.2
AES^{-1}	1,488	185,584	214.0

(b) Resource usage of infrastructure

	Slices	LUTs	Device utilisation
PDRC	172	247	1.6%
PIRC	118	221	1.1%
Access Controller	22	42	0.2%
Empty Slot	213	403	2%

Summing up, the overall implementation cost is acceptable given the potential save of FPGA area through the use of DPR.

7.8 Conclusion and Future Work

In this chapter, we presented a complete modelling and synthesis flow for DPR systems based on the modelling framework OSSS+R. Using an adaptive car audio system we demonstrated how a designer can efficiently design such systems without having to deal with the implementation details of DPR. Using the abstraction mechanism of polymorphism, reconfiguration can easily be expressed and captured already on application level. OSSS+R models are synthesised to RTL models using the synthesis tool *Fossy*. Using the Xilinx EAPR flow, we were able show that the overhead introduced by the DPR infrastructure is acceptable.

In future we will extend *Fossy* to support the synthesis of reconfigurable objects with Named Contexts [20]. We are also planning to integrate a more flexible approach for the implementation of the communication infrastructure between processes and reconfigurable objects.

Acknowledgements Supported by the DFG project PolyDyn, as part of the Priority Programme 1148, *Reconfigurable Computing Systems*.

References

1. Amicucci, C., Ferrandi, F., Santambrogio, M., Sciuto, D.: Sycers: a SystemC design exploration framework for soc reconfigurable architecture. In: International Conference on Engineering of Reconfigurable Systems & Algorithm (ERSA), pp. 63–69 (2006)
2. Asano, K., Kitamichi, J., Kuroda, K.: Dynamic module library for system level modeling and simulation of dynamically reconfigurable systems. J. Comput. **3**(2), 55–62 (2008)
3. Bellows, P., Hutchings, B.: JHDL—an HDL for reconfigurable systems. In: Proc. IEEE Symposium on FPGAs for Custom Computing Machines (1998)
4. Brito, A.V., Kuhnle, M., Hübner, M., Becker, J., Melcher, E.U.K.: Modelling and simulation of dynamic and partially reconfigurable systems using SystemC. In: ISVLSI '07: Proc. of the IEEE Computer Society Annual Symposium on VLSI, pp. 35–40 (2007)
5. Brunzema, C., Grabbe, C., Grüttner, K., Hartmann, P.A., Herrholz, A., Kleen, H., Oppenheimer, F., Schallenberg, A., Stehno, C., Schubert, T.: Osss 2.0—a library for synthesisable system-level models in SystemC (2007). http://system-synthesis.org

6. Fältman, I., Hast, M., Lundgren, A., Malki, S., Montnemery, E., Rångevall, A., Sandvall, J., Stamenkovic, M.: A hardware implementation of an MP3 decoder. Technical report, LTH Sweden (2003). Digital IC-Project
7. FOSSY synthesiser. http://www.system-synthesis.org
8. Grimpe, E., Timmermann, B., Fandrey, T., Biniasch, R., Oppenheimer, F.: SystemC object-oriented extensions and synthesis features. In: Forum on Design Languages FDL '02 (2002). http://odette.offis.de
9. Hedberg, H., Lenart, T., Svensson, H.: A complete MP3 decoder on a chip. In: Proc. of the IEEE International Conference on Microelectronic Systems Education (MSE'05) CCCD, Department of Electroscience, Lund University (2005).
10. IEEE Standards Association Standards Board: IEEE Std 1666–2005 Open SystemC Lang. Reference Manual (2005). http://www.systemc.org
11. Kriaa, L., Adriano, S., Vaumorin, E., Nouacer, R., Blanc, F., Pajaniardja, S., Coussy, P., Martin, E., Heller, D., Tabet, F., Fouilliart, A.: SystemC'mantic: A high level modeling and co-design framework for reconfigurable real time systems. In: Forum on Specification and Design Languages (2005)
12. Lee, T., Derbyshire, A., Luk, W., Cheung, P.: High-level language extensions for run-time reconfigurable systems. In: Proc. IEEE International Conference on Field-Programmable Technology (FPT), pp. 144–151 (2003)
13. Luk, W., McKeever, S.: Pebble: A language for parametrised and reconfigurable hardware design. In: Proc. International Conference on Field Programmable Logic and Applications (FPL), pp. 9–18 (1998)
14. McGregor, G., Lysaght, P.: Self controlling dynamic reconfiguration: A case study. In: Proc. International Conference on Field Programmable Logic and Applications (FPL), pp. 144–154 (1999)
15. Pelkonen, A., Masselos, K., Cupák, M.: System-level modeling of dynamically reconfigurable hardware with SystemC. In: Proc. of International Symposium on Parallel and Distributed Processing (Reconfigurable Architectures Workshop), pp. 174–181 (2003)
16. Raabe, A., Hartmann, P.A., Anlauf, J.K.: ReChannel: Describing and simulating reconfigurable hardware in SystemC. ACM Trans. Des. Autom. Electron. Syst. **13**(1), 1–18 (2008). doi:10.1145/1297666.1297681
17. Radetzki, M.: Synthesis of digital circuits from object-oriented specifications. PhD thesis, Carl von Ossietzky University Oldenburg, http://odette.offis.de (2000)
18. Robertson, I., Irvine, J.: A design flow for partially reconfigurable hardware. In: Trans. on Embedded Computing Syst., pp. 257–283 (2004)
19. Schallenberg, A., Oppenheimer, F., Nebel, W.: Designing for dynamic partially reconfigurable FPGAs with SystemC and OSSS. In: Proc. Forum on Design and Specification Languages (FDL), pp. 440–451 (2004)
20. Schallenberg, A., Oppenheimer, F., Nebel, W.: OSSS+R: modelling and simulating self-reconfigurable systems. In: Proc. International Conference on Field Programmable Logic and Applications (FPL), pp. 177–182 (2006)
21. Schallenberg, A., Herrholz, A., Hartmann, P.A., Oppenheimer, F., Nebel, W.: Osss+r: A framework for application level modelling and synthesis of reconfigurable systems. In: Proceedings—Design and Automation Conference in Europe (DATE 2009) (2009)
22. Semia, L., Sato, K., Micheli, G.D.: Resolution of dynamic memory allocation and pointers for the behavioral synthesis from C. In: Proc. Design Automation and Test in Europe DATE'00, pp. 312–319 (2000)
23. Tiensyria, K., Qu, Y., Zhang, Y., Cupak, M., Rynders, L., Vanmeerbeeck, G., Masselos, K., Potamianos, K., Pettisalo, M.: SystemC and OCAPI-xl based system-level design for reconfigurable systems-on-chip. In: Forum on Specification and Design Languages, pp. 428–439 (2004)
24. Xilinx, Inc.: Early access partial reconfiguration user guide (UG208). Xilinx, Inc (2006). http://www.xilinx.com/

Part III
Design Methods and Tools—Optimization and Runtime Systems

Chapter 8
Design Methods and Tools for Improved Partial Dynamic Reconfiguration

Markus Rullmann and Renate Merker

We dedicate this chapter to Prof. Wunsch on the occasion of his 85th anniversary.

Abstract In current FPGAs the overhead associated with partial dynamic reconfiguration limits the application of this method in system design. We review the origins of this overhead and present a novel approach to solve this problem. We introduce the reconfiguration state graph which is used to describe dynamic reconfiguration for individual resources and to assess reconfiguration cost. We present new method to map reconfigurable modules to resources such that the reconfiguration cost are small. The method can be applied to both digital circuits and dataflow graphs. We demonstrate that we can exploit the trade-off between resource requirements and reconfiguration cost by a unique high-level synthesis tool. We further discuss how our methodology can be integrated into a design flow for efficient runtime reconfigurable systems.

8.1 Introduction

Reconfigurable computing architectures provide a combination of high data processing throughput, similar to ASICs, and the flexibility of a software processor. In such architectures an array of functional units provides the resources to perform many operations in parallel, thus enabling high throughput. The function of the resources and the data transfer between resources are programmable, hence functionality is customized during deployment, after device fabrication. Whereas in reconfigurable computing systems with one static configuration any anticipated functionality must be provided statically, in systems with partial dynamic reconfiguration

Markus Rullmann · Renate Merker
Technische Universität Dresden, Dresden, Germany, e-mails: markus.rullmann@gmx.de, renate.merker@tu-dresden.de

M. Platzner et al. (eds.), *Dynamically Reconfigurable Systems*,
DOI 10.1007/978-90-481-3485-4_8,

more efficient realizations are possible because the functionality can be adapted at runtime to the requirements.

The design of reconfigurable systems-on-a-chip (RSoC) often follows the principles shown in Fig. 8.1. The application consists of a number of tasks. The tasks are partitioned into hardware tasks (HW tasks) and software tasks (SW tasks). In the final system implementation, the software tasks constitute the software program which is run on the RSoC's CPU. The hardware tasks are implemented as reconfigurable modules, which are loaded into the reconfigurable areas of the RSoC during runtime. This method is called partial dynamic reconfiguration. The partitioning of the tasks is crucial for the system performance and the requirements of reconfigurable resources. The reconfiguration between different reconfigurable modules induces a high runtime overhead. In order to prevent frequent dynamic reconfiguration we introduced multimode reconfigurable modules. A multimode module can perform the computation for different tasks without reconfiguration.

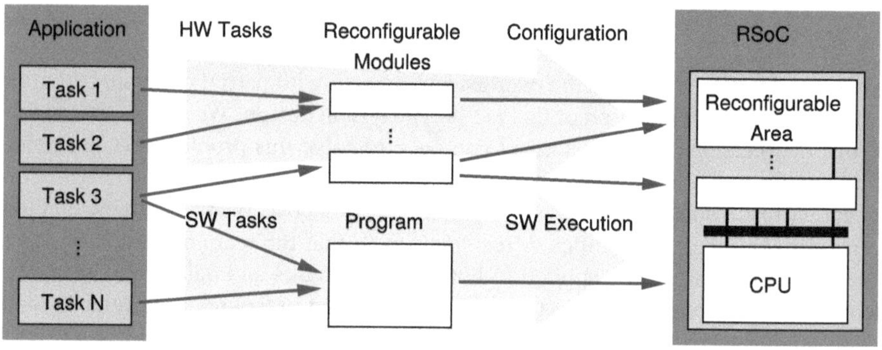

Fig. 8.1 Application partitioning into HW tasks and SW tasks. The tasks are run on the RSoCs reconfigurable area and the CPU.

We investigate the problems of reconfiguration overhead from the device architecture point of view. Consistently with current methodologies we assume that the design functionality is partitioned into modules, but we make two important extensions: (1) modules are not reconfigured completely but based on individual resources. We call this *fine grain reconfiguration*. (2) one module can provide several functions for an application, implemented in a *multimode circuit*.

In this chapter we describe new models, methods and tools that reduce the overhead associated with the dynamic reconfiguration of reconfigurable modules. The methods use a new high-level synthesis (HLS) approach to generate reconfigurable modules that execute the HW tasks. At first we motivate our approach by a short description of current limitations of partial reconfiguration in Sect. 8.2. In Sect. 8.3 we explain general reconfigurable module architecture that is produced by our HLS tool. The models used in the reconfiguration cost assessment are introduced in Sects. 8.4 and 8.5. In Sect. 8.6 we describe our HLS approach. Experimental results obtained with our HLS tool are provided in Sect. 8.7. Finally in Sect. 8.8, we present our work in a broader context of general system design issues. Here we describe how our methods should be integrated into an FPGA system design flow and give further references to related work.

8.2 Motivation

Partial dynamic reconfiguration in FPGAs is usually associated with a module based design approach, see Fig. 8.1. At first, the designer defines a *reconfigurable area* on the device. Second, he implements the reconfigurable tasks as modules that can be loaded on the reconfigurable area. At runtime the resources in the reconfigurable area are reconfigured to enable different modules.

Using standard methodology, the reconfiguration cost of the implementation depends on the size of the reconfigurable area, cf. Fig. 8.2. Each reconfiguration is performed by loading a partial *bitstream* into the device. The configuration bitstream itself is composed of *configuration frames* that contain any data needed to configure the entire reconfigurable area. A configuration frame is the smallest reconfigurable unit in a device; the size of a frame and configurable logic that is associated with each frame depends on the FPGA device. Because the standard bitstreams contain all data for a reconfigurable area, the size of these bitstreams is large, typically hundreds of kilobyte. The *configuration port* of the device has only a limited bandwidth. Together, this leads to configuration times in the order of some hundred microseconds. As a conclusion, configuration data becomes often too large for on-chip storage and frequent reconfiguration leads to considerable runtime overhead.

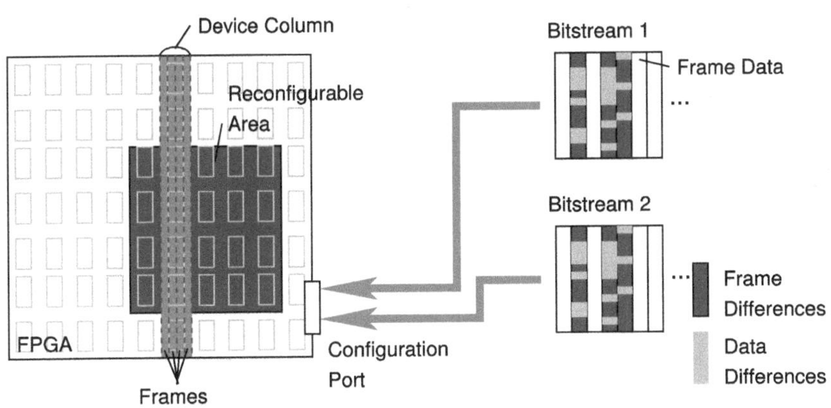

Fig. 8.2 Illustration of module-based partial reconfiguration.

If the properties of reconfiguration data are analyzed in detail, it can be observed that the data does not differ completely between reconfigurable modules: (1) some of the reconfiguration frames are equal in two designs and (2) the data in two frames that configure the same part of the device frequently exhibit only a few different bytes (see frame/data differences in Fig. 8.2). This has implications on the device reconfiguration at runtime. After the initial device configuration, the reconfigurable area is always in a known configuration state. When a new configuration must be established on this area, a new bitstream is used to program the associated device resources. In the ideal case, the reconfiguration programs only the device resources that need a different configuration. Currently the granularity of the reconfiguration is limited by the configuration frame size. For an efficient partial reconfiguration, only configuration frames that contain new configuration data are used to program the reconfigurable area. This is only possible if the current configuration and the frame-based differences are known at runtime.

In the latest FPGA generations (Virtex4, Virtex5) the size of configuration frames has been reduced considerable, which enables a more fine-granular reconfiguration. Also, the bandwidth of the configuration port has been increased, which enables faster reconfiguration. Nevertheless the drawback of existing design flows persists: reconfiguration overhead depends solely on the reconfigurable area, but not on the contents of the reconfigurable modules.

The configuration data themselves are the result of the circuit design and the place and route tools. The configuration data of two modules can become very similar if the initial design exhibits a similar circuit structure and the tools place and route the circuits similarly. In this chapter we describe how the similarity between reconfigurable modules can be increased in order to reduce the differences in configuration data and therefore reduce reconfiguration cost.

8.3 Reconfigurable Module Architecture and Partitioning

The HW tasks are implemented in the reconfigurable modules. In these modules, the tasks are executed in parallel to the CPU and other modules in the RSoC. Therefore, each module needs its own local execution control that runs the task within the module. The HW task execution is started by the software program running on the CPU. The software also receives data and status information from the HW task. The HW task functionality is realized by using computational resources and a state machine that is implemented in the module. The state machine creates a sequence of control signals for the resources in order to execute a task.

Here we describe the logical architecture of a reconfigurable module. Our high-level synthesis tool automatically generates modules with such an architecture. The modules consist of a datapath unit, a control unit, control memories, a bus interface and additional I/O interfaces. The general structure is shown in Fig. 8.3.

The control unit operates as a state machine that is split into a state control memory, which holds the sequence of states and a datapath control memory which stores the control sequence for the datapath. The datapath unit contains registers as storage elements for variables, operations to process data, and multiplexers to control the dataflow on the datapath connections. The control signals of these units are driven

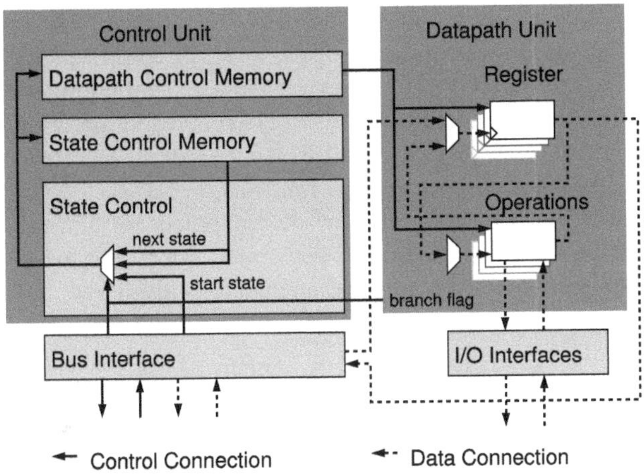

Fig. 8.3 Reconfigurable module architecture used by the HLS tool.

from the datapath control memory. The operations can realize either math and logic functionality or they are used as an interface to external I/O. External I/O can realize bus master accesses and access to FIFOs, memories and other periphery.

The amount of reconfiguration can be chosen if the module is partitioned into static and dynamic sub-modules accordingly. With existing methods, the whole module would be implemented as one monolithic reconfigurable module. Instead we propose to keep the bus and I/O interfaces static and to reconfigure the contents of the control memory and the resources of the datapath independently. Thus reconfiguration can be used for a subset of resources, depending on the configuration differences between modules.

8.4 Reconfiguration State Graph

The current configuration and the partial reconfiguration of an FPGA must be managed at runtime. A reconfiguration state graph (RSG) [6] is used to model the partial reconfiguration. The RSG defines the configurations and the reconfiguration formally. The RSG describes the different configurations available and for each reconfiguration it can be decided what resources must be reconfigured. It provides a framework for reconfiguration management and reconfiguration overhead assessment.

The RSG is defined as a digraph $G(\mathscr{N}_\mathsf{T}, \mathscr{E}_\mathsf{T})$ where the set $\mathscr{N}_\mathsf{T}$ of nodes i represents the reconfigurable modules and the set $\mathscr{E}_\mathsf{T}$ of edges $e = (i, j)$ represents the reconfiguration from reconfigurable module i to reconfigurable module j. With $\mathbf{d} : \mathscr{N}_\mathsf{T} \mapsto \mathscr{D}^m$ for each reconfigurable module i a configuration $\mathbf{d}(i) = (\mathrm{d}(i)_1, \ldots, \mathrm{d}(i)_m)$ is given. The set $\mathscr{D}$ denotes possible configurations of a resource. The elements $\mathrm{d}(i)_k, k = \{1, \ldots, m\}$ describe the configurations of the smallest independent reconfigurable resources k in a device. The reconfiguration $\mathbf{r} : \mathscr{E}_\mathsf{T} \mapsto \{0,1\}^m$ describes for each edge $e = (i,j) \in \mathscr{E}_\mathsf{T}$ which resources k in a device must be reconfigured in order to change a current configuration $\mathbf{d}(i)$ to a new configuration $\mathbf{d}(j)$. Hence, if $\mathrm{d}(i)_k \neq \mathrm{d}(j)_k$ (i.e. reconfiguration of resource k is necessary) $\mathrm{r}(e)_k = \mathrm{r}((i,j))_k = 1$ and if $\mathrm{d}(i)_k = \mathrm{d}(j)_k$ (i.e. reconfiguration of resource k is not necessary) $\mathrm{r}(e)_k = \mathrm{r}((i,j))_k = 0$.

The reconfiguration overhead can now be computed on the basis of the RSG. We assume that each reconfiguration is performed once. The average number of resources that are reloaded if reconfiguration occurs is given by:

$$c_{\mathsf{rc}} = \frac{1}{|\mathscr{E}_\mathsf{T}|} \sum_{e \in \mathscr{E}_\mathsf{T}} \sum_{k=1}^{m} \mathrm{w}_1(k)\mathrm{r}(e)_k, \tag{8.1}$$

where the function $\mathrm{w}_1(k)$ yields the cost for the reconfiguration of element k. We assume that the reconfiguration time is proportional to the weighted sum of reconfigured resources and therefore c_{rc} is called average reconfiguration time.

The RSG model is further illustrated in Example 8.1.

8.5 Module Mapping and Virtual Architecture

The RSG model describes only how the reconfiguration overhead is affected by the configuration data in $\mathbf{d}$. For any module functionality there exist many possible realizations, where each yields a different configuration $\mathbf{d}$. We have developed a method to map the original HW tasks to the reconfigurable modules such that the differences between module configurations are minimized. As a result, the average reconfiguration time is reduced, too.

We observe that functionality of a module is given as a structural representation until the module is finally translated to binary configuration data. It is not possible to describe the structural representation directly as a configuration $\mathbf{d}$, because the functionality is not directly related to fixed resources. Here, we introduce a model that enables us to provide a configuration $\mathbf{d}(i)$ for any structural representation of a module i. First, we define the structural representation formally as a digraph and then we map the digraph to a *virtual architecture* (VA) in order to derive the configuration $\mathbf{d}(i)$. For any such mapping we can therefore compute the reconfiguration overhead and thus can optimize the mapping accordingly, without creating bitstreams for different mappings.

A module $i \in \mathscr{N}_\mathsf{T}$ is represented by a digraph $G_i(\mathscr{N}_i, \mathscr{E}_i)$ where the set $\mathscr{N}_i$ of nodes defines the functions used by a module and the set $\mathscr{E}_i$ of edges defines the data transfer between the functions. The resource configuration required by a node is assigned by the function $\mathrm{l} : \mathscr{N}_i \mapsto \mathscr{D}$. Note that the digraph can describe structural representations at several levels in the design flow, e.g. it can describe dataflow graphs, synthesized digital circuits etc. The digraphs $G_i, i \in \mathscr{N}_\mathsf{T}$ are called *input graphs*.

The input graphs of all modules i are mapped to a VA. A VA is defined as a digraph $G_A(\mathcal{N}_A, \mathcal{E}_A)$. The nodes $\mathcal{N}_A$ denote reconfigurable resources and the edges $\mathcal{E}_A$ denote reconfigurable interconnect. Furthermore we define an allocation $a : \mathcal{N}_i \mapsto \mathcal{N}_A$, which maps the node of any input graph to a resource in the virtual architecture. As a by-product the edges from the input graphs are mapped to the VA, too. The most important feature of the allocation is that the different input graphs are mapped to a common context, and hence, their configuration can be compared to each other.

Now we specify for a module i the configuration of the VA, i.e. the configuration of the resources $n \in \mathcal{N}_A$ and of the interconnects $e \in \mathcal{E}_A$ that realizes the module on the VA. The configuration of these $m = |\mathcal{N}_A| + |\mathcal{E}_A|$ elements is given for module i by $\mathbf{d}(i) = (\mathrm{d}(i)_1, \ldots, \mathrm{d}(i)_m)$ as follows:

- The configuration of resource $n_k \in \mathcal{N}_A, k \in \{1, \ldots, |\mathcal{N}_A|\}$ is specified by $\mathrm{d}(i)_k = \mathrm{l}(n_i)$ if a node $n_i \in \mathcal{N}_i$ exists with $\mathrm{a}(n_i) = n_k$, otherwise $\mathrm{d}(i)_k = 0$.
- The configuration of interconnect $e_k \in \mathcal{E}_A, k \in \{|\mathcal{N}_A| + 1, \ldots, m\}$ is given with $\mathrm{d}(i)_k = 1$, if an edge $e_i \in \mathcal{E}_i$ is mapped to an edge $e_k \in \mathcal{E}_A$, otherwise $\mathrm{d}(i)_k = 0$.

In summary, we provided a formal definition of the module's structural representation and an abstract reconfigurable architecture model. An allocation describes how the module is mapped to the architecture model. In addition, the input graphs, the allocation, and the VA define the configuration in the RSG model. Thus, we can calculate the reconfiguration cost. The model is illustrated in Fig. 8.4. In order to reduce reconfiguration cost, we are interested in an allocation that minimizes reconfiguration cost.

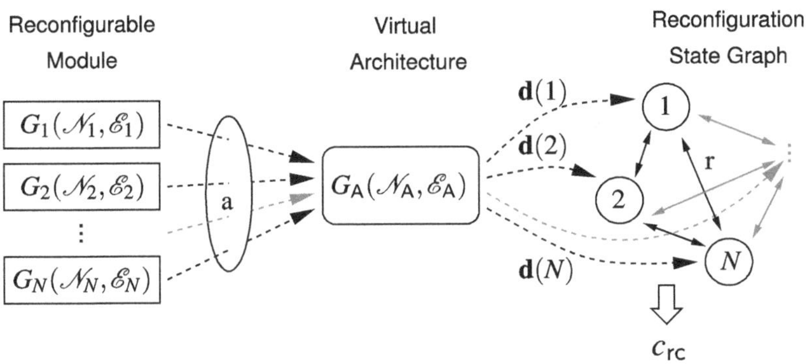

Fig. 8.4 Mapping of the structural representations of modules to a VA and its relationship to the RSG model.

Example 8.1. Consider three modules 1, 2, and 3 represented by the input graphs G_1, G_2, G_3 shown in Fig. 8.5. The nodes $n_i \in \mathcal{N}_i$ of the input graphs are labeled with their respective configuration $\mathrm{l}(n_i) = f_i$, e.g. node n_1 requires a configuration $\mathrm{l}(n_1) = f_1$ of the resource in order to realize the required functionality.

The VA graph is depicted in Fig. 8.5, too. The VA $G_\mathsf{A}(\mathscr{N}_\mathsf{A}, \mathscr{E}_\mathsf{A})$ is given by the elements $\mathscr{N}_\mathsf{A} = \{n'_1, n'_2, n'_3, n'_4\}$ and $\mathscr{E}_\mathsf{A} = \{e_5, e_6, e_7\}$.

For each node n_i in the input graphs G_1, G_2, G_3 the allocation to a node n'_k of the VA is shown, i.e. $\mathrm{a}(n_i) = n'_k$. The allocation of an edge $e_i = (n'_{i_1}, n'_{i_2}) \in \mathscr{E}_i$ results directly from the allocation of the nodes n'_{i_1}, n'_{i_2}. For example, the edge $(n_1, n_2) \in \mathscr{E}_1$ is allocated to edge $e_6 = (\mathrm{a}(n_1), \mathrm{a}(n_2)) \in \mathscr{E}_\mathsf{A}$ of the VA.

The configuration $\mathbf{d}(i)$ of the VA that realizes the functionality required by module i is depends on the allocation of nodes $\mathscr{N}_1$ and edges $\mathscr{E}_1$. The configuration $\mathbf{d}(i) = (\mathrm{d}(i)_1, \ldots, \mathrm{d}(i)_7)$ describes the configuration of the resources and interconnects in the following order: $n'_1, n'_2, n'_3, n'_4, e_5, e_6, e_7$. For example the allocation of module 1 yields the configuration $\mathbf{d}(1) = (\mathrm{l}(n_1), \mathrm{l}(n_2), \mathrm{l}(n_3), 0, 0, 1, 1) = (f_1, f_2, f_3, 0, 0, 1, 1)$.

The RSG model related to the input graphs G_1, G_2, G_3 contains the modules $\mathscr{N}_\mathsf{T} = \{1, 2, 3\}$ and the reconfigurations between the modules $\mathscr{E}_\mathsf{T} = \{(1,2), (1,3), (2,1), (2,3), (3,1), (3,2)\}$, cf. Fig. 8.5. The reconfiguration $\mathbf{r}((i,j))$ is derived from the configurations $\mathbf{d}(i), \mathbf{d}(j)$ of the modules i, j. Consider the configurations $\mathbf{d}(1) = (f_1, f_2, f_3, 0, 0, 1, 1)$ and $\mathbf{d}(2) = (f_5, f_6, 0, f_4, 1, 1, 0)$: the related reconfiguration yields $\mathbf{r}((1,2)) = (1,1,1,1,1,0,1)$ because the configuration of all elements k differs except for $k = 6$. The edge e_6 is allocated by both configurations.

The reconfiguration for the full RSG is as follows: $\mathbf{r}((1,2)) = \mathbf{r}((2,1)) = (1,1,1,1,1,0,1), \mathbf{r}((1,3)) = \mathbf{r}((3,1)) = (1,1,1,1,1,0,0), \mathbf{r}((2,3)) = \mathbf{r}((3,2)) = (1,1,1,1,0,0,1)$. If we assume a unit weight for all reconfigurable elements, i.e. $\mathrm{w}_1(k) = 1, k \in \{1, \ldots, 7\}$, then the reconfiguration cost (Eq. (8.1)) evaluate to:

$$c_\mathsf{rc} = \frac{1}{6}(6+6+5+5+5+5) = 5\frac{1}{3}.$$

8.6 High-Level Synthesis of Reconfigurable Modules

In this section we describe our methodology that enables us to compute allocations that lead to minimal reconfiguration cost. More specifically we implemented the methodology into a high-level synthesis (HLS) tool. The tool receives the functionality which must be implemented in the reconfigurable modules as ANSI-C like source code. The source code is compiled into control dataflow graphs (CDFG) first, one for each C-function. The CDFGs are considered as input graphs to our allocation problem. The tool performs four essential HLS steps: (1) for each node in the CDFG an appropriate resource type is chosen, (2) scheduling assigns an execution time to each node, (3) each node is allocated to a resource instance in the datapath which is described as a VA, and (4) architectural synthesis creates the submodules according to the module architecture, i.e. the control unit and the datapath unit.

HLS provides the ideal abstraction level to describe reconfigurable modules. From the designer's point of view, it is much easier to use C-like descriptions of the algorithm: they can be modified more easily and they can be integrated into system-

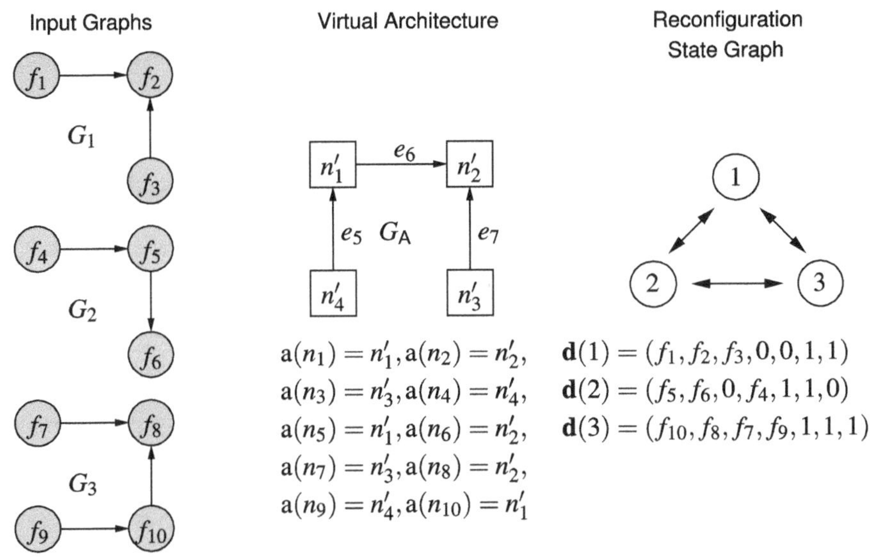

Fig. 8.5 Input graphs G_1, G_2, G_3 for Example 8.1. The input graphs are mapped to the VA G_A provides a common reference for the image graphs.

level simulations. It is also possible to adapt the hardware/software partitioning late in the design process because functions can be moved to the software processor or to hardware modules with little effort. On the tool side, the high-level descriptions provide a great deal of freedom to map the functionality to a datapath. In our work we exploit this freedom in order to generate reconfigurable modules with small reconfiguration overhead. More specifically, we investigated several methods to choose the resource types in HLS step 1, and we implemented a resource allocation method that takes advantage of our VA model to solve HLS step 3.

Both reconfiguration overhead and resource overhead in HLS is avoided by resource sharing. In intra-module resource sharing, several CDFG nodes of one module are mapped to the same resource instance in the VA, thus reducing resource overhead. In inter-module resource sharing, several CDFG nodes of multiple modules are mapped to the same resource instance in the VA. As a result, the resources are used in several modules and are not reconfigured between those modules.

In the following we describe the HLS steps 1 and 3 in more detail.

8.6.1 Resource Type Binding

In HLS step 1, Resource type binding can either enable or disable the reuse of VA resources in step 3. In extension to the input graphs defined previously, the CDFG nodes represent fixed operations or variables. An operation (or variable) can only be executed (or stored) on selected resource types. During resource type binding, one resource type is chosen for each CDFG node. In the resource instance binding step, the allocation can only be chosen such that each node is allocated to a resource instance that is of the previously specified type. The resource type binding defines the potential reuse of nodes and edges in the VA because it enables or disables resource sharing possibilities between nodes. We investigated several different type binding strategies in our research in order to see how much the strategy affects the final results. We assume that many nodes can be bound to resource types of different complexity, which effects resource sharing: e.g. an addition could be bound to a simple adder-resource or to a complex ALU-resource. The different type binding strategies are stated below:

(a) *Minimum Cost Resource Type*. Here, we choose a resource type for each node independently with the objective to select the least costly one, e.g. in terms of FPGA resources.
(b) *Minimum Number of Resource Types*. The resource types are chosen such that the number of different resource types becomes minimal. As a result, it is possible that there are fewer resource instances in the VA because resources can be shared more often. For instance two nodes may realize two different functions, but if they are mapped to the same resource type, they may share a resource instance in the datapath. The number of resource types can be minimized either over each task independently or for all tasks at once.
(c) *Minimum Number of Interconnect Types*. The data transfers, indicated by edges in the CDFG, are mapped to interconnects in the VA. Although the exact interconnect is not known during type binding, it is already possible to determine if two edges may share an interconnect or not. Two edges can share an interconnect if the they are mapped to the same interconnect type. The interconnect type is defined by the resource types were the source node and the drain node of an edge are mapped to. The minimization of interconnect types targets specifically the reuse of interconnect in the datapath. As above, the number of resource types can be minimized either over each task independently or for all tasks at once.

The effect of the different strategies for resource type binding will be discussed for the benchmarks provided in Sect. 8.7.

8.6.2 Resource Instance Binding

After a resource type has been selected for each operation and the operation has been determined in HLS step 2, the operations must be allocated to specific resource instances. In step 3, the HLS tool allocates the CDFG nodes to resource instances and derives the datapath interconnect to realize the data transfer between nodes. This step is based on the transformation of the input graphs to the VA presented in Sect. 8.5. In our tool, we employ a heuristic optimization method that is based on simulated annealing [7] in order to find an allocation that minimizes a cost function c. The cost function is a weighted sum of several cost parameters of the datapath. Thus the datapath can be optimized to achieve minimal reconfiguration cost (for resources and interconnect), resource use, interconnect overhead, and dataflow multiplexers.

Dataflow multiplexers are introduced into the datapath in order to realize resource sharing. Assume that multiple nodes of one CDFG are allocated to the same resource. However, the data supplied to the resource originates from different resources. For each computation, a dataflow multiplexer connects the output of the resource which provide the data to the shared resource input. The dataflow multiplexers are controlled by the control unit. The node allocation defines the resource sharing and thus the interconnect structure and the dataflow multiplexers.

Our HLS tool can generate multimode circuits in order to reduce reconfiguration cost. For a set of tasks it can be chosen, which tasks are implemented in the same reconfigurable module. Hence we can take advantage of inter-module resource sharing within one configuration, which reduces resource overhead in multimode circuits, and between different configurations, which reduces overhead for dynamic reconfiguration.

The resource instance binding step works as follows: The scheduling algorithm provides information how many resource instances are required. This defines the number of resources in the VA. Further, the scheduling algorithm provides an initial solution which assigns each CDFG node to a node in the VA. The initial solution is modified iteratively by the simulated annealing algorithm in order to improve the initial solution.

In simulated annealing a current solution is modified iteratively. At first, the current solution is slightly modified to gain a new solution. Then, the cost function for the new solution is calculated. Depending on the state of the algorithm and the cost of the new solution, the new solution is either accepted and becomes the current solution or the new solution is discarded.

In our tool, the new solution is derived from the current solution by a random permutation of the allocation. Subsequently, the interconnect structure in the VA is derived as well as the dataflow multiplexers. The permutation must observe the constraints imposed by the scheduling: Any resource instance can only be allocated by one CDFG node at any cycle at runtime.

For the permutation, a node $n \in \mathscr{N}_i$ is selected randomly and the allocation $\mathrm{a}(n) = r$ is changed to a randomly selected resource r', i.e. $\mathrm{a}(n) = r'$. Vice versa any node $n' \in \mathscr{N}_i$ which is already allocated to the resource r' and which is in conflict with the new allocation of n, will be allocated to the previous allocation of n, i.e. $\mathrm{a}(n') = r$. Thus, starting from a valid initial solution we permute the solutions iteratively such that each solution remains valid.

The cost function used in the simulated annealing algorithm is composed of several components. The basis for all components is the allocation to the VA and the RSG model. In the following we derive the relevant computations. The resource cost for a module $i \in \mathscr{N}_\mathsf{T}$ is calculated from the input graph G_i and the allocation a (cf. Sect. 8.5).

Here we assume that the resource instances n_k in the VA are either in use ($\mathrm{d}(i)_k = 1$) or unused ($\mathrm{d}(i)_k = 0$), similar to the interconnect configuration. Each resource instance and each interconnect is associated with a cost factor w_2 that represents either the number of used FPGA slices for a resource instance or the bus width of an interconnect. Hence, the resource cost for a module i are given by:

$$c_{\mathsf{res}}(i) = \sum_{k=1}^{m} \mathrm{w}_2(k)\mathrm{d}(i)_k. \tag{8.2}$$

The resource cost $c_{\mathsf{res}}(i)$ do not include the cost for the dataflow multiplexers. A single dataflow multiplexer switches between all edges running into the same input of node $n \in \mathscr{N}_{\mathsf{A}}$. For a module i, the set $\mathscr{E}_{\mathsf{A},n}$ of such edges are given by $\mathscr{E}_{\mathsf{A},n} = \{e_k \in \mathscr{E}_{\mathsf{A}} : e_k = (n', n), n' \in \mathscr{N}_{\mathsf{A}} \wedge \mathrm{d}(i)_k = 1\}$. The resource cost caused by the dataflow multiplexers are now computed as:

$$c_{\mathsf{mux}}(i) = \sum_{n \in \mathscr{N}_{\mathsf{A}}} \mathrm{w}_3(|\mathscr{E}_{\mathsf{A},n}|), \tag{8.3}$$

where $\mathrm{w}_3(x)$ yields the resource cost of an x-to-1 multiplexer.

The reconfiguration cost are already defined in Eq. (8.1). The overall cost function for the simulated annealing algorithm is given by:

$$c = \frac{1}{|\mathscr{N}_{\mathsf{T}}|} \sum_{i \in \mathscr{N}_{\mathsf{T}}} c_{\mathsf{res}}(i) + \frac{1}{|\mathscr{N}_{\mathsf{T}}|} \sum_{i \in \mathscr{N}_{\mathsf{T}}} c_{\mathsf{mux}}(i) + c_{\mathsf{rc}}. \tag{8.4}$$

8.6.3 Control Generation

The scheduling of nodes from the input graphs is computed in HLS step 2. Together with the allocation of those nodes, the tool generates the contents of the datapath control memory and the state control memory. The resources in large datapaths are often not used in every control state, i.e. the datapath control memory can be underutilized. Therefore we combined our tool with the approach described in Chap. 15. We implemented a greedy algorithm that translates the contents of the datapath control memory to multi-context tables. In [18] we have shown that this method can reduce the storage overhead for datapath control information significantly. We further proposed two possible extensions to current FPGA architectures that enable a very efficient and yet flexible integration of multi-context tables into FPGAs.

In this section we have provided a brief overview of a HLS tool for improved partial dynamic reconfiguration. We have illustrated that there exists a large space for optimizations that increase the similarity of reconfigurable modules. The aim of these optimizations are—besides conventional optimization targets time and area—the reduction of reconfiguration overhead.

8.7 Experiments

In this section we discuss the efficiency of our HLS approach on a set of benchmarks. The benchmark set has been implemented with different combinations of type binding and instance binding methods presented in the previous section. The results obtained depend on the combination of the HLS steps 1 and 3. This allows us to analyze the quality of results for each combination as well as the costs in terms of tool runtime. First we describe the experimental setup and then we discuss the results and draw some conclusion concerning the design of reconfigurable modules.

8.7.1 Experimental Setup

In Sect. 8.6 we described several options to perform resource type binding. The resource instance binding can also be performed with different objectives: with the help of the weight functions w_1, w_2, and w_3 we are able to set up different cost functions that are used by the simulated annealing algorithm as a cost function. The different objectives are used to optimize the HW task implementations for different scenarios within a common framework. The implementation scenarios describe how the modules are implemented in the RSoC. Thus we can compare non-reconfigurable and reconfigurable solutions.

Resource Type Binding Methods The resource type binding methods discussed in Sect. 8.6.1 (a)–(c) are used in our experiments as follows:

1. Minimum cost resource type,
2. Minimum number of resource types for each module individually,
3. Minimum number of resource types and interconnect types for each module individually,
4. Minimum number of resource types over all modules,
5. Minimum number of resource types and interconnect types over all modules,

where the cost for interconnect types is not optimized independently but in combination with the cost for resource types.

Resource Instance Binding Methods Further we investigated several optimization targets during HLS step 3 that were combined with the different type binding methods. The different optimization targets for the instance binding are as follows:

1. Minimum average resource and interconnect cost of the tasks individually,
2. Minimum resource and interconnect cost for all tasks merged into one datapath,
3. Minimum average resource reconfiguration cost,
4. Minimum average resource and interconnect reconfiguration cost.

Implementation Scenarios In this work we compare the non-reconfigurable and reconfigurable implementation of HW tasks, both for existing methods and for our new methodology. The following implementation scenarios are considered:

- A: *static, parallel implementation.* Classic implementation, where each HW task is implemented as an individual module, which is placed statically on the device. The HW tasks can be executed concurrently.
- B: *static, sequential implementation.* Our method allows that several HW task are implemented in a multimode module. The tasks can share resources but can not operate in parallel.
- C: *reconfiguration without reuse of resources.* Classic module based reconfiguration. Each HW task is assigned to an individual module which is completely dynamically reconfigured.
- D: *reconfiguration with reuse of resources.* In this scenario our new reconfiguration model is applied. It is assumed that only those resources and interconnect within the datapath are reconfigured that are different between configurations.

With the data obtained from our benchmarks we want to investigate, which resource type and which resource instance binding strategies lead to the best solution for a scenario. Further, we will show that the scenarios B and D, which are available through our methodology, are superior to previous concepts A and C.

Benchmark Characteristics The chosen benchmarks consist of several task sets. Each task set contains tasks that might be used in a real reconfigurable system. The tasks within one set are assumed to be reconfigured against each other. Thus the tasks provide a good example on how our methodology can be employed in practice. This kind of tasks can be found in many similar work on HLS. Here with give a short summary of the tasks functionality and complexity by (number of tasks, total number of nodes for all tasks). The benchmark ADPCM (2 tasks, 280 nodes) contains an ADPCM encoder and decoder from the MediaBench suite [8]. EDGE (3 tasks, 422 nodes) contains three different Sobel edge detection filters: a combined horizontal and vertical filter, a horizontal only, and a vertical only filter. JPEG_DCT (2 tasks, 613 nodes) consists of tasks that perform an integer based forward discrete cosine transform (DCT) and a task for the backward transform. Both tasks are also taken from MediaBench. The JPEG_DCT represents the most complex task set in terms of operations per input graph. Finally the RGB_YUV (2 tasks, 84 nodes) describes a color conversion from RGB color space to the YUV color space and vice versa, this function is used in many image and video coding applications.

8.7.2 Benchmark Results and Discussion

Our HLS tool has been used to implement the HW tasks of the benchmarks according to the different scenarios, by using different resource type and instance binding methods. Fig. 8.6 shows the results obtained for our benchmarks using the scenarios A–D. For each scenario we present the results for the best overall combination of resource type and resource instance binding method. For each benchmark, the results are labeled on the x-axis as follows: *scenario: resource type binding method, resource instance binding method.*

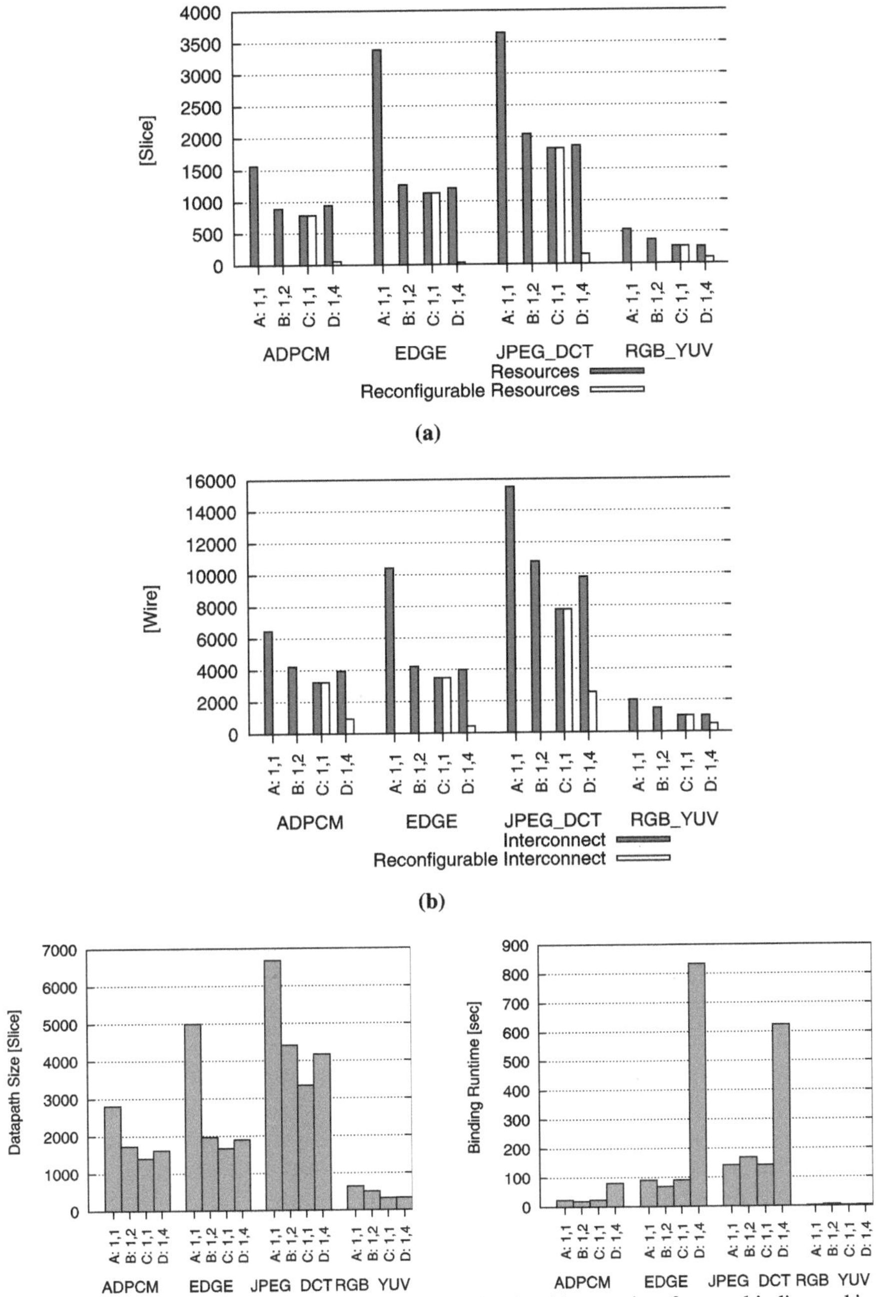

(c) Cost for resources and multiplexers.

(d) Algorithm runtime for type binding and instance binding.

Fig. 8.6 Comparison of results obtained with selected binding methods for scenarios A–D.

In Fig. 8.6(a) the average amount of module resources used for operations and storage in the datapath of a module is shown. The bright bars show the average amount of reconfigurable resources in each module. Similarly, the average number of interconnect wires used in the datapath of a module are depicted in Fig. 8.6(b). Here, the bright bars show the average amount of reconfigurable interconnect. Thus, the average resource cost $\frac{1}{|\mathcal{N}_\mathsf{T}|}\sum_{i\in\mathcal{N}_\mathsf{T}} c_{\mathsf{res}}(i)$ and the average reconfiguration cost c_{rc} are shown separately for resources and interconnect in Fig. 8.6(a, b). The average datapath size in terms of resources, which includes both operation/storage and multiplexer cost is shown in Fig. 8.6(c). The tool runtime for the type and instance binding algorithms is depicted in Fig. 8.6(d).

In our experiments we found that the most straightforward resource type binding method (1) achieves the best overall results in the final implementation for all scenarios. While the type binding method has a considerable effect on the resource sharing possibilities, the differences are negligible in the final datapath implementation. We suspect that there are always good resource sharing possibilities, because the high-level description uses only a limited set of different operations. We found that the resource instance binding method has a much more severe effect. For the classic scenarios A, C we choose instance binding method (1) because in both scenarios, the modules are implemented independently. For scenario B we found that instance binding method (2) performs best, because only in this case the resource sharing between tasks that are implemented in one module is exploited. Finally for scenario D we could show that instance binding method (4) performs best in terms of reconfiguration cost, because both resources and interconnect reconfiguration are targeted by the optimization.

The scenarios A and C represent the conventional approaches to implement HW tasks on FPGAs. In scenario A, a static configuration contains all reconfigurable module, which may lead to a high resource and interconnect overhead but enables a parallel execution of tasks. As shown in our benchmarks, the resource and interconnect requirements are reduced drastically if the modules are dynamically reconfigured. At the same time, reconfiguration of all resources and interconnects used by a module causes a high overhead.

The newly introduced scenario B, which implements static, merged datapaths provides an attractive trade-off between the scenarios A and C. Scenario B requires much less resources than scenario A because resources are shared between HW tasks. Further scenario B causes no reconfiguration cost, but scenario C employs partial reconfiguration of the module. Nevertheless, scenario B requires more resources than scenario C, because the flexibility is gained with additional operations and dataflow multiplexers, which result in a datapath with more, temporarily unused resources. Furthermore, it is interesting to note that the differences in the resource allocation for scenarios B and C are small, cf. Fig. 8.6(c).

In scenario D, the implemented datapaths are optimized for maximum similarity and hence, for minimal reconfiguration cost in terms of resource and interconnect resources. Scenario D demonstrates how much the datapaths can be optimized, such that they differ in only few resources and interconnects. For resources, the differences are less than 10% and for interconnects, the differences are less than 26% in most cases, which is a significant reduction compared to the full reconfiguration of the datapath in scenario C. However, in scenario D the resource and interconnect cost were not included in the optimization. Therefore, the datapaths are slightly more costly in terms of resources and interconnects. Actually, the scenarios C and D represent two extremes between area optimization (scenario C) and reconfiguration optimization (scenario D). With the flexible weights w_1, w_2, w_3 in the cost function, we are able to generate intermediate solutions that meet the needs of the overall system.

The runtime of the binding algorithms shown in Fig. 8.6(d) is comparable for the scenarios A–C, but the runtime for scenario D is much higher. The resource type binding has been performed with method (1) in all scenarios. Obviously the resource instance binding requires a much higher optimization effort, when the result is optimized for resource and interconnect reconfiguration cost. From our experience we believe that this optimization is still much more efficient in terms of runtime and results, than a similarity extraction of the final netlist. E.g. benchmark JPEG_DCT contains a total of 613 nodes in the input graphs for which quality binding must be found. However, at netlist level the similarity extraction must be performed for 2×1800 slices, which is much more complex.

The performance of the datapath implementations is very similar. The task execution cycles are equal in all scenarios because the same scheduling is used. The maximum clock frequency differs slightly between the scenarios, but usually less than 10%. However, our optimization does not target the critical path delay directly, it reduces the complexity of dataflow multiplexers instead.

Although the examples presented here have only a limited number of tasks, we discuss the development for an increased number of tasks. As we merge more tasks into one module (scenario B) it is likely that the increase in operation resources is small. However, because the dataflow in the tasks is different, more flexibility in interconnect is needed and the overhead in interconnect and dataflow multiplexers increases. Likewise, if the datapath implementation is optimized for low reconfiguration cost we expect that for more reconfigurable modules the average reconfiguration cost increase, because an efficient inter-module resource/interconnect sharing can not be achieved for many tasks at the same time. Our predictions are supported by an analysis in [11] and the fact that FPGAs, which target maximum flexibility, contain highly reconfigurable resources and very flexible interconnect routing. As a general rule, in FPGAs about 90% silicon area are used for interconnect and only 10% for reconfigurable logic.

Here we suggest the following strategy for a balanced use of our new methodology. HW tasks that are frequently reconfigured against each other should be merged in one reconfigurable module or optimized for low reconfiguration cost. HW tasks or the reconfigurable modules that are not frequently reconfigured must not be optimized for low reconfiguration cost. Thus, the implementation depends on the overall execution behavior of the application. With our methodology it is possible to use the reconfigurable area more efficiently and to reduce the penalty of runtime reconfiguration for the most critical parts of the application.

8.8 System Design for Efficient Partial Dynamic Reconfiguration

In this chapter we have presented a concise model to describe reconfiguration on the level of individual resources and interconnect. The cost model is based on the reconfiguration state graph. We introduced the model of a virtual architecture that allows us to assess reconfiguration cost for any structural representation of hardware tasks. The benefits of the model have been demonstrated on several examples, which have been implemented by our high-level synthesis tool. Finally, we describe a possible design flow that takes advantage of our methodology.

The general idea of our method, the reduction of reconfiguration overhead by reducing the differences between reconfigurable modules, is reflected throughout our proposed design flow for reconfigurable systems. In this section we summarize the steps of our design flow and provide references to further work.

In the system design phase, there must be decided how the functionality of the tasks is partitioned into HW tasks and SW tasks. The characteristics of the application and of the reconfigurable HW tasks provide information on how many reconfigurable resources are required and what kind of runtime management should be used. Our tools can aid the decision which HW tasks could be integrated into the same reconfigurable module and provide information on the size of the reconfigurable area.

Tasks that depend on each other can not run concurrently in a system. Therefore they can be either integrated into the same reconfigurable module or in different reconfigurable modules that are configured successively. If those tasks are executed frequently they should be integrated into the same reconfigurable module because then dynamic reconfiguration is avoided. Of course, this is only possible if the multimode module fits on the reconfigurable area. More information on the partitioning problem is given in Chap. 9. An alternative approach is described in Chap. 4: In hyper-reconfigurable hardware it is assumed that a sequence of configurations is known. Next, the sequence is partitioned into hypercontexts which contain reconfigurable resources and resources that are static within the hypercontext. The concept can be interpreted in our HLS context as follows: The sequence of configurations is similar to the sequence of control data supplied to the datapath. For such a sequence it can be derived with the methods described in Chap. 4, which parts of the sequence should be grouped into a reconfigurable module in order to minimize reconfiguration cost.

The RSoC is usually managed at runtime by an operating system (OS), which has been adapted to support dynamic reconfiguration. Examples for such systems are described e.g. in [4, 20, 19, 10]. In [1] we have demonstrated a video-based, realtime region-of-interest-detection application. The application demonstrates the capabilities of our HLS tool and the integration of HW tasks and the ReconOS OS (cf. Chap. 13). The application contains a HW task that runs as an independent threat, parallel to the software application. The HW task sends continuously data to the software application via the ReconOS API. As a hardware platform we use the ESM, cf. Chap. 3.

For design entry, the two major methods are HLS and synthesis from register transfer level (RTL) code. During this design phase it is possible to increase the reconfigurable module similarity significantly by special design practices. A brief overview of our HLS method is presented in this chapter, more details can be found in [16, 15]. We also explored possibilities to increase similarity during RTL design. Here, the designer can describe digital circuits that have an intended similarity [17, 12]. Expert knowledge of the device architecture and synthesis is necessary to achieve good results. Other synthesis methods that are used to reduce reconfiguration cost are described e.g. in [3, 9, 2].

The similarity information, which is required later in the module implementation phase can be provided either by the synthesis tool or it can be derived after synthesis. As discussed before, the HLS tool directly generates the similarity information. In addition we have developed a tool that is able to extract the similarity from generated netlists after synthesis [13].

In an FPGA design flow, the synthesized netlists are mapped to device resources before place and route. During the mapping the netlist elements are assigned to device specific resources, e.g. logic blocks (Slices). In this mapping several netlist elements (logic and interconnect) can be assigned to the same device resource, which may destroy the module similarity or invalidate similarity information. We have described a mapping tool [14] that is able to take advantage of the similarity instead. The mapping tool treats all netlists of reconfigurable modules at the same time and thus can retain the similarity directly. The tool takes the reconfiguration cost model into account in order to improve the mapping result.

For our methodology, existing place and route tools must be extended in order to take advantage of the similarity information. The tools must observe the following constraints: nodes that are allocated to the same resources in the VA must be placed on the same device resource later on. Similarly, edges that are allocated to the same interconnect in the VA must be routed using the same switch box configuration. With existing tools this can only be realized to a limited extend, e.g. by using the guide mode in the Xilinx ISE tools. This method has been used in [17, 12].

Finally, the placed-and-routed reconfigurable modules are transcribed into bitstreams that contain the binary programming data for the device. Because we use partial reconfiguration, the bitstreams contain only data that is relevant to adapt the reconfigurable area to the new module. In the established EAPR design flow, the bitstreams for the module contain the full configuration of a predefined reconfigurable area. The authors in [5] describe a tool that can produce bitstreams which remove the frames from the bitstreams which are static in all modules. We have described another method to create partial bitstreams in [15]. There, the reconfiguration bitstreams are chosen such that minimal reconfiguration time or minimal storage of configuration data is ensured. The method is based on the RSG model described above.

Acknowledgements Supported by DFG grant ME 1625/3-3, project *Development of Methods and Tools for the Minimization of Reconfiguration Cost*, as part of the Priority Programme 1148, *Reconfigurable Computing Systems*. We would like to thank all members of the Priority Programme

1148 for the fruitful discussions at the project meetings and for the close collaboration within the programme.

References

1. Angermeier, J., Majer, M., Teich, J., Braun, L., Schwalb, T., Graf, P., Hubner, M., Becker, J., Lubbers, E., Platzner, M., Claus, C., Stechele, W., Herkersdorf, A., Rullmann, M., Merker, R.: Spp1148 booth: Fine grain reconfigurable architectures. In: International Conference on Field Programmable Logic and Applications (FPL 2008), Heidelberg, Germany, p. 348 (2008)
2. Aravind, D., Sudarsanam, A.: High level—application analysis techniques & architectures—to explore design possibilities for reduced reconfiguration area overheads in FPGAs executing compute intensive applications. In: Parallel and Distributed Processing Symposium, 2005. Proceedings. 19th IEEE International, pp. 158–158 (2005)
3. Boden, M., Fiebig, T., Reiband, M., Reichel, P., Rulke, S.: Gepard—a high-level generation flow for partially reconfigurable designs. In: IEEE Computer Society Annual Symposium on VLSI (ISVLSI'08), pp. 298–303 (2008)
4. Brebner, G.: A virtual hardware operating system for the xilinx 6200. In: Field-Programmable Logic, Smart Applications, New Paradigms and Compilers. LNCS, vol. 1142, pp. 327–336. Springer, Berlin (1996)
5. Claus, C., Müller, F.H., Zeppenfeld, J., Stechele, W.: A new framework to accelerate Virtex-II pro dynamic partial self-reconfiguration. In: IEEE International Parallel and Distributed Processing Symposium, 2007. IPDPS 2007, pp. 1–7 (2007)
6. Heron, J., Woods, R., Sezer, S., Turner, R.: Development of a run-time reconfiguration system with low reconfiguration overhead. J. VLSI Signal Process. **28**(1–2), 97–113 (2001)
7. Kirkpatrick, S., Gelatt, C.D., Vecchi, M.P.: Optimization by simulated annealing. Science **4598**, 671–680 (1983)
8. Lee, C., Potkonjak, M., Mangione-Smith, W.H.: Mediabench: A tool for evaluating and synthesizing multimedia and communications systems. In: Proceedings of the 30th International Symposium on Microarchitecture (MICRO-30), Research Triangle Park, USA, pp. 330–335 (1997)
9. Moreano, N., Borin, E., de Souza, C., Araujo, G.: Efficient datapath merging for partially reconfigurable architectures. IEEE Trans. Comput.-Aided Des. Integr. Circuits Syst. **24**(7), 969–980 (2005)
10. Nollet, V., Coene, P., Verkest, D., Vernalde, S., Lauwereins, R.: Designing an operating system for a heterogeneous reconfigurable soc. In: Proceedings of the International Conference on Parallel and Distributed Processing Symposium (IPDPS 2003), Nice, France (2003)
11. Rullmann, M.: Models, design methods, and tools for improved partial dynamic reconfiguration. PhD thesis, Technische Universität Dresden (2009, to appear)
12. Rullmann, M., Merker, R.: Design and implementation of reconfigurable tasks with minimum reconfiguration overhead. In: Dynamically Reconfigurable Architectures Workshop at 19th International Conference Architecture of Computing Systems (ARCS 2006), Frankfurt/Main, Germany, pp. 132–141 (2006)
13. Rullmann, M., Merker, R.: Maximum edge matching for reconfigurable computing. In: Reconfigurable Architectures Workshop at 13th IEEE International Parallel & Distributed Processing Symposium (IPDPS 2006), Rhodes, Greece (2006)
14. Rullmann, M., Merker, R.: A reconfiguration aware circuit mapper for fpgas. In: IEEE International Parallel & Distributed Processing Symposium—IPDPS 2007, 14th Reconfigurable Architectures Workshop (2007)
15. Rullmann, M., Merker, R.: A cost model for partial dynamic reconfiguration. In: Najjar, W., Blume, H. (eds.) International Conference on Embedded Computer Systems: Architectures, Modeling and Simulation (IC-SAMOS), pp. 182–186 (2008)

16. Rullmann, M., Merker, R.: Synthesis of efficiently reconfigurable datapaths for reconfigurable computing. In: International Conference on Field-Programmable Technology 2008 (ICFPT'08) (2008)
17. Rullmann, M., Siegel, S., Merker, R.: Optimization of reconfiguration overhead by algorithmic transformations and hardware matching. In: Workshop RAW 2005 at the 19th IEEE International Parallel and Distributed Processing Symposium, pp. 151–156 (2005)
18. Rullmann, M., Merker, R., Hinkelmann, H., Zipf, P., Glesner, M.: An integrated tool flow to realize runtime-reconfigurable applications on a new class of partial multi-context fpgas. In: International Conference on Field Programmable Logic and Applications (FPL 2009, to appear), Prague, Czeck Republic (2009)
19. Steiger, C., Walder, H., Platzner, M.: Operating systems for reconfigurable embedded platforms: online scheduling of real-time tasks. IEEE Trans. Comput. **53**(11), 1393–1407 (2004)
20. Walder, H., Platzner, M.: Online scheduling for block-partitioned reconfigurable devices. In: Design, Automation and Test in Europe Conference and Exhibition, pp. 290–295 (2003)

Chapter 9
Dynamic Partial Reconfiguration by Means of Algorithmic Skeletons—A Case Study

Norma Montealegre and Franz J. Rammig

Abstract A digital hardware system which implementation does not fit in a FPGA device can be placed into the FPGA. The system with inherent parallelism must be partitioned into hardware modules to be executed in different time slots using partial reconfiguration. That is the so called temporal partitioning and temporal placement. The partitioning of the system can be done using standard patterns used in parallel systems. Algorithmic skeletons are common parallelization patterns which encapsulate parallelism, communication and synchronization. They help to avoid concentrating in unnecessary details about the underlying implementation of parallelism. Algorithmic skeletons seem to be promising as a methodology for the design of partial reconfigurable systems. In this chapter, the design of a speech recognition front-end is described to show the feasibility of using algorithmic skeletons in the design of reconfigurable systems. A speech recognition front-end is a digital signal processing device used to transform an audio signal into feature vectors used for Automatic Speech Recognition or storage of semantic audio information. This device does not fit in the fabric of the FPGA of the used development board. After it was redesigned using a developed library of algorithmic skeletons, the use of dynamic partial reconfiguration has made possible to fit the device into the FPGA. This chapter demonstrates how algorithmic skeletons allow to simplify and to speed up the development of partial reconfigurable systems.

9.1 Introduction

Algorithmic Skeletons can be seen as meta-functions or *high order functions*. They can be used in applications that already have a parallel solution or are adequate to be

Norma Montealegre · Franz J. Rammig
Heinz Nixdorf Institute, Paderborn, Germany, e-mails: `norma@uni-paderborn.de`, `franz@uni-paderborn.de`

M. Platzner et al. (eds.), *Dynamically Reconfigurable Systems*,
DOI 10.1007/978-90-481-3485-4_9, © Springer Science+Business Media B.V. 2010

parallelized. Algorithmic Skeletons were coined in the mid 1980s by Murray Cole in his Dissertation [3]. Afterwards, other publications followed like [5, 2, 16, 17, 4, 1].

Algorithmic skeletons can be implemented and collected into a*library*, which can be used by the system developer or programmer. Algorithmic skeletons were already employed in parallel computers to execute parallel algorithms. FPGAs can also execute parallel algorithms. Therefore, applications implemented in FPGAs can be described by means of algorithmic skeletons too.

Algorithmic skeletons offer a structured means in parallel computation, therefore, they can be helpful for partitioning a system into independent parts which can be executed in different time slots and be placed as a dynamic module in a FPGA. This procedure needs partial reconfiguration support by the FPGA and a *communication infrastructure* among parts of the partitioned system. A library of algorithmic skeletons for FPGAs should consider to implement this infrastructure so that the programmer concentrates on the algorithms and gives some hints of the structure, but needs not to spend much effort in how the algorithm will be executed in the FPGA.

The paradigm of *stream parallelism*, where a stream of data is waiting to be processed, is taken as input in two algorithmic skeletons: pipe and farm, see Fig. 9.1. Signal processing applications have also streams of data as input. Such applications can be taken as examples to prove the usefulness of algorithmic skeletons in the design of reconfigurable systems. A *Channel Vocoder Analyzer* is a filter-bank used as a front-end for automatic speech recognition. A very detailed implementation of this device can be found in [14]. It can be described in terms of a farm of pipes. A farm of pipes is a so called *high order skeleton* because a pipe is passed as argument to a farm skeleton. In consequence, it is an adequate example to prove the feasibility of applying algorithmic skeletons in the development of complex systems in FPGAs. With the implementation of an application using algorithmic skeletons to support dynamic partial reconfiguration is intended to prove the feasibility of the methodology exposed by Florian Dittmann [7, 6] and to go a step further in the implementation started by Stephan Frank [11, 9].

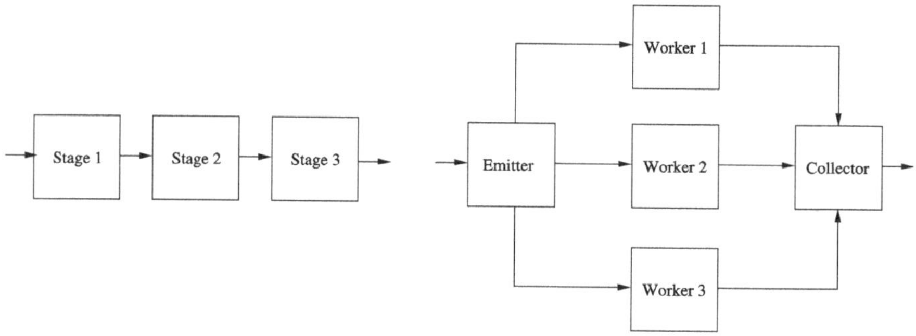

Fig. 9.1 *Pipe* and *farm* skeleton types.

In the next section, an overview of the overall system is presented. It deals with how to use algorithmic skeletons in dynamic reconfigurable systems. It follows a description of a implemented library. Afterwards, the implementation of the Channel Vocoder Analyzer on a FPGA will be presented as a partial reconfigurable system implemented with algorithmic skeletons. This implementation solves the problem of reduced resources available in the FPGA for such a complex application.

9.2 Overview of the Overall System

For the implementation of a partial reconfigurable system, the FPGA is divided into a *base region* and *partial reconfigurable regions* (PRRs)[19], see Fig. 9.2. The base region contains logic which does not change during partial reconfiguration. This logic can include modules which interact with peripheral devices or modules which control the partial reconfiguration. Partial reconfigurable regions contain logic which can be reconfigured independently from the base region or other reconfigurable regions. Each partial reconfigurable region has at least one, but usually multiple partially reconfigurable modules, which can be loaded into the reconfigurable region. The shape, size and placement of each partial reconfigurable region can be defined through constraints. Those regions are instantiated as components at the top module of the design.

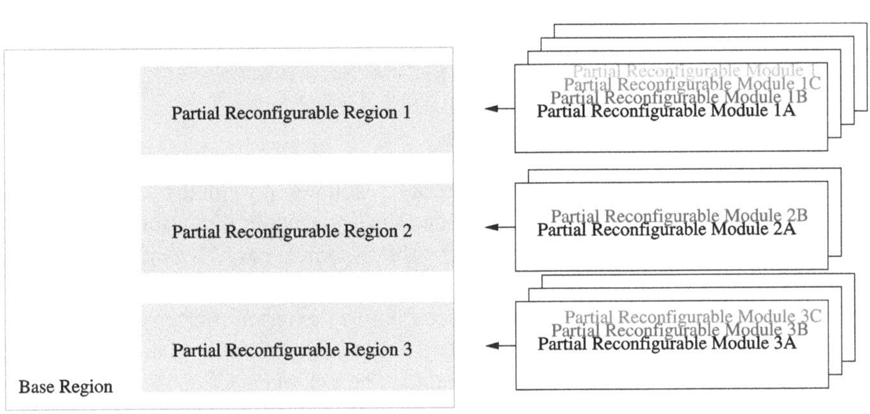

Fig. 9.2 *Base region* and *partial reconfigurable regions* in a partial reconfigurable FPGA.

Partial reconfigurable modules to be reconfigured in a determined partial reconfigurable region can constitute the modules of a skeleton. This schema implies the existence of some *intermediate memory* so that the modules can communicate and transfer information. Modules take data to process from memory and leave results in memory too. In this way, modules which are reconfigured can use the results of prior modules. Such memory can either be internal or external to the reconfigurable device.

If the amount of data transfer among partial reconfigurable modules is small, the intermediate memory can be implemented inside the FPGA and placed in the base region, see Fig. 9.3. In this case *internal memory* is organized for each skeleton to enable their modules to communicate each other. This approach has been taken in [11] and a pattern database has been developed. Once a pattern is elected from the database, the wires among reconfigurable regions are placed by the tool PaReTo-FAS after the user has given a set of constraints. This wires constitute the infrastructure for the communication among skeleton modules and intermediate memory. The communication protocol should be implemented for each skeleton. For an example of such an implementation, please see [11].

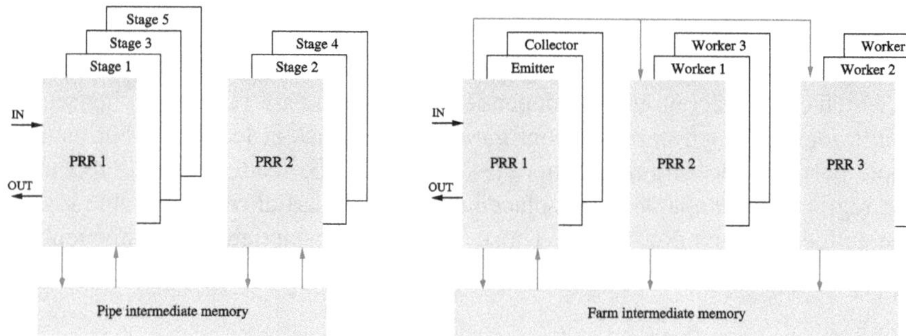

Fig. 9.3 Skeletons implemented with *intermediate memory inside the FPGA*. A pipe implementation (left) and a farm implementation (right) [11].

Many practical real world applications require a huge amount of information to be processed. It means that internal memory resources provided by the reconfigurable device could be insufficient for needs of intermediate memory. Such a request can be solved with the help of *external Random Access Memory* (RAM), see Fig. 9.4. In this case, a *memory controller* is required. It can be placed in the base region of the FPGA, acting as an interface between external memory and the system. DDR SDRAM and SRAM memory are found in prototyping boards, then, their respective controllers should be implemented. The Erlangen Slot Machine [15] dispose of external banks of SRAM memory, which are directly connected to the FPGA in its Baby Board. This SRAM memory could be used to implement the intermediate memory, among partial reconfigurable modules implemented as slots in the Virtex II 6000 Xilinx FPGA available in the board.

In Fig. 9.4, inputs and outputs from peripherals have respective controllers, which leave data to be processed in external memory. Emitter and collector from a farm skeleton, take and leave data from external memory. This two last considerations in the implementation of algorithmic skeletons help to construct a *generalized skeleton module*. In the case of a farm skeleton: emitter, workers and collector, take and leave data from external memory. They do not pass data directly each other. Modules of

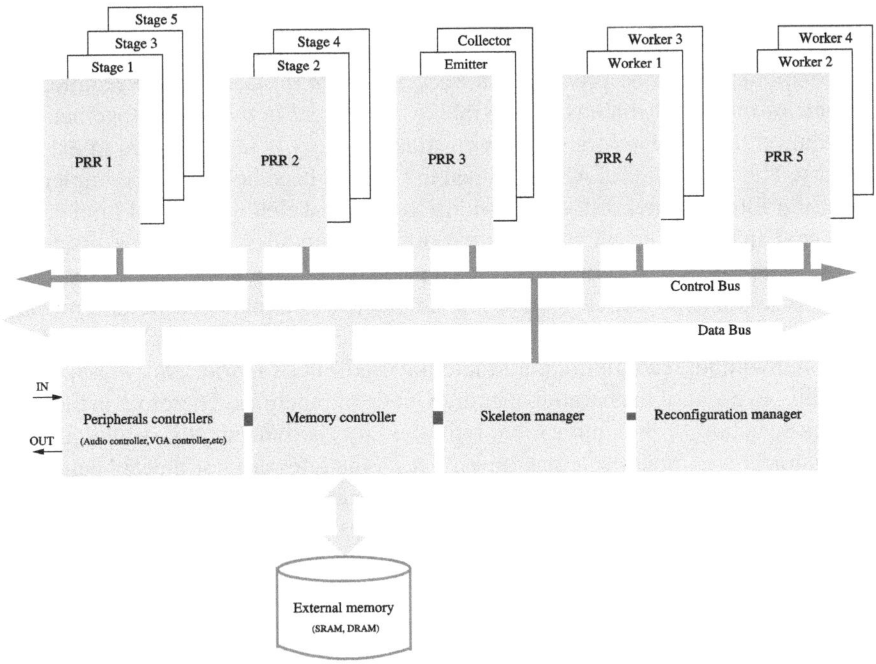

Fig. 9.4 Skeletons implemented with *intermediate memory outside the FPGA.*

a skeleton take data from and leave data in external memory in the way determined by the behavior of the skeleton. The generalization of modules makes easier dealing with a bigger amount of skeletons or using high order skeletons.

The *communication protocol* and the *data flow* among modules of a skeleton are very important tasks. Therefore, a *skeleton manager* is needed for organizing the flow of data according to the behavior of the implemented skeleton, group of skeletons or high order skeleton. Besides, it is important to schedule the reconfiguration of partial reconfigurable modules into the partial reconfigurable regions according to the requirement of the implemented skeletons. Because this task is skeleton dependent, it should also be carried on by the skeleton manager. Therefore, the reconfiguration manager deals only with the scheduling of the reconfiguration port [8] and the reconfiguration process itself.

Communication lines in each skeleton module are provided for the implementation of the communication protocol. A determined logic can be provided which such lines through a wrapper or a template VHDL to be used in the implementation of the determined logic. Such control lines are connected to the skeleton manager through the *control bus*. Depending to the algorithmic skeleton the partial reconfigurable module belongs to, the skeleton manager will signal the memory controller to put the correct data in the *data bus* or to store the data present in the same bus into external memory. Once the processing of a partial reconfigurable module is finished or its assigned processing time is passed, the skeleton manager will send to the reconfiguration manager the signal of which partial reconfigurable region should be reconfigured with which partial reconfigurable module.

The application to be presented in Sect. 9.4 is *data intensive* and requires big amounts of memory, which is impossible to implement in the FPGA together with the required logic. Therefore, the intermediate memory is implemented in external memory. The tool PaReToFAS developed in [11] has been helpful in the implementation of a filter channel of the application as a pipe skeleton. The tool implements the shared intermediate memory for the different modules of the pipe inside the FPGA, which is suitable for small transfer data rates. Since the application deals with a large amount of speech data, it is demanded to implement the intermediate memory externally as recommended in [11]. On the other side, a single channel implemented without reconfiguration fit into the used FPGA fabric. This wakes up the possibility to profit of the parallel operation mode of pipelines. Therefore in the farm of pipes structure of the application, only the farm is dynamically reconfigurable. The channels are implemented as pipes, but its modules are not independently reconfigurable. The whole pipe is configured at a time.

In order to implement such a paradigm, a *script description template* which implements the structure of parallel modules with reconfigurable modules only at the outer skeleton is introduced. The design method is based on a *library of algorithmic skeletons* which has been programmed in Python. The library has to be used in a template program, which is also written in Python. This means that the program has to describe the global structure of the system in terms of the algorithmic skeletons available in the library. The template of such a program is very intuitive. The script generates the VHDL top module of the entire design, which is able to be synthesized and implemented in a FPGA using partial reconfiguration. The methodology followed by the scripts can also enhance the functionality of PaReToFAS in a future step.

9.3 Library of Algorithmic Skeletons

This section describes the implementation of a *library of algorithmic skeletons* to be useful for parallel applications which hardware descriptions are written in VHDL.

9.3.1 Algorithmic Skeletons for Defining the Structure of the Design

Each algorithmic skeleton is implemented as a *class*. Then, algorithmic skeletons can be represented in the memory of the program as *instances*. The collection of algorithmic skeletons implemented as classes constitutes a library. This library is implemented in Python. Python is a scripting language, therefore it is possible to use scripting constructions to describe the structure of a hardware system in terms of algorithmic skeletons. The designer, who is unfamiliar with Python, is provided with a template for each skeleton. In the template it is necessary to give just the names of the VHDL modules to be used.

The system designer implements his/her designs as VHDL modules and decides the type of skeleton or skeletons to be used. This will define the structure of the overall system. The designer enters the name of the VHDL modules and the type of skeleton into the template. Then, the implemented *code generator* produces the top VHDL module, which connects properly the components reflecting the desired algorithmic behavior.

Code listing 9.1 shows an example template for three VHDL modules which use a pipe skeleton. Line 1 calls the library of algorithmic skeletons. Line 2 assigns a VHDL module to each stage of the pipe. Line 5 instantiates the pipe skeleton. Line 6 generates the VHDL top module. Line 7 writes the top module into a file.

Code Listing 9.1 Template for the use of a *pipe skeleton*.

```
1 from algoskels import *
2 stages = [vhdl_entity('stage1.vhd'),
3           vhdl_entity('stage2.vhd'),
4           vhdl_entity('stage3.vhd')]
5 pipe = vhdl_pipe(stages)
6 top = vhdl_top_module('pipe_top',[pipe])
7 top.write()
```

Code listing 9.2 shows an example template for three VHDL modules which use a farm skeleton. Line 1 calls the library of algorithmic skeletons. Line 2 assigns a VHDL module to each worker of the farm. Line 5 instantiates the farm skeleton. Line 6 generates the VHDL top module. Line 7 writes the top module into a file.

Code Listing 9.2 Template for the use of a *farm skeleton*.

```
1 from algoskels import *
2 workers = [vhdl_entity('worker1.vhd'),
3            vhdl_entity('worker2.vhd'),
4            vhdl_entity('worker3.vhd')]
5 farm = vhdl_farm(workers)
6 top = vhdl_top_module('farm_top',[farm])
7 top.write()
```

9.3.2 Algoskels

The developed library of algorithmic skeletons is called *algoskels.py*. Initially, it implements two types of algorithmic skeletons: pipe and farm. It consists of the following 4 classes:

- vhdl_entity—this class contains methods for lexical, syntax and semantic analysis of the input VHDL files.
- vhdl_pipe—this class includes all methods required for the Pipe skeleton.
- vhdl_farm—this class includes all methods required for the Farm skeleton.
- vhdl_top_module—this class contains functions responsible for the automatic generation of the VHDL top module.

The first task of the library is to analyze the given VHDL files. It extracts a list of the port names, their directions, and their types. This information is used to bind the VHDL components in the top module in the way, defined by the algorithmic skeleton these VHDL components belong to.

The library allows to implement *high order skeletons*. A high order skeleton is a skeleton which is composed of elements described also in terms of some skeleton. A skeleton which is composed of some other skeletons must be instantiated only after all skeletons it includes have been instantiated before. That is because any compound skeleton uses VHDL files which must be previously generated. There is no limit on how many compound skeletons are used.

For extending the library, it is necessary to program a class for a further skeleton. Such class can be instantiated, in the same way as the pipe or farm classes. Further examples of skeletons and their possible applications please refer [14, 3, 2, 5]

9.3.3 Algorithmic Skeletons for Partial Reconfigurable Systems

In the case of a farm skeleton, implemented with external intermediate memory, emitter and collector can be placed in the base region and workers can be organized in partial reconfigurable regions. Its top level may have at least two component instances, one for the base region and one for the partial reconfigurable region interconnected through bus macros.

Code Listing 9.3 Representation of the *analyzer* using a *farm skeleton*.

```
from algoskels import *
workers = [vhdl_entity('base_region.vhd'),
           vhdl_entity('pr_region.vhd')]
farm_PR = vhdl_farm(workers, mode='PR')
top = vhdl_top_module('farm_PR_top',[farm_PR])
top.write()
```

Code listing 9.3 shows a template script which produces the top module of a partial reconfigurable farm skeleton. The partial reconfigurable region may contain the workers as reconfigurable modules. The output values of each worker can be obtained and be given to the modules implemented in the base region before this partial reconfigurable module will be replaced by the next one. The top module of this skeleton has two component instances called *base_region* and *pr_region*. Thus, all workers of the full farm skeleton operate in sequence, one by one, at different time.

Adequate inter-module communication via *bus macros* and a proper placement of the modules must be established too. Therefore, the generated top module includes a template for the constraints: AREA_ GROUP, AREA_GROUP RANGE, MODE and LOCs. The constraints themselves have to be configured manually since they are application and FPGA dependent.

9.4 Application Scenario: Channel Vocoder Analyzer

The *Automatic Speech Recognition* process basically consists of the functional stages listed below [12].

1. Front-end analysis
2. Pattern matching
3. Decision making

The *Channel Vocoder Analyzer* represents the front-end analysis stage in the Automatic Speech Recognition. It processes the input speech signal and constantly produces feature vectors—also named spectral vectors-, which are to be processed by the Pattern Matching stage. A method for implementing the front-end analysis stage is a bank of filters. The sound signal is processed independently by each filtering channel. Each channel has a band-pass FIR filter, a rectifier, a low-pass anti-aliasing FIR filter, and a bit rate reducer also called decimator. The full set of filters measures the short-term spectrum of the audio signal. The so called *spectral vectors* are produced at the output of the Channel Vocoder Analyzer. Each component of every spectral vector represents the magnitude of the signal—the square root of the power of the signal—within the corresponding frequency band. The frequency band of each band-pass filter is determined by the Bark scale [13], which perceptual division of the sound frequency band gives narrower band-pass frequencies to lower frequencies, see Fig. 9.5.

Every channel of the analyzer has as modules a band-pass filter, rectifier, low-pass filter and decimator. They are applied to the samples as a composed function, which can be structured like a pipe skeleton. The structure of the entire analyzer with 16 channels can be represented as a *farm of pipes*. Code listing 9.4 shows the scripting template corresponding to the analyzer. Line 2 creates a list of pipes. Line 3 gives the number of channels to be used. Line 4 and 8 instantiate automatically the 16 channels. It is to be noticed that the band-pass filters of the channels are different. Therefore, the workers of the farm skeleton implement different functions. With line 9 to 11 the top module of the bank of filters is created. The underlying methods of the library produce proper interconnections for every skeleton instance only if the skeletons are instantiated in the bottom-up order. It means, first the pipe skeletons are instantiated and then the farm skeleton.

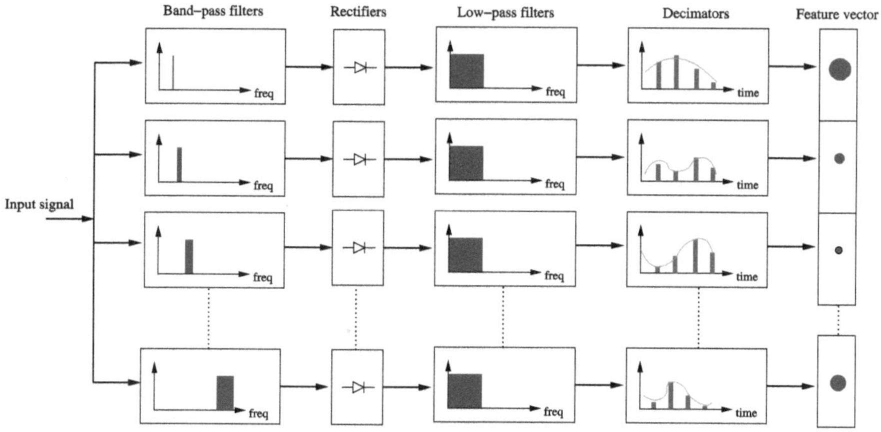

Fig. 9.5 Channel vocoder analyzer implemented with a *bank of filters*.

Code Listing 9.4 Channel vocoder analyzer described as a *farm of pipes*.

```
from algoskels import *
pipes = []
NUM_OF_CHANNELS = 16
for i in range(NUM_OF_CHANNELS):
  bpname = 'bandpass'+i+'.vhd'
  pipe = vhdl_pipe(bpname,'rect.vhd','lowpass.vhd','dec.vhd')
  pipe_top = vhdl_top_module('pipe_top'+i+'.vhd',[pipe])
  top.write()
  workers.append( vhdl_entity('pipe_top'+i+'.vhd') )
farm_of_pipes = vhdl_farm(workers)
top = vhdl_top_module('farm_of_pipes_top',[farm])
top.write()
```

The Channel Vocoder Analyzer has been implemented in the *Xilinx Virtex-4 FX12 FPGA* installed in the *ML403 prototyping board* developed by AVNET [20]. This board offers a great variety of useful peripheral devices like a DDR SDRAM memory, a SRAM memory, an audio chip-set, a video digital to analog converter, and a flash card reader among others. The Erlangen Slot Machine is another option as target platform, please see Chap. 3.

It has been decided to have as input a n seconds wide audio signal taken from a microphone and to visualize the computed feature vectors on a display. For this objective, some of the on-board devices needed to be used. In order to communicate with these on-board devices, additional hardware modules had to be designed and implemented in the same FPGA. The prototype receives the input sound from a microphone and produces a visual representation of the feature vectors on the attached VGA video display, see Fig. 9.6.

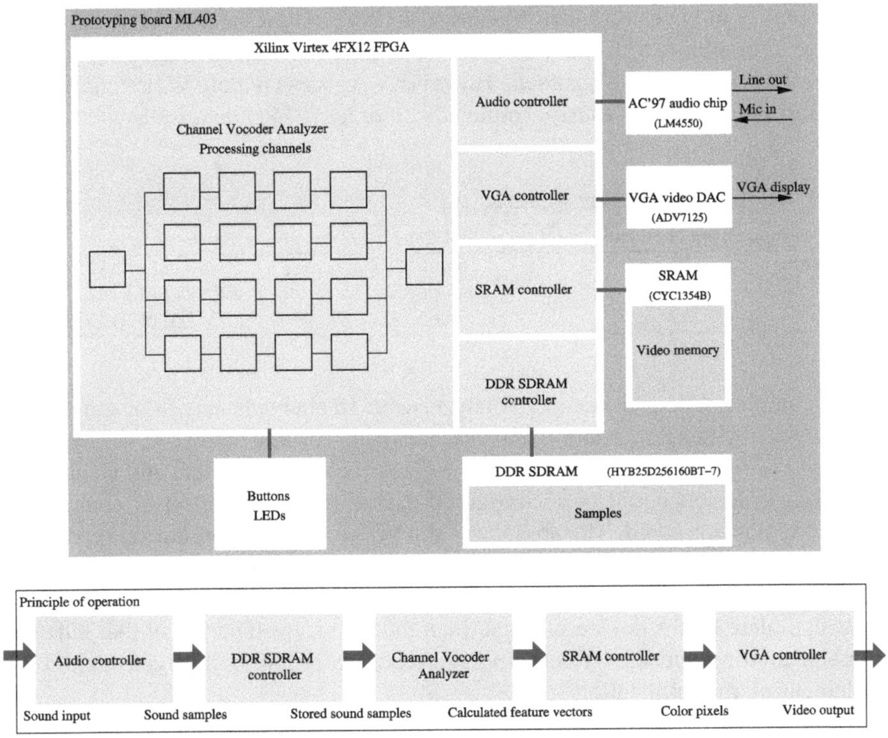

Fig. 9.6 The *prototype* implementation of the channel vocoder analyzer (above) and its *principle of operation* (below).

The input sound stream from the external microphone is processed by the on-board audio chip AC'97. The operation of this device is controlled by an *audio controller* which receives sound samples and sends them to the *DDR SDRAM controller*. After a sequence of sound samples has been stored in the external on-board DDR SDRAM memory, the processing channels of the analyzer start to operate. The DDR SDRAM controller reads every sample from the stored sequence in the same order as received from the audio controller and sends it to the inputs of the channels of the analyzer. The processing channels of the analyzer process the sound samples and produce the feature vectors. These values are stored in the on-board SRAM memory. This on-board memory is controlled by the *SRAM controller*. The SRAM memory is read constantly by the *VGA controller*, which controls the external on-board VGA digital to analog converter device, connected to the VGA video display.

After the complete design had been developed, the synthesized netlist of the complete prototype *did not fit into the Virtex-4 FX12 FPGA device*. In order to analyze the need of resources for different design parameters, the number of processing channels of the analyzer were varied. Respective results of the hardware synthesis are presented in Table 9.1. The number of required slices is shown in percent relative to the number of slices provided by Virtex-4 FX12. These results show that there would be a possibility to fit the entire design in the FPGA fabric if the analyzer has no more than 3 processing channels. However, to process feature vectors according to the Bark scale requires to have implemented *at least 16 channels*.

Table 9.1 Results of the hardware synthesis.

Number of processing channels	DSP48s	Slices
16 (required)	16 out of 32	30444 out of 5472 (556%)
4	4 out of 32	6969 out of 5472 (127%)
3	3 out of 32	5337 out of 5472 (97%)

The complete Channel Vocoder Analyzer with 16 channels may fit in some other FPGA device with a larger fabric. An alternative approach would be to use *dynamic reconfiguration of the entire FPGA device*. That means, controllers and a subset of processing channels could be reconfigured each time a new subset of channels is required to be configured. The absence of the SDRAM controller during reconfiguration can produce that all data contained in the external SDRAM memory be lost each time the total reconfiguration happens. It is because the reconfiguration time of the complete FPGA device is longer than the auto refresh period of the SDRAM memory. In consequence, dynamic *partial* reconfiguration is the right method for implementing this application.

Our intention is to study *dynamic partial reconfiguration*. This means that at a point of time only a subset of processing channels are loaded in the configuration memory of the FPGA. After they processed a determined number of sound samples, a next subset of channels is loaded and so on. This methodology makes profit of the inherent parallelism of the design. It has been decided to implement the Audio, VGA, SRAM and DDR SDRAM controllers as static modules placed in the base region. All single processing channels of the analyzer are implemented as partial reconfigurable modules placed in partial reconfigurable regions. The number of reconfigurable regions could be 3, as the results of the hardware synthesis show. However, the reconfiguration manager needs also resources in the FPGA, therefore it is possible to have only 1 or 2 partial reconfigurable regions.

Code listing 9.5 shows the implementation of a reconfigurable farm of pipes. In lines 2 to 8, the 16 pipes are created automatically. This modules can be taken as partial reconfigurable modules which can be used for producing partial bit-streams. Lines 9 to 13 produce the top module required to implement a design with a base region and one partial reconfigurable region. In the partial reconfigurable region the 16 created pipe modules will be reconfigured, one at a time slot. The base region will place memory, audio and VGA controllers and the reconfiguration manager. Figure 9.7, shows the placement of hardware in this two regions.

According to the *modular partial reconfiguration design flow* from Xilinx [19], the area for each reconfigurable region must be defined. Almost the whole area of the *right part of the fabric of the FPGA* is allocated for the reconfigurable region, see code listing 9.6.

Code Listing 9.5 Channel vocoder analyzer described as a *partial reconfigurable farm of pipes*.

```
from algoskels import *
pipes = []
NUM_OF_CHANNELS = 16
for i in range(NUM_OF_CHANNELS):
    bpname = 'bandpass'+i+'.vhd'
    pipe = vhdl_pipe(bpname, 'rect.vhd', 'lowpass.vhd', 'dec.vhd')
    pipe_top = vhdl_top_module('pipe_top'+i+'.vhd',[pipe])
    top.write()
workers = [vhdl_entity('base_region.vhd'),
           vhdl_entity('pr_region.vhd')]
farm_PR = vhdl_farm(workers, mode='PR')
top = vhdl_top_module('farm_PR_top', [farm_PR])
top.write()
```

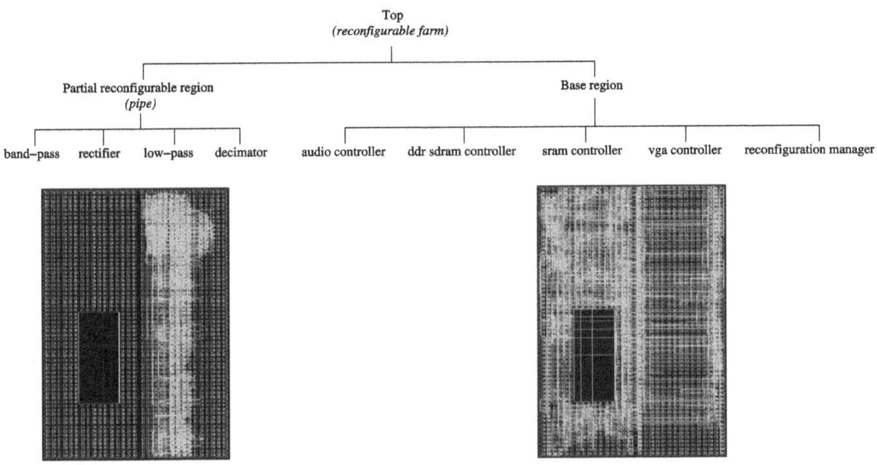

Fig. 9.7 *Partial reconfigurable region* (left) and the *base region* (right) of the *reconfigurable farm of pipes*. The circuit implements the dynamic reconfigurable design of the channel vocoder analyzer. The black area corresponds to a PowerPC processor not used in this application.

The reconfigurable region contains exactly one processing channel described as a pipe skeleton per time period. The average amount of data to be processed for a n seconds speech sampled at a frequency of 8000 $\frac{\text{samples}}{\text{second}}$ is n∗8000 samples. After the samples are processed, produced feature vectors are composed of 16 components. Therefore, in the worst case of no decimation present, the amount of data at the output of the bank of filters of 16 channels is $n * 8000 * 16$ two byte data, because each sample is a 16 bit one. In consequence, a bus macro must supply a 16 bit signal for the incoming sample values and 16 bit signal for the outgoing spectral values.

Code listing 9.7, shows the instantiation of eight bit bus macros. As the samples are composed of 16 bits, two 8 bits bus macros are instantiated. The right to left bus macros are responsible to transfer samples from the DDR SDRAM con-

Code Listing 9.6 Constraints for the *placement* of the partial reconfigurable region.

```
#components placement
INST "base_inst" AREA_GROUP=AG_static_1;
INST "reconfigurable_inst" AREA_GROUP=AG_tile_a;
#constrains for the partial reconfigurable region
AREA_GROUP "AG_tile_a" RANGE=SLICE_X26Y0:SLICE_X45Y125;
AREA_GROUP "AG_tile_a" RANGE=FIFO16_X2Y0:FIFO16_X2Y15;
AREA_GROUP "AG_tile_a" RANGE=RAMB16_X2Y0:RAMB16_X2Y15;
AREA_GROUP "AG_tile_a" RANGE=DSP48_X0Y0:DSP48_X0Y31;
AREA_GROUP "AG_tile_a" MODE=RECONFIG;
```

Code Listing 9.7 Bus macros.

```
#right to left bus macros (from base to PR region)
INST "Y_channels_to_static_1" LOC = SLICE_X24Y116;
INST "Y_channels_to_static_2" LOC = SLICE_X24Y114;
#left to right bus macros (from PR region to base)
INST "X_static_to_channels_1" LOC = SLICE_X24Y122;
INST "X_static_to_channels_2" LOC = SLICE_X24Y120;
INST "clk_8KHz_static_to_channels" LOC = SLICE_X24Y118;
```

troller in the base region to the processing channel in the partial reconfigurable region. In the VHDL top design description, the output signal *X* of the component instance *base_inst* is connected to the bus macros *X_static_to_channels_1* and *X_static_to_channels_2*. The outputs of these bus macros are connected directly to the inputs of the component *reconfigurable_inst*. The left to right bus macros are responsible to transfer feature vectors from the processing channel of the partial reconfigurable region to the SRAM controller in the base region. In the VHDL top design description, the output signal *Y* of the component *reconfigurable_inst* is connected to the bus macros *Y_channels_to_static_1* and *Y_channels_to_static_2*. The outputs of these bus macros are connected directly to the inputs of the component *base_inst*. Synchronization is achieved connecting clk_8KHz_static_to_channels to the clock inputs.

Once the system has been structured and we have the VHDL and constraint files, the partial reconfiguration design flow of Xilinx is applied in order to produce the bit-streams. It is possible to generate several partial bit-streams in parallel on several remote computers using the facility provided by Part-E [18].

The FPGA should trigger the reconfiguration of the next processing channel by itself, after the present channel finishes processing. This task is executed by the reconfiguration manager, which has been implemented as a hardware module placed on the base region of the FPGA. The reconfiguration manager loads the configuration memory of the FPGA with the required bit-stream at the required time slot. All configuration bit-streams are stored on the attached flash memory card. The implemented reconfiguration manager is controlled by the signal *start*. If this signal is asserted, the reconfiguration manager communicates with the external flash controller of the prototyping board. It starts the reconfiguration process taking as a bit-stream the one indicated by a counter which is incremented each time a new bit-stream is loaded.

The reconfiguration time of a channel is 0.5 seconds. The processing of all 16 channels, for a 2 seconds speech data stream has been of approximately 30 seconds.

9.5 Conclusion

It has been shown that the methodology for implementing dynamic partial reconfiguration by means of algorithmic skeletons is feasible. The methodology demonstrated to be helpful in structuring the design. The design used as demonstrator, has only one partial reconfigurable region. Whenever the FPGA area size allows to implement more than one partial reconfigurable region and many skeletons are implemented in the same FPGA, a skeleton manager is required. The skeleton manager would have the task of implementing the logic for communication among modules, to control the access to the intermediate memory in the case of the use external memory and to schedule reconfiguration of skeleton modules when required. In this way the reconfiguration manager could only concentrate in managing the reconfiguration port of the FPGA, maybe implementing interesting scheduling strategies like preemption during reconfiguration [10].

This application considered to implement the reconfiguration manager as a hardware module inside the FPGA. However some other prototyping boards, like the Erlangen Slot Machine [15], offer an external reconfiguration manager implemented in an extra Spartan 2E FPGA, see Sect. 3.5. The design methodology can be easily transferred to the Erlangen Slot Machine. For this purpose, the partial reconfigurable modules corresponding to the modules of an algorithmic skeleton can easily be implemented on the FPGA, provided the Virtex II FPGA of the ESM is divided into slots. Therefore, each slot can represent a module and the partial reconfiguration can be slot-based. Nevertheless, the hardware modules for driving audio and video should still be implemented on the FPGA. SRAM memory blocks and the DRAM socket, which is intended to provide as much RAM as needed for the tasks running on the main FPGA, could be used as intermediate memory for the skeleton modules, in our application, for storing audio stream data and calculated feature vectors.

Acknowledgements This work is supported by the German Research Fundation under grant *RA 612/9-3*, Project "Temporal Partitioning and Temporal Placement in the Design of Reconfigurable Computing Systems", part of the Priority Programm 1148 "Reconfigurable Computing Systems".

References

1. Caarls, W., Jonker, P.P., Corporaal, H.: Algorithmic skeletons for stream programming in embedded heterogeneous parallel image processing applications. In: IEEE International Parallel and Distributed Processing Symposium (2006)

2. Campbell, D.K.G.: Towards the classification of algorithmic skeletons. Technical report. University of York (1996)
3. Cole, M.: Algorithmic skeletons: structured management of parallel computation. PhD thesis, University of Glasgow (1989)
4. Cole, M.: Bringing skeletons out of the closet: a pragmatic manifesto for skeletal parallel programming. Parall. Comput. **30**, 389–406 (2004)
5. Darlington, J., Field, A.J., Harrison, P.G., Kelly, P.H.J., Sharp, D.W.N., Wu, Q., While, R.L.: Parallel programming using skeleton functions. Conference on Parallel Architectures and Languages (1993)
6. Dittmann, F.: Algorithmic skeletons for the programming of reconfigurable systems. In: Software Technologies for Embedded and Ubiquitous Systems. Springer, Berlin (2007)
7. Dittmann, F.: Methods to exploit reconfigurable fabrics. Making reconfigurable systems mature. PhD thesis, University of Paderborn (2007)
8. Dittmann, F., Frank, S.: Hard real-time reconfiguration port scheduling. In: Design, Automation, and Test in Europe. EDA Consortium (2007)
9. Dittmann, F., Frank, S., Oberthuer, S.: Algorithmic skeletons for the design of partially reconfigurable systems. In: 15th Reconfigurable Architectures Workshop, IEEE (2008).
10. Dittmann, F., Weber, E., Montealegre, N.: Implementation of the reconfiguration port scheduling on the Rrlangen slot machine. In: International Symposium on Field-Programmable Gate Arrays, ACM/SIGDA (2009)
11. Frank, S.: Einsatz algorithmischer Skelette zur Generierung partiell dynamisch rekonfigurierbarer Schaltungen. Diploma thesis. University of Paderborn (2007)
12. Furui, S.: Digital Speech Processing, Synthesis, and Recognition. Dekker, New York (2001)
13. Gold, B., Morgan, N.: Speech and Audio Signal Processing. Processing and Perception of Speech and Music. Wiley, New York (2000)
14. Kataev, V.: Application scenarios for algorithmic skeletons in the design of partial reconfigurable systems. Diploma thesis. University of Paderborn (2009)
15. Majer, M., Teich, J., Ahmadinia, A., Bobda, C.: The Erlangen slot machine: a dynamically reconfigurable fpga-based computer. J. VLSI Signal Process. Syst. **46**(2), 15–31 (2007)
16. Pelagatti, S.: Structured Development of Parallel Programs. Taylor & Francis, London (1997)
17. Rabhi, A.F., Gorlatch, S.: Patterns and Skeletons for Parallel and Distributed Computing. Springer, Berlin (2003)
18. Weber, E., Dittmann, F., Montealegre, N.: Part-E—A tool for reconfigurable system design. In: International Conference on Reconfigurable Computing and FPGAs, IEEE (2008)
19. Xilinx: Early Access Partial Reconfiguration User Guide. Xilinx Inc. (2006)
20. Xilinx: ML401/ML402/ML403 Evaluation Platform. User Guide. Xilinx Inc. (2006)

Chapter 10
ReCoNodes—Optimization Methods for Module Scheduling and Placement on Reconfigurable Hardware Devices

Ali Ahmadinia, Josef Angermeier, Sándor P. Fekete, Tom Kamphans, Dirk Koch, Mateusz Majer, Nils Schweer, Jürgen Teich, Christopher Tessars, and Jan C. van der Veen

Abstract Placement and scheduling are recognized as the most important problems when exploiting the benefit of partially reconfigurable devices such as FPGAs. For example, dynamically loading and unloading modules onto an FPGA causes fragmentation, and—in turn—may decrease performance. To counteract this effect, we use methods from algorithmics and mathematical optimization to increase the performance and present algorithms for placing, scheduling, and defragmenting modules on FPGAs. Taking communication between modules into account, we further present strategies to minimize communication overhead. Finally, we consider scheduling module requests with time-varying resource demands.

10.1 Introduction

The increasing performance of reconfigurable hardware devices leads to numerous interesting applications; many examples such as video acceleration for driving assistance (Chap. 18) can be found in other chapters of this book. On the other hand, while systems that involve such devices become more and more complex, there is a growing need for powerful design methods and tools. In particular, the available hardware resources should be used efficiently to satisfy the demands required. For example, modules loaded onto an FPGA should be packed densely without wasting much of the FPGAs area, leaving as much space as possible for further tasks.

Ali Ahmadinia · Josef Angermeier · Dirk Koch · Mateusz Majer · Jürgen Teich
Hardware/Software Co-Design, Department of Computer Science, University of Erlangen-Nuremberg, Erlangen, Germany, e-mails: A.Ahmadinia@ed.ac.uk, angermeier@cs.fau.de, dirk.koch@cs.fau.de, teich@cs.fau.de

Sándor P. Fekete · Tom Kamphans · Nils Schweer · Christopher Tessars · Jan C. van der Veen
Department of Computer Science, Braunschweig University of Technology, Braunschweig, Germany, e-mails: s.fekete@tu-bs.de, tom@kamphans.de, n.schweer@tu-bs.de, j.van-der-veen@tu-bs.de

M. Platzner et al. (eds.), *Dynamically Reconfigurable Systems*,
DOI 10.1007/978-90-481-3485-4_10,

The main idea for our research was to combine expert knowledge from the computer engineering and algorithms/mathematics, to provide optimization methods for the design and operation of reconfigurable hardware devices, in particular, FPGA-based systems. It turned out that powerful methods from algorithm theory and mathematical optimization produce a significant increase in efficiency and performance. In this chapter, we summarize the results of a 6 year lasting, fruitful cooperation.

The rest of this chapter is organized as follows: Modern generations of FPGAs allow for partial reconfiguration. If the sequence of modules to be loaded is unknown beforehand, repeated insertion and deletion of modules leads to progressive fragmentation of the available space, making defragmentation an important issue. In Sect. 10.2,[1] we address this problem. We show that defragmenting an FPGA corresponds to solving a two-dimensional strip-packing problem. Problems of this type are NP-hard in the strong sense, and previous algorithmic results are rather limited. Based on a graph-theoretic characterization of feasible packings, we develop a method that can solve defragmentation instances of practical size to optimality. We also discuss a simple strategy for dealing with online scenarios, called *least-interference fit* (LIF) and compare LIF with the best offline solution.

In Sect. 10.3,[2] we discuss the problem of communication-aware module placement in array-like reconfigurable environments, such as the Erlangen Slot Machine (ESM; see Chap. 3). Bad placement of modules may degrade performance due to increased signal delays and wasted chip space for the reconfigurable multiple bus. We present ILP formulations that address both of these problems; both ILPs can be used stand-alone or as building blocks for more involved mathematical models.

In Sect. 10.4,[3] we propose a new method for defragmenting the module layout of a reconfigurable device, enabled by a novel approach for dealing with communication needs between relocatable modules and with inhomogeneities found in commonly used FPGAs. Our method is based on dynamic relocation of module positions during runtime, with only very little reconfiguration overhead—in contrast to Sect. 10.2 where defragmentation happens during idle times of the whole FPGA. The objective is to maximize the length of contiguous free space that is available for new modules. We describe a number of algorithmic aspects of good defragmentation, and present an optimization method based on tabu search. Experimental results indicate that we can improve the quality of module placement by roughly 50% over static layout. Among other benefits, this improvement avoids unnecessary rejection of modules.

Finally, we consider in Sect. 10.5 scheduling for modules whose resource requests (i.e., area on the FPGA) changes over time. We consider slot-based reconfigurations and assume that a modules requests a certain number of slots. So we

[1] A subset of the results of this paper appeared as a Distinguished Paper in the proceedings of ERSA'05 [45], the full version was published in TVLSI [22]. Portions reprinted, with permission, from Transactions on Very Large Scale Integration (VLSI) Systems, © [2008] IEEE.

[2] A proceedings version was presented at FPL 06 [19]. Portions reprinted, with permission, from Proc. of 16th Internat. Conf. on Field Programmable Logic and Applications, © [2006] IEEE.

[3] A proceedings version was presented at FPL 08 [21]. Portions reprinted, with permission, from Proc. of 18th Internat. Conf. on Field Programmable Logic and Applications, © [2008] IEEE.

are dealing with a two-dimensional strip-packing problem, where the width of the strip is given by the number of slots and the height of the strip corresponds to the time axis. However, in contrast to classical strip-packing solutions, we are allowed to postpone a module's requests for more space by pausing the execution of the module. We present an ILP and compare some heuristics for this problem.

10.2 Offline and Online Aspects of Defragmenting the Module Layout of a Partially Reconfigurable Device

One of the cutting-edge aspects of modern reconfigurable computing is the possibility of *partial* reconfiguration of a device: Ideally, a new module can be placed on a reconfigurable chip without interfering with other running tasks. Clearly, this approach has many advantages over a full reconfiguration of the whole chip. Predominantly it lessens the bottleneck of reconfigurable computing: reconfiguration time.

On the other hand, partial reconfiguration introduces a new complexity: management of the free space on the FPGA. In the 2D model this is an NP-hard optimization problem. There has been a considerable amount of work to solve this problem computationally. However, due to its computational complexity most recent work has focused on the online setting or on the 1D area model (see [41] for a recent survey).

Management of free space and scheduling of arriving tasks are the core components of an operating system for reconfigurable platforms (see Fig. 10.1). In all previous work these components use simple online strategies for the placement problem. The use of these strategies leads to fragmentation of the free space, as modules are placed on and removed from the chip area. This leads to situations where a new module has to be rejected by the placer because there is no free rectangle that could accommodate the new module, even though the total free space available would be more than sufficient (see [14] for further discussion).

We propose a different placer module. Instead of just relying on online strategies our placer has an additional offline component: the *defragmenter*. Recurring idle times can be used to optimally defragment the FPGA chip area.

This optimal defragmentation has two goals. One is to maximize the available contiguous free space. The other comes from the FPGA device we use. For example, the Xilinx Virtex-II series does not admit full two-dimensional partial reconfiguration. Instead, configuration can be performed only columnwise: While a column is reconfigured, all other modules that use this column have to be stopped, because reconfiguration interferes with the running tasks in a non-trivial way. So the other goal of the offline defragmenter is to free as many columns as possible. This way the next modules placed by an online placer will not interfere with other modules. In the following, we assume that the FPGA consists of W columns and H rows of *configurable logic blocks* (CLBs). Reconfiguration takes place on the column level, taking c units of time to configure one column of CLBs.

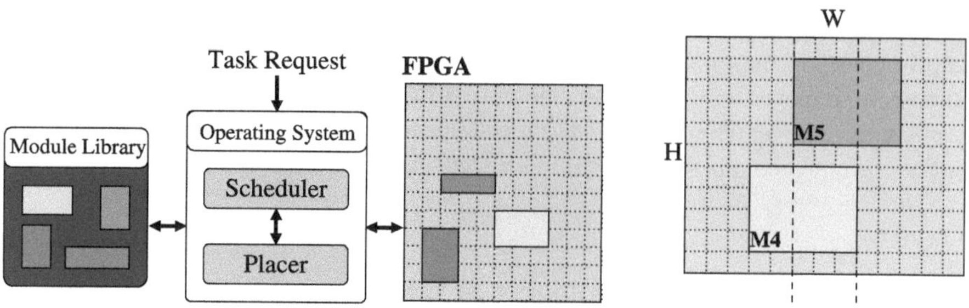

Fig. 10.1 Left: A schematic overview of an operating system for reconfigurable computers. Relocatable, presynthesized modules that are constrained to a rectangular layout are stored in a module library. As requests for tasks arrive, a module capable of running the task is selected, scheduled and eventually placed on the FPGA. Right: An FPGA of width $W = 13$ and height $H = 11$ CLBs. Assume that module M4 of width $w_4 = 5$ and height $h_4 = 4$ is located at position $(3, 1)$. If module M5 of same width and height is placed at position $(5, 6)$ the resulting overlap is 3 columns as indicated by the dashed lines. Consequently M4 is interrupted for $3c$ time units [22] © [2008] IEEE.

On this FPGA we execute a certain set of tasks $T = \{t_1, t_2, \ldots\}$. Each task is performed by a module j (i.e., a relocatable presynthesized digital circuit) that occupies a rectangular area of size (w_j, h_j) on the FPGA. As a consequence, placing module j on the FPGA takes time cw_j. The set of all modules is given by $M = \{m_1, m_2, \ldots\}$.

In an offline setting we simultaneously seek for a feasible schedule for the tasks (i.e., each task i is assigned a starting time s_i), a configuration schedule for the modules (each module j is assigned a reconfiguration time c_j), and a feasible placement of the modules on the FPGA (for each module j its location $x_i \in [0, W - w_i)$ and $y_i \in [0, H - h_i)$ has to be determined). Among all feasible solutions we select one that minimizes the *makespan*, i.e., the completion time of the last task. This alone is an NP-hard optimization problem, as it contains two-dimensional packing as a subproblem. At the same time, this problem is closely related to scheduling problems. (See [35] for an overview of classical one-dimensional scheduling problems.)

Columnwise reconfiguration has the drawback that reconfiguration of one column interrupts all modules using this column for the reconfiguration time c. Experimental evidence suggests that a placement strategy should take this interference into account. This suggests a simple heuristic called *least interference fit* (LIF): new modules are placed in consecutive columns that are used by as few other modules as possible. Just like other heuristics, LIF can be used in both offline and online scenarios. As we showed in [1], LIF seems to perform better than other simple heuristics based on bin packing.

Defragmentation as described above can be regarded as the two-dimensional strip packing problem. This, we take a closer look at this classic NP-complete optimization problem. As it turns out, for currently relevant numbers of modules, optimal placements can still be computed, using a cutting-edge algorithm for higher-dimensional packing.

10.2.1 Two-dimensional Strip Packing

The *Strip Packing Problem* (SPP) is to minimize the width W of a strip of fixed height H such that all rectangles fit into a rectangle of size $W \times H$. The corresponding decision problem is the *Orthogonal Packing Problem* (OPP); to decide whether a given set of rectangles can be placed within a given rectangle of size $W \times H$. Being a generalization of the one-dimensional problem 3-PARTITION, the OPP is NP-complete in the strict sense, and so the SPP is NP-hard [24].

Dealing with an NP-hard problem (often dubbed "intractable") does not mean that it is impossible to find provably optimal solutions. While the time for this task may be quite long in the worst case, a good understanding of the underlying mathematical structure may allow it to find an optimal solution in reasonable time. A good example of this type can be found in Grötschel [26], where the exact solution of a 120-city instance of the Traveling Salesman Problem is described, showing that the right mathematical tools and sufficient computing power may combine to explore search spaces of tremendous size.

Higher-dimensional packing problems have been considered by a great number of authors, but only few of them have dealt with the exact solution of general two-dimensional problems. See [16] for an overview. To our knowledge there is only one work that tries to solve SPP to optimality [38]. See also [23] for online packing.

In [16, 17, 20, 43], a different approach to characterizing feasible packings and constructing optimal solutions is described. A graph-theoretic characterization of the relative position of the boxes in a feasible packing (by so-called *packing classes*) is used, representing d-dimensional packings by a d-tuple of interval graphs (called *component graphs*); see Fig. 10.2. This factors out a great deal of symmetries between different feasible packings, it allows to make use of a number of elegant graph-theoretic tools, and it reduces the geometric problem to a purely combinatorial one without using brute-force methods like introducing an underlying coordinate grid. Combined with good heuristics for dismissing infeasible sets of boxes [15], a tree search for constructing feasible packings was developed. This

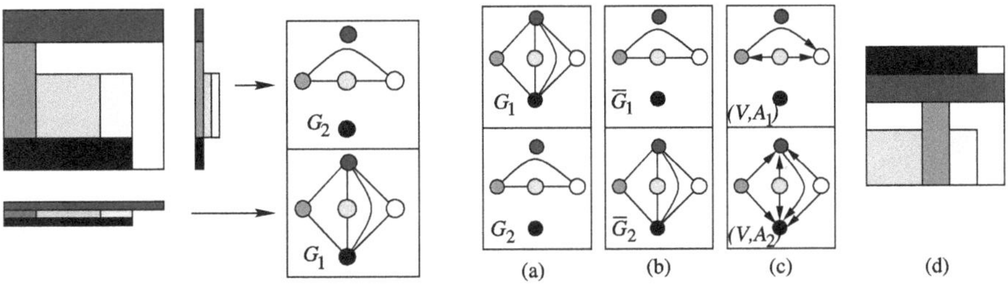

Fig. 10.2 Left: Projecting the boxes of a feasible packing in 2-dimensional space onto the coordinate axes converts the one 2-dimensional arrangement into 2 one-dimensional ones. These projections define interval graphs. Right: (a) A two-dimensional packing class. (b) The corresponding comparability graphs. (c) The transitive orientations. (d) A feasible packing corresponding to the orientation [22] © [2008] IEEE.

exact algorithm has been implemented; it outperforms previous methods by a clear margin.

10.2.2 Defragmentation Approach and Computational Results

We have used our implementation for the OPP as a building block for our defragmentation algorithm (see Algorithm 10.1). We have used simple lower and upper bounds to restrict the search. Suppose I denotes the set of indices of the modules currently present on the FPGA; then a lower bound for the number of columns W_L used by all modules after defragmentation is given by $W_L = \lceil (\sum_{i \in I} w_i h_i)/H \rceil$. The upper bound is computed as the minimum of the three shelf-packing heuristics next-fit-decreasing, first-fit-decreasing and best-fit-decreasing [5]. Based on these bounds the defragmentation algorithm performs a binary search until an optimal solution is found. We have benchmarked our code against a set of 10 instances, see Table 10.1. On an Intel Pentium IV clocked at 3 GHz, the running time for computing the optimum for these instances was less than 0.5 s for each scenario.

Algorithm 10.1 A binary search algorithm for determining an optimal module layout.

```
DefragmentModuleLayout()
LB := CalculateLowerBound()
UB := CalculateUpperBound()
while LB ≠ UB do
        W := LB + ⌊(LB+UB)/2⌋
        if SolveOPP(W) then
LB := W
        else
UB := W
```

10.2.3 Online Packing

A key motivation for considering partial reconfiguration is the necessity to develop fast and effective strategies for dealing with new and changing demands. In those cases, decisions have to be made that are based on incomplete information, as the system does not know what further requests will arise in the future.

In the context of dynamic partial reconfiguration, we can again describe module placement as a two-dimensional packing problem: Rectangular module requests arrive one at a time and need to be placed on the chip area.

One-dimensional online packing is well studied; for example the classical strategies FirstFit, BestFit, etc., can be used as online strategies [39]. Much less is known about two-dimensional online packing, and hardly anything for cost functions aris-

Table 10.1 Results for ten different scenarios, based on Fig. 10.3. The next columns show the number of placed modules, the total free space, the maximal free rectangle, the number of free columns before defragmentation, and after defragmentation. The last two columns show results for the online scenario, comparing LIF to the optimum. * without M11; ** including M11 (LIF fails to place M11) [22] © [2008] IEEE.

Scenario	$\|I\|$	Free space	Before defragmentation		After defragmentation		Online	
			Max. rectangle	Free columns	Max. rectangle	Free columns	LIF edges	OPT edges
A	11	30	7×1	0	2×11	2	12*	11* or 13**
B	9	52	2×8	0	4×11	4	9	9
C	9	70	3×7	0	6×11	6	6	6
D	9	42	4×4	0	3×11	3	10	9
E	6	83	6×8	0	6×11	6	4	3
F	6	54	8×2	0	4×11	4	4	3
G	5	76	6×4	2	6×11	6	1	1
H	6	53	3×11	3	4×11	7	3	2
I	5	87	9×6	1	7×11	7	1	1
J	6	42	3×8	0	3×11	3	4	3

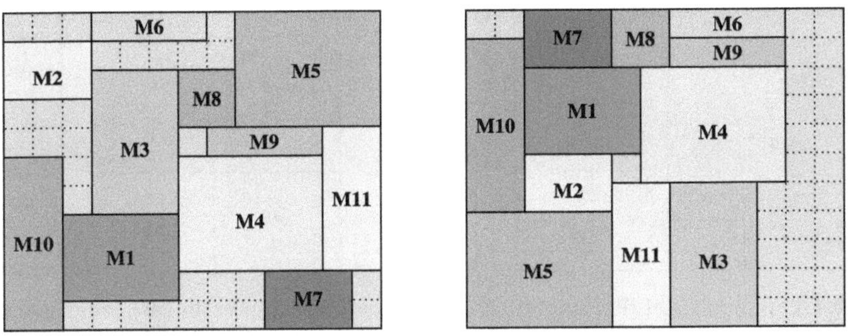

Fig. 10.3 Left: The FPGA before defragmentation. Even though the remaining free space is 30 reconfigurable units (CLBs), the maximal free rectangle of dimension 7×1 has only 7 CLBs. Note that there is no free column. Right: The same FPGA after defragmentation. The remaining free space is 30 CLBs. Now the maximal free rectangle is of dimension 2×11 has 22 CLBs. The number of free columns is 2 [22] © [2008] IEEE.

ing from column-wise reconfiguration as on Xilinx Virtex-II chips, where the reconfiguration cost arises by the interruption of existing modules that use one of the same columns as the newly placed modules. LIF seems to be a reasonable strategy for dealing with an online scenario. Before we argue that it is indeed a good strategy, let us note the following:

Theorem 10.1. *Consider a sequence of modules that is placed in some arbitrary permutation, resulting in an arrangement of modules. Then the total number of module interruptions is independent of the order in which the modules were placed* [22].

This means that we should aim at minimizing the total number of edges of G_1, which is precisely what LIF is trying to do in a greedy fashion.

Theorem 10.2. *Consider a set of modules, all having the same width $w_i = w$, that are to be placed on a chip area of sufficient height. Then LIF is a 1-competitive online algorithm, i.e., optimal at each step* [22].

Moreover, our approach allows it to compute an arrangement of modules that is optimal with respect to reconfiguration cost, thus making it possible to perform ex-post comparison of online strategies like LIF with an optimal solution:

Theorem 10.3. *The set of module placements that incurs minimal total configuration cost T correspond precisely to those feasible packings that have smallest possible number of edges in the interval graph G_1* [22].

As our OPP algorithm is already based on a tree search that makes use of the interval graphs, minimizing this edge number requires only a simple extension to our tree search. Using the same benchmark instances as before, we compared LIF with an optimal placement; see Table 10.1 and Fig. 10.4. It is clear to see that LIF is near-optimal for most instances.

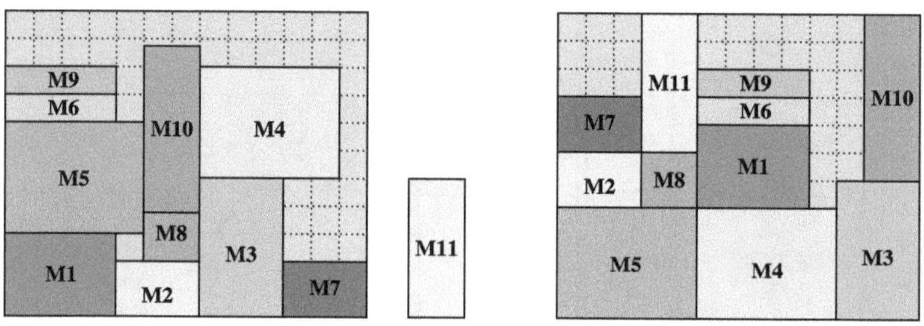

Fig. 10.4 Left: Applying LIF to scenario A results in insufficient space for the last module. Right: An optimal solution for scenario A: a feasible placement for all modules with minimum total overlap [22] © [2008] IEEE.

10.3 Minimizing Communication Cost for Reconfigurable Slot Modules

When trying to fully exploit the enormous practical potential of dynamically reconfigurable devices such as FPGAs, a crucial issue is intermodule communication, especially in the context of module placement. Existing techniques for FPGAs [47, 40, 44]) tend to be very slow, and do not generate high-quality placements. In addition, they cannot cope with problems arising from dynamic placement and routing; high run-times are also due to the fine-grain view of communicating modules.

In the online (dynamic) case where requests are generated at run-time, many approaches have been proposed [6, 48, 2, 3]. All these approaches still lack a practical proof of concept because only little research has been spent on dynamically routing signals between dynamically placed modules such as automatic circuit switching (e.g., [4]) or dynamic networks on a chip (e.g., [37, 10]).

For the offline (static) case, Teich et al. [42] showed that the problem of optimally placing a set of independent modules in space and time is a 3D strip packing problem and reduced the search space using *packing classes*; see Sect. 10.2.1. This approach has been extended in to include temporal precedence constraints between communicating modules for finding legal temporal placements [18]. However, the concepts have yet to be verified on real hardware, mainly due to the lack of FPGA-based platforms that allow free placement of 2D modules in time, or with great restrictions about how modules can communicate with each other [46].

The first restriction can be solved using the FPGA-based *Erlangen Slot Machine* (ESM), see Chap. 3. The architecture is slot-oriented and allows reconfiguring modules in slots of either static or dynamically adjustable width. For inter-module communication, it provides four main methods: direct communication using bus-macros between adjacently placed modules, SRAMs or BlockRAMs for shared memory communication, a dynamic signal switching communication architecture called *Reconfigurable Multiple Bus* (RMB), and an external crossbar; see Sect. 3.4 and Fig. 10.5.

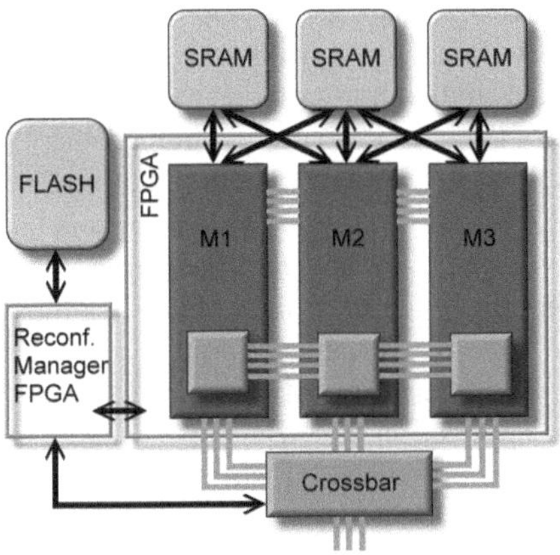

Fig. 10.5 Schematic overview of the ESM architecture with four methods for inter-module communication: direct communication via bus-macros, SRAMs for shared memory communication, the Reconfigurable Multiple Bus (RMB), and an external crossbar.

This architecture gives rise to the following problems: Given a set of $n \leq s$ communicating modules; find a placement that minimizes either (a) the number of segments m, or (b) the maximal number of segments a signal must cross from a source to a sink slot. The second objective means minimization of the maximal delays.

10.3.1 Mathematical Model

Our goal is to allocate the modules of an application to the slots of the ESM, such that the number of parallel RMB bus segments or the maximal length of a connection between connected modules is minimized. Given an array-like architecture (such as the ESM) that has s identical slots. Some of the slots may be occupied by other applications. Consequently let $S \subseteq \{1,\ldots,s\}$ denote the set of available slots. Each slot has width w. There may be some extra space between two neighbouring slots j and $j+1$; this extra space is denoted by e_j. The distance between the centers of two slots $j < l$ is given by $d_{jl} = d_{lj} = |l-j|w + \sum_{i=j}^{l-1} e_i$.

We have to place an application consisting of $n \leq |S|$ modules. Each of the modules is capable of executing a certain task. Some of these modules can have placement restrictions, e.g. they may require certain pins that are available in the first or last slot only. Let $P_i \subseteq S$ denote the set of slots into which a module i can be placed. Each module may wish to communicate with other modules; in the sequel we concentrate on connections via the RMB only, after removing those that are dealt with by other means. Each of these communication links can occupy one or more RMB bus segments. The implementation of the RMB on the ESM supports unidirectional communication only. Therefore, the communication graph is a directed graph $G = (V, A)$ with weight function $t : A \to \mathbb{N}$, indicating the number of bus segments necessary for realizing an edge on the RMB.

In this more formal setting, the task of placing modules on the ESM in order to minimize one of the two objectives given above reduces to the problem of placing the vertices of the communication graph on the integer points of the line. These problems are variations of the classical optimization problem called *minimum bandwidth problem* (MBW); see [12] for a recent computational study.

We present two ILP models that map modules to slots such that the number of parallel bus segments is minimized and the maximum distance between any two connected modules is minimized subject to a restricted number of parallel bus segments. The ILPs use two kinds of variables: Variable $x_{ij} \in \{0,1\}$ indicates whether a module $i \in \{1,\ldots,n\}$ is placed in slot $j \in \{1,\ldots,S\}$ or not. If a slot j is not available for the current application, the variables x_{ij} are fixed to zero for all modules i. If a module i must not be placed in slot j (i.e., $j \notin P_i$), x_{ij} is fixed to zero as well. We also have variables $0 \leq x_{ijkl} \leq 1$. Even though these variables are not binary, they can take only values in $\{0,1\}$: If two modules i and k that are connected by an edge ik are placed in slots j and l, respectively, x_{ijkl} is set to 1; otherwise it is set to 0.

10.3.1.1 Minimizing the Number of Parallel Segments

Each of the modules i has to be placed in any of the available slots. On the other hand, each of the slots j can hold at most one module. This can be expressed by the following two assignment constraints

$$\sum_{j\in S} x_{ij} = 1, \quad i \in M \qquad \sum_{i=1}^{n} x_{ij} \leq 1, \quad j \in S. \tag{10.1}$$

In this section we aim at minimizing the maximum number of bus segments in use. In order to count the number t_s of parallel segments crossing the border of slots s and $s+1$, we could formulate a constraint like $t_s = \sum_{jl\in A:j\leq s<l}\sum_{i=1}^{n}\sum_{k=1}^{n} t_{jl}\times x_{ij}x_{kl}$. Unfortunately, this equation is quadratic in x. As x_{ij} and x_{kl} are binary variables, this constraint can be linearized. To do so we introduce variables x_{ijkl} and add the constraint

$$x_{ijkl} \geq x_{ij} + x_{kl} - 1. \tag{10.2}$$

Equation (10.2) is nothing but a logical AND. x_{ijkl} is set to one if and only if neither of x_{ij} and x_{kl} is set to zero. Replacing the product in the quadratic equation by the new variables results in

$$t_s = \sum_{jl\in A:j\leq s<l}\sum_{i=1}^{n}\sum_{k=1}^{n} t_{jl}x_{ijkl}. \tag{10.3}$$

Adding the equation

$$t_s \leq T \qquad s \in S \setminus \max_{j\in S} j \tag{10.4}$$

we can now formulate the integer linear program

$$\begin{aligned}
&\text{Minimize } T\\
&\text{subject to } \quad (10.1), (10.2), (10.3), \text{and } (10.4)\\
&\qquad x_{ij} = 0 \quad \text{for } i \in \{1,\ldots,n\},\ j \notin S, && (10.5)\\
&\qquad x_{ij} = 0 \quad \text{for } i \in \{1,\ldots,n\},\ j \notin P_i, && (10.6)\\
&\qquad x_{ij} \in \{0,1\}, \quad 0 \leq x_{ijkl} \leq 1. && (10.7)
\end{aligned}$$

10.3.1.2 Minimize Segment-Constrained Bandwidth

Using the ILP presented in the previous section, we are able to compute the minimal number parallel segments in use, thereby minimizing the space overhead caused by the RMB. Given this minimal number of segments or at least the number T of parallel bus segments available, the ILP presented in this subsection finds a placement of the modules that minimizes the length L of the longest edge between any two connected modules. This number L can be determined by adding the linear inequality

$$L \geq d_{jl}x_{ijkl} \quad ik \in A, j, l \in S \tag{10.8}$$

to the ILP described above. Given this additional constraint, the ILP can be formulated as Minimize L subject to (10.1), (10.2)–(10.4), and (10.5)–(10.8).

10.3.2 Case Study and Results

We applied our ILPs to an ESM-based implementation of the classical Pong video game. This video game consists of four modules for user input, racket position calculation, ball position calculation, and video interface. The communication graph has four vertices and five edges. The first ILP yields a solution with 58 parallel segments. Even though this solution is optimal, it has unnecessarily long connections. Applying the second ILP with T set to 58 results in a placement where the length of the longest edge has been reduced to 2; see Fig. 10.6.

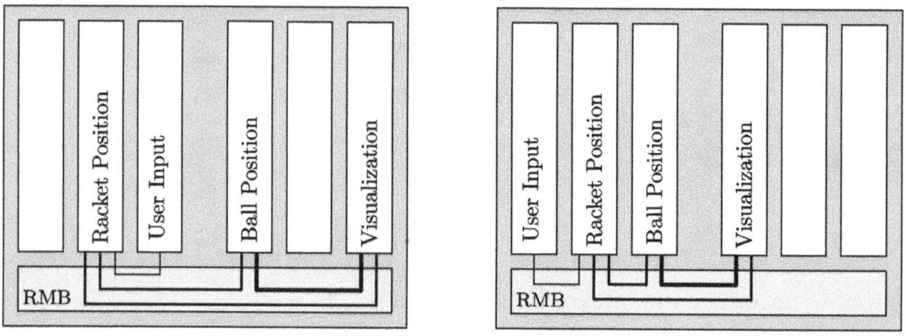

Fig. 10.6 Left: An optimal solution of the first ILP for the Pong example. The thin, medium, and heavy lines connecting the modules represent 4, 20, and 38 segments, respectively. The maximum number of parallel segments is 58. Note that the length of the longest connection is 4. Right: An optimal solution of the second ILP. The solution is restricted to a minimum number of 58 parallel segments. The length of the longest connection is now 2 [19] © [2006] IEEE.

Our ILPs work well for small applications like the Pong game in less than a tenth of a second. To demonstrate the ability of our approach to tackle much larger problems, we randomly generated a large number of artificial benchmark instances. We assume an application consisting of n modules; all of these modules require some kind of communication with other modules. The results are shown in Table 10.2.

Table 10.2 Left: Results for the first ILP. n is the number of vertices, $|E_x|$ are the numbers of edges of the three types of communication graphs considered. After performing 7 runs (AMD Athlon64 X2 3800+, Linux, CPLEX 10.0), time/s is the arithmetic mean taken over five runs, deleting the best and the worst result. Right: Results for the second ILP [19] © [2006] IEEE.

n	$\|E_1\|$	Time/s	$\|E_2\|$	Time/s	$\|E_3\|$	Time/s
6	8	0.04	8	0.04	10	0.05
8	10	0.14	10	0.13	13	0.15
10	12	0.45	13	0.34	16	0.46
12	14	0.82	16	0.96	20	1.21
14	16	1.80	18	2.55	23	3.72
16	18	4.40	21	2.54	26	18.15

n	$\|E_1\|$	Time/s	$\|E_2\|$	Time/s	$\|E_3\|$	Time/s
6	8	0.14	8	0.15	10	0.19
8	10	0.89	10	0.89	13	1.43
10	12	3.90	13	5.25	16	10.96
12	14	30.06	16	44.64	20	80.95
14	16	153.74	18	573.27	23	775.71
16	18	2655.44	21	2812.81	26	>3600.00

10.4 No-break Dynamic Defragmentation of Reconfigurable Devices

FPGAs suffer from a significant area overhead, a higher power consumption, or a speed penalty as compared to ASIC solutions [34]. Partial runtime reconfiguration is an applicable technique to overcome these issues. In order to take more benefit from runtime reconfiguration, such systems should be able to provide the reconfigurable resources in a very flexible way to the modules. Most related work is based on the assumption that a reconfigurable area is used exclusively by one partially reconfigurable module (e.g. [36]) at a point of time. Hagemeyer el al. [27] introduce a system that provides a bus-based communication for integrating up to 16 modules. Koch et al. [32] partition the reconfigurable area into 60 tiles, each capable of connecting a tiny 8-bit module over a so-called ReCoBus.

For such systems, an efficient resource management is necessary. One problem that has to be solved at runtime is the fragmentation due to the time variant execution of modules. For FPGAs available from the Xilinx Inc., the tiles will be vertically aligned column by column, as shown in Fig. 10.7. Accordingly, a module requiring multiple tiles to implement its logic will demand a consecutive adjacent set of tiles without any gaps. This problem is discussed in this section.

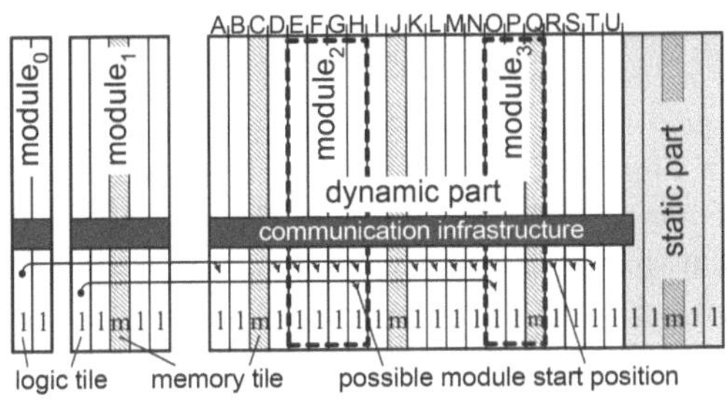

Fig. 10.7 A dynamically reconfigurable system shares some logic tiles l and memory tiles m among a set of modules within the dynamic part of the system. Some modules require a memory tile at a fixed offset with respect to the start position within the modules [21] © [2008] IEEE.

At first glance, this problem has a striking resemblance to one of the classical problems of computing: *Dynamic storage allocation* considers a memory array and a sequence of storage requests of varying size, looking for an assignment of each request to a contiguous block of memory cells, such that the length of each block corresponds to the size of the request. This allocation is static in space: after a block has been occupied, it will remain fixed until the corresponding data is no longer needed and the block is released. This can result in fragmentation of the memory array, making it hard or even impossible to store new data. A large variety of methods and results for allocating storage such as FirstFit, BestFit or different Buddy system have been proposed (e.g., [30, 11, 28, 29]). Newer approaches that use cache-oblivious structures for allocating space in memory hierarchies (e.g., [8, 9]). There are three notable differences between the dynamic allocation of modules to a reconfigurable device and dynamic storage allocation: Using pointers for creating virtual contiguous free blocks is not an option for the placement of modules, the reconfigurable device may–in contrast to uniform memory—contain inhomogeneities, and it is possible to relocate modules on a reconfigurable device during runtime.

There is a certain amount of related work from within the FPGA community [7, 25, 13, 31, 33]. These papers do not consider the algorithmic implications and how the relocation capabilities can be exploited to optimize module layout in a fast, interruption-free, no-break fashion or the problem description differs in some points.

10.4.1 Model and Problem Description

When modules are relocated for defragmentation, we have to distinguish between moving only the module configuration, and the configuration together with the internal state. In the first case, we just make a copy of the reconfiguration data to the new position and start the next computation on the module at the new position. In the second case, source and target have to be interrupted to copy the state. If we allow overlapping regions for the defragmentation, the interruption time will dominate the reconfiguration process, because we have to copy the routing information and logic settings in addition to the state. As a consequence, we will prevent our defragmentation algorithms to use overlapping regions to place modules.

An additional aspect is that FPGAs typically provide logic tiles l and memory tiles m, as shown in Fig. 10.7. The placement of a module on the FPGA must fit exactly to the particular module. This restricts the possible module start positions. For example, module_1 in Fig. 10.7 can be placed only at the positions A, H, and O.

We consider a device that allows allocating modules in a contiguous manner on an array L of length ℓ; modules will be denoted by $M_1, \ldots, M_n$. A module M_i placed in the array occupies a contiguous interval, L_{M_i}. Modules are placed such that $L_{M_i} \cap L_{M_j} = \emptyset$ for $i \neq j$; that is, two modules do not overlap; see Fig. 10.8.

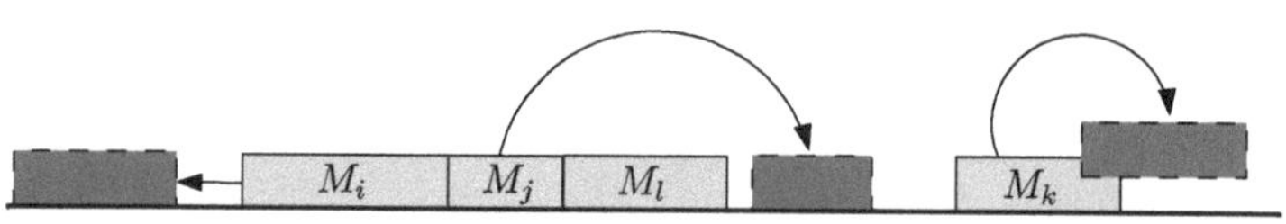

Fig. 10.8 A module corresponds to a set of columns on an FPGA. Each module occupies a contiguous block. The moves of M_i and M_j are allowed, the move of M_k is forbidden, because the current and the target position overlap [21] © [2008] IEEE.

Modules placed in the array divide L into sections that are occupied by a module and sections that are not occupied; the latter are called *free intervals*. Reconfiguration allows us to relocate a module M_i of size m_i from interval L_{M_i} to a new position within a free interval L_M of size m within the array, provided that the following two conditions are fulfilled: $L_M \cap \bigcup_{k=1}^{n} L_{M_k} = \emptyset$ and $m \geq m_i$. The first condition implies that L_M and L_{M_i} are not allowed to intersect and ensures that the new position is not occupied by any other module. The second condition ensures that there is sufficient free space to provide a new position for the module.

The Maximum Defragmentation Problem (MDP) asks for a sequence of relocation moves, such that the resulting connected free interval is as large as possible. We distinguish between the *homogeneous* MDP, where every cell in the array is equivalent, and the *heterogeneous* MDP for FPGAs with heterogeneities. Clearly, the heterogeneous MDP is more difficult.

10.4.2 Problem Complexity and Moderate Densities

We state two results: one for deciding whether one contiguous free block can be formed, and one for the maximization version of the (homogeneous) defragmentation problem. Both can be shown by a reduction from the *3-Partition Problem* [24].

Theorem 10.4. *The Maximum Defragmentation Problem with free intervals $F_1, \dots, F_k$ is strongly NP-complete* [21].

Proving NP-completeness for the decision version of a problem makes it interesting to consider approximating the size of the maximal constructible free intervals.

Theorem 10.5. *There is no approximation algorithm for the maximum defragmentation problem with an approximation factor polynomially bounded in the input size of the problem, unless $P = NP$* [21].

A special case of the homogeneous MDP can be solved with linear computing time and at most $2n$ moves: We define the density $\delta := \frac{1}{\ell}\sum_{i=1}^{n} m_i$. If the density is small enough (i.e., Eq. (10.9) holds), the total free space can always be connected.

$$\delta \leq \frac{1}{2} - \frac{1}{2\ell} \cdot \max_{i=1,\dots,n} \{m_i\}. \tag{10.9}$$

Theorem 10.6. *Algorithm* 10.2 *connects the total free space with at most $2n$ moves and uses $O(n)$ computing time* [21].

Algorithm 10.2 LeftRightShift

Input: A array L with n modules $M_1, \dots, M_n$ such that (10.9) is fulfilled.
Output: A placement of $M_1, \dots, M_n$ such that there is only one free interval at the left end of L.
for $i = 1$ **to** n **do** Shift M_i to the left as far as possible.
for $i = n$ **to** 1 **do** Shift M_i to the right as far as possible.

10.4.3 A Heuristic Method

As a consequence of the hardness and inapproximability results we focus on developing heuristic approaches for the MDP. There is a bound on the number of steps needed by any algorithm that constructs a maximum free interval, even in the homogeneous version:

Theorem 10.7. *There is an instance of the maximum defragmentation problem such that any algorithm needs at least $\Omega(n^2)$ steps to solve it* [21].

We implemented a standard tabu search. In every iteration all homogeneous modules M_i are moved to the left end and the right end of the free intervals that are greater than or equal to m_i. All inhomogeneous modules are moved to any feasible position. Each move is evaluated by a fitness function that divides the size of the maximal free interval by the size of the total free interval. The move yielding the configuration with the highest fitness is chosen. Ties are broken by choosing the first one. The resulting configuration is added to the tabu list. If the current solution is the best one found so far it is stored.

We performed experiments on two different arrays, both having $\ell = 94$ slots. The first one without heterogeneities, the second one with heterogeneities (Virtex-II). Moreover, we compared our heuristic to a simple greedy approach that moves every module to the most promising position. For the density ranging from 0.3 to 0.9 we performed 100 runs for each value and took the average value of the *number of free intervals* and the *size of the maximal free interval*. The results are show in Fig. 10.9.

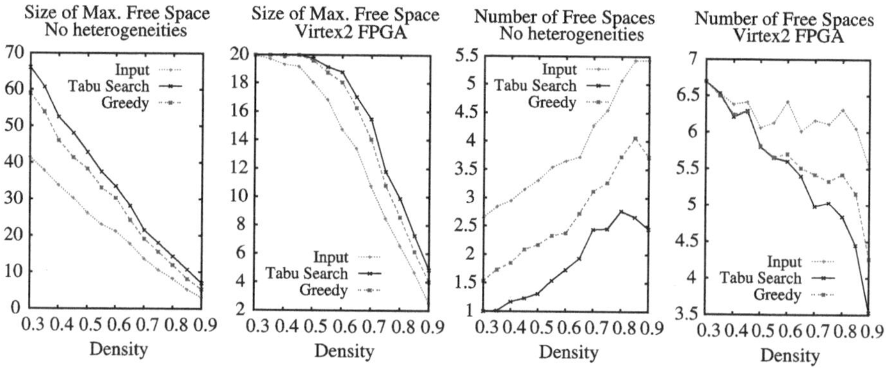

Fig. 10.9 (Left to right) Size of the maximal free interval before and after defragmentation, using our heuristic and a simple greedy approach in an array with no heterogeneities, size of the maximal free interval before and after defragmentation of the Virtex-II FPGA, number of free intervals before and after defragmentation in an array with no heterogeneities, number of free intervals before and after defragmentation of the Virtex-II FPGA [21] © [2008] IEEE.

10.5 Scheduling Dynamic Resource Requests

Finally we consider the problem of scheduling modules whose resource requests (i.e., space on an FPGA) may vary over time. This may be, for example, a router module that needs more resources if the traffic increases. In contrast to the preceding sections, we assume that a module occupies a certain number of full slots on the FPGA. Furthermore, we assume that requests come in time slots of a fixed size. Now, scheduling a sequence of modules with time-varying resources corresponds to strip packing (see Sect. 10.2.1) where the width of the strip is the number of slots on the FPGA and the height of the strip corresponds to the time axis. Thus, we will use height and time synonymously. However, we are allowed to *delay* a request; that is, we may stretch the modules along the time axis; see Fig. 10.10.

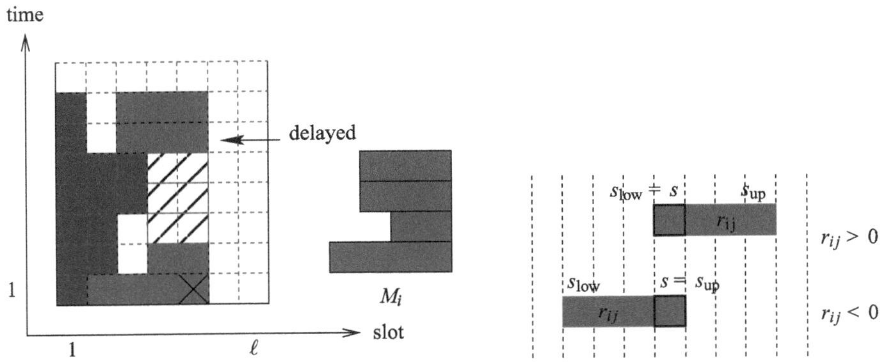

Fig. 10.10 Left: We can place the module M_i at the position marked by × (the *base slot* of M_i), if we *delay* the third request. That is, the second request stays on the FPGA until the third request can be fulfilled. Right: Occupancy Constraints: If $x_{stij} = 1$, the request (i, j) occupies r_{ij} slots left or right to s—depending on $\text{sgn}(r_{ij})$.

10.5.1 An ILP

We want to place n modules on an FPGA with ℓ slots. The resource requests of each module, M_i, is given as a sequence of *requests*, (i, j), of length r_{ij}, $r_{ij} \in \mathbb{Z}\backslash\{0\}$, where r_{ij} denotes the number of slots requested by M_i after running $j - 1$ time steps. We assume that every module occupies a base slot and extends to the left ($r_{ij} < 0$) or to the right ($r_{ij} > 0$) of the base slot. Let m_i denote the length of the request sequence for module M_i.

For the ILP, we introduce three kinds of variables: The *assignment variables*, x_{stij}, the *occupancy variables*, y_{stij}, and the *usage variables*, u_t. The first ones specify when and where a request is scheduled. More precisely, setting x_{stij} to 1 indicates that request (i, j) of module M_i is scheduled in slot s at time t (i.e., the slots $s, \ldots, s + r_{ij} - \text{sgn}(r_{ij})$ are occupied at time t), where s is the base slot of module M_i. For every (i, j) there is exactly one x_{stij} with $x_{stij} = 1$. Usually (i.e., if $|r_{ij}| > 1$) a request occupies more than one slot when executed. Moreover, if the request $(i, j+1)$ is delayed, (i, j) remains on the FPGA for more than one time unit.

To keep track of the occupied slots, we set y_{stij} to 1, if slot s is occupied by request (i, j) at time t. The usage variables simply specify which time steps are used.

Clearly, the size of the FPGA, ℓ, and the number of modules, n, strongly determine the size of an ILP and, in turn, the time needed to solve it. In addition, we assume that an upper bound, T, on the number of time steps is given. The closer this bound is to the optimum, the smaller the resulting ILP.

Assignment Constraints: There can be at most one request per slot and time and—vice versa—each request must be scheduled exactly once. The following constraints express these conditions:

$$\begin{aligned} &\sum_{i=1}^{n}\sum_{j=1}^{m_i} x_{stij} \leq 1 \quad \forall t = 1,\ldots,T,\ s = 1,\ldots,\ell \\ &\sum_{t=1}^{T}\sum_{s=1}^{\ell} x_{stij} = 1 \quad \forall i, j. \end{aligned} \tag{10.10}$$

Boundary Constraints: Next, we ensure that a request does not exceed the FPGA's boundary by forcing the corresponding assignment variables to be zero.

$$\begin{aligned} &\forall i, j,\ t = 1,\ldots,T,\ s = s_{\text{low}},\ldots,s_{\text{up}}: \quad x_{stij} = 0 \\ &\text{where } s_{\text{low}} := \begin{cases} \ell - r_{ij} + 2, & r_{ij} > 0 \\ 1, & r_{ij} < 0 \end{cases} \\ &\text{and } s_{\text{up}} := \begin{cases} \ell, & r_{ij} > 0 \\ -r_{ij} - 1, & r_{ij} < 0 \end{cases} \end{aligned} \tag{10.11}$$

Base-Slot Constraints: Each request of a module must be scheduled in the same base slot:

$$\sum_{t=1}^{T}\sum_{s=1}^{\ell} s x_{stij} - \sum_{t=1}^{T}\sum_{s=1}^{\ell} s x_{sti0} = 0 \quad \forall i, j. \tag{10.12}$$

For every (i, j), there is exactly one s, t such that $x_{stij} = 1$. Thus, summing up $s \cdot x_{stij}$ over s and t for fixed i and j yields the base slot of module M_i.

Order Constraints: Now we ensure that request (i, j) of module M_i is not scheduled before request $(i, j-1)$ is finished:

$$\sum_{t=1}^{T}\sum_{s=1}^{\ell} t x_{stij} - \sum_{t=1}^{T}\sum_{s=1}^{\ell} t x_{stij-1} > 0 \quad \forall i, j > 0. \tag{10.13}$$

Similar to the base-slot constraints, summing up $t \cdot x_{stij}$ over s and t for fixed i and j yields the time step where request (i, j) is scheduled.

Occupancy Constraints: If $x_{stij} = 1$, the request (i, j) occupies r_{ij} slots at time t. Thus, we set the appropriate occupancy variables as follows:

$$\begin{aligned}&\forall i=1,\ldots,n,\ j=1,\ldots,m_i,\ s=1,\ldots,\ell,\ t=1,\ldots,T,\\&s'=s_{\text{low}},\ldots,s_{\text{up}}:\quad x_{stij}-y_{s'tij}\leq 0,\\&\text{with } s_{\text{low}}=\begin{cases}s, & r_{ij}>0\\ \max\{1,s+r_{ij}+1\}, & r_{ij}<0\end{cases}\\&s_{\text{up}}=\begin{cases}\min\{\ell,s+r_{i'j'}-1\}, & r_{ij}>0\\ s, & r_{ij}<0.\end{cases}\end{aligned}\tag{10.14}$$

Exclusive Constraints: Now, we can ensure that requests do not overlap by allowing at most one occupancy variable for a fixed slot and a fixed time to be 1.

$$\forall t=1,\ldots,T,\ s=1,\ldots,\ell:\quad \sum_{i=1}^{n}\sum_{j=1}^{m_i} y_{stij}\leq 1.\tag{10.15}$$

Delay Constraints: If a request, $(i,j+1)$, is delayed, the preceding request (i,j) remains on the FPGA until $(i,j+1)$ is scheduled. Thus, if $y_{stij}=1$ either $y_{s(t+1)ij}$ must be 1 or $(i,j+1)$ is scheduled at time $t+1$; that is, there is an s' such that $x_{s'(t+1)i(j+1)}=1$ holds. The following constraints keep track of delayed requests. $\forall i=1,\ldots n,\ j=1,\ldots,m_i,\ s=1,\ldots,\ell,\ t=1,\ldots,T-1$:

$$y_{stij}-y_{s(t+1)ij}-\sum_{s'=1}^{\ell} x_{s'(t+1)i(j+1)}\leq 0.\tag{10.16}$$

Usage Constraints: Last, we introduce some constraints that define our usage variables. Let u_t be 1, if at least one x_{stij} is 1 or if u_{t+1} is 1.

$$\forall t=1,\ldots T, s=1,\ldots,\ell,\ i=1,\ldots,n,\ j=1,\ldots,m_i:$$

$$u_t - x_{stij}\geq 0\tag{10.17}$$

$$\text{and}\quad \forall t=2,\ldots,T:\quad u_{t-1}-u_t\geq 0.\tag{10.18}$$

Objective Function: To minimize the makespan, we use the following ILP:

$$\min\sum_{i=1}^{T} i\,u_i\quad \text{subject to Eqs. (10.10)–(10.18)}$$

$$x_{stij}\in\{0,1\},\qquad y_{stij}\in\{0,1\},\qquad u_t\in\{0,1\}.$$

10.5.2 Heuristic Methods

We implemented several heuristics for our problem. For comparison, a simple First-Fit with and without delaying requests, and two more elaborated heuristics, BestFit and TabuSearch. Our methods pack the given modules in a semi-infinite strip. The width of the strip is given by the number of slots on the FPGA, the height of the strip corresponds to the time axis. See Fig. 10.11 for an example.

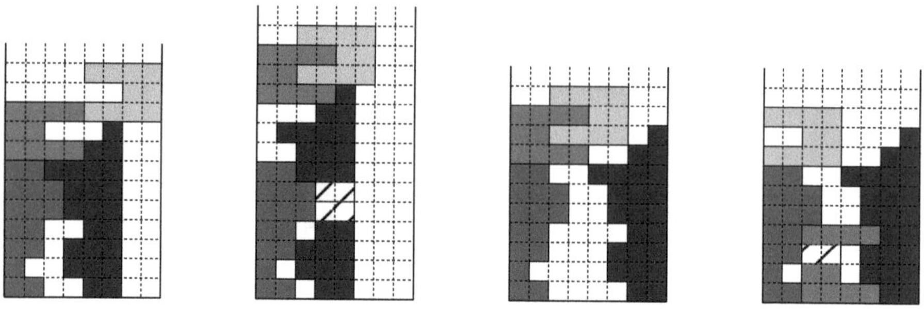

Fig. 10.11 (Left to right): Packings by FirstFit, FirstFit with delays, BestFit, and TabuSearch.

FirstFit. Probably the simplest heuristic is to place the modules, one by one, in a first-fit way: beginning with $s = 1$ and $t = 1$, we test for every position if the module that must be placed overlaps with already-placed modules. We choose the first position in which no overlap occurs. Note that we disregard the possibility of delaying requests.

FirstFit with delays. This method works the same as the method above, but allows the delaying of requests. That is, if for a certain start position the requests $0, \ldots, j-1$ fit into the strip without overlap, but request j does not fit in time step t', we search for the largest $j' < j$ such that request (i, j') fits in t' and delay every request $j'' = j' + 1, \ldots$ (i.e., we move them upwards in the strip); see Fig. 10.10.

BestFit. Similar to FirstFit with delays, we try to find a nonoverlapping position by testing every possible position. But now we do not choose the first feasible position, but we evaluate every position as follows: We separately count the unoccupied cells left and right to the placed module and take the minimum of the these two values as a score for the given position. For example, for the placement of M_i in Fig. 10.10, there are 4 unoccupied cells left to M_i and 14 unoccupied cells right to M_i, yielding a value of 4 for his placement. We choose the position that yields the minimal score and break ties by preferring the position with least number of delays.

To avoid that every module is placed on the left or right side (yielding a score of 0), we maintain an upper limit, $t_{\max}$, for the time. Before we place a new module, we increase $t_{\max}$ by $m_i/2$ and try to place the given module within the given time bound. If this is not possible, we increase $t_{\max}$ by m_i and try again.

TabuSearch. The idea of the heuristic is to try several BestFit runs. For every run, we choose a different order for the insertion of modules. Starting with the sequence $S = (1, \ldots, n)$, we swap two items of the sequence and compute the makespan that is achieved by BestFit. More precisely, we maintain a *swapping distance*, i, ranging from $i = 0$ to $n/2$. For a fixed i, we swap the items at positions j and $((j + i) \bmod n) + 1$ for $j = 1, \ldots n$, keeping track of the best makespan achieved so far, and accept the swap that achieves the best makespan known so far. A tabu list ensures that we do not swap an already accepted pair again, see Fig. 10.12.

Tabu Search

$S = (1, \ldots, n)$
for $i = 0$ **to** $n/2$
  $found = 0$
  for $j = 1$ **to** n
    if $(j, ((j + i) \bmod n) + 1)$
      are not in the tabu list **then**
     Swap items at positions j
      and $((j + i) \bmod n) + 1$ in S
     Calculate makespan of BestFit
     if makespan is the best so far **then**
       $found = j$
     Undo swapping
  if $found > 0$ **then**
    Swap positions $found$
     and $((found + i) \bmod n) + 1$ in S
    Store $(found, ((found + i) \bmod n) + 1)$
     in the tabu list

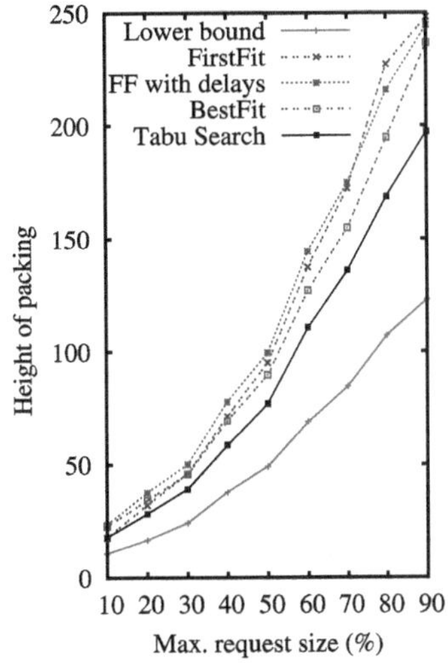

Fig. 10.12 Left: Tabu Search. Right: A comparison of different heuristics—FirstFit (without delays), FirstFit (with delays), BestFit, and TabuSearch—in settings with different densities (i.e., maximal value for a request). A lower bound (ratio of total area by number of slots) is shown.

To test our heuristics, we conducted a set of experiments. For an FPGA of width $\ell = 50$ and upper limits for the size of a request, $r_{\max}$, ranging from 10% to 90% we randomly generated sequences of 20 modules per sequence. Fig. 10.12 shows the mean value over 20 runs for every value of $r_{\max}$ and for every heuristic.

Acknowledgements This work was supported by DFG grants FE 407/8 and TE 163/14, project *ReCoNodes*, as part of the Priority Programme 1148. *Reconfigurable Computing Systems.*

References

1. Ahmadinia, A., Teich, J.: Speeding up online placement for XILINX FPGAs by reducing configuration overhead. In: Proc. IFIP Internat. Conf. VLSI-SOC, pp. 118–122 (2003)
2. Ahmadinia, A., Bobda, C., Bednara, M., Teich, J.: A new approach for on-line placement on reconfigurable devices. In: Proc. Internat. Parallel Distr. Proc. Sympos., p. 134 (2004)
3. Ahmadinia, A., Bobda, C., Fekete, S., Teich, J., van der Veen, J.: Optimal routing-conscious dynamic placement for reconfigurable devices. In: Internat. Conf. Field-Programm. Logic Appl., pp. 847–851 (2004)
4. Ahmadinia, A., Bobda, C., Ding, J., Majer, M., Teich, J., Fekete, S., van der Veen, J.: A practical approach for circuit routing on dynamically reconfigurable devices. In: Proc. 16th IEEE Internat. Workshop Rapid System Prototyping (RSP), pp. 84–90 (2005)
5. Baker, B.S., Schwarz, J.S.: Shelf algorithms for two-dimensional packing problems. SIAM J. Comput. **12**(3), 508–525 (1983)
6. Bazargan, K., Kastner, R., Sarrafzadeh, M.: Fast template placement for reconfigurable computing systems. IEEE Des. Test Comput. **17**, 68–83 (2000)

7. Becker, T., Luk, W., Cheung, P.Y.: Enhancing relocatability of partial bitstreams for run-time reconfiguration. In: Proc. 15th Annu. Sympos. Field-Programm. Custom Comput. Mach., pp. 35–44 (2007)
8. Bender, M.A., Demaine, E.D., Farach-Colton, M.: Cache-oblivious B-trees. SIAM J. Comput. **35**, 341–358 (2005)
9. Bender, M.A., Fineman, J.T., Gilbert, S., Kuszmaul, B.C.: Concurrent cache-oblivious B-trees. In: Proc. 17th Annu. ACM Sympos. Parallel. Algor. Architect., pp. 228–237 (2005)
10. Bobda, C., Ahmadinia, A., Majer, M., Teich, J., Fekete, S., van der Veen, J.: DyNoC: A dynamic infrastructure for communication in dynamically reconfigurable devices. In: Internat. Conf. Field-Programmable Logic and Applications, pp. 153–158 (2005)
11. Bromley, G.: Memory fragmentation in buddy methods for dynamic storage allocation. Acta Inform. **14**, 107–117 (1980)
12. Caprara, A., Salazar-Gonzalez, J.J.: Laying out sparse graphs with provably minimum bandwidth. INFORMS J. Comput. **17**, 356–373 (2005)
13. Compton, K., Li, Z., Cooley, J., Knol, S., Hauck, S.: Configuration relocation and defragmentation for run-time reconfigurable systems. IEEE Trans. VLSI **10**, 209–220 (2002)
14. Diessel, O., ElGindy, H., Middendorf, M., Schmeck, H., Schmidt, B.: Dynamic scheduling of tasks on partially reconfigurable FPGAs. IEEE Proc. Comput. Dig. Tech. **147**, 181–188 (2000)
15. Fekete, S.P., Schepers, J.: New classes of lower bounds for the bin packing problem. Math. Program. **91**, 11–31 (2001)
16. Fekete, S.P., Schepers, J.: A combinatorial characterization of higher-dimensional orthogonal packing. Math. Oper. Res. **29**, 353–368 (2004)
17. Fekete, S.P., Schepers, J.: A general framework for bounds for higher-dimensional orthogonal packing problems. Math. Methods Oper. Res. **60**, 311–329 (2004)
18. Fekete, S.P., Köhler, E., Teich, J.: Optimal FPGA module placement with temporal precedence constraints. In: Proc. Design Automat. Test Europe (DATE), pp. 658–665 (2001)
19. Fekete, S.P., van der Veen, J.C., Majer, M., Teich, J.: Minimizing communication cost for reconfigurable slot modules. In: Proceedings of 16th International Conference on Field Programmable Logic and Applications (FPL06), pp. 535–540 (2006)
20. Fekete, S.P., Schepers, J., van der Veen, J.: An exact algorithm for higher-dimensional orthogonal packing. Oper. Res. **55**, 569–587 (2007)
21. Fekete, S.P., Kamphans, T., Schweer, N., Tessars, C., van der Veen, J.C., Angermeier, J., Koch, D., Teich, J.: No-break dynamic defragmentation of reconfigurable devices. In: Proc. 18th Internat. Conf. Field Programm. Logic Appl., pp. 113–118 (2008)
22. Fekete, S.P., van der Veen, J.C., Ahmadinia, A., Göhringer, D., Majer, M., Teich, J.: Offline and online aspects of defragmenting the module layout of a partially reconfigurable device. IEEE Trans. VLSI **16**, 1210–1219 (2008)
23. Fekete, S.P., Kamphans, T., Schweer, N.: Online square packing. In: Proc. 20th Algorithms and Data Structure Symposium (WADS), Banff, Alberta, Canada, pp. 302–314
24. Garey, M.R., Johnson, D.S.: Computers and Intractability: A Guide to the Theory of NP-Completeness. Freeman, New York (1979)
25. Gericota, M.G., Alves, G.R., Silva, M.L., Ferreira, J.M.: Run-time defragmentation for dynamically reconfigurable hardware. In: New Algorithms, Architectures and Applications for Reconfigurable Computing (2005)
26. Grötschel, M.: On the symmetric travelling salesman problem: solution of a 120-city problem. Math. Program. Study **12**, 61–77 (1980)
27. Hagemeyer, J., Kettelhoit, B., Koester, M., Porrmann, M.: Design of homogeneous communication infrastructures for partially reconfigurable FPGAs. In: Proc. Internat. Conf. Engineering of Reconf. Systems and Algor., Las Vegas, USA (2007)
28. Hirschberg, D.S.: A class of dynamic memory allocation algorithms. Commun. ACM **16**, 615–618 (1973)
29. Knowlton, K.C.: A fast storage allocator. Commun. ACM **8**, 623–625 (1965)
30. Knuth, D.E.: The Art of Computer Programming: Fundamental Algorithms, 3rd edn. vol. 1, Addison–Wesley, Reading (1997)

31. Koch, D., Ahmadinia, A., Bobda, C., Kalte, H.: FPGA architecture extensions for preemptive multitasking and hardware defragmentation. In: Proc. IEEE Internat. Conf. Field-Programmable Technology, Brisbane, Australia, pp. 433–436 (2004)
32. Koch, D., Beckhoff, C., Teich, J.: ReCoBusBuilder—a novel tool and technique to build static and dynamically reconfigurable systems for FPGAs. In: Proc. 18th Internat. Conf. Field Programm. Logic Appl., pp. 119–124 (2008)
33. Koester, M., Kalte, H., Porrmann, M., Ruckert, U.: Defragmentation algorithms for partially reconfigurable hardware. Int. Fed. Inf. Process. Publ.-IFIP **240**, 41 (2007)
34. Kuon, I., Rose, J.: Measuring the gap between FPGAs and ASICs. IEEE Trans. CAD Integr. Circuits Syst. **26**, 203–215 (2007)
35. Lawler, E.L., Lenstra, J.K., Rinooy Kan, A.H.G., Shmoys, D.B.: Sequencing and scheduling: Algorithms and complexity. In: Graves, S.C., Rinnooy Kan, A.H.G., Zipkin, P.H. (eds.) Logistics of Production and Inventory. Handbooks in Operations Research and Management, vol. 4, pp. 445–522. North–Holland, Amsterdam (1993)
36. Lysaght, P., Blodget, B., Mason, J., Young, J., Bridgford, B.: Invited paper: Enhanced architecture, design methodologies and CAD tools for dynamic reconfiguration of Xilinx FPGAs. In: Proc. 16th Internat. Conf. Field Programm. Logic Appl., pp. 1–6 (2006)
37. Majer, M., Bobda, C., Ahmadinia, A., Teich, J.: Packet routing in dynamically changing networks on chip. In: Proc. Internat. Parallel Distr. Proc. Sympos. (IPDPS), p. 154 (2005)
38. Martello, S., Monaci, M., Vigo, D.: An exact approach to the strip-packing problem. INFORMS J. Comput. **15**(3), 310–319 (2003)
39. Seiden, S.S.: On the online bin packing problem. J. ACM **49**(5), 640–671 (2002). doi:10.1145/585265.585269
40. Shi, J., Bhatia, D.: Performance driven floorplanning for FPGA based designs. In: Internat. Sympos. Field Programm. Gate Arrays, pp. 112–118 (1997)
41. Steiger, C., Walder, H., Platzner, M.: Operating systems for reconfigurable embedded platforms: Online scheduling of real-time tasks. IEEE Trans. Comput. **53**, 1393–1407 (2004)
42. Teich, J., Fekete, S., Schepers, J.: Compile-time optimization of dynamic hardware reconfigurations. In: Internat. Conf. Parallel Distr. Proc. Techn. Appl., pp. 1097–1103 (1999)
43. Teich, J., Fekete, S.P., Schepers, J.: Optimal hardware reconfiguration techniques. J. Supercomput. **19**, 57–75 (2001)
44. Tessier, R.G.: Fast place and route approaches for fpgas. PhD thesis, Massachusetts Institute of Technology, Department of Electrical Engineering and Computer Science (1999)
45. van der Veen, J., Fekete, S., Majer, M., Ahmadinia, A., Bobda, C., Hannig, F., Teich, J.: Defragmenting the module layout of a partially reconfigurable devices. In: Plaks, T.P. (ed.) Proc. Internat. Conf. Engin. Reconf. Syst. Algor., pp. 92–101 (2005)
46. Walder, H., Platzner, M.: Online scheduling for block-partitioned reconfigurable devices. In: Proc. Design Automat. Test Europe (DATE), pp. 290–295 (2003)
47. Yang, X., Wang, M., Kastner, R., Ghiasi, S., Sarrafzadeh, M.: Fast placement approaches for FPGAs. ACM Trans. Des Automat. Electr. Syst. **8**, 316–333 (2003)
48. Yuh, P.H., Yang, C.L., Chang, Y.W.: Temporal floorplanning using the T-tree formulation. In: Proc. Internat. Conf. Comput. Aided Design, pp. 300–305 (2004)

Chapter 11
ReCoNets—Design Methodology for Embedded Systems Consisting of Small Networks of Reconfigurable Nodes and Connections

Christian Haubelt, Dirk Koch, Felix Reimann, Thilo Streichert, and Jürgen Teich

Abstract Automotive, avionic or body-area networks are systems that consist of several communicating control units specialized for certain purposes. Typically, constraints regarding reliability, availability but also flexibility are imposed on these systems. In this chapter, we will present the *ReCoNets* approach for increasing reliability and flexibility of such systems by solving the hardware/software codesign problem online. A ReCoNet allows to migrate tasks implemented in hardware or software from one node to another. Typically, it consists of a network of communicating Field-Programmable Gate Arrays (FPGAs) and CPUs. Moreover, if a sufficient number of hardware/software resources is not available, the migration of functionality from hardware to software or vice versa is initiated by the system itself. For supporting such flexibility, new design methods as well as services integrated in a distributed operating system for networked embedded systems are revealed. Besides the formal definition of methods and concepts providing several *self-x* properties such as self-healing, self-adaptiveness and self-optimization, a ReCoNet demonstrator is presented hosting a driver assistance application.

11.1 Introduction

More and more complex embedded systems consist of several control units connected via a network. Each control unit is specialized to provide a certain functionality. With the introduction of reconfigurable hardware (commonly FPGAs) as a

Christian Haubelt · Dirk Koch · Felix Reimann · Jürgen Teich
Hardware/Software Co-Design, Department of Computer Science, University of Erlangen-Nuremberg, Erlangen, Germany, e-mail: haubelt@cs.fau.de, koch@cs.fau.de, felix.reimann@cs.fau.de, teich@cs.fau.de

Thilo Streichert
Daimler AG, Group Research & Advanced Engineering, Sindelfingen, Germany, e-mail: thilo.streichert@daimler.com

M. Platzner et al. (eds.), *Dynamically Reconfigurable Systems*,
DOI 10.1007/978-90-481-3485-4_11,

centric part of the control units, a new dimension of flexibility can be implemented in such systems with the help of *runtime reconfiguration*. At the *micro level*, this includes reconfigurable hardware and software tasks on a single control unit called *ReCoNode*, while the reconfiguration may also involve a *macro level* where tasks are *migrated* among different ReCoNodes. Such a reconfigurable network, in the following also called a *ReCoNet*, can automatically react to environmental changes such as faults in a ReCoNode, network topology changes, or varying workload scenarios. This is established by a decentralized decision making to decide which task is executed *where*, hence, answering the question on which node (macro level) and at which placement position on a particular reconfigurable device (micro level) a task is executed. Additionally, this includes to decide *how* a task is executed, i.e., in software or in hardware.

Beside these self-healing and self-adapting properties, self-optimizing strategies allow for improving the reliability of a ReCoNet. For example, instead of using *Triple Modular Redundancy* (TMR) with a fixed, redundant task binding, a ReCoNet may adapt communication, task binding, or even the implementation style (hardware or software) of a task according to the currently available resources or load of the network components. Then, in case of a fault, tasks may migrate to other intact resources at runtime. Obviously, in order to guarantee fault tolerance, decentralized methods for self-healing, self-adaptiveness, and self-optimization are required.

While different levels of granularity have to be considered in the design of fault-tolerant and self-adaptive reconfigurable networked embedded systems, we will put focus onto the system level in this chapter. In particular, we will focus on topology changes like node or link defects and integration of new nodes. Central to the self-x properties of a ReCoNet is a methodology for *online hardware/software partitioning* (cf. [28]) which describes the procedure of binding functionality onto resources in the network at runtime. In order to allow moving functionality from one node to another and execute it either on hardware resources or in software, we will introduce the concepts of *task migration* and *task morphing*. Both, task migration as well as task morphing where the implementation style of a task changes from hardware to software or vice versa depending on load or timing constraints, require *hardware* and/or *software checkpointing* mechanisms and an extended design flow for providing an application engineer with such new design methods.

The remainder of this chapter is structured as follows: In Sect. 11.2, a formal model of fault-tolerant and self-adaptive reconfigurable networked embedded systems is introduced. Section 11.3 is devoted to operation system services implementing the self-healing, self-adapting, and self-optimizing capabilities of a ReCoNet in a distributed manner. This includes online hardware/software partitioning, checkpointing as well as methods for task migration and task morphing. After this, in Sect. 11.4, a methodology for designing ReCoNet systems is presented. In particular, the reliability analysis and optimization will be discussed. Finally, in Sect. 11.5, a ReCoNets demonstrator capable of compensating link or node failures by dynamically adapting and optimizing the routing and task binding will be presented for an automotive driver assistance system.

11.2 System Model

A ReCoNet is specified by a so-called *network model* which separates functionality from the network architecture.

Definition 11.1 (Network Model). The entire network model $M(G^{tg}, G^{sca}, E^m)$ consists of a *topology graph* G^{tg}, a set of *sensor-controller-actuator chains* G^{sca}, and a set of mapping edges E^m.

The topology graph models the ReCoNet architecture (resources and their connections) available to implement the intended functionality of the ReCoNet.

Definition 11.2 (Topology Graph). The *topology graph* $G^{tg}(V^{tg}, E^{tg})$ consists of a set of vertices V^{tg} modeling the network nodes and a set of edges $E^{tg} \subseteq V^{tg} \times V^{tg}$ modeling the connections between nodes. The set of vertices V^{tg} can be partitioned into *sensor nodes* $s \in S^{tg}$, *computational nodes* $n \in N^{tg}$, and *actuator nodes* $a \in A^{tg}$.

We assume the computational nodes, also called ReCoNodes, to be hardware/software reconfigurable, e.g., they are composed of an FPGA together with at least one CPU. ReCoNodes may integrate tasks implemented in software or hardware at run-time. Additionally, these FPGA-based nodes contain dedicated analog hardware for driving sensors and actuators which leads to a certain heterogeneity in the network. In particular, a sensor node $s \in S^{tg}$ must not be directly connected to an actuator node $a \in A^{tg}$ and vice versa. This restricts the set of edges to $E^{tg} \subseteq S^{tg} \times N^{tg} \cup N^{tg} \times N^{tg} \cup N^{tg} \times A^{tg}$.

Exemplarily, Fig. 11.1(b) shows a network topology with four *computational nodes* $n_i \in N^{tg}$, *sensors* $s_i \in S$, *actuators* $a_i \in A$, and communication links represented by the edges between the nodes n_i. Note that some sensors and actuators are not connected to all nodes in the network, thus building a heterogeneous network structure.

Similar to the ReCoNet architecture, the application is modeled by a set of sensor-controller-actuator chains composed in a single graph termed *sensor-controller-actuator chain graph*.

Definition 11.3 (Sensor-Controller-Actuator Chain Graph). The *sensor-controller-actuator chain graph* $G^{sca}(V^{sca}, E^{sca})$ consists of a set of vertices V^{sca} representing task of the application and a set of edges E^{sca} modeling data dependencies. The set of vertices V^{sca} can be partitioned into sensor tasks $t \in T^s$, controller tasks $t \in T^c$, and actuator tasks $t \in T^a$.

As sensor task always provide data for controller tasks and controller tasks send control information to actuator tasks, the set of edges can be restricted to $E^{sca} \subseteq T^s \times T^c \cup T^c \times T^c \cup T^c \times T^a$.

An example of a sensor-controller-actuator chain graph consisting of two sensor tasks, two controller tasks, and a single actuator task is shown in Fig. 11.1(a). In a

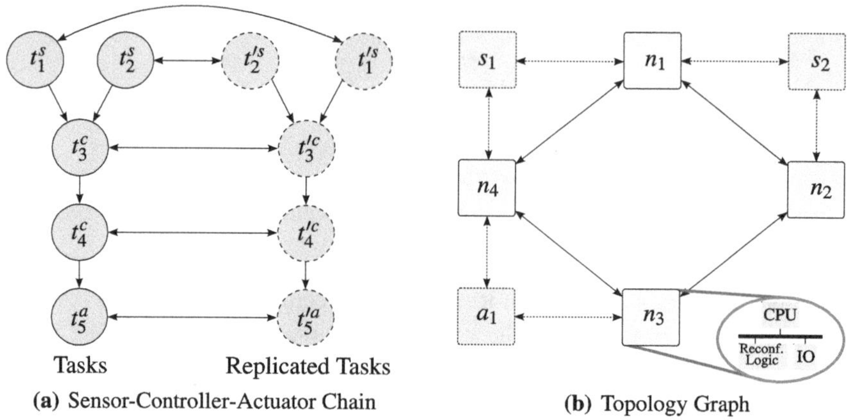

Fig. 11.1 (a) An application is modeled by a set of so-called *sensor-controller-actuator* chains. Each vertex represents a task or a replica task that both will be bound onto (b) the nodes of the *topology graph* modeling the ReCoNet architecture. Each ReCoNode $n \in N^{tg}$ is assumed to consist of a reconfigurable logic as well as at least one CPU.

ReCoNet, tasks might be replicated for fault-tolerance reasons and are called *replica tasks* $t_i^{\prime\{s,c,a\}}$.

Finally, the set of mapping edges which describes the binding possibilities of the tasks onto the network nodes. Due to the heterogeneity caused by the sensor nodes s_i and actuator nodes a_i in the ReCoNet architecture, the binding of sensor tasks t_i^s and actuator tasks t_i^a is restricted. In particular, a sensor task t_i^s is only allowed to be bound onto a computational node $n_j \in N$ if s_i is a direct predecessor of n_j, i.e., $(s_i, n_j) \in E^{tg}$. Analogously, an actuator task t_i^a is only allowed to be bound onto a computational node n_j if a corresponding actuator node a_i is a direct successor of n_j. It is assumed that all controller tasks t_i^c may run on each computational node n_j.

11.3 A Distributed Operating System Architecture for Networked Embedded Systems

All mechanisms for establishing a fault-tolerant and self-adaptive reconfigurable network have to be integrated in a distributed operating system infrastructure which is shown in Fig. 11.2, cf. [26]. While the reconfigurable network forms the physical layer consisting of reconfigurable nodes and communication links, the top layer represents the application that will be dynamically bound on the physical layer. This binding of tasks to resources is determined by an online partitioning approach that involves three basic and novel mechanisms: (1) *dynamic rerouting* of communication data, (2) *hardware/software task migration*, and (3) *hardware/software morphing*. Dynamic rerouting is necessary because messages will have to be sent between tasks that can migrate at runtime. The task migration mechanisms are required for moving tasks from one node to another while the hardware/software morphing allows a dynamic binding of tasks to either reconfigurable hardware resources or a CPU. The task migration and morphing mechanisms require in turn efficient hardware/software checkpointing methods such that states of tasks will not get lost. The basic network services for addressing nodes, detecting link failures and sending/receiving messages are discussed in [10].

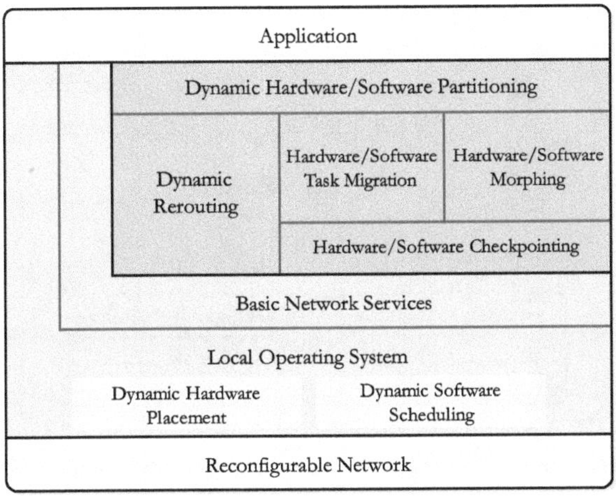

Fig. 11.2 Operating system layers of a fault-tolerant and self-adaptive network (ReCoNet). In order to abstract from the hardware, this local operating system is assumed to run at each node. On top of this local OS, basic network services are defined and used by the application to establish the intended fault tolerance and self-adaptiveness.

Beside the network services on the macro level, the hardware reconfiguration management on a particular ReCoNode has to be considered on the micro level. This includes the definition of communication architectures capable to efficiently integrate multiple partially reconfigurable modules into a system at runtime. Additionally, it is not only required to decide which node has to host which tasks within a ReCoNet (macro level), but also, in the case of hardware tasks, it is necessary to decide the placement position (FPGA) on a particular ReCoNode (micro level). This includes further management services including defragmenting the reconfigurable area. The reconfigurable area may get fragmented because different hardware tasks with individual resource requirements may be started and terminated over time. Consequently, the free resources on a ReCoNode alone are not sufficient for deciding if a hardware task can be placed on a ReCoNode or not. These problems have been solved in cooperation with the project ReCoNodes, cf. Chap. 10.

11.3.1 Self-healing and Self-adaptiveness

In the context of self-adaptiveness, we aim at integrating new tasks into a running ReCoNet. These tasks can be either platform-independent and can be executed on each ReCoNode in the network or they can be executed by a subset of all nodes in the network due to, e.g., I/O-interfaces or some inhomogeneity. While *self-adaptiveness* treats changes in the entire functionality of a network in general, *self-healing* treats the problem of maintaining all currently running functionality to keep running correctly in case of topology changes in the network. All in all, the treatment of these changes have to happen in limited time. Thus, we propose a two-step strategy consisting of a so-called *fast repair phase* and an *optimization phase* (see Fig. 11.3).

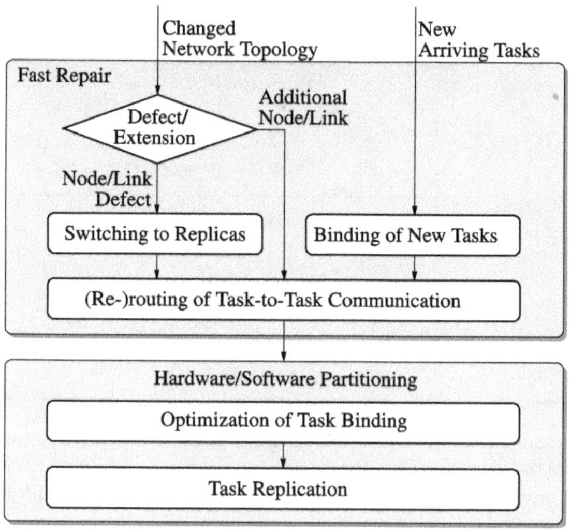

Fig. 11.3 If a link or node defects, the *fast repair* phase activates replicated tasks and reestablishes the communication routes. If the network is extended, all nodes collectively determine new communication routes between tasks hosted on the new nodes and the tasks in the network (*fast repair*). If a new task arrives, a node with enough computational resources is searched to bind the task onto it. The optimization phase optimizes the binding of tasks and creates new replicas.

In case of a node or link defect, the fast repair phase activates replicated tasks and reestablishes the communication between the tasks. If new tasks arrive and need to be bound onto nodes in the network, the same strategy as used for placing replicas is applied (cf. 11.3.1.2). Then, the communication between all tasks in the network is established if it does not exist yet. Afterwards, in the optimization phase, the problem of online hardware/software partitioning is solved with respect to different objectives, e.g., the overall computational load, the reliability, etc. When the hardware/software partitioning phase finishes, replicas need to be placed in order to tolerate future defects.

In the following two subsections, we present one investigated algorithm for online hardware/software partitioning followed by a method for task replica placement.

11.3.1.1 Discrete Diffusion-Based Task Binding

The first strategy to online hardware/software partitioning works with the objective to distribute tasks in the network such that the *computational load* between the hardware resources and the computational load between the software resources are equally balanced. For this purpose, the class of diffusion algorithms seems to be the most appropriate one in the context of a ReCoNet. In particular, it runs in a distributed manner in the network, allows balancing the load with local knowledge only, and may perform load exchanges with all neighbors simultaneously. In this context, we propose a discrete version of a diffusion algorithm in the following, cf. [25].

The algorithm moves load entities along the edges in the network to other nodes. Characteristic to a diffusion-based algorithm, cf. [2, 7], is that iteratively each node is allowed to move any size of load to each of its neighbors. Communication is only allowed along edges $e \in E^{tg}$. The quality of such an algorithm may be measured in terms of the number of iterations that are required in order to achieve a balanced state and in terms of the amount of load moved over the edges of the graph.

Definition 11.4 (Continuous Diffusion Algorithm). A local iterative load balancing algorithm performs iterations on the nodes of $v \in V^{tg}$ determining load exchanges between adjacent computing nodes. On each computing node $v_i^{\mathrm{c}} \in V^{tg}$, the following iteration k is performed:

$$\begin{aligned} ycont_e^{k-1} &= \gamma \cdot w_i^{k-1} - \gamma \cdot w_j^{k-1} \quad \forall e = (v_i^{\mathrm{c}}, v_j^{\mathrm{c}}) \in E^{tg} \\ xcont_e^{k} &= xcont_e^{k-1} + ycont_e^{k-1} \quad \forall e = (v_i^{\mathrm{c}}, v_j^{\mathrm{c}}) \in E^{tg} \\ w_i^{k} &= w_i^{k-1} - \sum_{e=(v_i^{\mathrm{c}}, v_j^{\mathrm{c}}) \in E^{tg}} ycont_e^{k-1} \end{aligned} \tag{11.1}$$

The real-valued load entity $ycont_e^k$ is calculated in each iteration k which is sent via edge $e \in E^{tg}$. This amount of load is a fraction γ of the load difference of two adjacent nodes connected by edge e. x_e^k is the amount of load sent via edge e until iteration k. w_i^k is the load of node i after the k-th iteration. If arbitrary real-valued load portions may be sent at each iteration k, it has been shown in [2] that the iteration converges to the average load $\bar{w}$. The number of iterations needed to obtain a certain error bound may be large and is in general not known a priori.

A slight modification of the iteration scheme above that works with changing values of γ in each iteration k has shown that the convergence speed can be drastically improved to exactly $K-1$ iterations [7]. Simply choose $\gamma = \frac{1}{\lambda_k}$ in the k-th iteration of Eq. (11.1) where λ_k, $1 \le k \le K-1$ denotes an arbitrary numbering for the K non-zero eigenvalues of L. L is called the *Laplacian*-matrix of the network and is defined as $L = D - B$ where D contains the node degrees as diagonal entries and B is the adjacency matrix of the network. Hence, $\gamma = \gamma_k = \frac{1}{\lambda_k}$, and in each iteration k, a node v_i^{c} adds a flow of $\frac{1}{\lambda_k}(w_i^{k-1} - w_j^{k-1})$ to the flow of edge $e = (v_i^{\mathrm{c}}, v_j^{\mathrm{c}})$, choosing a different eigenvalue for each iteration. An equally balanced load distribution is obtained after exactly $K-1$ iterations.

In order to apply a diffusion algorithm in applications where it is not possible to migrate a real-valued part of a task from one node to another, an extension is required. With our following extension, two problems will be handled:

- First of all, it is not advisable nor realistic to exchange a real-valued fraction of load between network nodes. This would imply that tasks need to be split and distributed to multiple nodes. This might produce a lot of data traffic in the network in addition to the inter-task communication. Moreover, the implementation of such mechanisms seems to be quite complex.
- Since the eigenvalue-based diffusion algorithm is an alternating iterative balancing scheme, it could occur that negative loads are assigned to computing nodes. It is obvious that such a balancing scheme does not work if tasks are migrated immediately in each balancing round.

Therefore, a novel discrete diffusion scheme is proposed which overcomes these problems. From now on, let $ycont_e^k$ be the real-valued continuous flow on edge $e \in E^{tg}$ which is determined by the continuous diffusion algorithm in iteration k (cf. Definition 11.4). Also, let $ydisc_e^k$ be the discrete flow on edge e determined by the discrete diffusion algorithm. Analogously, all variables of the continuous diffusion algorithm are extended with the index *cont* and all variables of the discrete version are extended with *disc*. Then, the discrete diffusion algorithm tries to optimize the following objective: Minimize the maximal load difference between software resources and hardware resources separately:

$$\max_{i,j} |wdisc_i^{SW} - wdisc_j^{SW}|$$

$$\max_{i,j} |wdisc_i^{HW} - wdisc_j^{HW}|$$

In order to fulfill these objectives, the discrete diffusion algorithms works in each iteration k with $k \in \{1, \ldots, K-1\}$ as follows:

In the first step of an iteration, the real-valued continuous flow $ycont_e^k$ on all edges for all nodes is computed. In the next step, each node tries to fulfill this real-valued continuous flow for its incident edges. Thus, it sends or receives tasks, respectively. Here, another optimization problem is encountered where a certain number and size of tasks of one node has to be chosen, in order to keep the optimality of the real-valued flow. This is an instance of the knapsack problem which is known to be NP-complete. Therefore, the discrete diffusion algorithm randomly selects tasks to be sent via one edge as long as the discrete flow $ydisc_e^k$ does not exceed the continuous flow or no more task remains on the node:

$$ydisc_e^k \leq ycont_e^k + \Delta_e^{k-1} \quad \text{with } \Delta_e^0 = 0 \tag{11.2}$$

In this equation, an error Δ_e^k which remains from the load exchange of the previous iteration is already respected. This error was caused by not fulfilling the optimality condition of the real-valued flow. This error Δ_e^k that occurred in the current iteration step can be computed as follows:

$$\Delta_e^k = ycont_e^k + \Delta_e^{k-1} - ydisc_e^k \quad \forall e = (v_i^c, v_j^c) \in E^{tg} \tag{11.3}$$

From here on, the discrete diffusion starts with a next iteration until the last iteration step $K-1$. In order to minimize the final error Δ_e^K, the error of the current iteration step is respected in the next iteration step, see (11.2). After the last iteration step, the remaining error Δ_e^{K-1} is minimized in one additional adjustment step in which nodes exchange tasks according to the error after $m-1$ iterations:

$$\Delta_e^K = \Delta_e^{K-1} - y_e^{adj} \tag{11.4}$$

y_e^{adj} denotes the flow in this adjustment step.

Compared to the continuous diffusion algorithm, it has been proven theoretically [25] that the proposed discrete version is always better concerning the congestion caused by load during the optimization. Moreover, in this monograph, the upper bounds for the deviation of the optimal load distribution and the discrete load distributions were theoretically deduced.

11.3.1.2 Replica Placement

According to the overall methodology presented in Fig. 11.3, task binding and replica placement are two subsequent steps with competing objectives. The task binding is performed with the objective to improve the performance of an application, i.e., reduce the traffic in the network (congestion) and balance the load on the network nodes such that the task response times and the overhead due to context switches are reduced. Due to the fact that the runtime behavior is of major interest, task binding is prioritized and executed first. Afterwards, a replica placement phase is started placing one replica per task onto the computing resources in the network. For this replica placement phase, the following *failure model* will be assumed:

Definition 11.5 (Failure Model). Only one node or communication link may fail simultaneously and another subsequent defect of such a resource occurs in a minimum arrival time greater than the repair and partitioning phase shown in Fig. 11.3.

Obviously, such a defect might result in a decomposition of a network into two or more disconnected parts. But for the presented replica placement algorithms, this assumption is very important and typically, the execution time of the online algorithm for repair and partitioning is much shorter than the time between two resource defects. Thus, this assumption does not reduce the applicability of the presented approach.

With the help of our failure model, it is possible to identify network regions in the topology graph G^{tg} that will under no circumstance decompose. Such components are called *biconnected components*: For the following examinations, let BCC_i be a set containing only the nodes of a biconnected component G_s: $BCC_i = V_s$. In networks with several biconnected components, so-called *articulation points* might occur which are critical to the reliability of the network: An articulation point is a vertex whose removal disconnects the graph, i.e., the graph is decomposed into disjoint components. Moreover, a *bridge* is an edge whose removal disconnects the graph, i.e., the graph is decomposed into disjoint components, cf. Fig. 11.4.

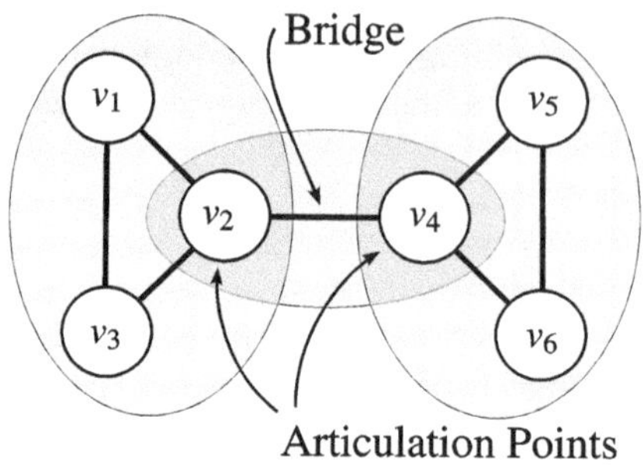

Fig. 11.4 Shown is a network with six nodes $v_1, \ldots, v_6$, with two articulation points v_2, v_4, one bridge (v_2, v_4), and three biconnected components given through the sets $\{v_1, v_2, v_3\}$, $\{v_4, v_5, v_6\}$, and $\{v_2, v_4\}$ and the corresponding edges.

After the biconnected components are identified in the network, each node searches for each task t_i it executes another node in the network which might be able to host its replica t_i'. This search can be simply implemented with a sort of depth-first search. Note that this depth-first search is also necessary for integrating new tasks in the network. A new task might be loaded to a node which is not able to execute the task because not enough computational resources are available. Therefore, the node starts to search for another node which is able to host the task t_i. In this search, the node sends a message $msg(t_i)$ to one neighboring node. The message $msg(t_i)$ consists of two ordered lists $msg(t_i).visited$ containing all visited nodes and $msg(t_i).backtracking$ with nodes which will be visited again if a search in a certain network direction was not successful. Additional task parameters are stored in $msg(t_i).constraints$ and contain the following information: (1) required resources like computational power or dedicated I/O components, (2) the current quality of the replica placement, and information about the current placement of the replica. Based on the information in $msg(t_i).constraints$, the receiver of the message $msg(t_i)$ can decide whether the replica will be accepted or not. If the receiver of $msg(t_i)$ accepts the new replica, it sends a message back to the former host node such that the task binaries can be transferred to the new host. In the following, several criteria are explained that allow for deciding whether a new replica placement is better than the current placement.

- *Max Rel* resembles the idea from [5] and tries to place tasks onto nodes with highest reliability as long as enough computational capacity is available and other resource constraints are not violated. The host node of the replica t_i' and its task t_i must not be the same. This strategy fails in the context of embedded systems as it does not account for data dependencies.
- *BCC Max1 Max Rel* places a replica in a biconnected component where the most adjacent tasks are located, i.e., tasks the replica communicates with af-

ter activation of the replica. Within this biconnected component, the replica is placed onto the most reliable node. As before, the host node of the replica t_i' and its task t_i must not be the same.

- *BCC Max2 Max Rel* places a replica in a biconnected component where the most tasks are located. In this case, data dependencies between tasks are not respected. Within this biconnected component, the replica is placed onto the most reliable node. Again, the host node of the replica t_i' and its task t_i must not be the same.
- *BCC Task Max Rel* places a replica t_i' in a biconnected component where the corresponding task t_i must not be located in. The idea is to distribute the tasks and replicas over the biconnected components such that one component with the entire functionality is still available after a defect. Within the biconnected component the replica is placed onto the most reliable node. If another biconnected component without the task t_i and a node with higher reliability is found, the replica t_i' will be placed onto the new node. If only one biconnected component exists, the replica is placed onto the node with the highest reliability in this component.

Note that these criterion can also be implemented neglecting the reliability values leading to *BCC Max1*, *BCC Max2*, and *BCC Task* criteria.

The entire replica placement approach is a greedy strategy which places a replica onto a network node if it improves the current binding. From this node, the search for a better node starts again. The decision whether the new network node is better or not is taken based on the above criteria. Note that the reliability values as well as the task binding are assumed to be static and known. Thus, a node which finds a better node for its replica will not be considered again. Thus, the algorithm has no cyclic behavior. Moreover, the approach runs asynchronously and concurrently in the network. For the termination of the algorithm, two possibilities exist: The algorithm runs until the next defect and tries to improve the replica placement all the time or the algorithm terminates after each replica migrated, resp. tried to migrate a certain number of times.

In the overall replica placement strategy, each node tries to find for each task t_i it executes another node for hosting a replica t_i'. Moreover, each node tries to improve the binding of the replicas it hosts. As an example, consider the network with a given task binding in Fig. 11.5. All tasks are assumed to consume half of the computational capacity of a node. Starting from the initial binding in Fig. 11.5(a), a node for the replica t_1' should be found using the BCC Max2 strategy, i.e., a replica will be placed in the biconnected component with the most tasks. For this purpose, our algorithm starts to search for another node which may host the replica. In this example, node v_3 is considered at first and has enough capacity to execute the replica t_1' in case of a defect. Thus, the replica is bound to v_3. This node starts again the search algorithm in order to improve the replica binding. The search algorithm considers v_2 but continues to v_4 because v_2 is also in a biconnected component with one task. Thus, the actual binding of the replica onto node v_3 would not be improved. By placing the replica onto node v_4, the replica is in a biconnected component with three tasks. Thus, the binding has been improved. Starting the search algorithm from

v_4 might lead to the situation shown in Fig. 11.5(d). On the way to node v_7, node v_5 is considered which is assumed to not improve the current binding. Afterwards, node v_8 is visited which has no free capacity for executing an additional task. Thus, the search stops at node v_7 which has enough capacity and may improve the current binding since four tasks belong to the same biconnected component.

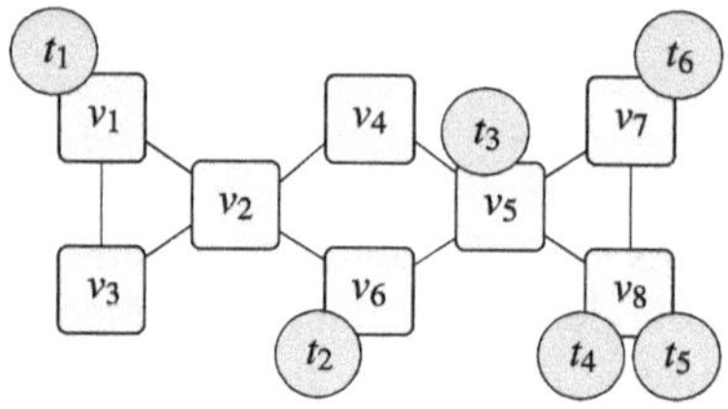

(a) initial task binding

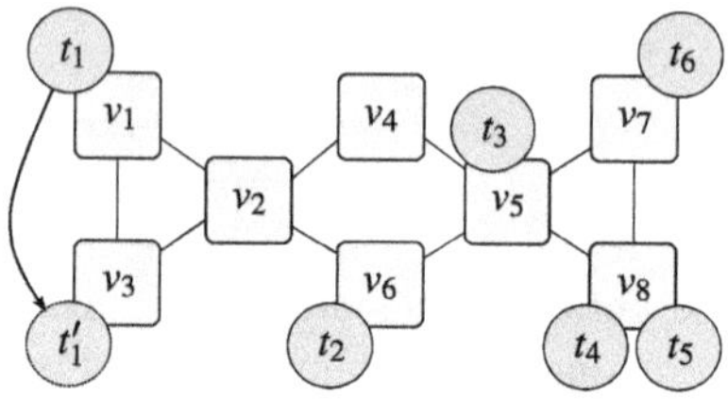

(b) using the search algorithm node v_3 is found for replica t_1' with one task in the same BCC

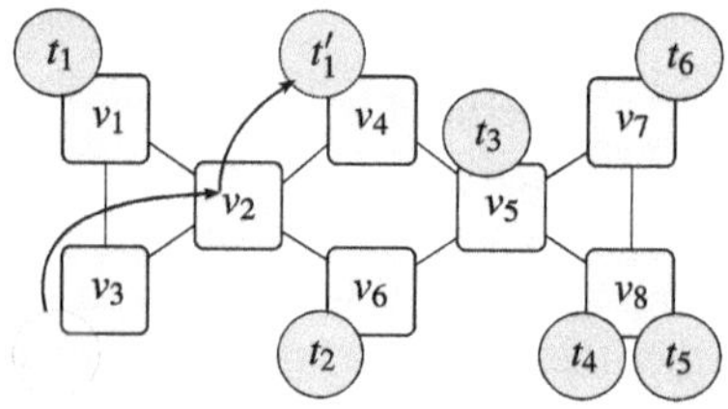

(c) next iteration: v_4 is found in a BCC with three tasks

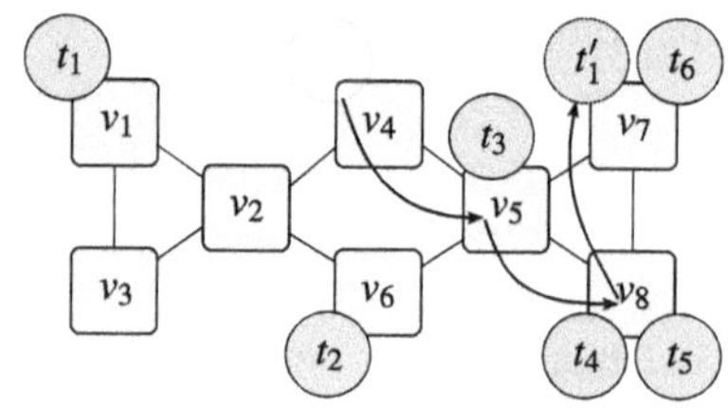

(d) next iteration: v_7 is found in a BCC with four tasks. Other nodes do not improve the binding of the replica. Thus, it will stay at node v_7.

Fig. 11.5 Exemplarily, four steps of BCC Max2 are shown until the replica for task t_1 is finally placed.

Extensive simulations have shown that our approach to replica placement may improve the reliability in terms of MTTF (Mean Time To Failure) by more than 20% compared to randomized replica placement strategies.

11.3.2 Hardware/Software Task Migration

In order to allow hardware/software partitioning at run-time, a mechanism for migrating software as well as hardware tasks between nodes is necessary. In case of software migration, two approaches can be considered:

- Each node in the network contains all software binaries, but executes only the assigned tasks. Unfortunately, this demands large memories on each node.

- Alternatively, binaries are transferred over the network and will be stored only on nodes that execute a task or host a replicated task. This migration of tasks demands a relocation of binaries in the memory, because nodes cannot reserve a certain part in the memory for a certain task. Due to this requirement, it has to be ensured that the generated binaries are location-independent.

Besides software functionality, it is desired to migrate functionality implemented in hardware between nodes in the reconfigurable network. Similar to the two approaches for software migration, two concepts for hardware migration exist:

- Each node in the network contains all hardware modules preloaded on the reconfigurable device. If a task implemented in hardware is bound to a certain node, the corresponding hardware module is enabled. This approach requires an unlikely huge amount of FPGA resources.
- For FPGAs supporting partial runtime reconfiguration, a more resource efficient approach can be applied. Configuration bitstreams of hardware modules are transferred between nodes and loaded to free resources on the FPGA. Comparable to location-independent software binaries, we demand that the configuration data is relocatable, too.

For speeding up the migration of hardware and software tasks, compression techniques can be used to hide some latency resulting from the relatively slow memories and the low network bandwidth. In [11], we accelerated the configuration process with a bitstream decompression accelerator that is capable to emit decompressed data at 400 MB/s while only requiring about 100 look-up tables. It was also shown that transferring software on an ESM (Erlangen Slot Machine) (see Chap. 3) can be improved with the help of a hardware accelerator [20] for decompressing binaries stored on the relatively slow flash memories.

11.3.3 Hardware/Software Morphing

Hardware/software morphing is required to dynamically change the implementation style of a task either to hardware or software on a ReCoNode. Naturally, not all tasks can be morphed from hardware to software or vice versa, e.g. tasks which drive or read I/O-pins. In [12], a propose a state machine-based approach. For this purpose, a task $t_i \in V^{sca}$ is assumed to be modeled by a finite state machine (FSM) $\mathcal{F} = (I, O, S, \delta, \omega, s_0)$, with input set I, output set O, state set S, state transition function δ, output function ω, and initial state s_0. In order to model hardware/software morphing formally, we define two mapping functions $M_{HW} : S \rightarrow S_{HW}$ and $M_{SW} : S \rightarrow S_{SW}$, where S_{HW} is the set of states of the hardware implementation whereas S_{SW} is the set of states of the software implementation of t_i, respectively. Note that, both, the hardware and software implementation, generally contain more states than the FSM model of t_i.

Definition 11.6 (Morph Point). A *morph point* $s \in S_m \subseteq S$ is a state $s \in S$ that has equivalent counterparts in both, the hardware and software implementation.

Furthermore, there exist inverse morph functions that permit to remap a state $s \in S$ from the refined hardware and software states:

$$\exists s \in S : M_{HW}^{-1}(M_{HW}(s)) = M_{SW}^{-1}(M_{SW}(s))$$

Basically, the morph process consists of three steps: At first, the state of a task has to be saved by storing a checkpoint (cf. Sec. 11.3.4). Then, the state *encoding* has to be transformed such that it can be loaded to the task with its new implementation style in the last step. This can be modeled by a so-called *morph FSM*.

Definition 11.7 (Morph FSM). Given a task $t_i \in V^{sca}$ modeled by a FSM $\mathcal{F}$, for each morph point $s_m \in S_m$ we extend the state set of the FSM by a corresponding *morph state* $s'_m \in S'_m$. A state machine FSM that can be morphed between hardware and software is called a *morph FSM* (MFSM). The corresponding MFSM $\mathcal{F}^M = (I', O', S', \delta', \omega', s'_0)$ can be derived as follows: $I' = I \times I_{morph}$; $I_{morph} = \{0, 1\}$ denotes an input signal indicating the intention to morph, $S' = S \cup S'_m$, $O' = O$, $\omega' = \omega$, $s'_0 = s_0$, and

$$\delta'(s', i') = \begin{cases} \delta(s', i) & \text{if } i' = (i, 0) \wedge s' \notin S'_m \text{ (normal operation)} \\ \delta(s', i) & \text{if } i' = (i, 1) \wedge s' \notin S_m \text{ (run until next morph point)} \\ s'_m & \text{if } i' = (-, 1) \wedge s' \in S_m \text{ (start morphing)} \\ s_m & \text{if } s' \in S'_m \text{ go back to morph point after morphing} \end{cases}$$

An example of a task FSM with the states $S = \{idle, run\}$ and one morph point $S_m = \{idle\}$ is presented in Fig. 11.6(a). The corresponding derived MFSM (Fig. 11.6(b)) contains the additional morph state $S'_m = \{idle'\}$. When the system decides to morph the task, this is indicated by the added input signal m. If this signal is active and if the state machine is in the morph point $idle$, the morphing takes place by entering the morph state $idle'$ and by translating internal state encodings to the state representation of the morphed domain. After finishing this process, the tasks continues its operation in the morphed implementation style, either hardware or software.

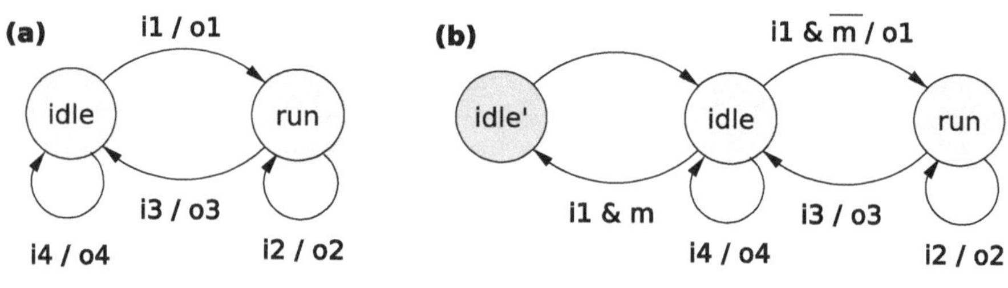

Fig. 11.6 Example of a task FSM (a) and the derived MFSM (b). A morph process starts from the *morph point idle* upon a morph request ($m = 1$). Then the state is translated into the states in the new implementation style (hardware or software) inside the *morph state idle'*.

A requirement to the interfaces of *morphable tasks* is that they have to be equivalent such that the surrounding system does not recognize the implementation style of the morphable task. Also, the transformation depends heavily on the implementation which especially leads to problems when transforming data types. While it is possible to represent numbers in hardware with almost arbitrary word width, current processors perform computations on 16 bit or 32 bit wide words. Thus, the numbers have to be extended or truncated. This modification causes again difficulties if numbers are presented in different number representations. The representation which can either be one's complements, two's complement, fixed point or floating point numbers, needs to be transformed, too. Additional complexity arises if functionality requires a sequential computation in software and a parallel computation in hardware. Exemplarily, assume that the computation of the parity of an 8 bit word is performed once in hardware and once in software. While the hardware module can evaluate the parity with an XOR-gate in parallel, the CPU needs to compute the parity by shifting and masking bits sequentially. Hence, during the computation of the parity, no transformable state exists. A contrary example is a sequential multiplier that requires several clock cycles in hardware but is atomic in software. Although the multiplication may require several cycles in the ALU of the processor, it is impossible to interrupt and extract the state (containing some intermediate values) of the multiplier. Due to these implementation dependent constraints, an automated morph-function generation only for bit vectors in the hardware that are interpreted as integers in the software is currently supported. The designer needs to give information about the possible morph states called *morph points* and together with the help of the automated insertion of checkpoints into hardware/software tasks, the hardware/software morphing becomes possible. Hardware/software morphing is discussed in more depth in [12].

11.3.4 Hardware/Software Checkpointing

Checkpointing is used during both, task migration and hardware/software morphing. Before suspending a task t, its context $C(t)$ must be saved and transferred together with its bitstream or binary, resp. to another node. This context is called a *checkpoint* and represents an image of the last error-free state of a module of computation from which it can restart in case of a fault. Consequently, it is used to update the state information between a task and its replica at certain times. It helps to reduce repair times by requiring computations from the last checkpoint at the side of a task replica after occurrence of a node fault. While software checkpointing has been a subject of former research, e.g., [6], we will point out how hardware checkpointing can be integrated into available designs.

Assuming an FSM model of operation of each hardware task t, a checkpoint FSM (CFSM) can be derived by making the internal state $C(t)$ accessible via the input and output. From a technical perspective, this comprises to extend a hardware module such that its internal state, represented by all flip-flop values, can be stored atomically to a safe memory. Basically, three possibilities for accessing the state of a hardware module have been examined:

1. *Scan Chain*: Similar to the boundary scan used in ASIC design, an extra scan multiplexer in front of each flip-flop allows to switch between a *regular execution mode* and *scan mode*. In the latter one, the registers are linked together to a shift register chain. The data inside the chain represents the checkpoint. For keeping the checkpoint consistent during the sequential checkpoint read and write processes, the chain is connected to a ring shifter and a state machine keeps track about the current shift position of the checkpoint.
2. *Scan Chain with Shadow Registers*: Each flip-flop of the original circuit will be duplicated and only these duplicated flip-flops are connected together in a chain for accessing the checkpoint. Data can be exchanged between the shadow register chain and the module flip-flops in a single clock cycle, while performing the state transfer concurrently with the module operation.
3. *Memory Mapping*: Alternatively, each flip-flop can be mapped directly into the address space of the system, hence allowing a CPU to access the checkpoint.

A tool was developed for automatically including any of the three state access mechanisms into a module at the netlist level, cf. [13].

11.4 Design and Synthesis of ReCoNets

As each task can run on different nodes in a ReCoNet, many decisions must be taken in the phase of building a new ReCoNet. In order to support designers in such a complex decision making, a new system synthesis approach is required.

11.4.1 Design Space Exploration

Our design space exploration searches for an ideal initial placement of tasks and replica tasks to increase the reliability of the system while minimizing the load on computational nodes and communication links. Besides reliability and load, other objectives like, e.g., monetary cost, area and power consumption have to be considered. Thus, the performed design space exploration is a multi-objective optimization problem, that aims to find a set of so called *Pareto-optimal* implementations. An implementation is Pareto-optimal, if it is better in at least one objective when compared to any other feasible implementation.

The used design space optimization approach performs the binding of tasks and shadow tasks as well as the placement of voters, if appropriate. It is based on the combination of a *Pseudo-Boolean* (PB) solver and a modern *Multi-Objective Evolutionary Algorithm* (MOEA). MOEAs are a *population*-based optimization heuristic taking advantage of the principles of biological evolution. The optimization is done iteratively in two alternating steps, the *variation* and *selection*. In the variation, new solutions, i.e., a new *generation*, is created from a set of existing solutions in the population. The design space exploration of dynamically reconfigurable systems has been integrated with approaches developed at the University of Paderborn, cf. Chap. 9 to determine optimal configurations and approaches developed at the University of Oldenburg, cf. Chap. 7 to simulate hardware reconfiguration at a cycle accurate level [4].

11.4.2 Dependability Analysis

To evaluate the reliability and safety benefits of the proposed ReCoNet methodology, an algorithm for an efficient analytical calculation of the *Mean Time To Failure* (MTTF) to quantify lifetime reliability and the *Mean Time To Unsafe Failure* (MTTUF) [3] to quantify safety has been developed, cf. [24]. While the MTTF denotes the expected time of the system to operate correctly, the MTTUF denotes the expected time that the system operates until a hazardous failure happens that the system cannot detect. That hazardous failure may result in a safety critical situation, thus, the MTTUF, although strongly related to the MTTF, is a safety measure. It has been shown, that these two measures do not deteriorate the optimization process as two other, arbitrarily chosen dependability measures would do, cf. [23].

Given the reliability of a system $R(\tau)$ and the safety $S(\tau)$, the MTTF calculates as $\int_0^\infty R(\tau)d\tau$ and the MTTUF as $\int_0^\infty S(\tau)d\tau$. To evaluate the reliability $R(\tau)$ and safety $S(\tau)$ of a ReCoNet with nodes V^{tg} and task assignment $\beta : V^{sca} \to V^{tg}$, the reliability analysis approach presented in [8] is used. Since a *fail-silent* [22] behavior is assumed in [8], an explicit introduction of the voting structures is needed.

If a task does not obey a fail silent behavior, additional fault detection mechanisms are required to enable its replica task to notice, if the task is faulty. This fault detection and, if possible, fault toleration is done by voters which are hardware resources or software tasks that aim to find a majority of k values delivered by n redundant task instances. Such voters are known as k-out-of-n-majority voting structures and are typically implemented as so called duplex voter (2-out-of-2-majority) or the well known *Triple Modular Redundancy* (TMR, 2-out-of-3-majority). The entire system synthesis approach is transparent for the designer, i.e., the system specification given by the designer should be as abstract as possible, while implementation details are hidden and derived automatically by an optimization process.

To evaluate a system including its implemented voting strategies, the structure function $\varphi : \{0,1\}^{|\alpha|+|V|} \to \{0,1\}$ with the Boolean vectors $\boldsymbol{\alpha} = (\boldsymbol{r}_1, \ldots, \boldsymbol{r}_{|V^{tg}|})$ and $\boldsymbol{V} = (\boldsymbol{v}_1, \ldots, \boldsymbol{v}_{|V^{sca}|})$ has to be calculated and represented as a *Binary Decision Diagram* (BDD) [1]. φ evaluates to 1 if and only if the system is operating properly under the given set of properly working resources and voters. Otherwise, φ evaluates to 0. Hereby, for each allocated resource $r \in \alpha$ and voter $v \in V$, the corresponding binary variables $\boldsymbol{r} = 1$ and $\boldsymbol{v} = 1$ indicate a proper operation while $\boldsymbol{r} = 0$ and $\boldsymbol{v} = 0$ indicate a defect, respectively. φ can be used to derive the reliability and safety, respectively, by the following recursive definition with $R_r(\tau)$ describing the specific reliability of a single resource node r over time τ:

$$\varphi(\tau) = R_r(\tau) \cdot \varphi|_{r=1}(\tau) + (1 - R_r(\tau)) \cdot \varphi|_{r=0}(\tau) \qquad (11.5)$$

Encoding the prerequisites for a working ReCoNet as presented in [9], the resulting BDD represents the ability of the system to cope with resource defects.

Using this analytically computed upper bound for the dependability, the replication strategies described in Sect. 11.3.1.2 were compared to the achievable MTTF_{max} by simulating them. It has been shown, that there is only a relatively small difference in the effectiveness of the different replica placement methodologies with regard to the known upper bound, while in [27], the relative difference between the methodologies seemed significant. With regard to the upper bound, the designer may choose the load balancing approach as well, since this approach is less than 5% worse than the BCC technique and offers a better load balancing of the resources.

11.5 Demonstrator

The previously described methods have been implemented and tested on the basis of a network consisting of four FPGA-based boards with a CPU and configurable logic resources. As an example, we implemented a driver assistant system that warns the driver in case of an unintended lane change and is implemented in a distributed manner in the network. As shown in Fig. 11.7, a camera is connected to node n4. The camera's video stream is then processed in basically three steps: (a) preprocessing, (b) segmentation, (c) lane detection. Each step is implemented as one task. The result of the lane detection is evaluated in a control task that gets in addition the present state of the drop arm switch. If the driver changes the lane without switching on the correct turn signal, an acoustic signal will warn the driver of an unintended lane change.

As depicted in Fig. 11.7, our prototype implementation of a ReCoNet consists of four fully connected FPGA boards. Each node is configured with a NIOS-II softcore CPU running MicroC/OS-II as a local operating system. The local OS supports multi-tasking through preemptive scheduling and has been extended by a message passing system. On top of this extended MicroC/OS-II, we implemented the different layers as depicted in Fig. 11.2. In detail, these are functions for checkpointing, task migration, task morphing and online hardware/software partitioning.

For fault-tolerance reasons, the ReCoNet uses a Point-to-point (P2P) communication protocol. In comparison to a bus, this will produce some overhead by the routing on the one side while omitting the problem of bus-arbitration. The routing allows us to deal with link failures by changing the routing tables in such a way that data can be sent via alternative paths. Besides fault tolerance, P2P networks have the advantage of an extreme high total bandwidth. In the present implementation, we set the physical data transfer rate of a single link to 12.5 Mb/s and measured a

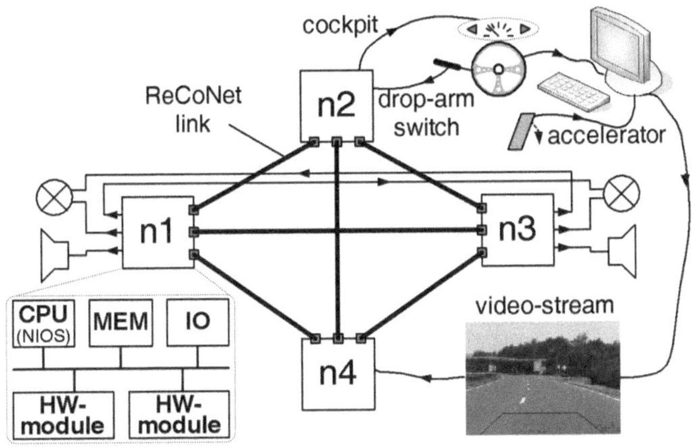

Fig. 11.7 Schematic composition of a ReCoNet demonstrator of an automotive driver assistance system: On the basis of four connected FPGA-boards, we implemented an operating system infrastructure which executes a distributed lane detection algorithm. This application warns the driver acoustically in case of an unintended lane change. Note that this ReCoNet is built upon heterogeneous nodes providing FPGAs from different vendors working seamlessly together [26].

maximum throughput of 700 kByte/s allowing to transfer the video stream in our driver assistant application.

Although the MicroC/OS-II has no runtime system that permits dynamic task creation, we enabled the software task migration by transferring binaries to other nodes and linking OS functions to the same address such that tasks can access these functions on each node. This methodology reduces the amount of transferred binary data drastically compared to the alternative that the OS functions are transferred either. Also, this methodology avoids implementing a complex runtime system and the operating system keeps tiny.

For integrating hardware modules at runtime, an according communication architecture is required on each RoCoNode. This includes (1) buses for shared memory communication (DMA, or register file access) and (2) dedicated point-to-point links for streaming data or I/O links. The communication architecture has material impact on the throughput, and hence, on the overall performance of the system. Furthermore, the system should not only provide one or more reconfigurable islands, as presented in prior work [21], but one fine-tiled resource area for hosting multiple modules in a flexible manner. In [18], it has been proven that the optimal size of such a *resource slot* should be in the rage of just 200–300 look-up tables for minimizing the logic overhead. For bus-based communication with runtime reconfigurable hardware, we developed the so-called *ReCoBus* [16, 19]. It uses an interleaving of bus signals and, hence, significantly reduces the amount of logic for implementation and provides high throughput. Beside the communication over the ReCoBus, reconfigurable modules may require links to I/O pins or point-to-point connections to other modules in the system or within the reconfigurable part of the system. This issue has been solved by reserving a horizontal set of wires that are aligned in parallel to the ReCoBus. These wires are routed in a similar regular fashion as the ReCoBus itself [17]. A tool for supporting designers to parameterize and integrate a ReCoBus into an FPGA-based systems has been implemented as prototype [14, 15].

In 2006, the ReCoNet demonstrator has been accepted for presentation at the Hanover Messe. Subsequently, it was also invited for presentation at ESOF 2006, the European Science Forum.

Acknowledgements This work was supported by DFG grant TE 163/10, project *ReCoNets*, as part of the Priority Programme 1148, *Reconfigurable Computing Systems*.

References

1. Bryant, R.E.: Graph-based algorithms for Boolean function manipulation. IEEE Trans. Comput. **35**(8), 677–691 (1986)
2. Cybenko, G.: Dynamic load balancing for distributed memory multiprocessors. J. Parall. Distrib. Comput. **7**, 279–301 (1989)
3. DeLong, T., Smith, D., Johnson, B.: Dependability metrics to assess safety-critical systems. IEEE Trans. Reliab. **54**(3), 498–505 (2005)
4. Dittmann, F., Rammig, F., Streubühr, M., Haubelt, C., Schallenberg, A., Nebel, W.: Exploration, partitioning and simulation of reconfigurable systems. Inf. Technol. **49**(3), 149–156 (2007)
5. Douceur, J.R., Wattenhofer, R.: Competitive hill-climbing strategies for replica placement in a distributed file system. In: DISC '01: Proceedings of the 15th International Conference on Distributed Computing, pp. 48–62. Springer, London (2001)
6. Elnozahy, E.N.M., Alvisi, L., Wang, Y.M., Johnson, D.: A survey of rollback-recovery protocols in message-passing systems. ACM Comput. Surv. **34**(3) (2002)
7. Elsässer, R., Frommer, A., Monien, B., Preis, R.: Optimal and alternating-direction loadbalancing schemes. In: Proc. of Euro-Par 99, Parallel Processing, pp. 280–290 (1999)
8. Glaß, M., Lukasiewycz, M., Streichert, T., Haubelt, C., Teich, J.: Reliability-aware system synthesis. In: Proc. of DATE '07, pp. 409–414 (2007)
9. Glaß, M., Lukasiewycz, M., Reimann, F., Haubelt, C., Teich, J.: Symbolic reliability analysis of self-healing networked embedded systems. In: Proceedings of the International Conference on Computer Safety, Reliability and Security (SAFECOMP 2008), Newcastle upon Tyne, UK, pp. 139–152 (2008)
10. Koch, D., Streichert, T., Dittrich, S., Strengert, C., Haubelt, C., Teich, J.: An operating system infrastructure for fault-tolerant reconfigurable networks. In: Proceedings of the 19th International Conference on Architecture of Computing Systems (ARCS 2006), Frankfurt/Main, Germany, pp. 202–216. Springer, Frankfurt (2006)
11. Koch, D., Beckhoff, C., Teich, J.: Bitstream decompression for high speed FPGA configuration from slow memories. In: Proceedings of International Conference on Field-Programmable Technology 2007 (ICFPT07), pp. 161–168. IEEE Press, Kokurakita (2007)
12. Koch, D., Haubelt, C., Streichert, T., Teich, J.: Modeling and synthesis of hardware/software morphing. In: ISCAS '07, New Orleans, CA, USA, pp. 2746–2749 (2007)
13. Koch, D., Haubelt, C., Teich, J.: Efficient hardware checkpointing—concepts, overhead analysis, and implementation. In: Proceedings of the 15th ACM/SIGDA International Symposium on Field-Programmable Gate Arrays (FPGA 2007), pp. 188–196. ACM, Monterey (2007)
14. Koch, D., Beckhoff, C., Teich, J.: ReCoBus-Builder—a Novel Tool and Technique to Build Statically and Dynamically Reconfigurable Systems for FPGAs. In: Proceedings of International Conference on Field-Programmable Logic and Applications (FPL 08), Heidelberg, Germany, pp. 119–124 (2008)

15. Koch, D., Haubelt, C., Teich, J.: Efficient reconfigurable on-chip buses for FPGAs. In: 16th Annual IEEE Symposium on Field-Programmable Custom Computing Machines (FCCM 2008), pp. 287–290. IEEE Computer Society, Palo Alto (2008)
16. Koch, D., Streichert, T., Haubelt, C., Teich, J.: Logic Chip, logic system and method for designing a logic chip. PCT/EP2008/007342 (2008)
17. Koch, D., Streichert, T., Haubelt, C., Teich, J.: Logic chip, method and computer program for providing a configuration information for a configurable logic chip. PCT/EP2008/007343 (2008)
18. Koch, D., Beckhoff, C., Teich, J.: Minimizing internal fragmentation by fine-grained two-dimensional module placement for runtime reconfigurable systems. In: 17th Annual IEEE Symposium on Field-Programmable Custom Computing Machines (FCCM 2009), pp. 251–254. IEEE Computer Society, Napa (2009)
19. Koch, D., Beckhoff, C., Teich, J.: A communication architecture for complex runtime reconfigurable systems and its implementation on spartan-3 FPGAs. In: Proceedings of the 17th ACM/SIGDA International Symposium on Field-Programmable Gate Arrays (FPGA 2009), pp. 233–236. ACM, Monterey (2009)
20. Koch, D., Beckhoff, C., Teich, J.: Hardware decompression techniques for FPGA-based embedded systems. ACM Trans. Reconfigurable Technol. Syst. **2**(2), 1–23 (2009)
21. Lysaght, P., Blodget, B., Mason, J., Young, J., Bridgford, B.: Invited paper: Enhanced architecture, design methodologies and CAD tools for dynamic reconfiguration of Xilinx FPGAs. In: Proc. of the 16th Int. Conf. on Field Programmable Logic and Application (FPL'06), pp. 1–6 (2006)
22. Powell, D., Bonn, G., Seaton, D., Verissimo, P., Waeselynck, F.: The delta-4 approach to dependability in open distributed computing systems. In: Symposium on Fault-Tolerant Computing, pp. 56–61 (1995)
23. Purshouse, R., Fleming, P.: On the evolutionary optimization of many conflicting objectives. IEEE Trans. Evol. Comput. **11**(6), 770–784 (2007)
24. Reimann, F., Glaß, M., Lukasiewycz, M., Haubelt, C., Keinert, J., Teich, J.: Symbolic Voter Placement for Dependability-Aware System Synthesis. In: Proceedings of the 6th International Conference on Hardware/Software Codesign and System Synthesis (CODES+ISSS), Atlanta GA, USA, pp. 237–242 (2008)
25. Streichert, T., Haubelt, C., Teich, J.: Distributed HW/SW-partitioning for embedded reconfigurable systems. In: Proc. of DATE '05, Munich, Germany, pp. 894–895 (2005)
26. Streichert, T., Koch, D., Haubelt, C., Teich, J.: Modeling and design of fault-tolerant and self-adaptive reconfigurable networked embedded systems. EURASIP J. Embed. Syst. **2006**, 1–15 (2006). doi:10.1155/ES/2006/42168
27. Streichert, T., Glaß, M., Wanka, R., Haubelt, C., Teich, J.: Topology-aware replica placement in fault-tolerant embedded networks. In: Proceedings of ARCS '08, pp. 23–37 (2008)
28. Streichert, T., Haubelt, C., Koch, D., Teich, J.: Concepts for self-adaptive and self-healing networked embedded systems. In: Würtz, R.P. (ed.) Organic Computing. Understanding Complex Systems, pp. 241–260. Springer, Berlin (2008)

Chapter 12
Adaptive Runtime System with Intelligent Allocation of Dynamically Reconfigurable Function Model and Optimized Interface Topologies

Lars Braun, Tobias Schwalb, Philipp Graf, Michael Hübner, Michael Ullmann, K.D. Müller-Glaser, and Jürgen Becker

Abstract Self-reconfiguration and adaptivity are important new concepts for reconfigurable hardware. The benefits include, for example, a reduction in power dissipation by sharing resources therefore requiring smaller reconfigurable chips. Idle applications are substituted on demand by actually needed functions. As a result, the number of possible functions controlled by such systems increases without raising the number of additionally required processing elements. Parallel tasks (functions) in hardware execute more efficiently compared with sequential microprocessors. The high performance of reconfigurable hardware and the possibility of hardware parallelization, help to overcome increasing problems of data processing with traditional microcontroller and microprocessors in future complex electronic systems. A relevant issue is the use of adaptive reconfigurable systems in real-time applications, which is one of the basic conditions for a variety of target applications. New approaches to create systems, which are able to manage their own configuration are called run-time systems. These systems use the flexibility of e.g. an FPGA by partially changing the configuration. Only the necessary functions are configured in the chip's memory. On demand one or more functions can be substituted by another while other parts stay operative. The ALadyn project targets exactly this kind of adaptive system approach and investigates in the physical hardware realization, system modeling and development tools.

Lars Braun · Tobias Schwalb · Michael Hübner · Michael Ullmann · K.D. Müller-Glaser · Jürgen Becker
Institut für Technik der Informationsverarbeitung (ITIV), University of Karlsruhe, Karlsruhe, Germany, e-mails: Braun@itiv.uni-karlsruhe.de, Schwalb@itiv.uni-karlsruhe.de, Huebner@itiv.uni-karlsruhe.de, Ullmann@itiv.uni-karlsruhe.de, Mueller-Glaser@itiv.uni-karlsruhe.de, Becker@itiv.uni-karlsruhe.de

Philipp Graf
FZI Forschungszentrum Informatik an der Universität Karlsruhe, Karlsruhe, Germany, e-mail: graf@fzi.de

M. Platzner et al. (eds.), *Dynamically Reconfigurable Systems*,
DOI 10.1007/978-90-481-3485-4_12,

12.1 Introduction

Complex systems regularly consist of mixed hardware/software parts due to performance reasons. Very often it is not required, that all in hardware realized modules of the system have to be performed in one point of time. In order to enable this, an adaptive run-time system with an intelligent process control for the hardware/software functional patterns is required. In this scenario, not all possible permutations consisting of the different functional patters have to be generated in advance. The novelty is that a Quality of Service (QoS) negotiation process with the run-time system leads to a dynamic on-demand and intelligent allocation and/or reallocation of functional patterns to the respective hardware resources. In this process not only the functional patterns are important for the system, also the communication patterns have to be considered. Also the most efficient communication infrastructure has to be provided dynamically by the run-time system. Possible realizations are e.g. a Network-on-Chip and the respective topologies (Star, Torus etc.), a bus system or a heterogeneous approach. The hardware/software realization of the topics mentioned above exploits reconfigurable hardware (FPGAs). Until today, no commercial tool or a toolset supports seamless design of run-time adaptive systems with the feature of the ALadyn approach. Due to this fact, within the ALadyn project a pioneering work in the area of tool development as well as the development of mechanisms for hardware reconfiguration was done. A meta-layer was developed and realized which enables the description of characteristics for the reconfiguration in the system specification. For this purpose, a hierarchical system model supporting and exploiting the different features of reconfigurable hardware architectures was developed with the goal for targeting a wide range of embedded applications. The applications can request functions using an API which then are realized physically as soft- or hardware in relation to the system status. The determination of suitable realizations of the functions with respect to the system status is performed with the intelligent distribution management system which realizes the QoS negotiation process as described above. On the lower layer of the system, local reconfigurable components (e.g. FPGAs) or dedicated processors (e.g. DSP) provide the physical realization of the requested functional patterns. A local run-time control system supports the dynamic behavior. In the case of the FPGA this system has to control e.g. the dynamic and partial reconfiguration. A schematic view of the ALadyn layers is shown in Fig. 12.1.

To show our ALadyn system we describe in Sect. 12.2 firstly the general aspects of partial and dynamic reconfiguration. The associated 1D and 2D Network-on-Chip is presented in Sect. 12.3. Thereby, the on demand system adaptation is shown in Sect. 12.4. After discussing the system modeling for the ALadyn architecture in Sect. 12.5, we introduce a model based tool chain in Sect. 12.6. We give detailed information on functionality and architecture of the model debugging in Sect. 12.7. The following section focuses on the linked model based test. We conclude and show relationships to other projects in Sect. 12.8.

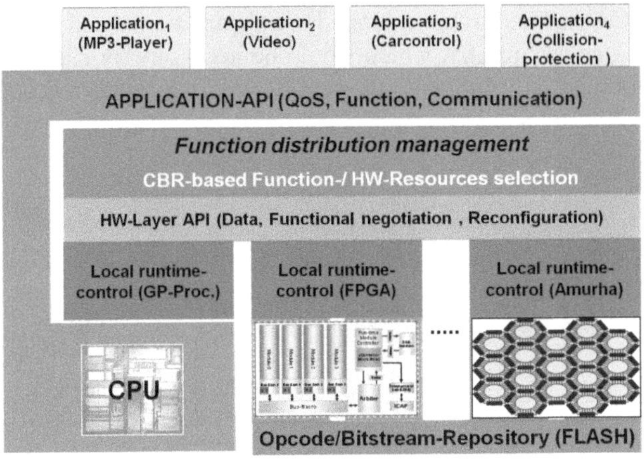

Fig. 12.1 ALadyn stack.

12.2 Partial and Dynamic Reconfiguration

12.2.1 One-dimensional Dynamic and Partial Reconfigurable System Architecture

Figure 12.2 shows a schematic representation of the entire system. The system consists of a runtime system (Run-Time Module Controller), which is based on a Xilinx MicroBlaze soft-core processor. Through its 32-bit RISC architecture it perfectly fits to the needs of the runtime system. An arbiter is connected to the MicroBlaze, which provides the connection via the FPGA internal bus system to the function modules. The arbiter is used to control the bus. The four modules are connected by a bus communication module (Bus Com) to the as macro implemented bus. With a fixed allocation of identification numbers, each module is addressable at any time. The number of four module slots was chosen in the first run of the system, because at the time of conception no estimation on the size of the entire system could be applied. Through the choice of four module slots it could be ensured to have enough space for the functions in each slot. The Bus Com modules serve as an interface for coupling functions to the FPGA bus, and they are completely identical for each slot. This allows a standardized interface definition for the design of function modules. The bus-macro modules, which are proposed in this approach were the first implementation in an LUT-based logic and represent an architecture-independent basis for all other physical realizations of this systems approach. In addition, this result is the basis for all hardware, which was realized within the concept of dynamic and partial reconfiguration within the so-called "Schwerpunktprogramm" [14]. Further, a decompression unit and an ICAP module are shown in Fig. 12.2. Connected to the microcontroller these modules are capable to read the reconfiguration data from an external flash memory, decompress it and send it to the Internal Configuration

Access Port (ICAP). Thereby an internal dynamic and partial self-reconfiguration is possible. An external circuitry, which would be necessary by using the SelectMap interface, is not required. As link to the outside world the system is connected to an external CAN controller.

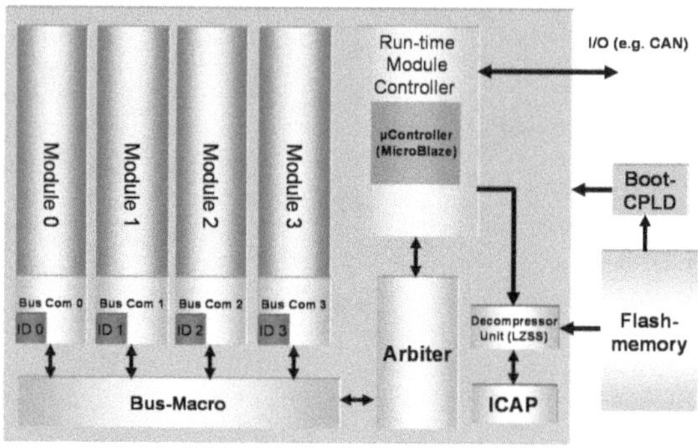

Fig. 12.2 Schematic representation of the 1-dimensional system approach for a runtime adaptive system.

12.2.1.1 LUT Based Communication Primitives

To establish a communication between modules busmacros are used. On the one hand this is necessary to specify an identifiable physical interface point for the reconfigurable area. Logic in this region can then use this interface to be connected. On the other hand, the busmacro is a mechanism to separate electrically the different reconfigurable regions. This prevents the violation of the principles of digital circuits. Short-circuits lead to increased power and the chip could possibly be damaged. Originally the busmacros proposed by Xilinx used TBUF elements. TBUF elements are Tristate drivers which drive TBUF lines. The disadvantage of these macros are its use of the limited TBUF resources, there are only four signal lines per CLB row available. Furthermore, on the new Xilinx FPGAs Tristate Buffers are no longer available which increases the need of an LUT-based implementation of the busmarcos. The basic idea of the new macro is to replace the TBUF elements by slices. This offers several advantages. On the one hand, the outputs of the macro external pins are assigned exactly on one side of the macro. This prevents undesirable wiring beyond the module boundaries. On the other hand, four slices, each with two Look-Up Tables are available per CLB row. Therefore, eight signal lines per CLB row can be realized. This is a doubling of the signal lines compared to the TBUF macros. In addition, the functionality of the macros can be extended by the functionalities, which are available at the slice resources. Thus, multiplexers can be realized in the Look-Up Tables or the flip-flops can be used to synchronize the transferred data. Another major advantage is that local wiring lines are used. These represent a much more flexible wiring resource and are available in a higher number. In the system of Fig. 12.2 in each slot four slices are placed to define the input and output pins of the macro. Figure 12.3 shows the structure of the macros. The output macro (Fig. 12.4) connects the outputs of the module to the input of the system. The basic structure of this macro is similar to the previous ones. However, each line has four sources within the slots and a sink in the static system area. Since the signal lines are interrupted by a slice in each slot, a combination of double and hex lines are used. The output of a slice is connected with a hex line. Then, a double line is added to connect the next slice. To determine the timing, the macros are analyzed with the timing analyzer which is part of the Xilinx ISE software. The critical paths of the macros have lag times of 5.651 ns and 6.139 ns. These values certainly contain additional delay of the test flip-flops, which were necessary to build a test system for the timing analyzer. If this delay time is subtracted, the macros themselves have delays of about 5–5.5 ns. However, it must be considered that the macros include no retiming elements. More information about the innovative "Busmacroarchitektur" is described in [14].

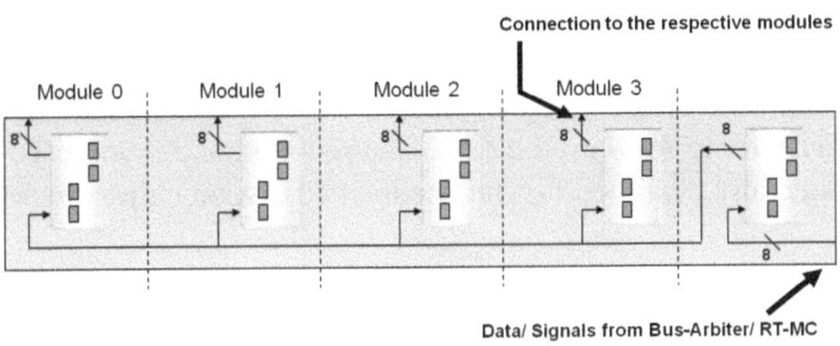

Fig. 12.3 Input macros from the arbiter to the modules.

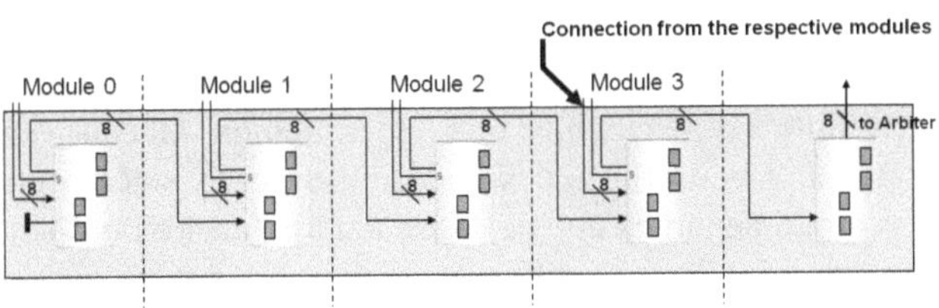

Fig. 12.4 Output macros from the modules to the arbiter.

As described in previous chapters, Field Programmable Gate Arrays (FPGA) of Xilinx allow the exploitation of dynamic and partial reconfiguration. For technical reasons, the Virtex-2 FPGA series only supports reconfiguration of complete columns. By manipulating the configuration bitstream it is possible to create variable rectangular reconfigurable areas. To perform a rectangular partial reconfiguration an appropriate communication structure to connect the different areas has to

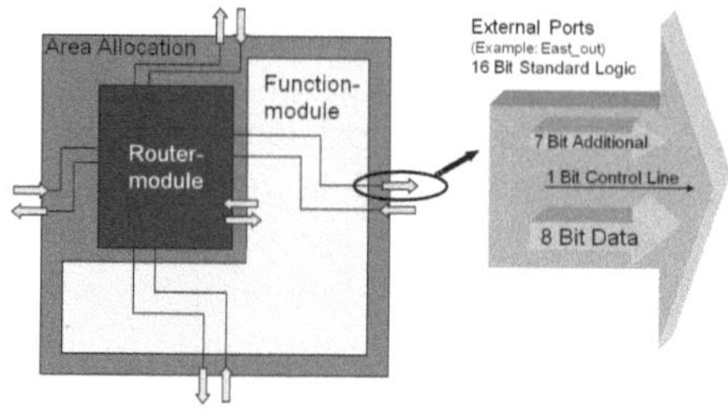

Fig. 12.5 Router module and functional unit.

be developed. In the following section the methods and the system integration for a two-dimensional partial reconfigurable system is described.

12.2.1.2 Physically Realized Two-dimensional System Approach

To achieve a solution the integrated network-topology has to been specified as first step. Since the FPGA is two-dimensional rectangular shaped a 2D mesh topology is selected. In the following two different mesh-based networks are developed. The first one uses a router to forward the data. The second one is switch-based, whereas the switch is specially designed for the needs of FPGA on chip network.

12.3 Network on Chip

12.3.1 Router Base Modules for the On-line Placement

In the previous chapter it was suggested that a functional unit connected to the router is added. The functional unit, which is connected to the router, can be structured quite simply, such as an adder or subtracter, but it can also contain more complex structures such as finite a state machine or a larger memory. Because of these different space utilizations, it is useful to design also differently sized router base modules. Their sizes should be large enough to contain the function but without wasting resources. Therefore initially four different module sizes were created which can be seen in image Sect. 12.6. In the picture the 1×1 base modules can be seen, as well as other modules of sizes 2×1, 2×2 and 1×2. It should be noted that all modules are multiples of the basic 1×1 module. Thereby each module can easily be substituted by any other module or any other module combination. The connection points are always on the same height, so that an inter-module communication gets never lost by exchanging or replacing the modules. Hence an existing connection between adjacent modules is guaranteed. It results in a kind of puzzle, in which at least one fitting piece (the 1×1 module) exists. Figure 12.7 shows the formalized routing strategy which is used in the presented router based network. The advanced

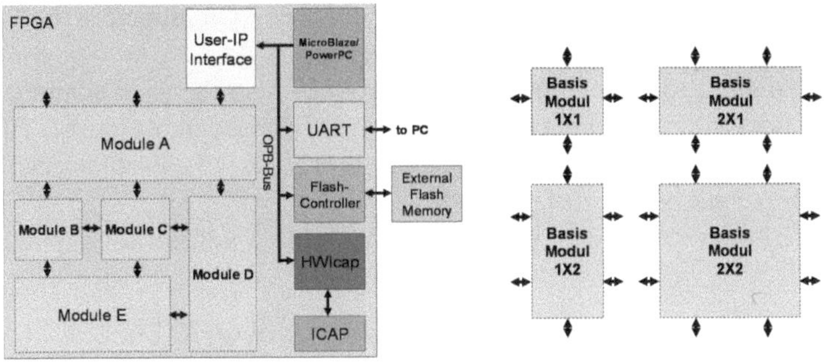

Fig. 12.6 Implemented sizes of the basis module.

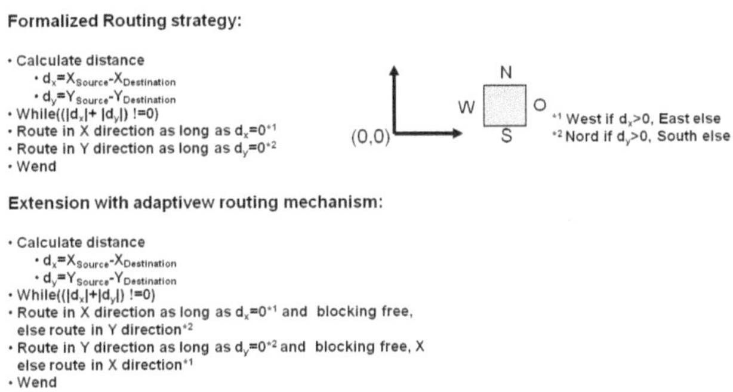

Fig. 12.7 Formalized routing strategy.

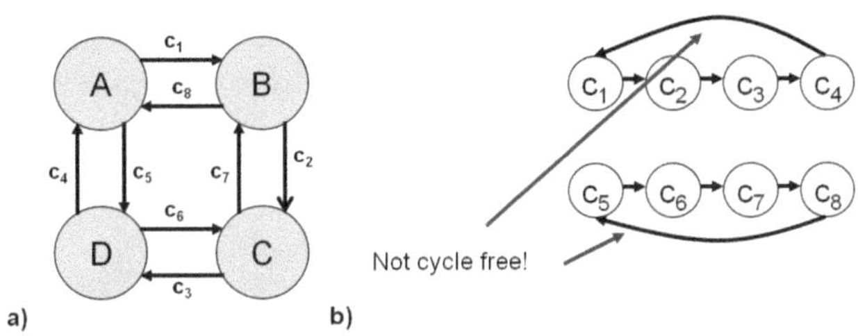

Fig. 12.8 (a) 2D-mesh topology; (b) Data Dependency Graph.

method of the XY-routing by the adaptive approach allows the forwarding of data packages in alternative directions, if a communication channel is blocked. In [4] it has been proved that the used method is non-blocking, because cycles in the Data Dependency Graph can be solved. Figure 12.8(a) shows the example of a 2D mesh topology of a network as a graph. The nodes of the graph describe the used router. The edges describe abstracted the possible paths in which a message packet can be forwarded. These edges are depicted in the corresponding Data Dependency Graph (Fig. 12.8(b)) as nodes. The nodes now represent the possible data paths through a router. It is clear to see that the resulting graph is not cycle-free, which would mean that this network topology is also not cycle-free. A solution for the non-cycle free graph problem is to remove edges. In real this is done by cutting connections targeted by default routing paths within the router. Also the used adaptive routing creates the possibility to produce cycles free Data Dependency Graphs. This is described in detail and mathematically proven in [4] and [13]. Further details of the router can be found in [15] and [1].

12.3.2 Switch for 2D Mesh Based NoC Approach

In this approach a switch is a module which changes the data flow path in the system. Thereby it is possible to choose different sinks for a source and hence to control the data path. This mechanism manages the switch and thereby the data stream which has to be routed via the switch. The goal is to minimize the resource utilization and to increase the speed of a switching mechanism. Another fact which argues for the use of switches is that the design is easy to implement on FPGAs. The network switch will be controlled by the configuration mechanism of the FPGA. Figure 12.9(a) shows a schematic of the switch.

12.3.2.1 Switch Layout

The switch will be designed using multiplexers as shown in Fig. 12.9(b). These multiplexers will be controlled by the outputs of LUTs. The innovation of the presented approach lies in the fact that the content of this register can be changed by a master without being connected directly with it. Hence it is possible to change the dataflow which is routed through the multiplexer by changing the outputs of the involved reconfigurable LUTs. To create larger switches it is possible to cascade the multiplexers.

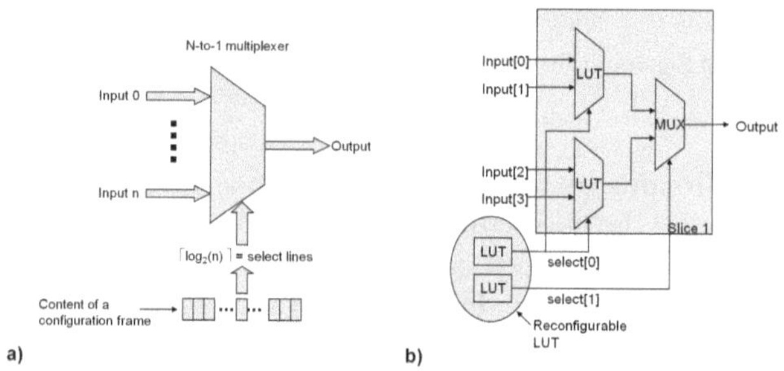

Fig. 12.9 (a) Schematically view of the switch structure; (b) Technologically switch design inside the CLB.

12.3.2.2 Controlling the Switch

There are several ways to change the content of the reconfigurable LUTs and thereby such a switch. If the system uses the ICAP in combination with a microprocessor such as MicroBlaze or the PowerPC, the LUTs can be reconfigured directly. It is also possible to control the switch utilizing the external reconfiguration port of an FPGA.

The approximate time the switch needs to change the dataflow depends on the speed of the internal ICAP interface. The following calculation is based on the assumption that the internal ICAP of a Virtex-2 FPGA is used. The configuration speed of this interface is 66 MHz with a bandwidth of 8 bit. One frame of a Virtex-2 Pro V2P100 will be written into the configuration memory within approximately 18 ms. This will also be the time the switch needs to change the dataflow. This enables to change the switches aligned in one frame with a frequency of 56 kHz. More information about the Switch for 2D Mesh based NoC is described in [3].

12.4 On Demand System Adaption

12.4.1 Physical on Line Routing of Communication Structures

Actual state-of-the-art systems require that communication structures are determined completely at design time. For this purpose, bus macros are used that are implemented fix in the system and cannot be moved around from their physical location. That is why modules can be placed at run-time only on predetermined positions, thus reducing the degree of freedom of the placement area. To avoid this problem, a methodology has been developed that takes care of the physical realization of the communication lines between the individual modules and the static area during run-time. Therefore a technique has been developed to route communications structures on the FPGA during run-time in order to react on the varying positions of the modules. In order to make use of the degrees of freedom described above the read-modify-writeback (RMW) (see [15]) technique was used. We assume that the position of the modules respectively the positions of the ports are known to the run-time system. Thus the run-time system has the information on the position of the start and end points of the communication lines that are to be established on the FPGA. Using these parameters the routing algorithm of the run-time system tries to find a suitable path for this point-to-point connection. Here the A*-algorithm showed to be well suited for this problem. The communication resources of the FPGA are constructed in a way that every end of a communication line is connected to one port of a switch box. Thereby a connection point describes a connection to a communication line of the FPGA. If a connection between two lines is to be established, the connection points of the two lines have to be connected in the switch box. Using this methodology it is possible to connect individual lines in order to establish a communication path based on individual components at run-time. This is done by using the routing templates that are configured to the adequate position on the FPGA. These routing templates include the information which points in a switch box have to be enabled. The information differs depending on the direction and the length the line is to be routed. Now that the path to be routed is known, the run-time system calculates the position of the routing templates to be placed in order to establish a communication link. Here, different routing templates for every orientation are used as well as 90 degree connections to change the orientation.

The method described above has been used to develop a system that enables routing from the static area to any point on the FPGA. Therefore routing templates for every orientation as well as 90 degree connections have been generated. The bit width of every single macro is 8 bit which is caused by the hardware structure of the FPGA. This limitation can be overcome by cascading the templates.

12.4.2 Physical On-line Routing of Parameterizable Filter Modules

For every functional unit, which is to be provided on FPGAs in state of the art systems, the according bit stream must be available. In systems in which a larger amount of modules are used, an accordingly higher number of bitstreams must be available in the extern memory. Now a technology has been designed which allows generating simple functions out of a set of basic blocks. These blocks are available to the run-time system in bitstream fragments and can be newly grouped according to the connecting rule provided by the system during run-time. Thus, it is possible to create a new module out of small single blocks. That way this module can be assimilated very flexibly to the requirements. As already mentioned, modules consist of single fine-granular sub-modules and are composed according to the connecting rule by the run-time system. Therefore these sub-modules must have defined connection points. These connecting points must be the same for all different sub-modules and must be identical whatever their function may be. This guarantees that the sub-modules are compatible and exchangeable and can be put together like a plugging system. To connect the single modules to each other and to realize any different types of necessary connections, additional routing blocks are placed between the modules. Via these modules it is possible to connect the exit of the sub-module to two succeeding modules. Sub-blocks like adders, subtractors, dividers and multipliers are available to the run-time system in form of bitstreams. Furthermore, a set of connecting macros is also available for the run-time system. The bitstream packer analyses the description of a given data flow graph and composes it by a mapping process using the bitstreams at hand. This bitstream generated during run-time according to the connecting rule is now written onto the FPGA via Read-Modify-Writeback (RMW). In Fig. 12.10 the test system is shown on a Virtes-2 Pro 30 FPGA. On the right there is the area, which contains the fixed modules of the system and is not altered during run-time. The connections marked in red are the routing templates used to connect the sub-modules. They are arranged in a way that the interfaces for the sub-modules are always in the same place. In-between the routing templates are the actual sub-modules. Because of the possibility to place the function modules freely within the given limits the method described offers a higher flexibility. Furthermore, the functional units can not only be adapted to the geometrical requirements in extension and form but also to the altering requirements from outside (e.g. the user). In contrast an alteration of the type of block within the model range is possible without further ado, as the bitstream fragments are identical for all blocks. Thus, a change from e.g. a Virtex-2 Pro 30 FPGA to a Virtex-2 Pro 100 or Virtex-2 6000 block, as it is used in the ESM (see Chap. 3), is possible without adapting the sub-modules.

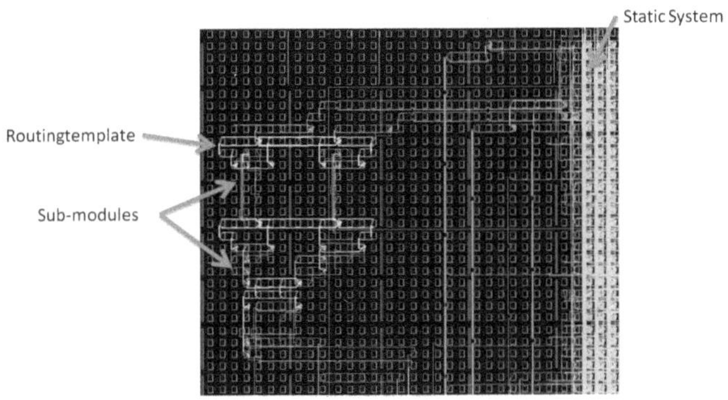

Fig. 12.10 Design of the test system shown in FPGA Editor.

12.5 System Modelling

We see a model based design process as a central point for application development, which uses graphical description, such as the Unified Modeling Language (UML) [17] or Statecharts [12] to specify the functionality and the run-time system. The development of graphical modeling languages is a continuation of the search for more powerful specification and design artifacts to meet the challenge of growing complexity in system design.

The requirements in a textual specification can, depending on project size and environment, be modeled and refined on the basis of different processes such as the V-model, the Rational Unified Process, or agile methods in the UML. Goal is an implementation model, which can automatically be transformed into an executable form. This can be a software implementation in classical central processing units as well as the configuration of a dynamically reconfigurable function unit. The same model should be reusable for different implementation alternatives and platform specific information is added automatically during the transformation step. Here we follow the Model Driven Architecture (MDA) specified by the OMG, which describes the way a platform-dependent model is generated in one or more transformation steps from platform-independent models. If transformation is automated, we gain the advantage that functional changes in the platform-independent model are propagated to specialized implementations and consistency problems can be avoided.

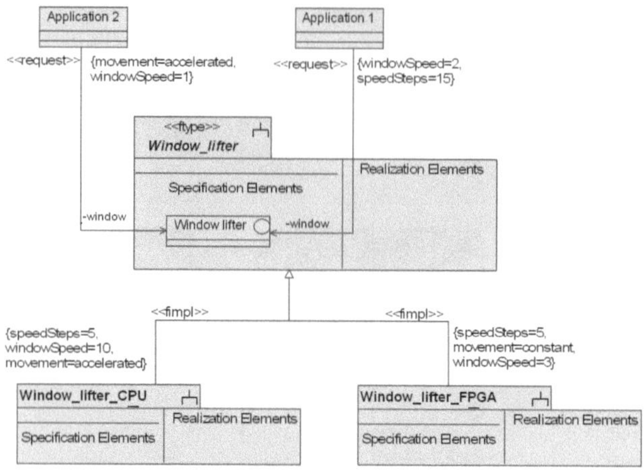

Fig. 12.11 Modeling of the realization alternatives and requirements.

The benefits for the ALadyn architecture are that functions can be developed as independent as possible from the implementation unit. The platform specific realization is selected during run-time. As part of this work an interface between application and the run-time system has been developed. The run-time system acts as a service provider to the application offering implementation alternatives on the basis of the required quality of service criteria, which can be accepted or rejected by the application. The interface on code level and thus the negotiation protocol has been defined in the first project phase and prototypically implemented. Figure 12.11 shows a model of realization alternatives using the example of a window lifter, which is managed by the ALadyn run-time system. This is done with an UML class diagram. The function type and its interface are defined by an abstract subsystem specification and its contained specification elements. Other subsystems contain implementing alternatives, and the relevant characteristics are assigned to the «fimpl»-stereotype. From the applications perspective requirements (stereotype «request») can be modeled, which are attached to the required feature characteristics [19, 20]. An application modeled in this way must be transformed into an implementation model, which then can be used to generate a system-specific run-time system. Figure 12.12(left) shows the basic approach to this transformation process. Basis of the modeling is the UML metamodel, based on which transformation rules, as shown in Fig. 12.12(right), can be created. This notation is characterized by the fact that it is easily understandable and maintainable, while powerful enough. The transformer REMIX, which converts the transformation rules has been implemented and can be equipped with suitable rules for the generation of the Aladyn run-time system.

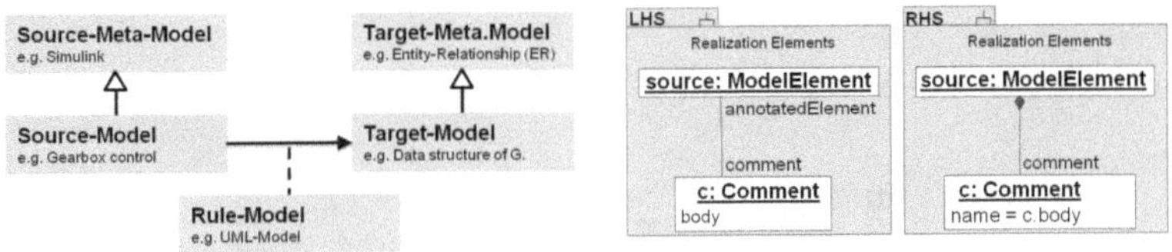

Fig. 12.12 Chain of transformation (left) and rule for transformation (right).

12.6 Tool Chain

One goal of the second project phase was the definition of appropriate tools for system design of the ALadyn architecture and classification of tools into a comprehensive tool chain [19, 20].

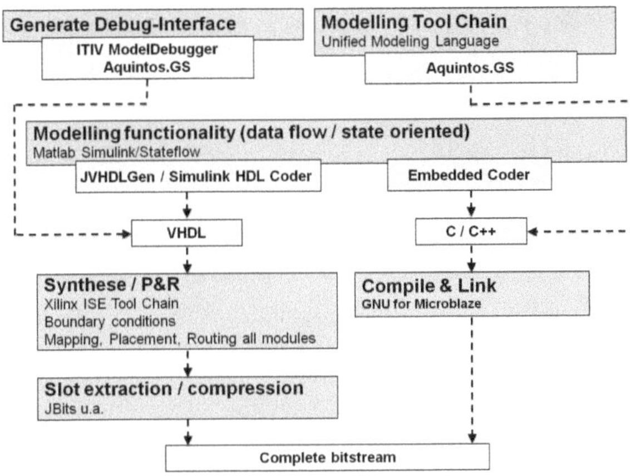

Fig. 12.13 Tool chain for a simple ALadyn system.

Figure 12.13 shows the tool chain, which has been used for the generation of prototypes [9]. It is based on a model based design. For function definition Matlab Simulink / Stateflow is chosen. This decision was driven by the available generators, which allow producing C-code and synthesizable VHDL [18]. The flow control and the associated scheduling strategies are specified in Unified Modeling Language (UML) class diagrams and with help of the Action Language. The code generator Aquintos.GS (http://www.aquintos.com) is used to generate the code from the class diagrams. VHDL models are extended with the described debugging interface and architecture, described in detail in Sect. 12.7. In the backend, the Xilinx ISE tool chain is used for generating bitstreams (http://www.xilinx.com). The free available GNU tools generate binary code for the MicroBlaze Softcore processors used in prototypes. From the generated bitstreams slots are extracted and compressed, and a bitstream for the complete configuration is created. This stream will be executed on our FPGA-based target system.

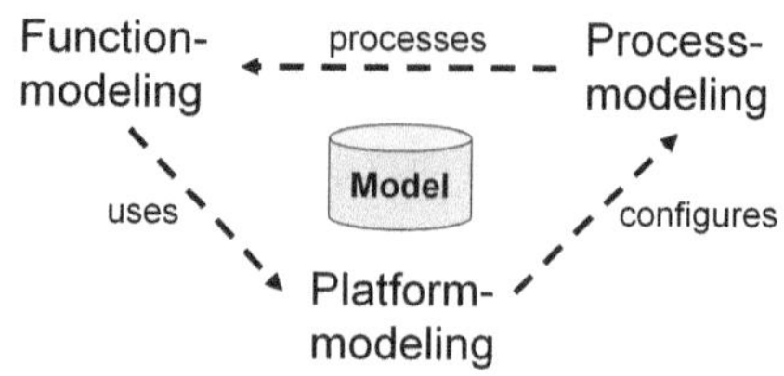

Fig. 12.14 Aspects of the overall model.

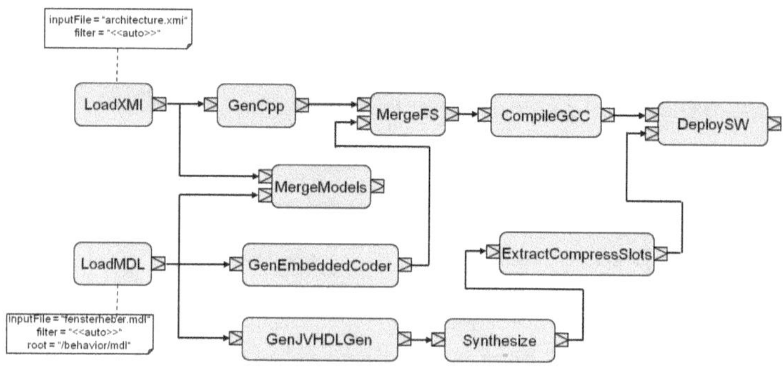

Fig. 12.15 Tool chain for the ALadyn-Architecture.

To extend the flexibility of the tool chain the experience gained in the areas of functional modeling and design process are systematized in the third phase. To achieve this aim, the tool chain can be modelled and is integrated along with a model of the target platform into an overall model. Figure 12.14 shows the relationships between the sub-models. The functional model utilizes the services of the execution units, which are specified in the platform modeling. The platform modeling also includes the numbers, parameters (size, speed, ...) of the execution units as well as the possible data flows in the form of 1-to-1-connections, buses and network interfaces. By using a subset of platforms over a consistent interface, a single functional model can be realized to different execution units. The process model aligns the individual work steps, such as communication with modeling tools, code generators, compilation, synthesize, configuration of distribution manager, function repository, enrichment of debugging interfaces etc. in a tool chain. The process model always refers to parts of the function model and processes them. The process model and its implementation will be influenced and configured by the available platform. Central to the claim of an integrated design environment for executable functional models is the integration of the functional model into the overall model. The Fig. 12.15 shows the designed tool chain for the ALadyn architecture. To allow multiple conjunctures of model parts a metamodel for the global model has been developed and implemented. The described overall model achieved the availability of the first description totally based on a graphical model for a high-level mixed configurable hardware and software system. The transition to implementation is completely automated, which leads to a dramatic improvement in productivity during the design phase.

12.7 Model Debugging

12.7.1 Problem

A model based design approach alone can not ensure a pure top-down approach in system design. Therefore, for the acceptance of an architecture it is essential, that the behavior of the designed system can be monitored, visualized and, where appropriate, influenced. The challenge is to create an approach which allows debugging of executable models across the boundaries of individual notations as well as across different execution units, which is especially important in the Aladyn architecture. As embedded systems always work in the context of an operating environment with sensors and actuators, a debugger approach can avoid the additional complexity and semantic equivalence problems by additional modeling of the associated environment and interfaces. Another aspect to consider is that in a completely model based designed approach, as it is the basis for the in Sect. 12.6 presented tool chain, a part of the behavioral semantics of the model is added only in the step of the model to code transformation. An interpretation of the model in a simulator, therefore, always implies doubts about the equivalence between interpretation and actual behavior.

12.7.2 Debugging Flow

Debugging a model based designed system can be understood as a reversal of the various transformation steps which lead from the model to an executable system. Figure 12.16 shows on the left side the relationships: The result of the design is a model, which can be converted by code generators to source code (C++/VHDL) and then compiled/synthesized to binary or bitstream format. This is in the next step executed on the target platform. In the system, the extraction of run-time information is the task of the On-Chip debugger. The hardware interface and architecture are discussed in Sect. 12.7.3. Based on the information at run-time on source code and/or signal level in further steps artefacts are obtained, which refer to elements of the model. These artifacts enrich the existing model, i.e. the metamodel is extended by meta classes, which permit a dynamic model part. The updating of this parts of the model is done by the Mapper, which conveys between drivers and model. The run-time information exists on different levels, which relate (at run-time static) on

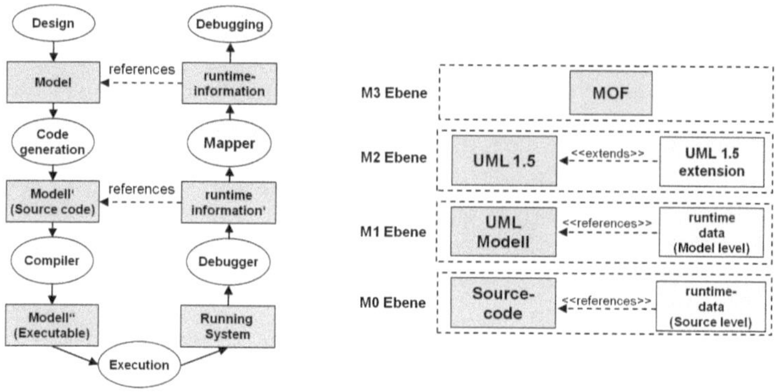

Fig. 12.16 Comparison: model levels (left) and OMG meta levels and run-time information (right).

model artifacts. The overall model is based on the most abstract level based on the UML 1.5 metamodel. The classification of our expansion in the OMG meta level hierarchy is shown in Fig. 12.16 [5].

For the user all these steps are automatically done, in the end the designed model can be directly debugged on the same level it has been designed. This debugging flow has been implemented in an own software development environment [8, 6, 7]. As an exemplary practical example a full design and debugging cycle has been realised for hierarchical state machines on microcontrollers and FPGAs.

12.7.3 Interface and Architecture

The extraction of run-time information on source code level for software is already possible with existing debuggers. Therefore, the widely used GDB debugger in the GNU compiler collection is introduced. The models executed on reconfigurable units are wrapped by a JTAG-compatible interface, as shown in Fig. 12.17. The interface allows accessing the on-chip registers, which allows reconstruction of the functional status. This interface is like the actual function automatically generated from the model and allows access through the serial signals tdi_i and tdo_o, which operate as shift registers. In addition, the system clock can be externally defined by the debug interface as needed for single-step processing. Hardware and software can be adapted transparently by addressing different memory areas using the JTAG interface. A special feature of the ALadyn system is that the execution unit is not known at design time and during run-time the function along with the system state can be migrated into a totally different type of execution unit. To enable a visualization of the system state in this environment, the distribution manager has to be instrumented in a way, that it is possible to track the function migration. With change of the type of execution unit, e.g. from CPU to FPGA, there must also be an on-the-fly change of the used driver component.

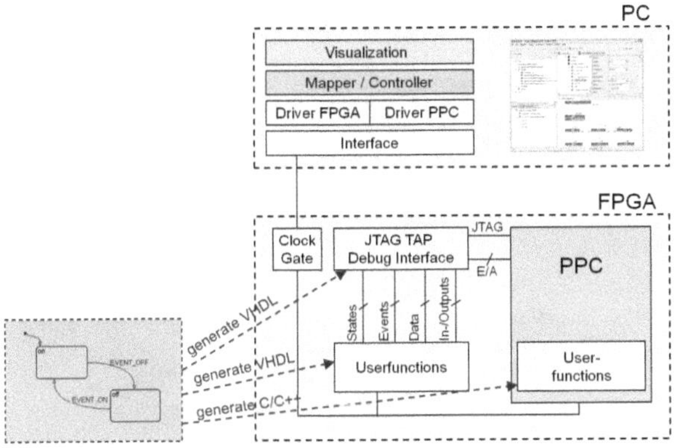

Fig. 12.17 Generated functionality and run-time interface.

Reconfigurable embedded systems also usually exist in the context of their environment, with which they interact having sensors and actuators. A slow down or even stopping of the system influences the system behavior. This characteristic limits the usefulness of classical debugging approaches, which stop the system clock for inspection. Therefore, it has been investigated to what extent real-time capability for simultaneous recording of the system process (trace) can be guaranteed, both for processor-like as well as for reconfigurable execution units.

According to the real-time perspective of an embedded microprocessor a software debugger component has been developed, which enables to record the system flow, changing data values, ... The recorded system run can later (post-mortem) be played in the Model-Debugger. Moreover, the above described run-time interface for VHDL-modules has been expanded for real-time recording of signals [21]. Figure 12.18 shows the designed modular on-chip architecture. This architecture also became part of the tool chain for system design and therefore can be easily integrated into the system. The design signals will in the first step be stored into a FIFO buffer. After the FIFO the amount of data can optionally be reduced by means of compression and coding algorithms. The reason for this is the limited internal memory on the chip and the limited bandwidth of external interfaces. The data after the compression can be either stored on-chip and later (in non real-time) transmitted or it can be directly transmitted to a PC. For faster data transfers additional optional interfaces are implemented, which can be used if they are supported by the system. Also optional is the implementation of a DDR-RAM-Controller, which enables to store more data on the system, if the system supports DDR-RAM.

Furthermore, trigger events can be set to define recording start- and stop-points. Also the clock and reset signal of the design can be controlled independently from the signals of the debugger. During run-time the signals can be modified and batch commands implemented, for different recording and transmitting scenarios, using the implemented controller and the interface to the PC. Moreover, in certain limits also the trigger conditions can be modified. This enables a wide range of possibil-

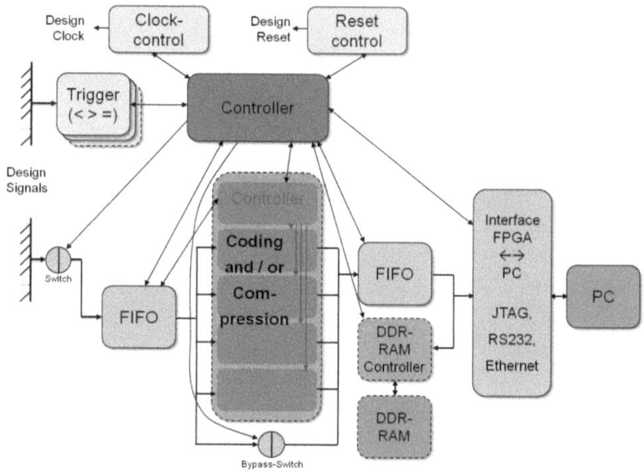

Fig. 12.18 On-Chip architecture for real-time debugging.

ities for real-time debugging of the embedded system without going through the implementation process again.

In the next step an approach was developed to integrate the debugging system in the two dimensional reconfigurable ALaydn architecture. The debugger works in the network as an independent function block and records the signals of the adaptive bus system. Thus, the architecture and function allocation as well as the functional units themselves can be debugged. By accessing the system using switch boxes and busmacros only a minimal influence on the target system is given, because the debugger acts as a normal function unit.

12.8 Test

In addition to debugging, tests play an important role for the quality assurance cycle during the design of embedded systems [16]. Based on the fundamental architecture of model based debugging in Sect. 12.7 also a model based test technique for heterogeneous systems has been developed [10], as shown in Fig. 12.19(left). In this context models are tested as functional models as well as used for test specification. The reason for a model based test approach is the same as for debugging; it raises abstraction level and therefore eases handling the design process of complex systems. One critical aspect is to be able to evaluate test cases against real time response, because of real time behavior of embedded systems.

Our test-framework allows the specification of test cases using UML use-case- and sequence diagrams. Use-case-diagrams structure the tests and sequence diagrams define stimuli and expected response of the system. Test cases can define participating objects and their messages including parameters, loops, control structures,

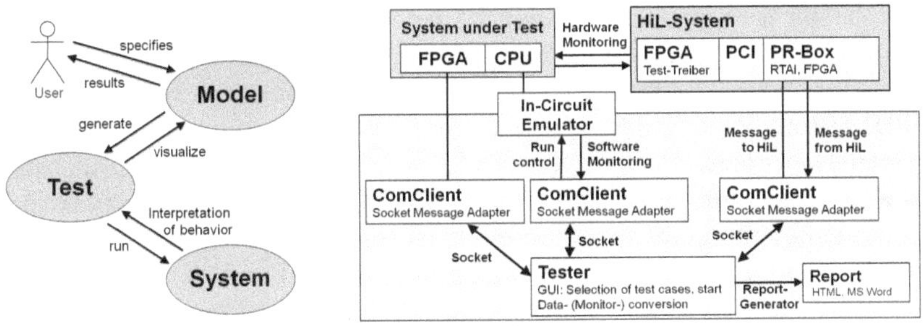

Fig. 12.19 Test approach (left) and overview of the test system (right).

inclusion of other collaborations and time constraints. The presented syntax applies to sent or received bus-messages using a Hardware-in-the-Loop (HiL) system as well as trace data collected from an in-circuit emulator or an on-chip debugger. Following the idea that relevant parts of the model should directly map into concepts on source code level, it is necessary to generate code of the overall model. Therefore, we also implemented the test approach into our code generator (see Sect. 12.5), to enable structural and behavioral code generation directly from UML models.

For a first implementation of this approach we used an embedded device consisting of a microcontroller and interfaces controlled by various peripherals. The setup is shown in Fig. 12.19(right). The system is monitored using a commercial Hardware-in-the-Loop and an in-circuit emulator. The collected data is mapped back to the model elements during the next step. Subsequently each required message, its parameters and the time constraints are translated into assertions the system trace is evaluated against. From the user's side, the tester is a GUI-based JAVA application that is used to select test cases from the model, run the test cases and generate HTML result reports.

The presented test approach supports white-box- and black-box-testing of software that is to be deployed on embedded processors. It can easily be extended for reconfigurable systems using a HiL-system for stimuli generation and the on-chip real-time debugger, as presented in Sect. 12.7.3, for tracing.

12.9 Conclusions

The DFG funded "Schwerpunktprogramm 1148 Rekonfigurierbare Rechensysteme" under the lead of Prof. Dr.-Ing. Jürgen Teich includes individual projects in the area of:

- Languages and Models
- Methodologies for design and development
- System analysis and evaluation
- Architectures and applications

Since 2003 the two projects ALadyn "Adaptives Laufzeitsystem mit intelligenter Allokation für dynamisch rekonfigurierbare Funktionsmuster und optimierte Interface-Topologien" in the area of design and development as well as the project AMURHA (Entwicklung und Synthese einer adaptiven multigranularen rekonfigurierbaren Hardwarearchitektur für dynamische Funktionsmuster) in the area of hardware architectures are under investigation at the Institut für Technik der Informationsverarbeitung (ITIV). The results of ALadyn were partially the basis for other projects within the "Schwerpunktprogramm" and vice versa, thanks to the deep collaboration with other projects ALadyn lead to an outstanding success of the project. In this section, the ALadyn project will be classified in the landscape of the different research projects.

In the area languages and models the project "PolyDyn" (Polymorphe Objekte für den Entwurf dynamisch rekonfigurierbarer FPGAs) targets the object oriented modelling approach for dynamic and partial reconfigurable hardwaresystems and is lead by Prof. Dr.-Ing. Wolfgang Nebel. The hardware description language SystemC-Plus exploits the methods of the object oriented programming in order to model the dynamic system behaviour of reconfigurable architectures. The efficient usage and the acceptance of the novel methodology "computing in time and space" is therefore also possible for users without deep hardware knowledge. Complex interactions and methods of the hardware components which can be found e.g. in the 2 dimensional reconfigurable system approach described above, can be developed by usage of the SystemC extension developed in the PolyDyn approach. More information about PolyDyn can be found in [11]. The area of design methodology, where as described also ALadyn belongs to, investigates in the hardware focused realization of strategies for an efficient application of reconfigurable hardware system architectures. The management of the reconfigurable area is a novel and necessary challenge in the research field of run-time reconfigurable architectures. The allocation of areas for hardware modules at run-time was under investigation of the project "ReCoNodes" (Optimierungsmethodik zur Steuerung hardwarekonfigurierbarer Knoten). The research group lead by Prof. Dr.-Ing. Jürgen Teich and Prof. Dr. Sandor Fekete investigated in the field of algorithms for efficient placement of hardware modules on dynamic and partial reconfigurable hardware. For this purpose, strategies form mathematics in the area of 1 and 2 dimensional packing are exploited while run-time in order to optimize the utilization of the hardware resources. The mathematical models of the approach for multidimensional optimization consider the parameters of the physical geometry of the target hardware, the modules to be places as well as the temporal relations of these modules amongst other important parameters. The used ILP (integer linear programming) approach delivers a possible optimized strategy for the placement of the hardware modules as well as the spatial distribution on the reconfigurable hardware. The description of the approach as well as an example scenario is published in [23]. Within the cooperation with the research group of Prof. Teich and Prof. Fekete, the described approach of the 2 dimensional reconfiguration is used to realize the approach of ReCoNodes physically. This approach would also offer novel possibilities for defragmentation or re-allocation of hardware modules in order to optimize the communication relations. Within the

"ReCoNodes" project, a rapid prototyping platform was developed. The Erlangen Slot Machine is a multifunctional platform enabling the flexible usage of the hardware resources. Within a cooperation the 2 dimensional reconfigurable hardware system as well as a visualization of the dynamic processes was implemented and demonstrated on this platform. The Erlangen Slot Machine is described in detail in the Chap. 3.

A special method for partitioning and placement, the temporal partitioning and placement, was developed within the research group around Prof. Dr. rer. nat. Franz Rammig in Paderborn. The developed algorithm enables to consider the temporal relations of functional reconfigurable blocks. In the research work the List scheduling algorithm was extended in that way, that run-time reconfiguration is supported. Further more, the method for 3 dimensional placement based on spectral analysis of task graphs was developed. Here the temporal relations (one of the dimensions) in taken into account for positioning and placing of the reconfigurable hardware modules. It is obvious, that this method can be exploited beneficially in the run-time system of the ALadyn approach. Further information about temporal partitioning can be found in [2]. The research project of Prof. Walter Stechele form the Technical University of Munich is engaged with the application of driver assistance systems in the automotive domain. Within the research project, reconfigurable image processing filters are used to optimize the hardware utilization of a FPGA based system. In a deep collaboration, the 2 dimensional reconfiguration approach of ALadyn was used to realize the run-time dynamic hardware approach. Prof. Stechele and his group were able to show the benefits of the adaptive system approach is this high performance application. More details can be found in [22].

ALadyn was perfectly integrated in the landscape of the different projects within the SPP 1148. Basic research enabled the development of real working hardware systems. Especially the development of the bus-macros finally enabled the break through of the 1 and 2 dimensional reconfigurable system approach [14]. With over 40 publications and 4 dissertations ALadyn generated outstanding success and valuable contribution to the academic world. Certainly this success was possible thanks to the deep and fruitful collaboration with other project partners within the program of emphasis.

Acknowledgements Supported by DFG grant MU 1323/6, project *ALadyn*, as part of the Priority Programme 1148, *Reconfigurable Computing Systems*.

References

1. Becker, J., Huebner, M., Paulsson, K., Thomas, A.: Dynamic reconfiguration on-demand: Real-time adaptivity in next generation microelectronics. ReCoSoc2005Montpellier, France (2005)
2. Bobda, C.: Synthesis of dataflow graphs for reconfigurable systems using temporal partitioning and temporal placement. PhD thesis (2003)

3. Braun, L., Goehringer, D., Perschke, T., Schatz, V., Huebner, M., Becker, J.: Adaptive real-time image processing exploiting two dimensional reconfigurable architecture. J. Real-Time Image Process. (2008). doi:10.1007/s11554-008-0095-8. http://dx.doi.org/10.1007/s11554-008-0095-8
4. Dally, W.J., Seitz, C.L.: Deadlock-free message routing in multiprocessor interconnection networks (5), 547–553 (1987). doi:10.1109/TC.1987.1676939
5. Graf, P., Müller-Glaser, K.D.: Dynamic mapping of runtime information models for debugging embedded software. In: 17th IEEE International Workshop on Rapid System Prototyping. Chania, Griechenland (2006)
6. Graf, P., Müller-Glaser, K.D.: Eine architektur für die modellbasierte fehlersuche in der software eingebetteter systeme. In: Modellierung 2006, Workshop: Modellbasierte Entwicklung von eingebetteten Fahrzeugfunktionen, Innsbruck, Österreich (2006)
7. Graf, P., Müller-Glaser, K.D.: Debugging modellbasiert entworfener software für eingebettete systeme. Informatik—Forschung und Entwicklung (2007)
8. Graf, P., Reichmann, C., Müller-Glaser, K.D.: Towards a platform for debugging executed uml-models in embedded systems. In: UML Modelling Languages and Applications: UML 2004. Springer, Lissabon (2004)
9. Graf, P., Reichmann, C., Müller-Glaser, K.D.: A model-based design approach for a dynamically configurable hardware-/software-architecture. 3rd Workshop on Object-oriented Modeling of Embedded Real-Time Systems (OMER 3) Paderborn, October 05 (2005)
10. Graf, P., Müller-Glaser, K.D., Reichmann, C.: Nonintrusive black- and white-box testing of embedded systems software against uml models. Rapid System Prototyping, 2007. RSP 2007. 18th IEEE/IFIP International Workshop on, pp. 130–138 (2007)
11. Grüttner, K., Grabbe, C., Oppenheimer, F., Nebel, W.: Object oriented design and synthesis of communication in hardware-/software systems with OSSS. In: Proceeding in SASIMI 2007, Hokkaido, Japan (2007)
12. Harel, D.: Statecharts: A visual formalism for complex systems. Sci. Comput. Program. **8**, 231–274 (1987)
13. Holzenspies, P., Schepers, E., Bach, W., Jonker, M., Sikkes, B., Smit, G., Havinga, P.: A communication model based on an n-dimensional torus architecture using deadlock-free wormhole routing. In: Proc. Euromicro Symposium on Digital System Design, pp. 166–172 (2003). doi:10.1109/DSD.2003.1231920
14. Huebner, M., Becker, T., Becker, J.: Real-time lut-based network topologies for dynamic and partial fpga self-reconfiguration. SBCCI04, Porto de Galinhas, Brasil (2004)
15. Huebner, M., Schuck, C., Kuehnle, M., Becker, J.: New 2-dimensional partial dynamic reconfiguration techniques for real-time adaptive microelectronic circuits. In: Schuck, C. (ed.) Proc. IEEE Computer Society Annual Symposium on Emerging VLSI Technologies and Architectures, p. 6 (2006). doi:10.1109/ISVLSI.2006.67
16. Lettrari, M., Klose, J.: Scenario-based monitoring and testing of real-time uml models. In: ≪UML≫ '01: Proceedings of the 4th International Conference on The Unified Modeling Language, Modeling Languages, Concepts, and Tools, pp. 317–328. Springer, London (2001)
17. Object Management Group (OMG): 2007 Unified Modeling Language (UML) Specification, Version 2.1 (2007)
18. Reichmann, C., Kühl, M., Müller-Glaser, K.D.: An overall systems design approach doing object-oriented modeling to code-generation for embedded electronic systems. In: FASE03, Warschau, Polen (2003)
19. Reichmann, C., Graf, P., Kühl, M., Müller-Glaser, K.D.: Automatisierte Modellkopplung heterogener eingebetteter Systeme. In: GI PEARL 2004, Eingebettete Systeme (2004)
20. Reichmann, C., Kühl, M., Graf, P., Müller-Glaser, K.D.: GeneralStore—A CASE-tool integration platform enabling model level coupling of heterogeneous designs for embedded electronic systems. In: Proceedings of the 11th IEEE International Conference on the Engineering of Computer-Based Systems. Springer, Brno (2004)
21. Schwalb, T., Graf, P., Müller-Glaser, K.D.: Architektur für das echtzeitfähige debugging ausführbarer modelle auf rekonfigurierbarer hardware. In: Methoden und Beschreibungssprachen zur Modellierung und Verifikation von Schaltungen und Systemen, Nico, Moser, p. 10 (2009)

22. Stechele, C.C.J.Z.F.M.W.: Using partial-run-time reconfigurable hardware to accelerate video processing in driver assistance systems. In: Proceedings of DATE 2007, Nice, France, April 16–20 (2007)
23. van der Veen, J., Angermeier, D., Göhringer, M., Majer, S.F.J., Teich, J.: Scheduling and communication-aware mapping of hw-sw modules for dynamically and partially reconfigurable soc architectures. In: In Proceedings of the Dynamically Reconfigurable Systems Workshop (DRS 2007), Zürich, Switzerland, March 15 (2007)

Chapter 13
ReconOS: An Operating System for Dynamically Reconfigurable Hardware

Enno Lübbers and Marco Platzner

Abstract In this chapter, we present the operating system ReconOS, which extends the concept of multithreaded programming to reconfigurable logic. ReconOS aims to provide hardware cores with the same services as the software threads of contemporary operating systems, thereby transferring the flexibility, portability and reusability of the established multithreaded programming model from software to reconfigurable hardware.

13.1 Introduction

The evolution of reconfigurable hardware devices has led from small logic-centric chips to powerful platforms combining microprocessor cores with dense logic fabrics. However, design methodologies for such configurable systems on chip have not kept up with the rise in complexity of reconfigurable hardware. In particular, there is little overlap between established programming models for embedded software and reconfigurable logic core design.

In this context, we are especially interested in the operating system layer. In the embedded systems domain, real-time operating systems provide the designer with a set of clearly defined objects and associated services, which are encapsulated in application programmer interfaces, e.g., the POSIX API [8]. Among these basic objects, we typically find *threads* and *processes* as units of execution, *semaphores* and related services for synchronization, and *mailboxes* and their derivatives for communication.

Enno Lübbers · Marco Platzner
Computer Engineering Group, Department of Computer Science, University of Paderborn, Paderborn, Germany, e-mails: `enno.luebbers@upb.de`, `platzner@upb.de`

Note: Parts of this book chapter have been previously published in [15, 11, 12].

M. Platzner et al. (eds.), *Dynamically Reconfigurable Systems*,
DOI 10.1007/978-90-481-3485-4_13, © Springer Science+Business Media B.V. 2010

The set of objects offered by an operating system together with the used scheduling policy can be considered a programming model. While this model is not comparable to formal models of computation, it does provide a designer with an established way of structuring an application. When going from a CPU-based system to a CPU/FPGA platform, it seems natural to simply extend the services offered by the operating system to customized hardware cores. Analogous to a software thread, a hardware core performing a specific task can be thought of as a *hardware thread.*

In this chapter we present ReconOS, an operating system for configurable systems on chip that extends the multithreaded programming model from software to reconfigurable hardware. ReconOS leverages standard operating system kernels, which allow for running existing code and facilitate the access to a variety of I/O devices. The ReconOS application programmer interface (API) essentially provides POSIX functions as one single development model for both *software* and *hardware* execution contexts. A designer is now able to map an application to a portable model that can be directly executed on a variety of CPU/FPGA execution platforms.

The remainder of this chapter is structured as follows: Sect. 13.2 reviews related work in operating system approaches for reconfigurable computers. The programming model used and implemented by ReconOS is explained in Sect. 13.3. Section 13.4 details the execution model for hardware and software threads, with Sect. 13.5 describing ReconOS implementations on two different existing software operating system kernels and the underlying hardware architecture. In Sect. 13.6, the performance of these implementations is evaluated and case studies are presented. Finally, Sect. 13.7 concludes the chapter and gives an overview of ongoing and future work.

13.2 Related Work

In the last decade, operating systems for reconfigurable computers have been researched from a number different angles. A large body of work has been focusing on single functions of future hardware operating systems. A prominent example is placement and scheduling of hardware tasks which has been studied under a variety of task and resource models as well as optimization objectives. Examples can be found in Jean et al. [9], Bazargan et al. [3], Teich et al. [24], Steiger et al. [23], Danne and Platzner [5], Danne et al. [6], and Pellizzoni and Caccamo [19].

Most of these works were either theoretical or, if experimental, evaluated their algorithms by simulation studies on synthetic workloads, given the absence of available hardware operating system implementations and accepted benchmarks. More recently, extensions of Linux appeared that promote the integration of software and hardware processes under the control of an operating system (OS).

Kosciuszkiewic et al. [10] built on top of an existing Linux operating system kernel and viewed so-called hardware tasks as a drop-in replacement for software tasks. These hardware tasks were executed on synthesized PicoBlaze softcore processors and did not exploit the fine-grained parallelism provided by FPGAs. In the described implementation, the interaction between software threads and hardware tasks was limited to FIFO communication. Bergmann et al. [4] encapsulated access to hardware modules into software wrappers, the so-called *ghost processes* which provide a transparent interface for interactions from the kernel and other processes. The authors considered sharing the same address space between hardware and software execution units as unsuitable. Technically, they used *processes* instead of *threads* to encapsulate hardware modules. For communication between software and hardware, FIFOs mapped to the Linux file system as well as dual-ported memory accessible from both software processes and a hardware process were used, as shown by Williams et al. [25]. So et al. [22] also modified and extended a standard Linux kernel with a hardware interface, providing conventional UNIX IPC mechanisms to the hardware using a message passing network. Again, communication between hardware and software processes was implemented by FIFOs and mapped to file system-based operating system objects.

A more closely related effort to ReconOS is the hthreads project [18]. In hthreads, hardware threads are managed by the operating system and are able to access various OS functions through a dedicated hardware thread interface, while sharing memory through a sophisticated inter-thread memory model [2]. hthreads is based on the POSIX pthreads programming model for both hardware and software threads and implements the OS components managing synchronization and task scheduling as hardware IP cores. In comparison to ReconOS, hthreads sacrifices the flexibility of a software operating system kernel for exceptionally low response time and jitter [1].

13.3 Programming Model

Two important design goals of the ReconOS programming model are portability, and—closely related—flexibility. ReconOS tries to (re)use much of the interface and functionality already present in established APIs, such as POSIX or the eCos kernel API. Consequently, most ReconOS programming model primitives or operating system objects are implemented by an operating system kernel running on the system CPU. ReconOS applications are typically crafted from the following operating system objects:

- *Threads* are the basic units of execution which make up an application. An application is partitioned into threads, which then communicate and synchronize using other operating system objects.
- *Semaphores* and *Mutexes* provide means for high-level *synchronization*; they can be used to sequentialize execution of threads, to protect critical code regions, or to manage exclusive access to shared resources.
- *Shared memory*, *message queues* and *mailboxes* are used for inter-thread *communication*. Generally, access to shared memory must be protected by synchronization primitives, as is necessary for any shared resource. Message queues and mailboxes occupy a special niche among the operating system objects—they provide both communication and synchronization at the same time.

The fact that all inter-thread activity is carried out using only these objects provides complete transparency within these interactions; a thread does not need to know whether its communication or synchronization partners are located in hardware or software—which, in turn greatly facilitates design space exploration with respect to the hardware/software partitioning. Also, as long as the interfaces to the respective operating system objects are supported, the interoperability and portability of threads can be easily maintained when moving to a different target platform. This applies both to different hardware target platforms, as well as to different host operating systems, as presented in [12]. An overview of the operating system objects and their related ReconOS and POSIX API calls, as used by hardware and software threads, respectively, is shown in Table 13.1.

Table 13.1 Overview of ReconOS API functions.

OS object	POSIX API (software)	ReconOS API (hardware)
Semaphores	sem_post()	reconos_sem_post()
	sem_wait()	reconos_sem_wait()
Mutexes	pthread_mutex_lock()	reconos_mutex_lock()
	pthread_mutex_unlock()	reconos_mutex_unlock()
Condition variables	pthread_cond_wait()	reconos_cond_wait()
	pthread_cond_signal()	reconos_cond_signal()
	pthread_cond_broadcast()	reconos_cond_broadcast()
Message queues/		reconos_mq_send()
mail boxes	mq_send()	reconos_mbox_put()
		reconos_mbox_tryput()
		reconos_mq_receive()
	mq_receive()	reconos_mbox_get()
		reconos_mbox_tryget()
Shared memory		reconos_write()
	**ptr = value*	reconos_write_burst()
		reconos_read()
	*value = *ptr*	reconos_read_burst()
Threads	pthread_exit()	reconos_thread_exit()
	pthread_create()	–
	sched_yield()	reconos_thread_yield()
	pthread_delay_np()	reconos_thread_delay()

To explain the fundamental principles of the ReconOS hardware/software programming model, we first present the programming of hardware threads under ReconOS before discussing the creation and termination of threads. ReconOS software threads are identical to regular threads of the host operating system both in concept and implementation. Since software threads are handled by the standard OS scheduler, they are independent from the ReconOS extensions.

13.3.1 Hardware Threads

Software threads have sequential execution semantics. To use an operating system service, a software thread simply calls the corresponding function in the operating system library. Hardware tasks, on the other hand, are inherently parallel. Mostly, there is no single control flow and, thus, no apparent notion of calling an operating system function. In particular, typical hardware description languages, such as VHDL, offer no built-in mechanism to implement *blocking calls*.

To present as unified a programming model as possible to the user, we rely on the following approach: We structure a hardware thread such that all interactions with the operating system are managed by a single sequential state machine. To this end, we have developed an operating system function library for VHDL. This library contains code implementing the system call signaling wrapped into VHDL procedures, e.g., `reconos_sem_wait()`. Together with the operating system interface (OSIF), a separate synchronizing logic module serving as the connection between the hardware thread and the OS, these procedures are able to establish the semantics of blocking calls in VHDL. A hardware thread thus consists of at least two VHDL processes: the synchronization state machine and the actual user logic. The state transitions in the synchronization state machine are always dependent on control signals from the OSIF; only after a previous operating system call returns, the next state can be reached. Thus, the communication with the operating system is purely sequential, while the processing of the hardware thread itself can be highly parallel. It is up to the programmer to decompose a hardware thread into a collection of user logic modules and one synchronization state machine. Besides the increased complexity due to the parallel nature of hardware, this process is no different from programming a software thread.

An example demonstrating this mechanism is illustrated in Fig. 13.1. In this example, the hardware thread receives a message into the local RAM, processes it, waits on a semaphore (SEM_READY), writes the result to shared memory, and then posts another semaphore (SEM_NEW). The OS synchronization state machine and the user logic communicate via the two handshake signals *run* and *done*.

13.3.2 Thread Creation and Termination

The creation of threads within the ReconOS programming model is almost identical for software threads (using `pthread_create()`) and hardware threads (using `rthread_create()`). A hardware thread takes the same scheduling and stack size parameters as a software thread. These are used for the hardware thread's associated delegate thread (see Sect. 13.4.3.1) and influence the hardware thread's priority when contending for access to operating system objects. Instead of a pointer to a software thread's executable code, a hardware thread is associated with a *core*, a data structure which is also passed to `rthread_create()`. A core represents the hardware thread's "executable circuit" which can be loaded into the FPGAs recon-

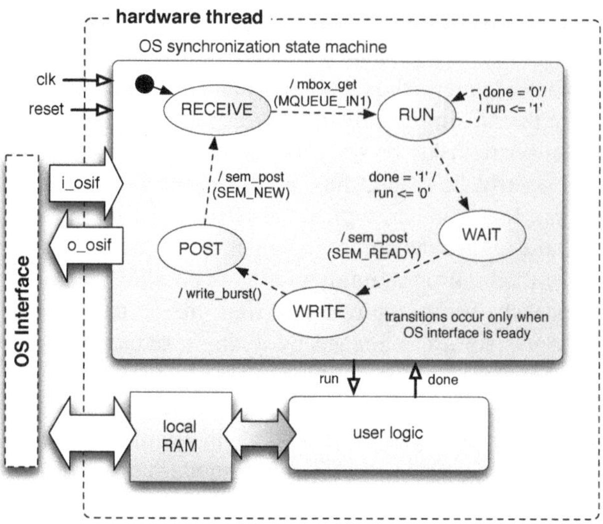

Fig. 13.1 Example of an OS synchronization state machine [11].

figurable fabric. Just as a section of machine code can be shared between multiple software threads, the same core can also be instantiated multiple times to form several hardware threads with the same functionality.

Cores can be loaded statically onto the FPGA together with the static hardware architecture (e.g., buses and peripherals) during system bootup, or be reconfigured dynamically into predefined slots when needed. If there are free slots when a new hardware thread is created, the corresponding core is loaded immediately; otherwise, it is marked for execution and will be loaded once a slot becomes free. The dynamic reconfiguration process is handled by a separate software thread, the *hardware scheduler*, which is described in Sect. 13.4.3.2.

Thread termination can either be initiated by the respective threads themselves—using `pthread_exit()` (within software threads) or `reconos_thread_exit()` (within hardware threads)—, or threads can be explicitly aborted using `pthread_kill()`.

13.4 Run-Time System

In order to support the programming model described in Sect. 13.3, a dedicated hardware/software run-time environment is needed to provide the connection between hardware threads and an existing operating system kernel. On the hardware side, a well-specified interface is required to manage the requests and responses of a hardware thread. On the software side, many of these requests need to be forwarded to the host operating system kernel, and their responses need to be relayed back to the hardware threads. Finally, to take advantage of the dynamic reconfiguration capabilities of modern platform FPGAs, the operating system needs to be extended to manage hardware multitasking. This section details the mechanisms employed by ReconOS to manage these tasks: the *operating system interface* (OSIF), the *delegate threads*, and the *hardware scheduler*.

13.4.1 Hardware Architecture

The ReconOS hardware architecture, shown in Fig. 13.2, consists of a main CPU executing the operating system kernel, one or more slots for hardware threads together with an operating system interface (OSIF) each, and two separate interconnection networks—usually buses—for accessing OS services and shared memory. Optionally, dedicated thread-to-thread communication channels can be established at design time between the slots. Such a configuration is typical for many signal processing applications that arrange filter stages in pipeline form, connected by hardware FIFOs (see Sect. 13.4.1.1).

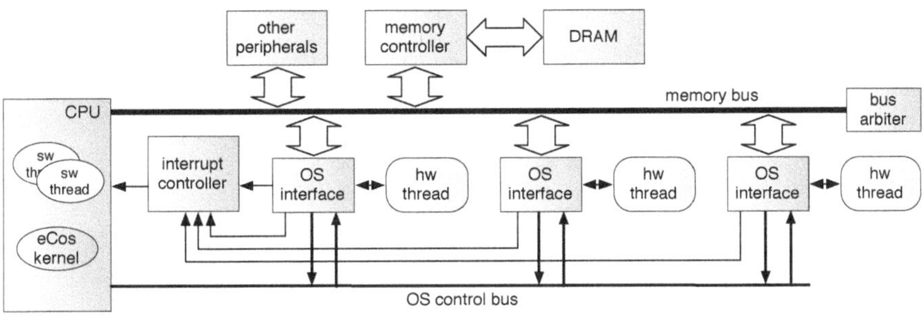

Fig. 13.2 ReconOS hardware architecture with three hardware threads [15].

The separation of control and data communications provides several benefits:

- OS control communications do not obstruct data communications on the memory bus, thus reducing its arbitration overhead and latency.
- Vice versa, memory communications, especially bursts, can not interfere with OS communications. This reduces the latency of OS calls, which is paramount to the use of ReconOS in real-time environments.
- OS interfaces for hardware threads that do not need direct access to system memory can be synthesized without the memory interface, thereby greatly reducing the area footprint.

13.4.1.1 The Operating System Interface

To be able to model hardware circuits executing on reconfigurable logic as threads, it is necessary to carefully define mechanisms for low-level synchronization and communication between the hardware circuitry and the operating system. In ReconOS, this is the task of the operating system interface (OSIF). Figure 13.3 gives an overview of the OSIF's structure and its interfaces. On one side, the OSIF connects to the hardware thread's OS synchronization state machine and local RAM. On the other side, the OSIF provides an interface to two bus systems, the system memory bus and an OS control bus. Further, the OSIF requires an interrupt line to the CPU's interrupt controller and features optional ports to connect to FIFO cores. The OSIF itself is built from several modules whose functions are described in the following.

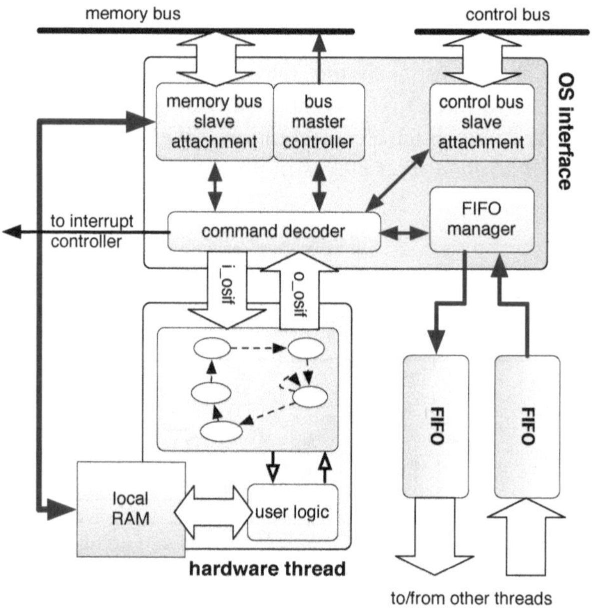

Fig. 13.3 OSIF overview and interfaces [13].

Thread Supervision and Control ReconOS provides hardware threads with a hardware API that comes in the form of a function library which specifies VHDL functions and procedures like `reconos_sem_post()` or `reconos_thread_exit()`. A designer can use these procedures inside the thread's OS synchronization state machine to sequentially call operating system functions, much like a software thread uses functions from the operating system's C-API. As a consequence, every state of the OS synchronization state machine may contain at most one VHDL system call. The VHDL procedures are purely combinational and communicate with the OSIF through a set of incoming and outgoing signals.

The mechanisms that govern the OS call request-response interactions between the OSIF and the hardware thread are controlled by the *command decoder* module. This module receives OS call requests from the hardware thread, decodes them and initiates the appropriate processes to fulfill that request. This may involve, for example, raising an interrupt with the system CPU, initiating a bus master transfer or feeding data into a FIFO.

Since the operating system executing on the CPU cannot process OS calls within one clock cycle, the OSIF needs a means to suspend state transitions of the thread's OS synchronization state machine. This is achieved by having the OS synchronization state machine routinely check input signals from the `osif2task` record before setting its next state. This way, the OSIF can block the part of the hardware thread that interacts with the operating system, which effectively implements the semantics of blocking calls in VHDL. A more detailed description of the mechanisms and procedures used to provide low-level synchronization between hardware threads and the OSIF can be found in [13].

OS Call Relaying OS services that are not provided by the OSIF directly (such as memory or FIFO accesses) are relayed to the OS kernel running on the CPU. Once the command decoder receives such a request from the hardware thread, it makes the command and associated arguments available to the CPU through the control bus, and raises an interrupt. This interrupt is forwarded to the software delegate thread associated with the hardware thread (see Sect. 13.4.3.1), which retrieves the command and arguments and executes the software OS call on behalf of the hardware thread. Any return values are sent back to the OSIF, which passes the values on to the hardware thread.

This mechanism provides maximum flexibility, since virtually every call that is possible from a software thread can now be requested by a hardware thread as well. However, OS call relaying comes with a considerable overhead which is quantified in Sect. 13.6. On every relayed OS call, the CPU needs to process an interrupt, switch to the associated delegate's context, and access the control bus registers before actually executing the call. During this time, the hardware thread's OS synchronization state machine remains suspended. Nonetheless, it must be noted that the parallel user processes inside the thread may continue their execution.

Data Communication Routing Due to the overhead involved in relaying OS requests to software, all high-throughput data communications should be handled in hardware without involving the CPU. In the ReconOS OSIF, this is realized in two variants, which provide the basis for any efficient, high-bandwidth thread-to-thread communication:

- *Bus master access*
  By utilizing the OSIF's memory bus interface, a hardware thread has direct access to any memory location and bus-connected peripheral in the system. Using the bus master controller (see Fig. 13.3), it is even possible to transfer bursts of data to and from memory. To request a burst write, the hardware thread must first store the data to transfer in the thread-local RAM. Then, the thread's OS synchronization state machine calls a `reconos_write_burst()` procedure. This prompts the bus master controller to initiate a memory bus transfer from the local RAM, which is mapped into the system memory space, to the target

address in main memory. Similarly, a thread can request a burst read transaction, which will place data from main memory in the local RAM.

- *Hardware FIFOs*
  While direct memory bus transfers represent an improvement over the indirect communication methods provided by the OS call relay technique, their performance can suffer considerably when several threads, the CPU, or other peripherals contend for the bus.
  To allow for bus-independent thread-to-thread data communication, the ReconOS run-time environment provides dedicated FIFO buffers implemented in hardware. Two threads connected by such a FIFO module can transfer data without interrupting the CPU or increasing bus load. When a hardware thread signals a pending read or write access to such a FIFO, the OSIF can pass the data directly to a hardware FIFO module. In the event of a write request to a full FIFO or a read request to an empty FIFO, the FIFO manager can also suspend the hardware thread's OS synchronization state machine, thus providing blocking *get/put* operations on FIFOs. Details on the performance of this message passing mechanism can be found in Sect. 13.6.

13.4.2 Hardware Multitasking

In the ReconOS execution environment, it is possible to synthesize multiple hardware threads to the same location in the FPGAs reconfigurable fabric. Using partial dynamic reconfiguration, the operating system is able to load hardware threads during run-time, effectively realizing hardware multitasking between different hardware threads. To this end, the area allocated to the execution of hardware threads is split into separate slots, each with its own OS interface. The partial bitstreams of the different cores are linked into the application's executable and are loaded onto the reconfigurable fabric under the control of a separate scheduling thread (see Sect. 13.4.3.2) running on the CPU. This mechanism is made available to hardware threads using the already established threading API.

To reduce the complexity of the reconfiguration logic and the overheads incurred by context saving and restoring in hardware, ReconOS in its current implementations employs a cooperative multitasking scheme for its hardware threads. Hardware threads cannot be preempted asynchronously, but use special OS calls to voluntarily yield their slot to other waiting threads. Typically, these preemption points will be inserted into a hardware threads wherever state information is minimal. Cooperative multitasking in ReconOS is described in detail in [14].

13.4.3 Software Architecture

To interface hardware threads with the host operating system kernel, the ReconOS software architecture employs *delegate threads* and a *hardware scheduler*. These concepts are common among all ReconOS implementations; the necessary adaptations to maintain portability across different host operating systems are illustrated through two examples in Sect. 13.5.

13.4.3.1 Delegate Threads

A fundamental assumption of the ReconOS programming model concerns the transparency of thread-to-thread communication and synchronization, regardless of the execution context (hardware or software) of the respective communication partners. This enables the designer to easily replace, for example, a software thread with a functionally equivalent hardware thread, allowing for rapid design space exploration with respect to the hardware/software partitioning.

To achieve this transparency in ReconOS, every hardware thread is associated with exactly one software thread, its *delegate*. The delegate is responsible for executing operating system calls on behalf of the corresponding hardware thread, making it appear as a software thread to the operating system kernel. Delegate threads are created as standard OS threads and passed additional parameters necessary to access the OSIF hardware. After creation, the delegate resets, initializes and starts the hardware thread. It then waits for an incoming OS request from the hardware to execute. The basic execution flow of a delegate thread is depicted in Fig. 13.4.

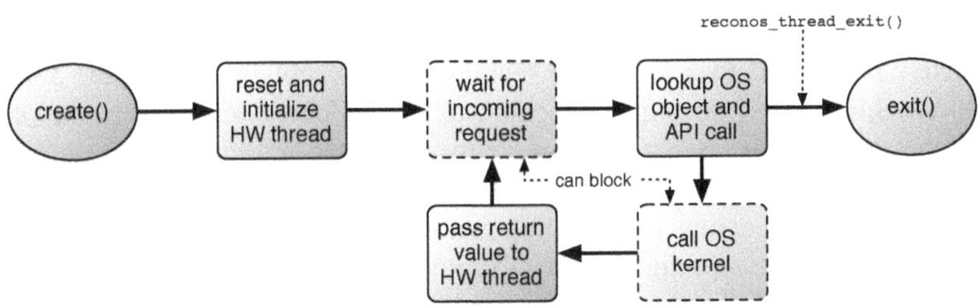

Fig. 13.4 Execution flow of a delegate thread [15].

To be able to map the OS objects referenced by the hardware thread to actual instances in the operating system kernel, the delegate thread maintains a table of object instances that are used by the hardware thread. Individual resources are represented towards the hardware thread as an index into this table. Hence, a single hardware thread description (or *core*) can be used for multiple instances in the system; giving different instances access to different resources is simply a matter of changing the delegate's OS object table. This mechanism is also a prerequisite for partial reconfiguration of hardware threads, when a core is relocated to a different slot.

13.4.3.2 Hardware Scheduler

All reconfiguration decisions a centrally managed by a high-priority software thread, the *hardware scheduler*. It communicates with the delegate threads through a central kernel data structure with synchronized access. Any thread ready to run—because of a `rthread_create()` call or a returning blocking function call—is marked as "waiting for a slot" in its kernel data structure. As mentioned in Sect. 13.4.2, a running thread can voluntarily yield its slot to other waiting threads. As soon as a delegate thread receives a OS interaction request from its hardware thread that is marked as "yielding", it notes this in the thread's kernel data structure and wakes up the hardware scheduler.

Upon invocation, the scheduler then checks for waiting threads and reconfigures them into any available free slot, or into a slot that is currently occupied by a yielding thread. If there are no free or yielding slots, a *yield request* is broadcasted to all running threads. This allows running threads to yield only when there are other threads waiting to avoid unnecessary reconfigurations.

A detailed description of the interaction between the hardware scheduler and the delegate threads as well as a quantitative analysis of scheduling overheads can be found in [14].

13.5 Implementation

13.5.1 Target Platforms

The principle of extending the multithreaded programming model to reconfigurable logic itself is not tied to a particular technology or architecture. Certain features of ReconOS, however, do require specific devices. For example, dynamic partial reconfiguration, which ReconOS employs for hardware multitasking, is as of now only available and supported on Xilinx Virtex-II, Virtex-II Pro, and Virtex-4 FPGAs.

The ReconOS extensions can be implemented on virtually any operating system which supports multithreading. As we will show in Sect. 13.6, we have currently implemented the ReconOS extensions on two different host operating systems, eCos and Linux. Depending on the actual choice of a host OS, its target platform requirements also dictate the platform for that particular ReconOS variant. For example, among the CPU families supported by eCos, only PowerPC processors can be found on Xilinx FPGAs, and only on a limited range of families. On the other hand, ReconOS/Linux can also be executed on soft core CPUs synthesizable to a wider range of target devices.

In essence, these requirements have led us to use Xilinx Virtex-II, Virtex-II Pro, and Virtex-4FX FPGAs for our prototypes. Besides off-the-shelf multi-purpose evaluation boards, we have also successfully run ReconOS on the Erlangen Slot Machine (see Sects. 13.6.1, 3 and 8.8).

13.5.2 Prototypes

Based on the hardware architecture shown in Fig. 13.2 and two OS kernels, we have created three ReconOS prototypes: ReconOS/eCos-PPC and ReconOS/Linux-PPC on a XC2VP30 FPGA, and ReconOS/Linux-MicroBlaze on a XC4VSX35 FPGA. The main features of these prototypes are listed in Table 13.2. Our implementation of the run-time environment uses the IBM CoreConnect bus topology available for Xilinx FPGAs. A 64-Bit high-throughput Processor Local Bus (PLB) connects the CPU and the OS interfaces to external SDRAM, which is used for both the operating system and shared thread memory. A separate Device Control Register (DCR) bus is used for control communications between the OS interfaces and the operating system kernel. Both the OS interfaces and the hardware threads run at the system's bus clock, which is 100 MHz for all prototypes.

Table 13.2 ReconOS prototype implementations.

Prototype (ReconOS/–)	eCos-PPC	Linux-PPC	Linux-MicroBlaze
Operating system	eCos	Linux	Linux
Based on kernel	eCos-VIRTEX4[a]	2.6-virtex[b]	2.6-nommu[c]
CPU	PowerPC 405	PowerPC 405	MicroBlaze 4.0
FPGA	XC2VP30	XC2VP30	XC4VSX35
CPU clock	300 MHz	300 MHz	100 MHz
PLB/DCR bus clock	100 MHz	100 MHz	100 MHz
MMU	No	Yes	No

[a][17]; [b][21]; [c][20].

A single OS interface requires 1147 slices which amounts to about 8.4%/7.5% of the logic resources of the employed XC2VP30 and XC4VSX35 FPGAs, respectively. Roughly two-thirds of the OS interface slices are taken up by the PLB bus interface, while the remaining logic is mainly used for the command decoder (267 slices) and the DCR bus interface (56 slices). The hardware architecture has been assembled and synthesized using Xilinx ISE 9.2, Xilinx EDK 9.2, and custom tools for automated OS interface and thread instantiation.

13.5.2.1 ReconOS/eCos

The eCos [7] real-time operating system provides a modular and configurable framework of operating system services. Application designers can select the necessary packages from the eCos repository and compile them into a library, which the final application is linked against. eCos is also configurable on a source code level. Using preprocessor macros, unneeded code is removed at compile time, resulting in small code sizes, which suits the targeted embedded segment. eCos is written in C/C++ and supports a range of target processor architectures, including the PowerPC 405, as found on Xilinx FPGAs of the Virtex-II Pro, Virtex-4 FX and Virtex-5 FXT families.

To transparently include ReconOS delegate threads in the eCos programming model, we have extended the eCos thread class to include additional information relevant to hardware threads, such as OSIF addresses, interrupt numbers, and OS object tables. Together with C wrappers for thread creation that are very similar to the eCos and POSIX API, reusing the existing kernel code allows ReconOS delegate threads (and, by extension, the associated hardware threads) to take advantage of all services provided by the eCos kernel.

Because eCos does not distinguish between user and kernel space but runs entirely in the processor's real mode, hardware access from user threads is greatly simplified. Although the delegate thread is logically part of the user application rather than the kernel, it can directly access the DCR bus to communicate with its corresponding OSIF. eCos also lets hardware and software threads share the same address space, since it disables the MMU, sacrificing memory protection and privilege management for a greatly simplified memory access model and higher performance. While unreasonable for larger-scale multiuser systems, this is entirely appropriate for small-footprint self-contained embedded systems, as targeted by eCos.

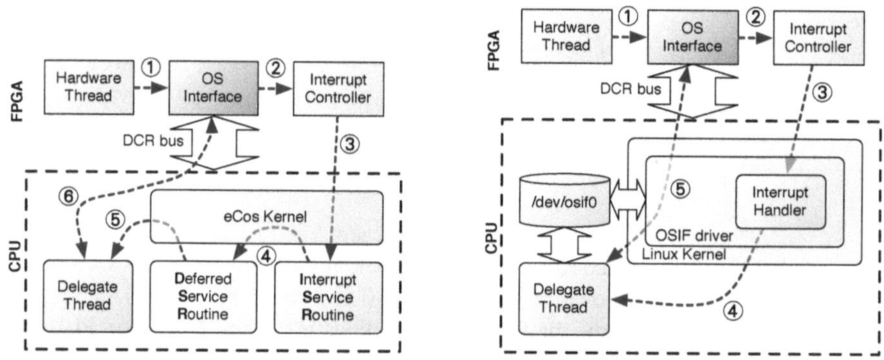

Fig. 13.5 Communication between hardware thread and delegate thread [12].

The sequence of events that is performed to relay an OS call from hardware to the eCos kernel is shown in Fig. 13.5(a). When a hardware thread uses a VHDL API call to request an operating system service, the respective VHDL procedure asserts certain handshake lines between the thread and its OSIF (1). Pending OS calls requested by the OSIF are signalled to the CPU's interrupt controller via a dedicated interrupt line (2). In eCos, interrupt processing is split into two steps to minimize interrupt latency. First, a very small *interrupt service routine* (ISR) is invoked (3), which executes in its own context, performs the necessary operations to enable reception of the next interrupt as quickly as possible, and marks the *deferred service routine* (DSR) (4) for execution. The latter is scheduled by the regular eCos scheduler so as not to interfere with the low-level interrupt processing. As the last step before the actual delegate thread is invoked, the DSR posts a semaphore (5) which the delegate is waiting on, essentially signaling an incoming request. The delegate thread then directly accesses the OSIF's registers via dedicated DCR access instructions to retrieve call parameters (6) and executes the requested eCos kernel function. Section 13.6 evaluates the timing overhead of this OS call sequence.

13.5.2.2 ReconOS/Linux

The Linux operating system is employed on a wider range of target architectures and therefore enjoys a wider adoption than eCos. The list of architectures includes, as the most interesting to us, the PowerPC 405 and the Xilinx MicroBlaze soft core. The latter widens the range of ReconOS targets to include FPGAs without an embedded CPU core. The MicroBlaze can be synthesized with or without an MMU. For our MicroBlaze prototype, we have opted for the omission of the MMU, which simplifies memory transfers between software and hardware threads.

While offering a wide set of configurable options, the Linux kernel does not allow to reduce its memory footprint as much as the eCos kernel. Absolute values on the size of the respective kernel images are difficult to obtain, as the code size greatly depends on the selected features, the target architecture, and the employed compiler. Also, an eCos kernel image already includes all necessary API implementations, the libc, and possibly a network stack. It can be expected, though, that a Linux kernel's size exceeds an equivalent eCos kernel by about an order of magnitude.

To communicate with its OSIF, a delegate thread needs access to the DCR bus. On a PowerPC system, this is accomplished through the `mtdcr` and `mfdcr` instructions, both of which are *privileged*. In Linux, user-space code, such as a delegate thread, typically cannot execute privileged instructions. To make the OSIF registers accessible to the delegate, we have thus implemented the low-level hardware access to the OSIF registers in a kernel driver, which publishes the registers through a device node, as depicted in Fig. 13.5(b). The hardware-independent code, such as the API wrappers and the delegate thread code, is implemented through a library that is linked with the user application.

Due to the separation of hardware-dependent and independent code, the sequence of events to relay an OS call from hardware to the Linux kernel differs from the one described for the eCos kernel. The signal assertions between hardware thread and OSIF (1) and the interrupt request to the system's interrupt controller (2) are identical. When a delegate thread needs to access its OSIF, it does so through filesystem accesses to the kernel driver's device node. In eCos, synchronization between the delegate thread and the OSIF was achieved through a dedicated semaphore. In Linux, this synchronization is implemented through read accesses blocking until an interrupt from the OSIF is registered (3). Only then is the blocking delegate thread resumed (4) and the read access translated into DCR operations (5). Write operations to an OSIF do not block. The timing overhead of this sequence is also analyzed in Sect. 13.6.

13.5.3 Debugging and Monitoring

The debugging of a multithreaded hardware/software system like ReconOS is a complex task. To aid application developers with debugging and optimization, ReconOS utilizes and extends the IBM CoreConnect bus functional simulation [26].

A hardware thread's VHDL description is integrated into a simulation testbench consisting of an OS interface and functional models of the memory and control buses. Because simulating an entire operating system booting and running on a complex microprocessor is rather time-consuming, we replace the CPU with a thread-specific stimulus file generated from a simple scripting language. When debugging, a hardware thread designer needs only to specify the expected timing and sequence of the thread's OS interactions, which are then automatically checked against the actual simulated thread behavior.

Debugging during run-time is also possible through a dedicated monitoring infrastructure. Intelligent probe cores collect data relevant to the functionality and performance of the hardware thread (or other critical modules such as the OS interface). The collected data is accumulated and pre-processed on a separate soft-core CPU, which allows a more refined tracking of the system's state than is possible with regular raw logic analyzers. It is also possible to feed the collected performance data back into the ReconOS execution environment and use it for run-time optimizations.

13.6 Experimental Measurements

The ReconOS programming and execution models have been experimentally verified using several prototypes based on the implementations described in Sect. 13.5. In the following, we present quantitative performance results of operations on operating system primitives on both host operating systems, which provides valuable pointers on the costs and overheads involved with individual thread interactions. Then we discuss another set of experiments that has been conducted to evaluate the throughput of the different communication mechanisms available to ReconOS threads. Lastly, we focus on more elaborate application case studies to analyze the impact of the operating system overheads on real-world implementations. The results demonstrate the applicability, feasibility, and portability of the proposed multithreaded programming model.

To enable quantitative measurements on the performance of operating system calls, we have run a set of benchmarks on the three prototype implementations listed in Table 13.2. The first set of experiments employs a set of synthetic threads analyzing the performance of timing critical OS calls. The mutex and semaphore primitives serve as representative examples, as most other supported API calls are either based on them or are not considered timing critical. The threads measure the raw execution time of single API calls to lock (unlock) a mutex or post (wait on) a semaphore, respectively, as well as a measure we call the *turnaround time*. The turnaround time is defined as the time it takes from one thread releasing a mutex (posting a semaphore), to the next thread acquiring a lock (receiving the semaphore).

The experiments have been run with different combinations of software and hardware threads. The results are shown in Fig. 13.6. While the synchronization overhead incurred by the operating system is not negligible, its impact on system performance remains within reasonable bounds, as application designers will usually use the reconfigurable fabric for data-centric parallel processing tasks, and implement control-intensive sequential tasks on the system's CPU.

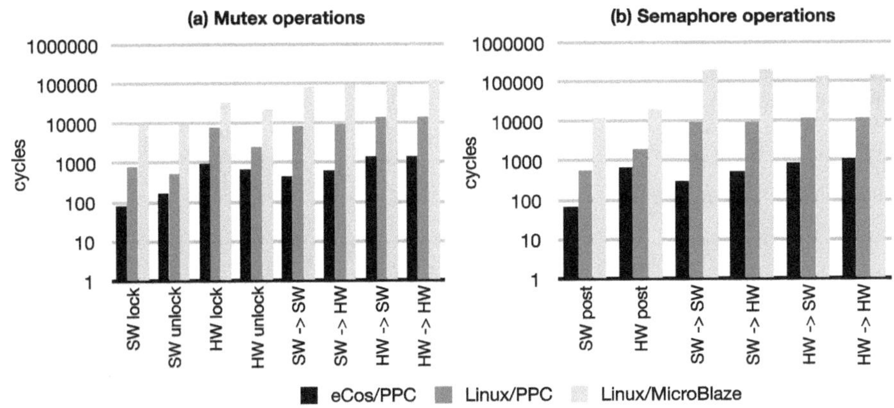

Fig. 13.6 Performance of ReconOS synchronization primitives.

In a second experiment, we have analyzed the attainable throughput for the communication primitives available to ReconOS threads. Two threads perform a sequence of data transfers, subsequently reading and writing data from and to main memory, as well as reading and writing data from and to a mailbox. Several configurations of the test have been run, using hardware and software threads, and with mailboxes mapped either to hardware FIFOs or to eCos software mailboxes. The throughput of ReconOS communication primitives for hardware threads depends primarily on the specifics of the hardware architecture (memory bus, hardware FIFOs), which is identical for the ReconOS/eCos and ReconOS/Linux prototypes. For this test, the ReconOS/eCos prototype employing a PowerPC processor and two hardware threads was selected.

The first hardware thread reads 8 kBytes of data from main memory into its local RAM. It then uses the ReconOS mailbox calls to transfer this data to the hardware FIFO, one 32-bit word at a time. Simultaneously, the second hardware thread reads from the hardware FIFO, also by using the ReconOS mailbox API. Once this data transfer is completed, the second thread writes the data back to main memory.

During the experiment, we have measured the times for reading and writing the data from and to main memory, and the times for writing and reading the data to and from the mailboxes. For comparison, we have also measured the times for data transfer between hardware and software threads using ReconOS' message queue primitives. Since software threads do not possess local memory, the memory read/write tests for software threads have been combined into a single *memcopy* test. The results are shown in Table 13.3.

While the hardware FIFOs only achieve 66% to 74% of the memory bus (PLB) in terms of raw throughput, one has to keep in mind that in order to transfer data

Table 13.3 Performance of ReconOS communication primitives.

Operation	With data cache		Without data cache	
	[μs]	[MB/s]	[μs]	[MB/s]
MEM→HW (burst read)	45.74	170.80	46.41	168.34
HW→MEM (burst write)	40.54	192.71	40.55	192.66
MEM→SW→MEM (memcopy)	132.51	58.96	625.00	12.50
HW→HW (mailbox)	61.42	127.20	61.42	127.20
SW→HW (mailbox)	58500	0.13	374000	0.02
HW→SW (mailbox)	58510	0.13	374000	0.02
SW→HW (message queue)	472.00	16.55	2166.79	3.61
HW→SW (message queue)	482.31	16.20	2160.69	3.62

All operations were run for 8 kBytes of data

from one thread to another, two memory transactions have to occur: first, the sending thread needs to write to shared memory, before the receiving thread can read the data. When using hardware FIFOs, reading and writing can occur concurrently. Considering this, an 8 kByte data transfer via hardware FIFOs is about 40% faster than a transfer of the same data via shared memory. Also, the transfer via mailboxes is implicitly synchronized, while two threads exchanging data via shared memory need explicit synchronization, e.g., via mutexes or semaphores.

13.6.1 Application Case Studies

The real-world implications of the ReconOS overheads on the overall system performance, and the feasibility of hardware/software system design based on the ReconOS programming model have been evaluated on a number of case studies covering several application domains:

- We have shown the benefits of ReconOS' incremental design methodology, as well as the integration of complex operating system services on a multithreaded image processing application [11], which streams image data from a workstation across an Ethernet network to an FPGA board, where it is processed in a chain of hardware and software threads. Similarly, another application accelerates cryptographic block ciphers modeled as hardware threads to encrypt and decrypt streamed video data. We have also used ReconOS to design a modular image processing library consisting of specialized hardware threads.
- ReconOS serves as a base for a multithreaded framework for Sequential Monte Carlo Methods [16], which can be used to track the state of a non-linear system through imprecise measurements. Using this framework, we have implemented a prototype for object tracking in video streams. Through ReconOS, the framework is able to adapt to changing data-dependant performance requirements by changing its HW/SW partitioning during run-time through partial reconfiguration of its hardware threads.

- To experimentally explore the portability and ease of use of the ReconOS programming model, we successfully combined the high-level synthesis (HLS) approach developed by the group of Prof. Merker (see Chap. 8) with ReconOS running on the Erlangen Slot Machine (see Chap. 3). In this prototype, a visual object tracking system has been realized, in which a hardware thread developed with the HLS tool communicates with software threads using the ReconOS execution environment.

In the following, we describe a general a ReconOS application which we routinely use for benchmarking architectural improvements of the run-time system. In this application, a multithreaded sorting algorithm has been implemented using ReconOS and mapped to different host operating systems and underlying hardware architectures.

A list of 2^{18} unsorted 32-bit integers is sorted, using a combination of bubble sort and merge sort; the basic concept is depicted in Fig. 13.7. First, the data is divided into 128 chunks, which are sorted individually using bubble sort. The resulting lists are then merged. To map this application onto our system, we divided it into two threads, one for the bubble sort routine, which has a software and a hardware implementation, and one for the merge operation, which is always performed in software. The threads communicate using shared memory and use message boxes for simultaneous synchronization and passing of buffer addresses. The application has been run on two prototype platforms, one running ReconOS/eCos on a PowerPC, the other running ReconOS/Linux on a MicroBlaze. Both systems use exactly the same application code for both software and hardware threads. Three tests have been performed: the first running the sort thread in software (SW); the second running the sort thread in hardware (HW); and the third running two sort threads concurrently, one in software, the other in hardware (SW+HW). The results of the measurements are shown in Fig. 13.7. In this figure, two times are given for each test and architecture, the first (bold) value denotes the time spent sorting, while the second corresponds to the merge time.

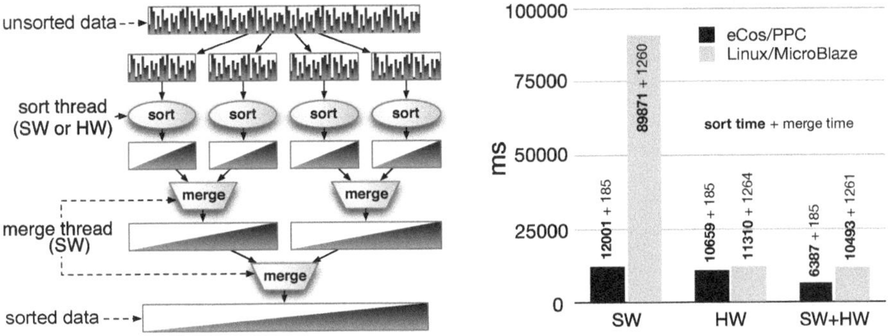

(a) Concept of sort application

(b) Execution times (in ms)

Fig. 13.7 Sort case study [15].

The first and last test, which perform (at least part of) the sorting routine in software, reveal, unsurprisingly, that the MicroBlaze processor performs the sort operation vastly slower than the PowerPC. However, when executing the sorting thread solely in hardware, both systems are almost on par. In this situation, the hardware thread interacts with the OS synchronization primitives infrequently enough so that the performance penalty due to additional software processing remains within acceptable limits. This is a typical scenario: an application designer will likely use the precious hardware resources for data-centric computations with relatively infrequent OS synchronization operations, and perform most control-dominated tasks inside software threads. Therefore, while the penalty incurred by the low-level synchronization and communication between delegate thread and OS interface is substantial for OS calls alone, the effect on overall application performance is marginal.

13.7 Conclusion and Outlook

In this chapter, we have presented the extension of the established multithreaded programming model to reconfigurable logic. Our operating system ReconOS facilitates rapid HW/SW design space exploration by providing transparent services for communication and synchronization, thereby promoting hardware cores from passive coprocessors to active peers in a multithreaded environment. The described multithreaded design process allows to accelerate critical system components through reconfigurable hardware without sacrificing portability, flexibility and reusability. Building on a portable execution model, ReconOS enables designers to exploit both fine-grained data parallelism as well as coarse-grained thread-level parallelism, and helps to take advantage of the rising complexity of modern reconfigurable devices.

Future work on ReconOS will focus on improved support of hardware vitalization, with emphasis on hardware thread scheduling and preemption techniques. Additionally, we will continue to work on supporting the synthesis of ReconOS threads from domain-specific application descriptions.

Acknowledgements This work was supported by DFG grant PL-471/2, project *ReconOS*, as part of the Priority Programme 1148, *Reconfigurable Computing Systems*.

References

1. Agron, J., Peck, W., Anderson, E., Andrews, D., Komp, E., Sass, R., Baijot, F., Stevens, J.: Run-time services for hybrid CPU/FPGA systems on chip. In: Proceedings of the 27th International Real-Time Systems Symposium (RTSS), pp. 3–12. IEEE Press, New York (2006)
2. Anderson, E., Peck, W., Stevens, J., Agron, J., Baijot, F., Warn, S., Andrews, D.: Supporting high level language semantics within hardware resident threads. In: Proceedings of the 17th

International Conference on Field Programmable Logic and Applications (FPL), vol. 1, pp. 98–103. IEEE Press, New York (2007)
3. Bazargan, K., Kastner, R., Sarrafzadeh, M.: Fast template placement for reconfigurable computing systems. IEEE Des. Test Comput. **17**(1), 68–83 (2000)
4. Bergmann, N.W., Williams, J.A., Han, J., Chen, Y.: A process model for hardware modules in reconfigurable system-on-chip. In: Proceedings of the Dynamically Reconfigurable Systems Workshop, 19th International Conference on Architecture of Computing Systems, vol. 81, pp. 205–214 (2006)
5. Danne, K., Platzner, M.: An EDF schedulability test for periodic tasks on reconfigurable hardware devices. In: ACM SIGPLAN/SIGBED Conference on Languages, Compilers, and Tools for Embedded Systems (LCTES), Ottawa, Canada (2006)
6. Danne, K., Mühlenbernd, R., Platzner, M.: Server-based execution of periodic tasks on dynamically reconfigurable hardware. IET Comput. Digit. Techn. **1**(4), 295–302 (2007)
7. eCosCentric: (2008). eCos, Website. http://ecos.sourceware.org/
8. IEEE, The Open Group: The Open Group Base Specifications Issue 6, IEEE Std. 1003.1, 2004 Edition (2004)
9. Jean, J.S.N., Tomko, K., Yavagal, V., Shah, J., Cook, R.: Dynamic reconfiguration to support concurrent applications. IEEE Trans. Comput. **48**(6), 591–602 (1999)
10. Kosciuszkiewicz, K., Morgan, F., Kepa, K.: Run-time management of reconfigurable hardware tasks using embedded Linux. In: Proceedings of the International Conference on Field-Programmable Technology (ICFPT), pp. 209–215. IEEE Press, New York (2007)
11. Lübbers, E., Platzner, M.: ReconOS: An RTOS supporting hard- and software threads. In: Proceedings of the 17th International Conference on Field Programmable Logic and Applications (FPL), vol. 1, pp. 441–446. IEEE Press, New York (2007)
12. Lübbers, E., Platzner, M.: A portable abstraction layer for hardware threads. In: Proceedings of the 18th International Conference on Field Programmable Logic and Applications (FPL). IEEE Press, New York (2008)
13. Lübbers, E., Platzner, M.: Communication and synchronization in multithreaded reconfigurable computing systems. In: Proceedings of the 8th International Conference on Engineering of Reconfigurable Systems and Algorithms (ERSA'08). CSREA Press (2008)
14. Lübbers, E., Platzner, M.: Cooperative multithreading in dynamically reconfigurable systems. In: Proceedings of the 19th International Conference on Field Programmable Logic and Applications (FPL). IEEE Press, Prague (2009)
15. Lübbers, E., Platzner, M.: ReconOS: multithreaded programming for reconfigurable computers. ACM Trans. Embed. Comput. Syst. **9**(1), 1–33 (2009)
16. Markus Happe, E.L., Platzner, M.: A multithreaded framework for sequential Monte Carlo methods on CPU/FPGA platforms. In: Proceedings of the 5th International Workshop on Applied Reconfigurable Computing (ARC). Springer, Berlin (2009)
17. Mind NV: Release of the eCos Port to the Xilinx Virtex4 ML403 board. Website (2008). http://www.mind.be/?page=ML403
18. Peck, W., Anderson, E., Agron, J., Stevens, J., Baijot, F., Andrews, D.: hthreads: a computational model for reconfigurable devices. In: Proceedings of the 16th International Conference on Field Programmable Logic and Applications (FPL), vol. 1, pp. 885–888. IEEE Press, New York (2006)
19. Pellizzoni, R., Caccamo, M.: Real-time management of hardware and software tasks for FPGA-based embedded systems. IEEE Trans. Comput. **56**(12), 1666–1678 (2007)
20. PetaLogix: Petalinux. Website (2007). http://developer.petalogix.com/
21. Secret Lab Technologies Ltd.: Linux on Xilinx Virtex. Website (2008). http://wiki.secretlab.ca/index.php/Linux_on_Xilinx_Virtex
22. So, H.K.H., Brodersen, R.W.: Improving usability of FPGA-based reconfigurable computers through operating system support. In: Proceedings of the 16th International Conference on Field Programmable Logic and Applications, pp. 349–354. IEEE Press, New York (2006)
23. Steiger, C., Walder, H., Platzner, M.: Operating systems for reconfigurable embedded platforms: online scheduling of real-time tasks. IEEE Trans. Comput. **53**(11), 1392–1407 (2004)

24. Teich, J., Fekete, S., Schepers, J.: Optimization of dynamic hardware reconfigurations. J. Supercomput. **19**(1), 57–75 (2000)
25. Williams, J.A., Bergmann, N.W., Xie, X.: FIFO Communication models in operating systems for reconfigurable computing. In: Proceedings of the 13th IEEE Symposium on Field-Programmable Custom Computing Machines (FCCM), pp. 277–278 (2005)
26. Xilinx Inc.: UG644: BFM Simulation in Platform Studio, User's Guide (2009). http://www.xilinx.com/support/documentation/sw_manuals/xilinx11/bfm_simulation.pdf

Part IV
Applications

Chapter 14
FlexiChaP: A Dynamically Reconfigurable ASIP for Channel Decoding for Future Mobile Systems

Matthias Alles, Timo Vogt, Christian Brehm, and Norbert Wehn

Abstract Future mobile and wireless communication networks require flexible modem architectures to support seamless services between different network standards. Hence, a common hardware platform that can support multiple protocols implemented or controlled by software, generally referred to as software defined radio (SDR), is essential. This chapter presents a family of application-specific instruction-set processors (ASIPs) for channel coding in wireless communication systems. Flexibility is provided by offering not only programmability but also dynamical reconfiguration within the ASIP pipeline. As a weakly programmable IP core, it can implement many channel decoding schemes for a SDR environment. It features binary convolutional decoding, turbo decoding for binary as well as duo-binary turbo codes, and LDPC decoding for current and upcoming standards. The ASIP consists of a specialized pipeline with 15 stages and a dedicated communication and memory infrastructure. A reconfigurable data shuffling allows for fast context switches, multi-standard support, and a efficient ASIP implementation.

14.1 Introduction

In recent years, we have seen the emergence of an increasing number of wireless protocols. Next generation mobile communication networks will feature new services, especially *multi-access and heterogeneous interoperability*, with increased data throughput. *Thus, flexibility becomes a dominant aspect for the transceiver.* To provide the flexibility for supporting seamless services between various wireless networks, there is a high demand for a common hardware platform that can support

Matthias Alles · Timo Vogt · Christian Brehm · Norbert Wehn
Microelectronic Systems Design Research Group, University of Kaiserslautern, 67663 Kaiserslautern, Germany, e-mails: `alles@eit.uni-kl.de`, `vogt@eit.uni-kl.de`, `brehm@eit.uni-kl.de`, `wehn@eit.uni-kl.de`

M. Platzner et al. (eds.), *Dynamically Reconfigurable Systems*,
DOI 10.1007/978-90-481-3485-4_14, © Springer Science+Business Media B.V. 2010

multiple protocols implemented or controlled by software, generally referred to as *software defined radio* (SDR).

The focus of this Chapter is put on *channel decoding in mobile and wireless communications systems in the context of SDR*. Channel decoding is the central processing task of the outer modem. Here convolutional codes (CC), binary turbo codes (bTC), duo-binary turbo codes (dbTC), and low-density parity-check (LDPC) codes are established techniques. The implementation complexity of the encoding algorithms for these codes is negligible. The decoding algorithms, however, have a very high computational complexity. In [16] it was shown that channel decoding contributes, depending on the implementation platform, about 40% to the total computational complexity of the physical layer of a UMTS or a WiFi 802.11a system. The challenge is to provide a flexible platform for channel coding that offers the needed computing power and a high efficiency w.r.t. area and power consumption.

Most of today's platforms for digital baseband processing support one or just a few standards, with an embedded channel decoder coprocessor engine especially designed for the respective outer modem. Such architectures are not suited for SDR since they do not provide the required flexibility for channel decoding.

Recently, platforms for SDR [16, 23, 26, 13, 12, 19] were presented. Most of these platforms are targeting the signal processing tasks in the inner modem, i.e., filtering, modulation and channel estimation. These algorithms have to process a huge amount of operations/samples, but have a high degree of data parallelism. Thus, these SDR platforms are multiprocessor systems with SIMD (single-instruction multiple-data) vector processing engines. Channel decoding substantially differs from the algorithms in the inner modem. They are non-standard signal processing algorithms with non-standard arithmetic and word widths. Here, the challenge is an efficient internal data management with a high utilization of the computational units.

In case of the SODA platform, which consists of an ARM processor and four *uniform* DSPs especially developed for SDR applications, one of these engines implements channel decoding solely [16]. Rather than using a pool of uniform processors for physical layer processing and reserving one of them for channel decoding only, we propose a heterogeneous platform in which channel decoding is performed by an ASIP that is optimized for this task. Thus, this ASIP can be considered as a *weakly programmable* processor that offers just enough flexibility and yields power efficiency. We will show that efficient utilization of application knowledge yields efficient and flexible architectures, as confirmed in a design study on a scalable reconfigurable channel decoder for future wireless handsets [15]. High performance combined with the advantages of processors, namely instruction level flexibility, are achieved by ASIPs, cf. Chap. 2.

In [20] the first ASIP targeting the channel coding domain of UMTS turbo codes was presented. The gain in processing speed of this ASIP was limited by the classical RISC (reduced instruction set computer) structure and its general load/store memory architecture. Total freedom in pipeline and memory architecture design gives room for further improvement. Moreover, it allows to add application specific *run-time reconfigurability* to the ASIP approach. Run-time reconfigurability yields a further performance boost allowing, e.g., fast context switches and a reduced instruction bit width.

An ASIP using this approach is presented in [22], but it only targets binary and duo-binary turbo codes with a maximum of eight states. Convolutional codes, turbo codes with more than eight states, and LDPC codes are not supported. The ASIP presented in this chapter, named FlexiChaP (*Flexi*ble *Cha*nnel Coding *P*rocessor), resolves this lack of code support and furthermore achieves a higher data throughput, as will be shown in Sect. 14.7. In detail the architecture was presented in [1, 29]. FlexiChaP represents a whole *family* of processors since it can be tailored to the application scenario by *design-time configurability*. It implements convolutional, turbo, and LDPC decoding for a vast number of existing and emerging standards in the field of mobile wireless communication systems. *Run-time reconfigurability* plays a major role in achieving the required flexibility as will be shown in Sect. 14.5.

14.2 Channel Codes

A careful analysis of channel codes incorporated in communication standards reveals that most of them are using convolutional codes, binary/duo-binary turbo codes, and LDPC codes. Although block sizes, polynomials, and coding rates differ both within one coding scheme as well as between the coding schemes, it is possible to find commonalities for convolutional and turbo decoders which can be exploited for the decoder's architecture. In contrast the decoding algorithm of LDPC codes varies substantially from convolutional and turbo code.

Table 14.1 summarizes relevant standards.

Table 14.1 Standards and channel codes.

Standard	Codes	States	Code Rates	Infobits	Throughput
GSM	CC	16, 64	1/4, 1/3, 1/2	Up to 876	Up to 12 kbit/s
EDGE	CC	64	1/4, 1/3, 1/2	Up to 870	Up to 384 kbit/s
UMTS	CC	256	1/4, 1/3, 1/2	Up to 504	Up to 32 kbit/s
	bTC	8	1/3	Up to 5114	Up to 2 Mbit/s
CDMA2000	CC	256	1/6, 1/4, 1/3, 1/2	Up to 744	Up to 28 kbit/s
	bTC	8	1/5, 1/4, 1/3, 1/2	Up to 20730	Up to 2 Mbit/s
HSDPA	bTC	8	1/2, 2/3, 3/4	Up to 5114	Up to 14.4 Mbit/s
LTE	bTC	8	1/3–9/10	Up to 6144	Up to 150 Mbit/s
DAB	CC	64	1/4	None	Up to 1.1 Mbit/s
DVB-H	CC	64	1/2, 2/3, 3/4, 5/6, 7/8	1624	Up to 32 Mbit/s
DVB-T	CC	64	1/2, 2/3, 3/4, 5/6, 7/8	1624	Up to 32 Mbit/s
DVB-RCT	dbTC	8	1/2, 3/4	Up to 648	Up to 31 Mbit/s
IEEE802.11a/g	CC	64	1/2, 2/3, 3/4	Up to 4095	Up to 54 Mbit/s
IEEE802.11n	CC	64	1/2, 2/3, 3/4	Up to 4095	Up to 450 Mbit/s
	LDPC	–	1/2, 2/3, 3/4, 5/6	Up to 1620	Up to 450 Mbit/s
IEEE802.16e	CC	64	1/2, 2/3, 3/4, 5/6	Up to 864	Up to 75 Mbit/s
(WiMAX)	dbTC	8	1/2, 2/3, 3/4	Up to 4800	Up to 75 Mbit/s
	LDPC	–	1/2, 2/3, 3/4, 5/6	Up to 1920	Up to 75 Mbit/s

14.2.1 Convolutional Codes

In convolutional codes forward error correction is enabled by introducing parity bits at the encoder. Convolutional encoding is based on a shift register process, which can also be interpreted as a Mealy automaton. Binary convolutional codes are characterized by a single binary input sequence (information sequence). They are fully specified by the constraint length $K_c = m+1$ with m the size of the shift register, a feedback polynomial G_{FB}, and generator polynomials G_i for the parity bits. In case of $G_{FB} = 0$ the code is *non-recursive*, otherwise *recursive*. Note that one output sequence can be equal to the input sequence: the systematic information. The number of output values per information bit defines the rate R of the code. The code rate is defined as $R = \frac{information\ bits}{information\ bits + parity\ bits}$. It can be adapted by the number of generator polynomials and the *puncturing* scheme. By puncturing, certain bits are removed from the coded bit stream and are not transmitted.

In mobile and wireless communications systems a large variety of binary convolutional codes with varying puncturing schemes are used. The constraint length typically varies between $K_c = 5 \ldots 9$, and the number of generator polynomials between 1 and 4. The polynomials themselves basically assume arbitrary values.

Convolutional codes are usually decoded using the Viterbi algorithm (VA) [27, 28]. It finds the most likely sequence of state transitions in a finite-state Markov chain and generates hard decision output values.

14.2.2 Turbo Codes

Turbo codes, proposed by Berrou et al. [3], are among the most advanced channel coding schemes. The idea behind these codes is to construct a powerful error control code out of simple building blocks. Two component codes are concatenated either in parallel or in serial, separated by an interleaver (INT) that rearranges the bits in a pseudo-random fashion. Typically, convolutional codes are used as component codes, see Fig. 14.1. The goal of the interleaver is to provide each component decoder with uncorrelated input data. The constraint length of the component codes

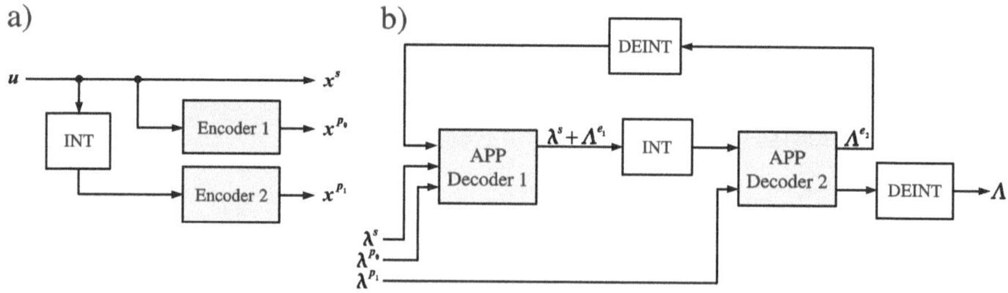

Fig. 14.1 (a) Berrou's turbo encoder and (b) turbo decoder structure.

typically is $K_c = 3 \ldots 5$. In binary turbo codes single bits enter the component encoders sequentially, where as in the case of duo-binary turbo codes two bits are encoded in parallel [4].

The ultimate goal of a channel decoder is to find the *a-posteriori probability* (APP) for each information bit. The turbo decoder approximates an APP decoder by decoding the component codes separately but cooperatively. A closed loop of component decoders, separated by an interleaver and deinterleaver, iterates several times, until a predetermined number of iterations is reached, or earlier if some other stop criterion is fulfilled [11]. The component decoders are soft-input soft-output (SISO) decoders that perform APP decoding. Turbo decoding typically uses the Maximum A-posteriori Probability (MAP) algorithm since it processes soft input values (intrinsic information) and it produces soft output values (extrinsic information). In contrast the Viterbi algorithm only generates hard decoded bits. An implementation friendly version of the MAP algorithm is the suboptimal Max-Log-MAP, since it works in the logarithmic domain. Here multiplications are substituted through additions. Both, Max-Log-MAP and Viterbi decoding, are based on the *add-compare-select* (ACS) recursion. For more details on the decoding algorithms refer to [27, 28, 3, 25].

14.2.3 LDPC Codes

LDPC codes [9] are linear block codes defined by a sparse parity check matrix H of size $M \times N$, see Fig. 14.2(a). For binary LDPC codes a valid codeword $\mathbf{x}$ has to satisfy $H\mathbf{x}^T = 0$ in a modulo-2 arithmetic. A column in H is associated to a codeword bit, a row corresponds to a parity check. A non-zero element in a row means that the corresponding bit contributes to this parity check. The complete code can be best visualized by a Tanner graph, a graphical representation of the associations between code bits and parity checks. Each column of H corresponds to a variable node (VN) and represents one code bit, each row of H corresponds to a check node (CN) and represents one parity check, respectively. Edges connect variable and check nodes according to the non-zero elements of the parity check matrix. Figure 14.2(b) shows the corresponding Tanner graph to the parity check matrix in Fig. 14.2(a) of an LDPC code with $N = 7$ variable nodes and $M = 3$

a)

$$H = \begin{pmatrix} 1 & 1 & 1 & 0 & 1 & 0 & 0 \\ 1 & 1 & 0 & 1 & 0 & 1 & 0 \\ 1 & 0 & 1 & 1 & 0 & 0 & 1 \end{pmatrix}$$

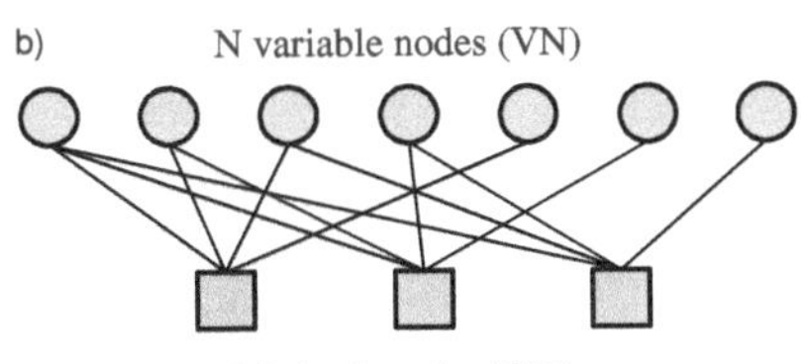

Fig. 14.2 Definition of an LDPC code: (a) parity check matrix (b) corresponding Tanner graph.

check nodes. The number of edges connected to a given node is called the node degree.

Like turbo codes, LDPC codes are decoded iteratively. Messages about the bit probabilities are transmitted via the edges between variable nodes and check nodes. The check nodes perform the parity check calculation, while the variable node perform an overall estimation of the decoded bit. The decoding process is stopped if all parity checks are satisfied or a fixed number of iterations has been performed.

Fully parallel decoder implementations instantiate all nodes of the Tanner graph. This approach lacks flexibility and is furthermore infeasible for high block lengths. LDPC decoders are therefore implemented in a partly parallel fashion, where only a subset of nodes in the Tanner graph is implemented as hardware units. The nodes of the Tanner graph are processed in a time multiplexed way on the functional units. Low latency and high throughput are obtained by exchanging many messages between variable and check nodes per clock cycle. A random connectivity between variable and check nodes poses big challenges for an efficient hardware implementation. Complex connectivity networks become mandatory to allow for a flexible and parallel message exchange, resolving occurring memory access conflicts. Thus, LDPC codes defined by standards are based on so called structured LDPC codes [5]. The matrices of these codes are composed of cyclically shifted identity matrices of size $P \times P$. Figure 14.3 shows the structured binary parity check matrix of the WiMAX code with $N = 576$ variable nodes, $M = 288$ check nodes, and $P = 24$.

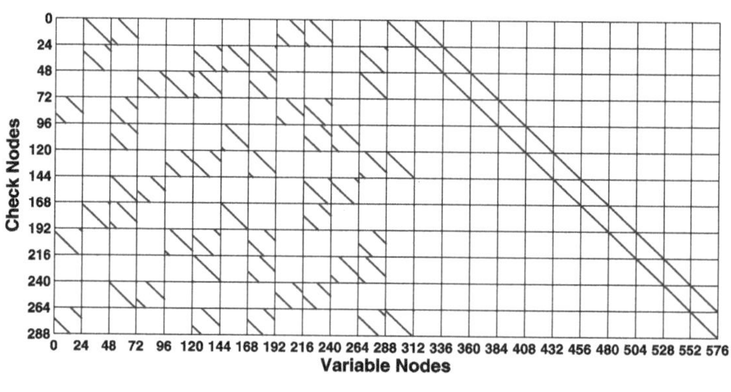

Fig. 14.3 A parity check matrix of the WiMAX code with $P = 24$.

These codes allow for an efficient implementation of the connectivity between up to P variable and P check nodes. Memory access conflicts are avoided and a low complexity logarithmic barrel shifter is sufficient as connectivity network for a given P. A flexible LDPC decoder has to support different submatrix sizes and varying node degrees. For more information on the decoding algorithms refer to [9, 18, 6].

14.3 Decoder Requirements

As one can see from Table 14.1, each standard has its own parameters for the channel encoder, such as different constraint lengths, generator polynomials, and, in case of turbo codes, different interleaving patterns. For LDPC codes many different parity check matrices have to be supported, depending on block size and code rate. Thus, the ASIP has to support a wide range of coding parameters.

In summary, the following specifications were derived which have to be provided by the ASIP to fulfill the flexibility requirements for SDR systems:

- combined decoder for CC, binary/duo-binary TC, and LDPC decoding,
- VA and Max-Log-MAP decoding for CC,
- support of $N = 16 \ldots 256$ states for CC,
- support of $N = 4 \ldots 16$ states for bTC,
- support of $N = 8 \ldots 16$ states for dbTC,
- arbitrary feedback and generator polynomials,
- rate flexibility by internal depuncturing of punctured codes for CC and TC,
- submatrix size flexibility for LDPC codes,
- check/variable node degree flexibility for LDPC codes,
- high throughput, low latency.

The throughput requirements strongly vary with the communication protocols and standards. In a mobile service like HSDPA, for instance, the maximum throughput for the turbo decoder is 14.4 Mbps. We target a throughput of at least the maximum of HSDPA. If the required throughput is not achieved by a single processor, e.g., in a WiFi system, FlexiChaP can be configured for a multiprocessor decoder system [30].

14.4 ASIP Design Methodologies

ASIPs allow for smooth trade-offs of implementation flexibility (in terms of software programmability) against hardware performance (throughput, energy, area). Application specific instructions offered by these processors can close the performance gap between traditional processors and dedicated hardwired solutions and improve the energy efficiency.

One can distinguish between software-driven top-down and architecture-driven bottom-up design methodologies for ASIPs. In the traditional top-down approach, one starts with a generic architectural processor template, e.g. a 5-stage RISC (reduced instruction set computer) pipeline, see Fig. 14.4(a). By profiling the target application on this processor the bottlenecks are identified and removed by inserting application specific instructions. This approach therefore can be considered as a software-driven design flow. In contrast, the bottom-up approach starts from the micro-architecture level, see Fig. 14.4(b). By a thorough analysis of the target algorithms and the required flexibility, functional blocks (indicated by triangles) are

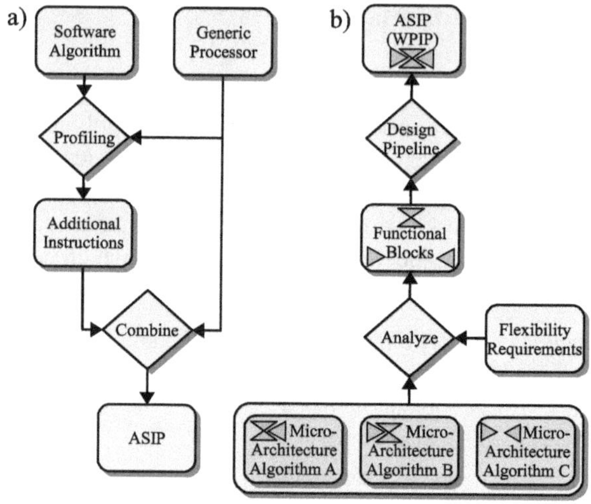

Fig. 14.4 ASIP design methodologies (a) top-down approach (b) bottom-up approach [2].

identified that need to be included into the processor pipeline. Thus, this approach can be considered as a architecture-driven design flow. All the flexibility and overhead in the processor that is not needed by the application is removed. Adapting bit widths, arithmetics, memory hierarchies etc. completely to the needs of the application yields best performance and energy efficiency. ASIPs designed in this manner share hardly any characteristic of classical RISC pipelines and can be considered as application specific building blocks enhanced with programmability. To emphasize this property these special processors are also called weakly programmable IP (WPIP).

14.5 Architecture

14.5.1 General Considerations

As pointed out in the previous sections, channel decoding for the various standards requires a high flexibility. However, since the ASIP is intended to only perform channel decoding, a "just enough policy" dominated the pipeline design. I.e., the FlexiChaP ASIPs were designed following the bottom-up approach as explained in the previous section. Dedicated architectures for the different decoding algorithms were intensively investigated w.r.t. commonalities in computation and memory organization. Table 14.2 shows the result of our analysis. The different building blocks and memories, and how they can be shared among the different algorithms are listed. It shows that many operations can be shared between convolutional and turbo decoding. LDPC decoding differs fundamentally from these trellis based decoding algorithms, but memory sharing is still applicable.

Table 14.2 Basic building blocks of the different decoding algorithms. Besides the common building blocks, the extrinsic and the survivor memory can be shared in a common decoder architecture.

	Viterbi	MAP	Turbo	LDPC
Branch metric calculation	x	x	x	
State metric calculation	x	x	x	
LLR calculation		x	x	
Traceback	x			
Variable node calculation				x
Check node calculation				x
Barrel shifter				x
Channel value memory	x	x	x	x
State metric memory	x	x	x	
Extrinsic memory			x	x
Survivor memory	x			

The ASIP was designed with an ASIP environment which does not impose any constraints on the instruction set, pipeline or memory structure. For this we used the ProcessorDesigner from CoWare [7]. The data path is modelled with LISA (language for instruction-set architectures). From the LISA description all needed tools and models can be generated (clock accurate C++ pipeline model, assembler, linker, debugger, synthesizable VHDL model).

Optimization for turbo and LDPC decoding was the primary goal since these are the most demanding tasks. We process one complete ACS recursion step of turbo code applications in a single clock cycle. Higher constraint lengths as they are common in convolutional codes are processed in a time multiplexed manner. To compute one recursion step per clock cycle, it must be possible to read channel values, to process branch and state metrics, and to store multiple state metrics or soft output values in parallel. This requires data parallelism within the pipeline, and a customized memory architecture with high memory bandwidth. Turbo codes do not use component convolutional codes with a constraint length K_c larger than 5. Hence, a maximum of 16 state metrics are computed in parallel. If K_c is larger than 5, the state metrics of a single trellis step have to be computed sequentially. Therefore a trellis step of a code with, e.g., 256 states needs to be divided into 16 partitions, each including 16 states or eight butterflies.

For LDPC decoding a whole submatrix can be processed per clock cycle. The maximum submatrix size is 27×27. In case bigger submatrices need to be supported, the parity check matrix has to be reordered to submatrix sizes of 27×27 or less.

The channel code structure is at least constant for a whole data block. The cost (w.r.t. area and energy) of specifying the code structure with each instruction is very high. On the other hand it is important to be able to switch within a few clock cycles from one channel code to another, for instance to support soft handover. Therefore the channel code configuration is not specified by the instructions but is kept *dynamically reconfigurable* within the ASIP, as explained later in Sect. 14.5.4.

The decoding algorithm itself is software defined, i.e., based on the special instruction set of the ASIP. Various windowing [8] and block termination schemes [24] must be supported for turbo, MAP, and Viterbi decoding. The software programmability yields flexibility, e.g., to adjust the acquisition length to the code structure and to the communication channel conditions. For LDPC decoding the node degree variability is achieved on the software level, whereas the submatrix size is reconfigurable.

14.5.2 Memory Concept

As already mentioned, flexible data routing and efficient data management are key for an efficient channel decoding. Thus the ASIP is based on a distributed memory concept, composed of various memories allocated to specific pipeline stages as depicted in Fig. 14.5.

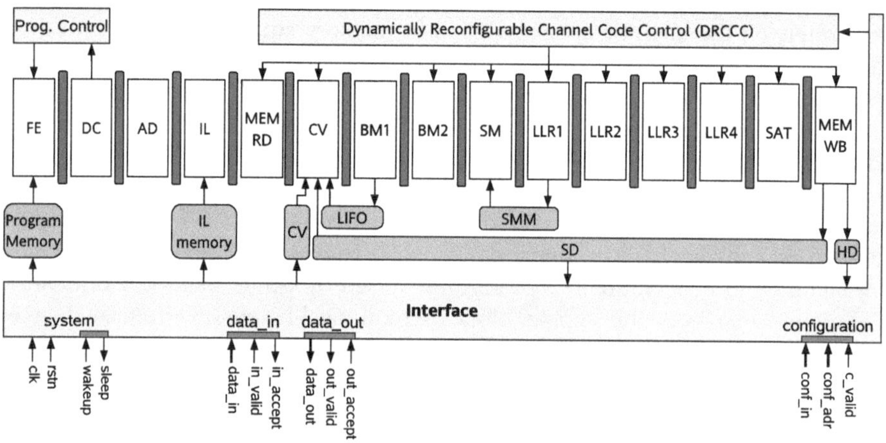

Fig. 14.5 General pipeline architecture for turbo/viterbi decoding, consisting of an interface, memories (a program memory, an interleaver memory, a channel value memory (CV), a last-in-first-out buffer (LIFO), a state metric memory (SMM), a hard decision output memory (HD), and a memory for a-priori or local survivor storage (AP/SUR)), a dynamically reconfigurable channel code control, a program control, and an instruction pipeline with 15 stages (Fetch (FE), Decode (DC), Address Generation (AD), Interleaver Access (IL), CV and AP/SUR Memory access (MEM), Branch Metric Calculation (BM1), Branch Metric Shuffling (BM2), State Metric calculation (SM), 4 stages for LLR calculation (LLR1 to LLR4), Extrinsic Calculation and Saturation (SAT), and Write Back (WB)) [29].

Besides the channel value (CV) memory for the input and the hard decision (HD) memory for the output data, two additional memories are reserved to store internal data only: one, a state metrics memory (SMM) for the storage of the state metrics during forward or backward recursion of the MAP algorithm, or for the storage of intermediate state metrics during Viterbi processing. It can store 128 words each 192 bits wide. Two, a last-in-first-out (LIFO) which serves as buffer for channel and a-priori values. The SD memory serves for internal data storage, namely the a-priori values for turbo decoding or the decision bits of the local survivors for the traceback of the Viterbi algorithm. For LDPC decoding the pipeline and memory organization looks different, since the main goal for the integration of LDPC codes was memory sharing between turbo and LDPC functionality. E.g. the SMM is used to store the channel/a-posteriori values of an LDPC code and it is furthermore accessed at different pipeline stages. Note that all memories are switched off when not accessed to reduce power consumption.

The instructions are able to access up to six memories in parallel. The addresses for these memory accesses are generated with the help of five dedicated address registers placed in the AD pipeline stage (e.g. WBA, RDA, CVA). The address registers are modified implicitly by the instructions, or explicitly by direct or indirect addressing coded in the instructions.

14.5.3 Pipeline

The data path consists of 15 pipeline stages. Figure 14.6 shows the pipeline with the hardware used during turbo and MAP decoding, and Fig. 14.7 depicts the hardware needed for LDPC decoding. While most of the logic of the turbo pipeline is reused for Viterbi decoding, almost no logic is shared with the LDPC functionality. When compared to a standard RISC pipeline, there are hardly any commonalities. Only the first two stages, instruction fetch (FE) and instruction decode (DC), are similar. The rest of the pipeline is completely tailored to the decoding algorithms. One essential operation for turbo and Viterbi decoding is the ACS operation. In Fig. 14.6 this operation is performed in the SM stage, in the so called butterflies. The inputs and outputs to/from these butterflies, however, have to be shuffled according to the code structure, i.e., the polynomials of the used convolutional code. The *Dynamically Reconfigurable Channel Code Control* (DRCCC) is responsible for applying the correct shuffling context to the data path. 16 states can be processed by the butterflies in binary mode in one clock cycle, and 8 states in duo-binary mode, respectively.

In case of more than 16 states, a load-store architecture is implemented: intermediate state metrics are loaded from the SMM to a pipeline register, processed, and then written back from the pipeline register to the SMM. A single trellis recursion with 256 states can thus be computed in 16 consecutive steps. Each step requires two clock cycles since the load and the write-back are executed in parallel on the dual-ported SMM. Furthermore, each step requires a different data shuffling.

For LDPC decoding it is necessary to support different submatrix sizes. Hence reconfigurable barrel shifters are used within the pipeline. The shifters have to be reconfigured if at all only after a block has been decoded.

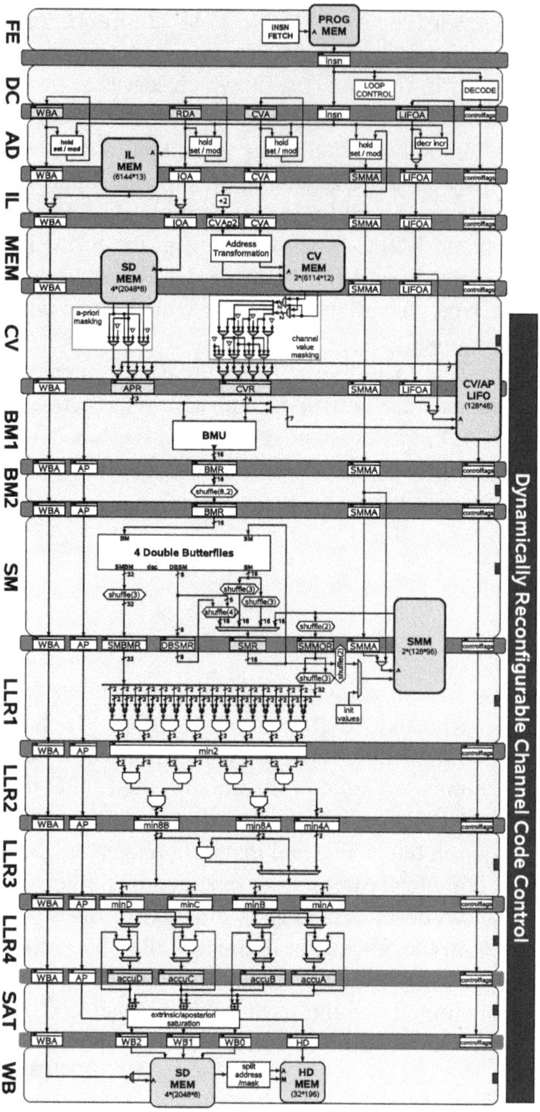

Fig. 14.6 Pipeline architecture as required for CC and TC decoding using the Max-Log-MAP algorithm [29].

14.5.4 Dynamically Reconfigurable Channel Code Control

The DRCCC unit defines the structure of the convolutional (component) code that is in use. It controls the internal data routing of the data path and the memory configuration. The DRCCC is look-up-table (LUT) based. It includes a *shadow* and a *working* configuration. The working registers hold the configuration that is actually

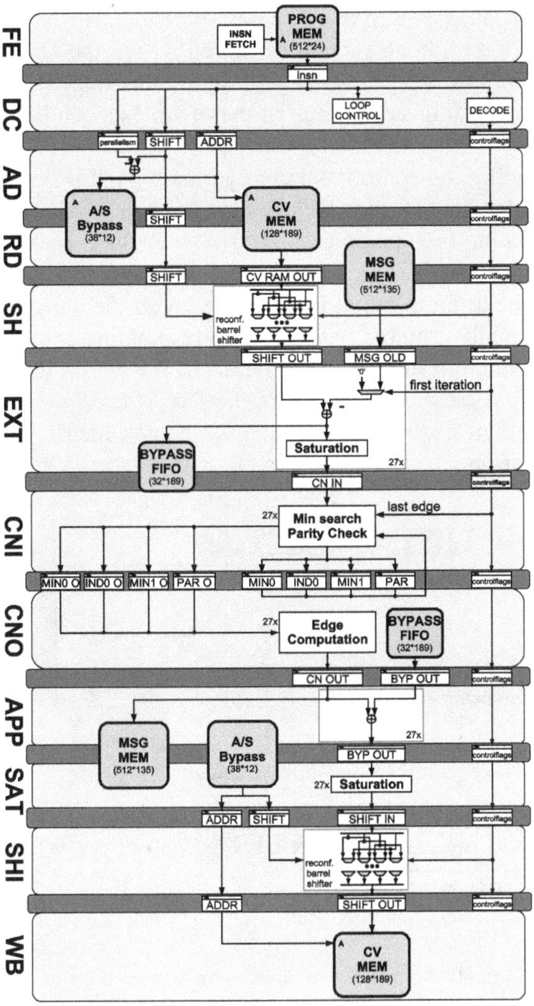

Fig. 14.7 Pipeline architecture as required for LDPC decoding. Pipeline stages are: fetch (FE), decode (DC), address (AD), read (RD), shift (SH), extrinsic computation (EXT), check node in (CNI), check node out (CNO), APP calculation (APP), saturation (SAT), shift inverse (SHI), write back (WB) [1].

in use, while the shadow registers allow for loading a new configuration without effecting the actual data processing. The content of the shadow registers is transferred to the working registers by a special instruction within one clock cycle, thus allowing a fast context switch between different codes.

Moreover, the DRCCC unit allows for efficient multi-context instructions. As already mentioned before, the state metrics of one recursion step for a convolutional (component) code with more than 16 states are computed in up to 16 processing steps in case of a code with 256 states. These different processing steps require, e.g., varying data routing within the data path pipeline. This routing is defined by a *context*. Eight different contexts can be stored within the DRCCC. These eight contexts are sufficient to decode a code with 256 states since the data routing does not differ for all processing steps. Each of these contexts can be freely assigned to the different steps.

The flow of loading the appropriate configuration parameters from the DRCCC to the pipeline is depicted in Fig. 14.8. A certain step number is associated with each data-manipulating instruction (e.g. step = 1 for *VA1*, step = 2 for *VA2*, and so forth in the assembler code example in Fig. 14.9(a)). This step number is used to look-up the context number that is associated with the current processing step. Afterwards, the actually required part of the corresponding context is loaded into a pipeline register, which in turn controls the data flow within the pipeline stage(s). The example in Fig. 14.8 shows the load of the *BM_allocation_context* for the control of the branch metric shuffling between the branch metric calculation and the state metric calculation.

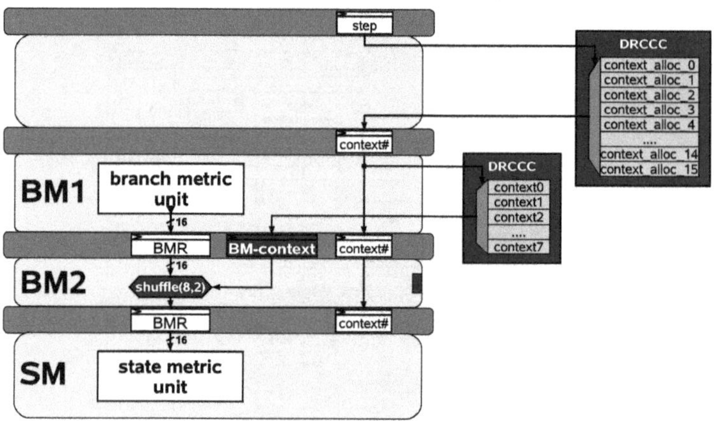

Fig. 14.8 Use of configuration contexts within the pipeline: each trellis instruction specifies a step number. Each step number is mapped to a certain context with the help of context_map. The retrieved context number finally specifies the context that is loaded into a pipeline register and used from there to control the data flow in the data path [29].

In total, 383 bits are sufficient to specify the whole channel code structure of a convolutional code. Due to the DRCCC, the instruction bit width is reduced from 68 to 24 bits which considerably saves area and energy.

14.5.5 Instruction Set

The processor has to be programmed in assembler in order to fully exploit the performance of the data path. Each instruction is 24 bits wide and can include besides the opcode additional address information for accessing the CV, SD or SM memories.

The instructions can be grouped in control, address modification, and algorithmic instructions.

The program flow control is implemented by hardware-loops, or zero-overhead-loops: a block of instructions is executed for a certain number of times defined by the instruction, or forever. The block starts after the *RPT* instruction and ends at a certain position in the program, defined by a label in the assembler program. A stack-based system allows to nest five of these loops. Furthermore, a branch into a subroutine is allowed.

Besides a no-operation instruction *nop*, a special instruction *PD* is defined to power down the processor's pipeline and to set it into a sleep mode. This sleep mode can only be left by applying the wakeup interrupt request through the interface of the ASIP. The content of the shadow registers of the DRCCC is transferred to the working registers by the *reconf* instruction.

The algorithmic instructions can be grouped into LDPC instructions, VA instructions, and MAP instructions for "fully parallel" and "partially parallel" processing of the state metrics. The "fully parallel" Log-MAP instructions can compute all state metrics of one recursion step in parallel. If the constraint length of the code is larger than five, the "partially parallel" instructions allow for sequential processing of a single trellis step. The example assembler code for Viterbi decoding of block with 254 infobits coded with $K_c = 7$ and $R = 1/3$ is shown in Fig. 14.9(a). The assembler code is independent of the feedback and generator polynomials. Those are defined in the DRCCC. For larger blocks basically only the number of repetitions of the loops has to be modified. Therefore it is possible to write very compact instruction code, resulting in a small program memory. Figure 14.9(b) shows an assembler code for LDPC decoding. This code exactly corresponds to the parity check matrix

a)

```
  .text

  nop
  reconf        ; reconfigure
; endless hardware loop
RPT ->END_MAIN FOREVER  st 0,CVA
  st 0,WBA
  st 0,RDA
  st AZS,(SMMA=0)
  st EQP,(SMMA=1)
  st (SMMA=0),SMR
  ; compute initial recursion step
  ; with four different contexts
  VA1 (CVA)+=3 #CVs=3,(SMMA=2)
  st (SMMA=1),SMR
  VA2 (SMMA=3)
  st (SMMA=1),SMR
  VA3 (SMMA=4)
  st (SMMA=1),SMR
  VA4 (SMMA=5)
; compute recursion steps
RPT ->END_REC #127
  st (SMMA=2)OS=1,SMR
  VA1 (CVA)+=3 #CVs=3,(SMMA=6)
  st (SMMA=5)OS=-1,SMR
  VA2 (SMMA=7)
  st (SMMA=3)OS=-1,SMR
  VA3 (SMMA=8)
  st (SMMA=4)OS=1,SMR
  VA4 (SMMA=9)
  st (SMMA=6)OS=1,SMR
  VA1 (CVA)+=3 #CVs=3,(SMMA=2)
  st (SMMA=9)OS=-1,SMR
  VA2 (SMMA=3)
  st (SMMA=7)OS=-1,SMR
  VA3 (SMMA=4)
  st (SMMA=8)OS=1,SMR
  VA4 (SMMA=5)
END_REC:
  st 254,RDA
  st0x80fe,WBA
  VAinitTB 0  ; Init Traceback
; Perform Traceback in a loop
RPT ->END_TB #127
  VATB
  VATB
END_TB:
  VATB
  PD               ; power down
  nop
END_MAIN:
```

b)

```
.text

1.subm 24                 ; reconfigure networks and
                          ; set submatrix size to 24
1.diag 0, s=1,  a=1  ; process submatrix
1.diag 0, s=6,  a=2  ; s determines shift offset
1.diag 0, s=11, a=8  ; a determines address
1.diag 0, s=4,  a=9
1.diag 0, s=23, a=12
1.diag 1, s=0,  a=13 ; last edge of check node

1.diag 0, s=18, a=1  ; new check node begins
1.diag 0, s=19, a=5
1.diag 0, s=5,  a=6
1.diag 0, s=22, a=7
1.diag 0, s=21, a=11
1.diag 0, s=0,  a=13
1.diag 1, s=0,  a=14 ; last edge of check node
 ...
1.diag 0, s=23, a=12
1.diag 1, s=0,  a=23 ; last edge of check node

1.pchk it=10              ; perform parity check
                          ; and start new iteration
                          ; if parity check fails
                          ; and less than 10 iter.
                          ; were processed
nop
PD                        ; power down
nop
```

Fig. 14.9 (a) Viterbi assembler code for a convolutional encoded block with 254 information bits, constraint length $K_c = 7$ equal to 64 states, rate $R = 1/3$ [29]. (b) LDPC assembler code snippet for WiMAX LDPC code of Fig. 14.3 ($N = 576$, $M = 288$, $R = 1/2$) [1].

in Fig. 14.3. The reconfigurable barrel shifters are initialized with *l.subm*, while a whole submatrix is being processed with *l.diag*.

14.6 ASIP Validation

Validation of the presented ASIP family is very challenging. Due to the high flexibility many use cases have to be considered when testing. E.g., the UMTS standard alone accounts for more than 5000 different block lengths that have to be tested. Furthermore, Monte Carlo simulations are preferable, since they show the bit and frame error rates (BER/FER) for given signal-to-noise ratios (SNR). Thus they show the overall system performance of the ASIP with the running program. Performing these simulations with the generated C++ pipeline simulation model or the VHDL model is infeasible. The decoding speed on standard PCs with more than 3 GHz is just several hundred payload bits per second. This speed is at least three orders of magnitude too low.

To overcome the aforementioned validation bottlenecks, a rapid prototyping environment was developed [2]. We chose the Xilinx ML507 evaluation platform [21] for rapid prototyping the ASIP, since this platform perfectly fits our requirements. It is a fully featured FPGA board with many interfaces and a hardwired IBM PowerPC 440 on the Virtex5 FPGA. The Xilinx Embedded Development Kit 10.1 (EDK) was used to integrate the prototyping platform.

Prior to running the ASIP three tasks have to be performed:

- A configuration has to be loaded into the ASIP.
- A program has to be loaded into the ASIP.
- Data has to be sent to the ASIP.

Due to the high flexibility of FlexiChaP, an intelligent interface is needed. Configuration data has to be generated, programs have to be loaded, and input data have to be generated and loaded into the ASIP. Furthermore, BER and FER have to be calculated after decoding. Performing all these tasks in software is the only way to offer the required flexibility. The PowerPC on the FPGA is used for this by communicating with the ASIP via the processor local bus (PLB). For this, a PLB wrapper has been created around the ASIP interface. Configuration and program are loaded in a memory mapped way into the ASIP, while for data transmission handshaking is used.

Figure 14.10 shows the system data flow, which is based on three components:

- A graphical user interface (GUI) running on the PowerPC [10].
- A baseband processing chain for performing BER and FER simulations running on the PowerPC.
- The FlexiChaP ASIP running in FPGA logic.

The GUI is used to set the system parameters, such as interleaver (e.g., UMTS or LTE), block size, code rate, signal-to-noise ratio (SNR), and many more. When all

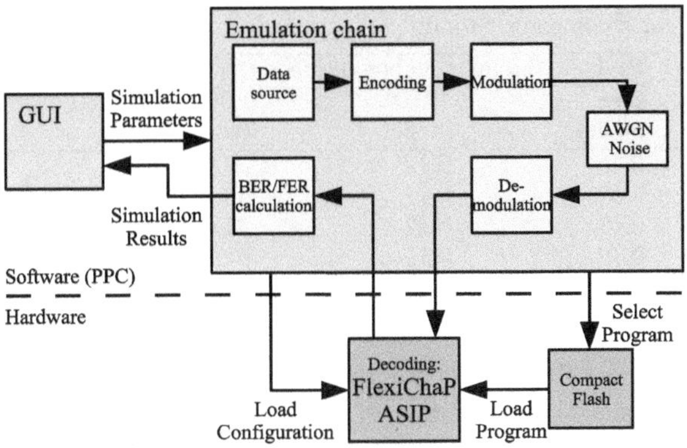

Fig. 14.10 Emulation system data flow [2].

parameters are set, the system chain starts its operation. It consists of a flexible baseband processing model with data generation, encoding, modulation (BPSK, QPSK, 8PSK, 16QAM, 64QAM or 256QAM), additive white Gaussian noise (AWGN), demodulation, and BER/FER calculation. Decoding is performed on the ASIP. The FlexiChaP software driver calculates the required configuration data and loads the selected program from an external Compact Flash into the ASIP. After decoding, the results are displayed on the GUI, but can also be sent to a PC, in order to check the results automatically. If these are not satisfying for a certain case, it is possible to apply the usual debugging methods.

The platform is used as a hardware/software development platform, that allows us to quickly validate the correct functionality of the ASIP in conjunction with the assembler programs. By using the GUI for controlling the whole system, the presented platform is able to run completely autonomously without any PC. Thus, it is suitable for customers of the ASIP that want to develop their own programs. The customers do neither need the LISA source code, nor the C++ pipeline simulation model, nor the VHDL code of the processor, but only the programming model.

In our setup, the hardwired PowerPC is clocked at 450 MHz, a double-precision floating point unit (FPU) runs at 150 MHz in FPGA logic, and the ASIP runs at 75 MHz. These clock frequencies were chosen because of the clock ratios that are constrained between all these components. Table 14.3 shows the resulting payload throughputs for the different validation techniques, i.e., C++ pipeline simulation model, VHDL RTL model, and the presented FPGA-based rapid prototyping platform. For the C++ and RTL simulations a standard PC with a Pentium4 CPU running at 3.2 GHz has been used.

For a UMTS turbo code with 882 information bits, the C++ model achieved a payload throughput of only 450 bits per second, since the ASIP uses mainly nonstandard data types, e.g. signed integers with a 6 bit quantization are used for the channel values. With this simulation speed it takes 22 days to simulate one million blocks, which is a necessity for tests at a FER level of 10^{-4}. This simulation corre-

Table 14.3 Throughput comparison for different evaluation techniques [2].

Code rate	Payload Throughput/kbps UMTS turbo code (882 info bits, 5 iterations)			
	C++ model	VHDL model	FPGA-based (soft-AWGN)	FPGA-based (hard-AWGN)
1/3	0.45	0.12	37.0–151.2	217.3–252.0
1/2	0.45	0.12	53.7–195.3	272.5–309.8
2/3	0.45	0.12	71.6–267.8	364.7–415.0
9/10	0.45	0.12	97.7–363.3	507.2–578.2

sponds to only a single SNR and a single configuration. The RTL simulation speed of 120 bits per second with ModelSim is another factor of three to four slower.

In contrast to these simulation methods, the prototyping platform achieves payload throughputs of up to 363.3 kbps with a software noise channel.

In order to further increase the throughput, the software noise generator is replaced by a high quality AWGN IP running in FPGA logic [31]. In this way, the throughput of the overall system can be drastically increased at the cost of flexibility regarding the supported SNRs (the AWGN IP only supports SNRs in steps of 0.1 dB). Accelerations of up to 600% for BPSK and up to 165% for 256QAM modulation are achievable with the hardware noise channel. Throughputs of more than 575 kbps are possible.

The obtained BERs/FERs from the platform are congruent with a turbo decoder software model that applies the same parameters. Thus, the prototyping platform successfully validated the ASIP when running a programm. Up to one million frames have been decoded per SNR with the software noise generator in just two to four hours instead of 22 days. Employing the hardware AWGN channel in this case gains another factor of five in throughput.

14.7 Results

As already mentioned, the FlexiChaP pipeline architecture was modelled in the LISA language [7]. This model is parameterizable to support the full set of codes as discussed before, or only a subset of the codes. Thus we provide design-time and run-time configurability (see Table 14.4).

Table 14.4 Run-time and design-time configurability of FlexiChaP.

	Design-time	Run-time
Code classes	x	
Code structure		x
Decoding algorithm	x	x
Memories	x	x

Various VHDL instances of the FlexiChaP pipeline were synthesized. A 65 nm low power low leakage standard cell library with worst case conditions is used as target technology. The area of the different ASIP instances without memories is listed in Table 14.5. The maximum clock frequency for FlexiChaP with full functionality is 400 MHz. If only a subset of codes is supported even higher clock frequencies are possible. The area overhead of the ASIP with bTC, dbTC, and VA flexibility is 46% compared to a dedicated 16-state bTC decoder ASIP, not considering the memories. LDPC functionality requires about as much logic as the support for these trellis-based decoding schemes. Since there is almost no logic reuse between TC/VA and LDPC decoding, the overall area with full functionality is approximately given by the sum of TC/VA functionality and LDPC functionality. The ASIP was also synthesized for a Xilinx Virtex4 FPGA. Here a maximum clock frequency of 132 MHz was achieved when only LDPC codes are supported.

Table 14.5 ASIC and FPGA synthesis results for various instances of the ASIP [1].

Functionality				ASIC (65 nm, 1.10V, 120°C)		FPGA (Xilinx xc4vlx80-12)	
bTC	dbTC	VA	LDPC	Size [mm^2]	Frequency [MHz]	Size [Slices]	Frequency [MHz]
X				0.074	450 (max)	4,207	135 (max)
X	X			0.089	415 (max)	5,494	117 (max)
X	X	X		0.109	400 (max)	7,012	109 (max)
			X	0.113	425 (max)	8,076	132 (max)
X	X	X	X	0.232	400 (max)	14,495	109 (max)

FlexiChaP with full functionality requires 0.39 mm^2 for the memories. Due to the high memory reuse for LDPC decoding still 0.31 mm^2 are needed when omitting LDPC support. Overall this sums up to 0.62 mm^2 with full functionality and 0.42 mm^2 without LDPC support. With this RAM configuration it is possible to decode binary turbo codes with up to 6144 information bits, duo-binary turbo and convolutional codes with up to 8192 information bits ($R = {}^1\!/_2$ for convolutional codes). LDPC codes with a block length of up to 3456 bits, a check node degree of up to 28 and up to 13824 edges are decodable. Thus, all LDPC codes in the WiMAX and WiFi standards are supported by the ASIP. Note that the memories can be tailored to the actual application scenario at design and run-time.

Table 14.6 Core payload throughputs of FlexiChaP for convolutional, turbo and LDPC codes [1].

Algorithm	Throughput/Mbps @ 400 MHz	Conditions
Viterbi	12–196	16–256 states
Binary MAP	1.6–170	4–256 states
Binary TC	17	4–16 states, 5 iterations
Duo-binary TC	17–34	8 or 16 states, 5 iterations
LDPC	27.7–257.0	10–20 iterations

The payload throughputs achieved by FlexiChaP for convolutional, turbo and LDPC decoding are listed in Table 14.6. For Viterbi and MAP decoding the throughput strongly depends on the number of states. If the number of states is higher than

16 in the binary case the load/store mechanisms as mentioned before result in a decreased throughput. The LDPC decoding throughput varies depending on the submatrix size and the code rate. The LDPC throughputs are valid for the WiMAX and WiFi standard.

Table 14.7 Comparison of different turbo decoder implementations for UMTS turbo codes [29].

	Implementation Technology	Size	Clock freq.	Cycles/ (bit*MAP)	Throughput @ 5 iter
Conf. RISC [20]	ASIC (130 nm)	104 kGE	133 MHz	9	1.4 Mbps
Clustered VLIW [14]	VirtexII-4000	517 slices	80 MHz	8	1 Mbps
XiRisc [17]	ASIC (130 nm)	6000 kGE	100 MHz	100	0.1 Mbps
SODA [32]	ASIC (180 nm)	1000 kGE	400 MHz	20	2 Mbps
ASIP [22]	ASIC (90 nm)	63 kGE	400 MHz	3.5	11.4 Mbps
FlexiChaP[a]	ASIC (65 nm)	53 kGE	400 MHz	2.35	17 Mbps
FlexiChaP[a]	Virtex4	7012 slices	109 MHz	2.35	4.6 Mbps

[a]Full functionality without LDPC.

Table 14.7 summarizes turbo decoder implementations for UMTS turbo code applications on different state-of-the-art target platforms. The total gate count of the ASIP core in the 65 nm standard cell technology without memories providing full functionality but LDPC support is 53 kilogate equivalents (kGE). Compared to the processors of [20] and [22] with 104 kGE and 63 kGE for the core's logic, respectively, FlexiChaP saves more than 15% of the area, but provides much higher flexibility. The size of FlexiChaP and the ASIP of [22] differ considerably from a platform like SODA. However, it is important to notice that the other processors provide a much higher flexibility whereas FlexiChaP and the ASIP of [22] are only intended for channel decoding. The clock frequencies listed are maximum values. They differ, among other things, because the target technology is not the same. FlexiChaP outperforms the other processor implementations even if they all run with the same clock frequency. This high performance is achieved due to the high internal memory bandwidth, the specialized data path pipeline, and the internal reconfigurable data shuffling and reordering mechanisms. The overhead of FlexiChaP in comparison to a fully dedicated solution for an eight state duo-binary turbo-decoder is less than 25%. This is mainly due to the fact that memory is the dominating part and the memory sizes are the same for a dedicated implementation and the FlexiChaP solution except for the program memory which is quite small. Moreover the basic building blocks of the data path pipeline like LLR and butterfly calculations do hardly differ from a dedicated solution. Considering the FPGA implementation, the processor presented in [14] is competitive in terms of area and throughput, but not in flexibility. The solution of [14] is only able to decode turbo codes with eight states and code rates of $1/3$ and higher, convolutional codes are not supported at all.

Postlayout analysis without LDPC support shows a clock frequency of 350 MHz and a size of 0.75 mm^2 in a 65 nm CMOS technology. First estimates for the core's logic power consumption are 100 mW. Fig. 14.11 depicts the ASIP layout with an

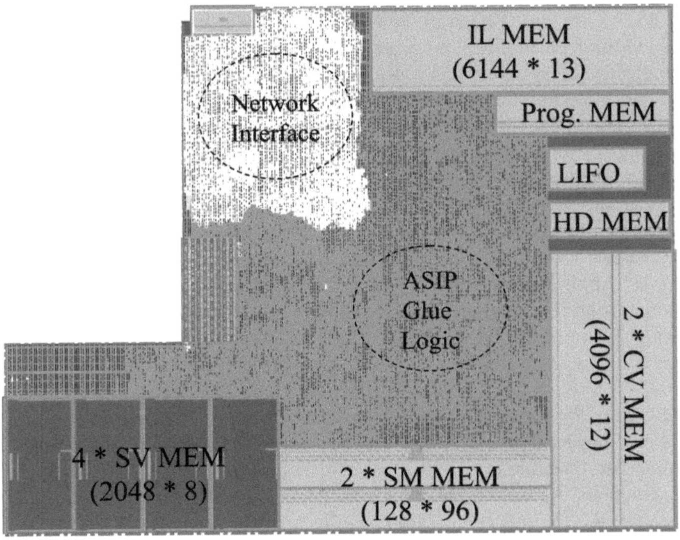

Fig. 14.11 ASIP layout after place and route. The total size is 0.75 mm^2 without the network interface [29].

additional Network-on-Chip (NoC) interface. This instance of FlexiChaP without LDPC support is currently in fabrication.

Acknowledgements This work was supported by DFG as part of the Priority Programme 1148, *Reconfigurable Computing Systems*.

References

1. Alles, M., Vogt, T., Wehn, N.: FlexiChaP: A reconfigurable ASIP for convolutional, Turbo, and LDPC code decoding. In: Proc. 5th International Symposium on Turbo Codes and Related Topics, Lausanne, Switzerland, pp. 84–89 (2008)
2. Alles, M., Lehnigk-Emden, T., Brehm, C., Wehn, N.: A rapid prototyping environment for ASIP validation in wireless systems. In: EDA Workshop 2009, Dresden, Germany (2009)
3. Berrou, C., Glavieux, A., Thitimajshima, P.: Near Shannon limit error-correcting coding and decoding: turbo-codes. In: Proc. 1993 International Conference on Communications (ICC '93), Geneva, Switzerland, pp. 1064–1070 (1993)
4. Berrou, C., Jezequel, M., Doullard, C., Kerouedan, S.: The advantages of non-binary turbo codes. In: Proceedings of Information Theory Workshop, Cairns, Australia, pp. 61–63 (2001)
5. Boutillon, E., Castura, J., Kschischang, F.: Decoder-first code design. In: Proc. 2nd International Symposium on Turbo Codes & Related Topics, Brest, France, pp. 459–462 (2000)
6. Chen, J., Dholakia, A., Eleftheriou, E., Fossorier, M.P.C., Hu, X.Y.: Reduced-complexity decoding of LDPC codes. IEEE Trans. Commun. **53**(8), 1288–1299 (2005)
7. CoWare, http://www.coware.com
8. Dawid, H.: Algorithmen und Schaltungsarchitekturen zur Maximum a Posteriori Faltungsdecodierung. PhD thesis, RWTH Aachen, Shaker Verlag, Aachen, Germany (1996). In German
9. Gallager, R.G.: Low-density parity-check codes. IRE Trans. Inf. Theory **8**(1), 21–28 (1962)

10. Genode FPGA Graphics (FX), http://www.genode-labs.com/products/fpga-graphics
11. Gilbert, F.: Optimized, highly parallel architectures for iterative decoding algorithms. PhD thesis, Microelectronic Systems Design Research Group, Department of Electrical Engineering and Information Technology, University of Kaiserslautern (2003). ISBN 3-936890-06-4
12. Glossner, J., Iancu, D., Moudgill, M., Nacer, G., Jinturkar, S., Schulte, M.: The Sandbridge SB3011 SDR platform. In: Joint IST Workshop on Mobile Future and the Symposium on Trends in Communications (SympoTIC '06), pp. ii–v (2006)
13. IMEC: Scientific Report 2006: Software Defined Radio Flexible Air Interface. www.microelektronica.be/wwwinter/mediacenter/en/SR2006/681340.html (2006)
14. Ituero, P., Lopez-Vallejo, M.: New schemes in clustered VLIW processors applied to turbo decoding. In: Application-specific Systems, Architectures and Processors, 2006. ASAP '06. International Conference on, pp. 291–296 (2006). doi:10.1109/ASAP.2006.48
15. Krishnaiah, G., Engin, N., Sawitzki, S.: Scalable reconfigurable channel decoder architecture for future wireless handsets. In: Proc. 2007 Design, Automation and Test in Europe (DATE '07) (2007)
16. Lin, Y., Lee, H., Woh, M., Harel, Y., Mahlke, S., Mudge, T., Chakrabarti, C., Flautner, K.: SODA: A low-power architecture for software radio. In: Proc. 33rd International Symposium on Computer Architecture (ISCA'06), pp. 89–101 (2006)
17. Lodi, A., Cappelli, A., Bocchi, M., Mucci, C., Innocenti, M., De Bartolomeis, C., Ciccarelli, L., Giansante, R., Deledda, A., Campi, F., Toma, M., Guerrieri, R.: XiSystem: a XiRisc-based SoC with reconfigurable IO module. IEEE J. Solid-State Circuits **41**(1), 85–96 (2006). doi:10.1109/JSSC.2005.859319
18. Mansour, M.M., Shanbhag, N.R.: VLSI architectures for SISO-APP decoders. IEEE Trans. Very Large Scale Integr. (VLSI) Syst. **11**(4), 627–650 (2003)
19. Matu, E., Seidel, H., Limberg, T., Robelly, P., Fettweis, G.: A GFLOPS vector-DSP for broadband wireless applications. In: Conference 2006, IEEE Custom Integrated Circuits, pp. 543–546 (2006). doi:10.1109/CICC.2006.320923
20. Michel, H., Worm, A., Münch, WehnN.: Hardware/software trade-offs for advanced 3G channel coding. In: Proc. 2002 Design, Automation and Test in Europe (DATE '02), Paris, France (2002)
21. ML507 Evaluation Platform, http://www.xilinx.com/products/devkits/HW-V5-ML507-UNI-G.htm
22. Muller, O., Baghdadi, A., Jezequel, M.: From parallelism levels to a multi-ASIP architecture for turbo decoding. IEEE Trans. Very Large Scale Integr. (VLSI) Syst. **17**(1), 92–102 (2009)
23. PACT XPP Technologies, www.pactcorp.com
24. Reed, M.C., Pietrobon, S.S.: Turbo-code termination schemes and a novel alternative for short frames. In: Proc. 1996 International Symposium on Personal, Indoor, and Mobile Radio Communications (PIMRC '96), Taipei, Taiwan, vol. 2, pp. 354–358 (1996)
25. Robertson, P., Hoeher, P., Villebrun, E.: Optimal and sub-optimal maximum a posteriori algorithms suitable for turbo decoding. European Transactions on Telecommunications (ETT) **8**(2), 119–125 (1997)
26. Stretch, http://www.stretchinc.com
27. Viterbi, A.J.: Error bounds for convolutional codes and an asymptotically optimum decoding algorithm. IEEE Trans. Inform. Theory **IT-13**, 260–269 (1967)
28. Viterbi, A.J., Omura, J.K.: Principles of Digital Communication and Coding. McGraw–Hill, New York (1979)
29. Vogt, T., Wehn, N.: A reconfigurable ASIP for convolutional and turbo decoding in an SDR environment. IEEE Trans. Very Large Scale Integrat. (VLSI) Syst. **16**(10), 1309–1320 (2008)
30. Vogt, T., Neeb, C., Wehn, N.: A reconfigurable multi-processor platform for convolutional and turbo decoding. In: Reconfigurable Communication-centric SoCs (ReCoSoC), Montpellier, France (2006)
31. Xilinx Inc., http://www.xilinx.com/ipcenter
32. Yuan, L., Mahlke, S., Trevor, M., Chaitali, C., Alastair, R., Krisztian, F.: Design and implementation of turbo decoders for software defined radio. In: Proc. IEEE 2006 Workshop on Signal Processing Systems (SiPS) (2006)

Chapter 15
Dynamically Reconfigurable Systems for Wireless Sensor Networks

Heiko Hinkelmann, Peter Zipf, and Manfred Glesner

Abstract In this chapter, we explore the design of a coarse-grained reconfigurable architecture for wireless sensor network nodes, which applies frequent runtime reconfiguration to obtain the required combination of high energy efficiency and programmability for the target domain. We particularly examine the effect of the reconfiguration overhead on total system efficiency, and propose a novel reconfiguration mechanism to reduce this overhead. Comparing the energy consumption, area, and performance of our architecture to processor and ASIC designs, our experiments show the low total energy consumption achieved, the low reconfiguration overhead, and the specific region of the architecture in the design space between processors and ASICs. In particular, large energy-savings of factor 2 to 6 and speed-ups of factor 6 to 14 compared to processors are obtained on average. Overall, our work shows the high suitability of frequent runtime reconfiguration for the design of very energy efficient but yet programmable embedded system architectures.

15.1 Introduction

Emerging technologies like ubiquitous computing, smart environments, wearable computing and—of particular interest for our research work—wireless sensor networks (WSN) all rely on the availability of small and efficient embedded system platforms in large quantities. While low area consumption and programmability are important cost factors, it is typically *energy efficiency* which constitutes the most critical design challenge for WSN nodes. Lifetimes of up to several years often need to be achieved with the energy budget of a small battery only. Processors typically cannot provide high energy efficiency for data processing on the nodes, whereas

Heiko Hinkelmann · Peter Zipf · Manfred Glesner
Institute of Microelectronic Systems, Technische Universität Darmstadt, Darmstadt, Germany,
e-mails: hinkelmann@mes.tu-darmstadt.de, zipf@mes.tu-darmstadt.de,
glesner@mes.tu-darmstadt.de

M. Platzner et al. (eds.), *Dynamically Reconfigurable Systems*,
DOI 10.1007/978-90-481-3485-4_15, © Springer Science+Business Media B.V. 2010

ASICs are usually too costly. Therefore, the development of new architectures for energy-constraint embedded systems becomes important for achieving the desired balance between high energy efficiency, small size, and flexibility.

The objective of our research work is to study specifically the suitability of a new type of reconfigurable architecture to provide this balance [16, 11]. Characteristic for this approach is the extensive use of *frequent runtime reconfiguration* of a small coarse-grained heterogeneous data path. Frequent runtime reconfiguration is however controversial: On the one hand, it allows efficient cycle based hardware reuse and hence keeps the data path small, flexible and cost-efficient. On the other hand, it is usually associated with a large energy and performance overhead. Yet, in contrast to this common experience, the presented results will show that the reconfiguration costs for our new architecture remain small even in the worst case. Thus, this architecture approach is shown to be indeed well suited to achieve the required combination of high energy efficiency, small size, and flexibility.

A key factor to achieve these low reconfiguration costs is a novel reconfiguration mechanism developed in the scope of this project, constituting an additional important result of our research work [14]. Due to its good scalability and easy adaptability, it is well suitable to be adopted for other reconfigurable architectures and application domains as well as for design automation tools for such systems.

15.1.1 Outline

The following sections of this chapter begin with a short review of WSNs and related work (Sect. 15.2). Then, the design of our reconfigurable architecture is introduced, including the datapath design (Sect. 15.3), the proposed reconfiguration mechanism (Sect.15.4), and the overall sensor node system (Sect. 15.5). Detailed analysis results evaluating the energy efficiency, area, and performance of our architecture in comparison to processor and ASIC reference designs are provided next (Sect. 15.6), followed by a short presentation of a developed prototyping platform (Sect. 15.7). The chapter concludes with a brief discussion of possible utilisation of the achieved results beyond our research project (Sect. 15.8) and a concluding evaluation (Sect. 15.9).

15.2 Motivation and Background

A typical WSN node (mote) consists of a small processing unit, sensors, a RF transceiver, and a battery for autonomous power supply [1]. For most applications, the motes must be small, cheap, and achieve long lifetimes up to several years. Energy consumption is therefore the most critical and limiting factor in WSNs today.

A node consumes energy mainly for wireless communication and for processing. A trade-off exists between both factors, since local data processing can be applied to reduce wireless communication demand and vice versa [25, 7]. Significant reduction of communication costs can be achieved, e.g., by reducing retransmissions through forward error correction [28] or by application-specific pre-processing of sensor data. Moreover, local data processing can be useful for smart functionality or required for providing security via encryption and authentication.

How costly local data processing is depends on the mote architecture. Most existing nodes use small processor architectures, e.g. [10, 21]. However, processors in general suffer from low energy efficiency, because low power consumption is outweighed by long execution times. Local data processing is therefore often too costly on such platforms and hence highly limited, representing a severe drawback.

ASIC architectures on the other hand provide best energy efficiency, often several orders of magnitude better than processors [3]. However, the large variety of WSN applications differs so much in functionality [7, 26], that individual ASIC designs would often not achieve the high volumes required to compensate their high design costs. A programmable platform is needed instead. The idea to use reconfigurable computing for designing more energy efficient programmable mote platforms was discussed, e.g., in [25]. Our own approach studies a new reconfigurable architecture concept for this purpose.

Our approach differs from previously presented reconfigurable architectures like [4, 5, 20, 22, 19] in several ways. First, we optimise our architecture for low energy consumption, and achieve substantially lower power values for the entire architecture of few mW only [16]. Second, from architectural perspective, dynamic reconfiguration is applied extremely frequently to enable efficient resource sharing of a small coarse-grain data path, possibly having a different configuration in every clock cycle. Hence, our architecture becomes much smaller than the pipeline or array structures of [4, 20, 22, 19] for example. Solely RaPiD [5] follows a somewhat similar approach. Yet, only two RaPiD cells (each 0.3 mm^2 when scaled to 130 nm) are already bigger than our complete architecture.

15.3 Design of a Reconfigurable Function Unit

Our objective to create a small, highly efficient and yet flexible mote architecture is realised by applying frequent dynamic reconfiguration to a small heterogeneous coarse-grain data path, while keeping the according reconfiguration overhead low. The purpose of dynamic reconfiguration is to allow efficient inter-task and intra-task resource sharing, thus requiring only few hardware resources. The data path is integrated as a *reconfigurable function unit* (RFU) into a processor and serves as energy-efficient accelerator for data processing of WSN functions [16, 11].

The RFU is designed flexible enough to support a large variety of WSN-typical applications. Yet, some general, *domain-specific* customisation of the RFU architecture to typical characteristics of the entire WSN domain can be made:

- Data processing is typically not based on continuous streams but rather on tasks, e.g. packet processing or reaction to events. Short phases of active computing will be followed by long phases in sleep mode.

- High throughput is typically not required, since data rates are low. Nevertheless, fast execution of tasks is of advantage. The earlier a node can finish its actual tasks, the sooner it can enter low power sleep modes.
- Limited accuracy is usually sufficient for data processing, resulting from the limited bit width of sampled sensor data. This affects favoured operand bit width and granularity for the RFU.

15.3.1 Functional Coverage of the RFU

The large majority of functions in the WSN domain is based on two arithmetic classes: either integer/fixed point arithmetic (DSP functions for sensor data processing), or finite field arithmetic (common for communication-related functions). As our main objective is the evaluation of different architecture concepts, it will be sufficient to concentrate our studies on one of the two arithmetic classes, without restricting the generality of the results.

The finite field arithmetic class has eventually been selected for this study, as it allows to support common functions that can be part of almost any kind of WSN application, like encryption, authentication, checksum calculation and forward error correction (FEC). The AES is part of common WSN standards [18] and used for granting privacy, integrity and authenticity. FEC like BCH coding [29] can avoid costly retransmissions [28]. Thus, total energy consumption is reduced if the energy invested in the computation of coding and decoding is outbalanced by larger energy savings in wireless communication. This trade-off between computation and communication costs can be optimised by selecting optimal code rates and lengths depending on the actual channel quality. The RFU allows to adapt these coding parameters quickly during runtime, by using its dynamic reconfigurability.

15.3.2 The RFU Data Path

According to the aforementioned design concepts and domain-specific requirements, a relatively small RFU data path with a task-based execution model is chosen [11, 16]. Neither deep pipelines nor parallel processing of multiple tasks will be required. Instead, frequent dynamic reconfiguration is utilised to execute tasks sequentially on a limited amount of reconfigurable hardware resources.

The data path is composed of different types of dedicated operator blocks of 8-bit granularity. The issue of arranging and interconnecting these heterogeneous resources flexibly has been solved by the modular top level data path structure shown in Fig. 15.1 [11]. It includes dedicated modules for multiply accumulate (MAC) operations, inversion, local storage, and a memory interface. Few operators per module are sufficient, because they can be reused flexibly by frequent dynamic reconfiguration. Three global buses serve as highly flexible top level interconnects between the

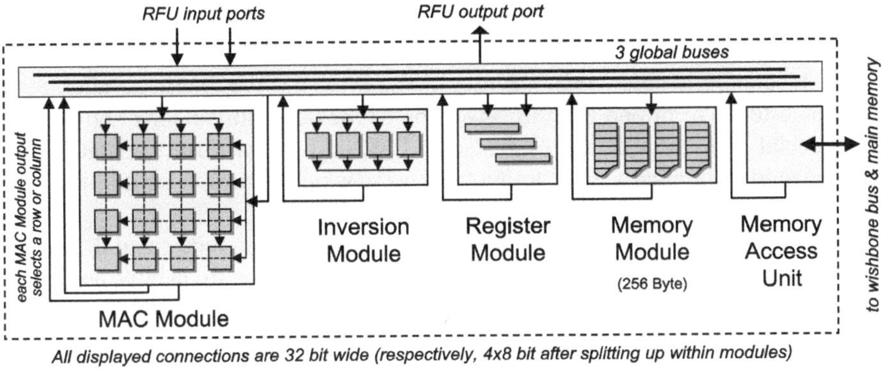

Fig. 15.1 Modular structure of the RFU data path.

modules and the RFU's inputs/outputs. Each module is sized to be able to process 4 bytes in parallel, in order to obtain a balanced throughput with the 32-bit global interconnect structures and avoid bottlenecks in the design.

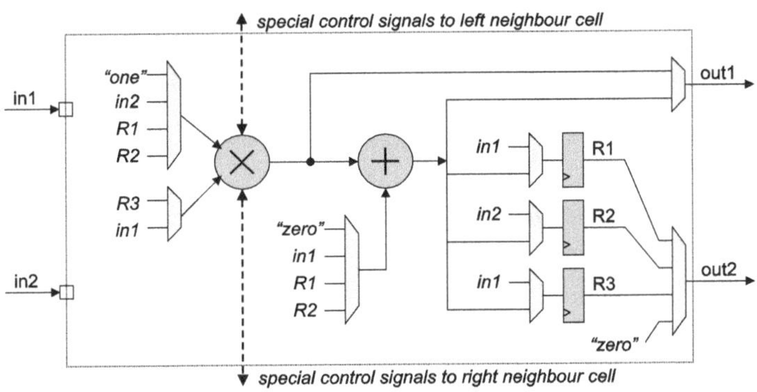

Fig. 15.2 Single cell of the MAC module.

The core of the data path in Fig. 15.1 is the MAC module, which combines all adder and multiplier operators. They are arranged internally in a 4×4 array of identical cells, which are illustrated in detail in Fig. 15.2 [16]. Each cell comprises one finite field (FF) multiplier, one FF adder, and three registers (8 bit granularity each). Multiplication and addition results can be routed immediately to other cells via output1 or stored temporarily in the local registers. The customised interconnect structure of the cells and the MAC array provides efficient support for common operation sequences like multiply-accumulate, addition of multiple operands, or instant sum-of-products. The array has a preferred data flow direction from top to bottom, but feedback and chaining of up to 16 cells are also possible. Horizontal concatenation of cells allows to configure larger operators (up to 32 bit) by using dedicated interconnects between the multipliers of a row.

The Inversion module comprises four operator blocks working in parallel, each being able to perform one 8-bit FF inversion or the very similar *sbox* and *inverse sbox* special operations of the AES standard [6]. A byte-wise shift operation has been integrated implicitly into the global buses, which allows to shuffle the four bytes of one bus arbitrarily.

The Register and Memory modules provide extra local storage capabilities like additional registers (e.g., for holding temporary variables or frequently used constants) and a versatile local memory of 256 bytes, which is configurable either as LUT or FIFO of selectable width and depth. The RFU also has its own interface to the main memory, allowing it to perform autonomous memory accesses quickly.

15.4 Dynamic Reconfiguration

Fast dynamic reconfiguration of the RFU's operators, memory blocks and interconnects is an essential part of regular task execution on the RFU, so these hardware resources can be reused flexibly for performing different computation steps in every clock cycle. Also, dynamic reconfiguration is applied for reusing the entire RFU for the execution of different tasks sequentially over time. For this fast and efficient *intra-task* and *inter-task* reconfiguration, a novel multi-layered reconfiguration mechanism has been developed [14].

15.4.1 Intra-task Reconfiguration

Intra-task reconfiguration of the RFU's 426 configuration bits is efficiently realised by *partial multi-context switching* and can be performed without causing latency. The actual configuration of all bits is defined through a set of multi-context configuration tables (MCT). Each MCT is driving one frame, i.e. the complete configuration bits of one RFU data path component (a module or a bus), resulting in the overall frame structure illustrated in Fig. 15.3 [14]. A different context can be selected in every clock cycle and for each MCT individually, which is controlled and synchronised via a central run control unit using a novel tag-matching mechanism. Unlike regular multi-context reconfiguration where contexts are only switched as a whole [30], the splitting into many small context tables allows to switch frame configurations individually and yet fully concurrently. A detailed description of the intra-task reconfiguration concept is given in the following.

15.4.1.1 The Multi-context Configuration Table

The MCT forms the basic building block of our reconfiguration mechanism [14]. Its structure is shown in Fig. 15.4. The MCT can hold several instances (= contexts)

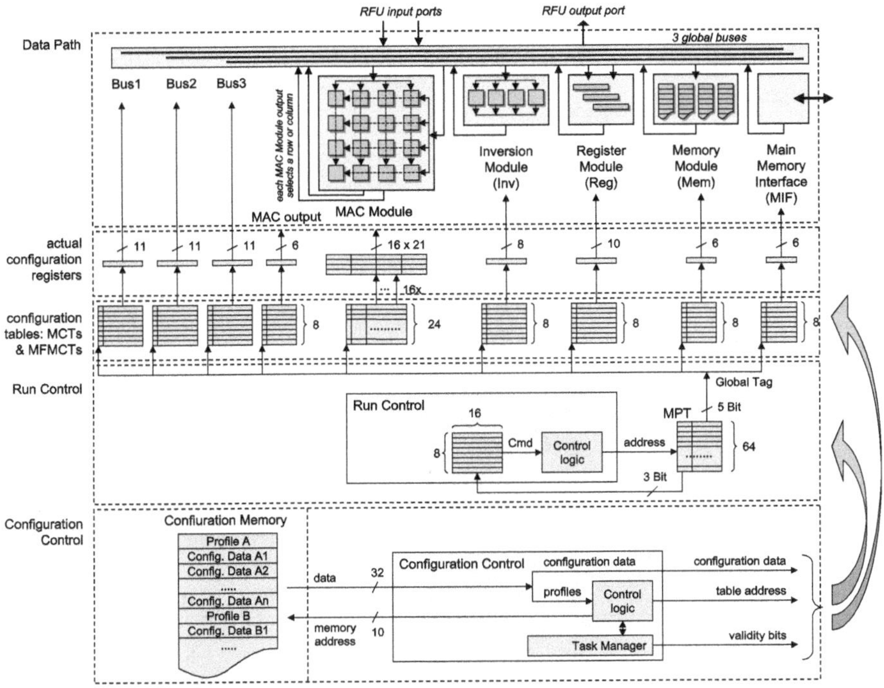

Fig. 15.3 The complete reconfiguration mechanism of the RFU.

of configuration data locally available. In each clock cycle, a different context can be selected and applied to the actual configuration register. The selection is done by using a *tag-matching* technique, which is illustrated in Fig. 15.5. Each context is extended by a tag field and one validity bit. A new context is selected for the configuration register if its associated tag matches a globally generated bit pattern (global tag) and if its validity bit is '1'. The given example shows the selection of context number two. The validity bits allow to keep configuration data of different tasks in a MCT without interference, by activating only those contexts belonging to the actual task. This is handled by a central task manager, which is explained later in Sect. 15.4.2.

A variant of the MCT, the multi-frame MCT, allows to drive multiple frames with identical structure by only one single multi-frame MCT, without restricting the possibility to assign different configurations to each frame concurrently (see [14] for a detailed explanation how this is solved). This has been applied for instance to drive the 16 identical frames of the 16 MAC cells, thus saving a considerable amount of memory compared to the hypothetical usage of 16 separate MCTs instead.

An important advantage of the MCT structure is its good scalability. The bit width of the configuration data and the number of contexts (table depth) can be chosen for each MCT individually, as can be seen in Fig. 15.3 and Table 15.1 for the RFU. In total, the RFU table structure comprises eight MCTs of different size, one

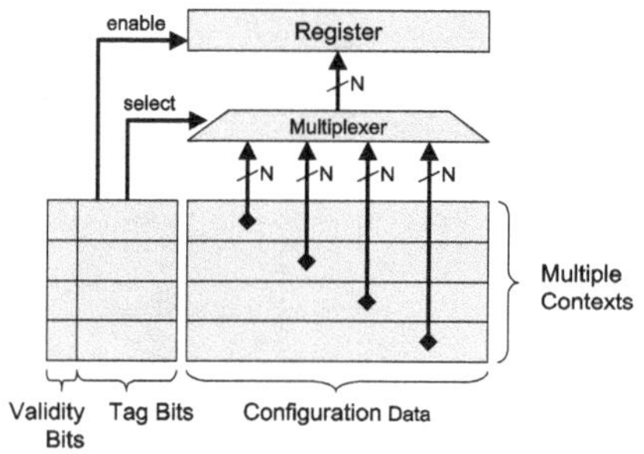

Fig. 15.4 Structure of the MCT.

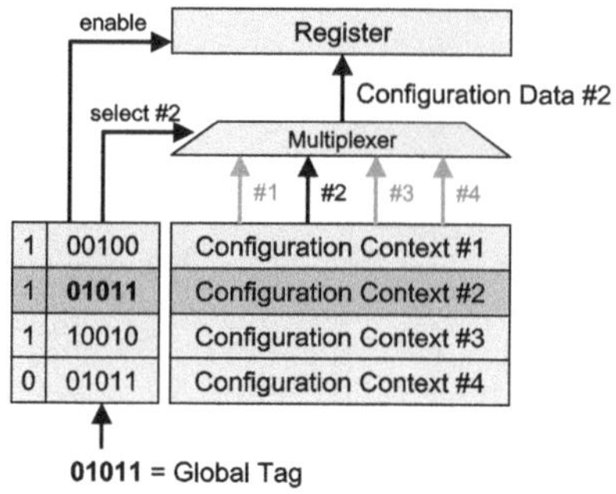

Fig. 15.5 Operating principles of the MCT.

multi-frame MCT driving the 16 frames of the MAC module, and one 21 bit wide register (not shown) that remains static during one task.

Table 15.1 Parameter settings for all tables.

Table name	MAC	MACo	Inv	Reg	Mem	MIF	Bus1	Bus2	Bus3	MPT	Cmds
Config. width	21	6	8	10	6	6	11	11	11	8	16
Table depth	24	8	8	8	8	8	8	8	8	64	8
Frames	16	1	1	1	1	1	1	1	1	–	–

15.4.1.2 Run Control via Tag-matching

The operation of all MCTs is controlled and synchronised through a central control signal, called the global tag. This global tag is generated from a different reconfigurable table being part of the RFU's run control unit. Each entry of this microprogram table (MPT) consists of a 3-bit control word and a 5-bit data value used as global tag, as shown in Fig. 15.6. In each clock cycle, the run control unit can address a different MPT entry and thus control the selection of different contexts in the MCTs. The order of MPT entries defines the sequence of consecutive operations over multiple cycles, including the possibility of jumps and loops inside the MPT by selecting one of eight reconfigurable command codes via the 3-bit control word in order to allow a better table utilisation for long sequences with repeating patterns.

One major advantage of using tag-matching is that it saves a considerable amount of memory compared to a direct addressing scheme of the MCTs: it requires to store only one small tag (5 bit) instead of addresses for all MCTs (would be 29 bit in total) in each MPT entry. Even if including the overhead for tag fields in all MCTs, tag-matching thus requires 60% less storage cells than direct MCT addressing [14].

Overall, the run control unit allows to control sequences of operations completely autonomously, thus unburdening the processor in its role as external control unit significantly. Operations on the RFU can be initiated by the processor in two ways:

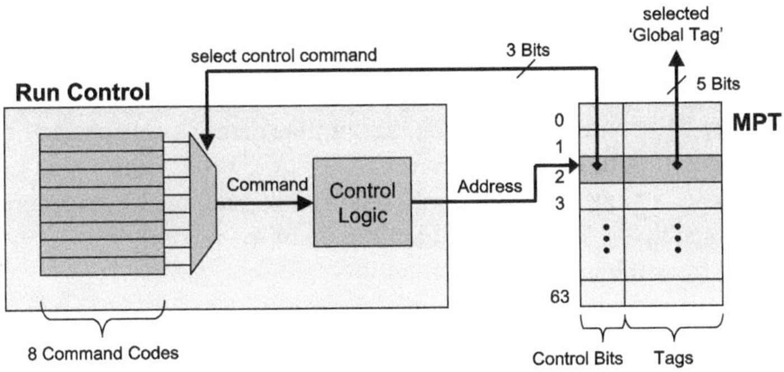

Fig. 15.6 Tag generation by the MPT and run control unit.

A single operation can be started via the single-cycle execution (ESR) instruction, which addresses a single MPT entry and hence selects a specific configuration for a single operation. The EMR instruction can be used to initiate sequences of operations over multiple cycles, which are then carried out autonomously by the RFU as described above. Thus, even very long sequences—up to the autonomous execution of an entire task lasting hundreds of cycles—can be started with a single EMR instruction. By the two instructions, intra-task reconfiguration (i.e., context switching) is practically transparent to a software programmer.

15.4.2 Inter-task Reconfiguration

Before a new task can be started on the RFU, all its required configuration data must be present in the MCTs and MPT to allow delay-free intra-task reconfiguration. The objective of the inter-task reconfiguration is therefore to reconfigure the MCTs and MPT prior to task execution by loading new content from an external memory to the tables. As this process causes non-negligible latency, the concept of reconfiguration *profiles* [14] is applied to minimise overhead. It ensures that just as many data as necessary is loaded, and maximises the throughput of configuration data through the memory interface (which is a bottleneck for reconfiguration latency). Each task has its own 32-bit profile, which specifies exactly how much data needs to be loaded for each table. A dedicated configuration control unit (Fig. 15.3) evaluates the profiles to assign all following configuration data automatically to the correct tables.

A revolving replacement strategy is used to load new configuration data to the MCTs and replace older data. A task manager keeps track of which configuration belongs to which task. If a recently executed task is started again, the task manager recognises if configuration data of this task is still (partly) present in the tables. In this case, it just activates the validity bits of this task's configuration contexts and deactivates the others, instead of loading the configuration data again from the memory. Thus, additional reconfiguration time is saved. In the best case, no reconfiguration is required for switching between two tasks.

The inter-task reconfiguration process is controlled fully autonomously by the RFU's configuration control unit. Hence, a complete reconfiguration can be initiated with a single reconfiguration instruction (CRT) by the processor, which just passes the start address of the profile as the only required parameter (like a unique task ID from user's perspective). Thus, all the complexity of the reconfiguration mechanism itself is hidden from the users.

15.5 General System Architecture

The RFU is embedded into a RISC processor, thus forming a hybrid computing core for our programmable sensor node platform. While the RFU's role is to execute data processing tasks at low energy costs, the processor mainly serves for performing control functions.

Two sensor node architectures have been designed accordingly, one based on the LEON2 32-bit RISC processor [8] and the other on Atmel's 8-bit ATmega AVR processor [2]. Figure 15.7 shows the tight integration of the RFU into the LEON2 processor's data path as a second function unit in parallel to the ALU, which is a well-known coupling concept for minimizing the communication costs between processor and reconfigurable unit [33, 24, 23]. The same integration concept has been implemented also for the ATmega processor.

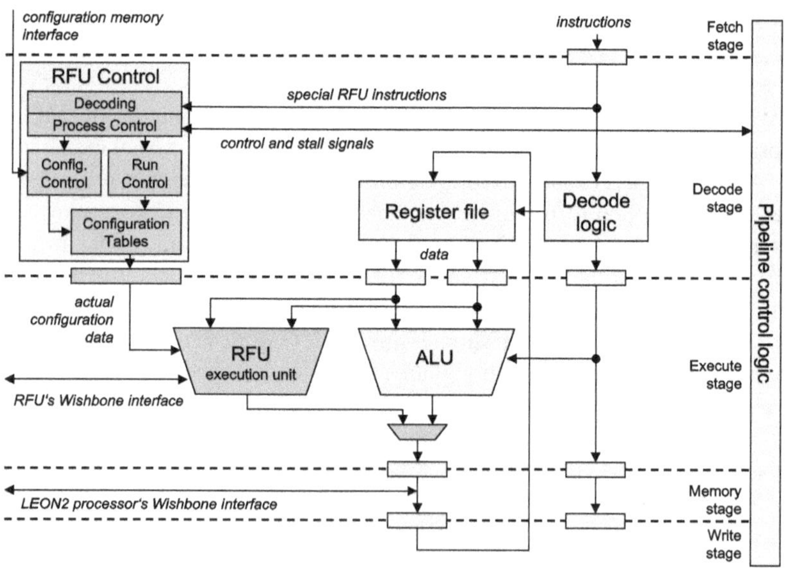

Fig. 15.7 Inner structure of the LEON2 core, showing the integration of the RFU.

As shown by Vassiliadis et al. [31], a few basic instructions are sufficient to let the processor control the RFU. In our case, these are two instructions for single-cycle and multi-cycle execution and one for configuration (ESR, EMR and CRT, see Sect. 15.4), which have been added to the SPARCV8 and AVR instruction sets respectively. The ESR instruction takes one clock cycle, whereas the duration of the EMR and CRT instructions is variable: Their execution times can range between two and several hundreds of clock cycles. During such multi-cycle instructions, the processor is not blocked but can perform regular instructions meanwhile. The next RFU instruction then resynchronises the processor and the RFU, simply stalling the CPU until the RFU has finished its previous operation.

Figure 15.8 illustrates the entire sensor node system for the LEON2 version [16]. Separate on-chip memories for data, instructions, and configurations are proposed (Harvard architecture style). The configuration memory is directly accessed by the RFU and serves as external repository for the configuration data of all tasks (see Sect. 15.4.2). The RFU also has its own master interface to the system bus, allowing it to access the main memory autonomously. This avoids a known bottleneck for such hybrid systems and unburdens the processor from data copying to and from the RFU.

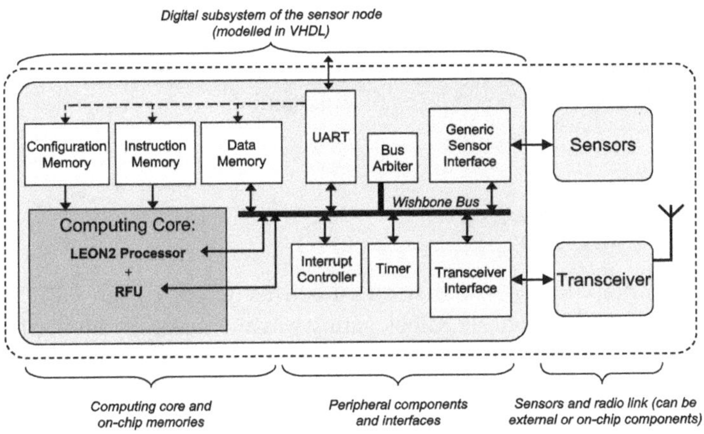

Fig. 15.8 General system architecture of our wireless sensor node platform.

15.6 Evaluation Results

The presented RFU architecture is evaluated regarding chip area, performance, and energy consumption in order to analyse how it compares to classical architectures like RISC processors and ASICs, which occupy well-known characteristic regions in a design space. Since our main objective—the identification of architecture-specific characteristics—implicates a fair and unbiased comparison on the same level of optimisation, comparing our VHDL model to real chip implementations is not feasible. All architectures have therefore been evaluated as VHDL models on the same level of synthesized gate netlists in a standard cell technology [16, 11].

15.6.1 Test Settings

The system in Fig. 15.8 is the basis for all three LEON2 based architectures, and a similar system has been designed for the ATmega architectures. The conventional processor platforms are designed without RFU, so all tasks are run in software only. The ASIC architectures are designed by extending these processor platforms with specific ASIC cores for all tested benchmark functions: one ASIC core for CRC calculations, one for AES de- and encryption, and one for BCH coding and decoding. All three ASIC accelerators are attached to the system bus as memory-mapped peripherals.

A set of benchmark tasks is defined that will be run on every architecture. Four tasks are included: (1) The computation of a 8-bit CRC checksum for a 128-bit packet [29]; (2) The decoding of a (15, 10, 3) BCH code word [29]; (3) The encryption of a 128-bit data block according to the AES standard [6]; (4) The key generation for the AES encryption task, which needs to be run prior to encryption. For a fair comparison, all software parts are written in hand-optimised assembler code for best possible efficiency on each architecture.

15.6.2 Synthesis Results

Synthesis was done in a 130 nm standard cell technology from UMC using the Synopsys Design Vision tool set. Clock gating was automatically inserted to include the effects of this important power reduction technique in our analyses.

Table 15.2 denotes the obtained area results, given as total cell area in μm^2. As can be seen, the RFU has a remarkably small size. It adds just 0.45 mm^2 to the system size, which increases from 2 mm^2 to 2.5 mm^2 for a LEON2-based sensor node with two 32 KByte SRAM on-chip memories [32], and from 1.7 mm^2 to 2.2 mm^2 for an Atmega equivalent. Importantly, this points out that the RFU is not dominating the total system size. The reason for the RFU's small size is found in the efficient reuse of hardware resources through inter-task and intra-task reconfiguration.

15.6.3 Evaluation of Energy Efficiency

While small area consumption is an important cost factor, the most critical design parameter for the WSN domain is energy efficiency. Therefore, the product of power

Table 15.2 Area results.

System component	Total cell area	Percentage
On-chip Memories [32]	1,680,000 μm^2	
Instruction Mem. 32 KB	840,000 μm^2	
Data Mem. 32 KB	840,000 μm^2	
LEON2 architecture	337,300 μm^2	
ATmega architecture	66,900 μm^2	
RFU	454,200 μm^2	100.0%
RFU data path	231,100 μm^2	50.9%
RFU config. tables	186,800 μm^2	41.1%
RFU control units	36,300 μm^2	8.0%
ASIC architecture	468,000 μm^2	100.0%
BCH core	260,800 μm^2	55.7%
AES core	199,500 μm^2	42.6%
CRC core	7,700 μm^2	1.7%

consumption and execution time needs to be investigated. The Synopsys PrimeTime tool is used for estimating power consumption at gate level based on the switching activity of the synthesized gate netlists [16].

15.6.3.1 Reconfiguration Overhead

Tables 15.3 and 15.4 show the resulting energy consumption and execution times of all benchmark tasks executed on the RFU, from which the impact of the reconfiguration costs on total energy efficiency and performance can be concluded. Assuming a worst case scenario in which a complete reconfiguration of the RFU would be required prior to each task, common experience would usually let us expect a very large overhead on total efficiency. However, our results prove that this is not the case for our architecture. The dynamic reconfiguration costs never reach a critical level, neither for latency nor energy consumption. This is a direct effect of the novel reconfiguration mechanism applied (see Sect. 15.4). Tables 15.5 and 15.6 summarise the required effort for inter-task and intra-task reconfiguration respectively [14].

15.6.3.2 Comparison of Alternative Reconfiguration Mechanisms

The latency overhead for dynamic reconfiguration of the RFU has also been analysed in [14] under the assumption that different common reconfiguration mechanisms were used. This comparison of our approach with common related work has been made with a simple frame addressing scheme (which is in similar form still the basic concept for typical FPGA reconfiguration today), an advanced frame addressing scheme with wildcard bits [9], and regular multi-context switching [30].

Table 15.3 Latency in [clock cycles].

	CRC	BCH	AES	
			enc.	key.
LEON2	930	2893	1474	538
AVR	1184	3978	3120	834
LEON2 RFU				
Execution	10	230	115	58
Reconfiguration	19	54	68	32
AVR RFU				
Execution	35	438	144	101
Reconfiguration	19	54	70	32
LEON2 ASIC	74	56	52	
AVR ASIC	136	77	369	

Table 15.4 Energy consumption in [nJ].

	CRC	BCH	AES	
			enc.	key.
LEON2	126.0	408.2	251.4	88.1
AVR	45.3	170.2	134.4	39.0
LEON2 RFU				
Execution	2.3	56.7	52.5	13.3
Reconfiguration	3.4	9.7	12.2	7.7
AVR RFU				
Execution	5.1	60.2	50.2	17.2
Reconfiguration	2.2	6.3	9.2	4.0
LEON2 ASIC	13.8	8.6	13.0	
AVR ASIC	14.0	9.0	43.5	

Table 15.5 Inter-task reconfiguration: the numbers denote how many configuration data entries need to be loaded to each table for a task.

Task	MAC	MACo	Inv	Reg	Mem	MIF	Bus1	Bus2	Bus3	MPT	Cmds
BCH	21	3	0	5	0	0	3	1	3	17	1
AES keygen	8	1	6	2	5	0	2	4	0	11	2
AES enc	22	5	5	6	6	0	6	6	2	36	2
CRC	3	1	0	2	0	0	1	0	2	3	1

Table 15.6 Intra-task reconfiguration activity: the numbers denote how often frame configurations change during one task.

Task	MAC1	MAC2	MAC3	MAC4	MAC5	MAC6	MAC7	MAC8	MAC9	MAC10	MAC11	MAC12	MAC13	MAC14	MAC15	MAC16	MACo	Inv	Reg	Mem	MIF	Bus1	Bus2	Bus3
BCH	71	71	71	71	26	26	20	0	0	0	0	0	0	0	0	0	3	0	5	0	0	2	2	3
AES keygen	24	24	0	0	24	0	0	0	24	0	0	0	24	0	0	0	1	25	3	24	0	2	22	0
AES enc	100	102	101	99	100	102	101	99	100	102	101	99	100	102	101	99	45	81	83	46	0	86	83	2
CRC	0	0	0	0	0	0	0	0	0	0	0	0	3	3	3	3	1	0	2	0	0	1	0	2

The summarised results in Table 15.7 denote pure reconfiguration latency. For tasks like the CRC requiring almost no intra-task reconfiguration (see Table 15.6), our mechanism provides comparable latencies to existing techniques, whereas in cases with high dynamic reconfiguration demand like the BCH and AES tasks, it achieves large improvements. When compared to execution times, it becomes clear that our mechanism is the only variant which achieves acceptably low costs for fast dynamic reconfiguration.

15.6.3.3 Reconfiguration Power

The power consumption of the RFU's main components is analysed over time, which is illustrated in Fig. 15.9 exemplarily for different phases of the AES task [16].

Table 15.7 Latency comparison: numbers of clock cycles required for reconfiguration by different mechanisms.

	Execution time	Frame addressing	Wildcard addressing	Multi-context	Ours
BCH	230	372	103	238	51
AES keygen	58	198	134	112	29
AES enc	115	2035	580	210	64
CRC	10	18	9	42	14

The energy consumption during each phase is proportional to the shaded area representing the power-latency-product. Alternating phases of configuration and execution can be observed, first for the key generation and then for the encryption.

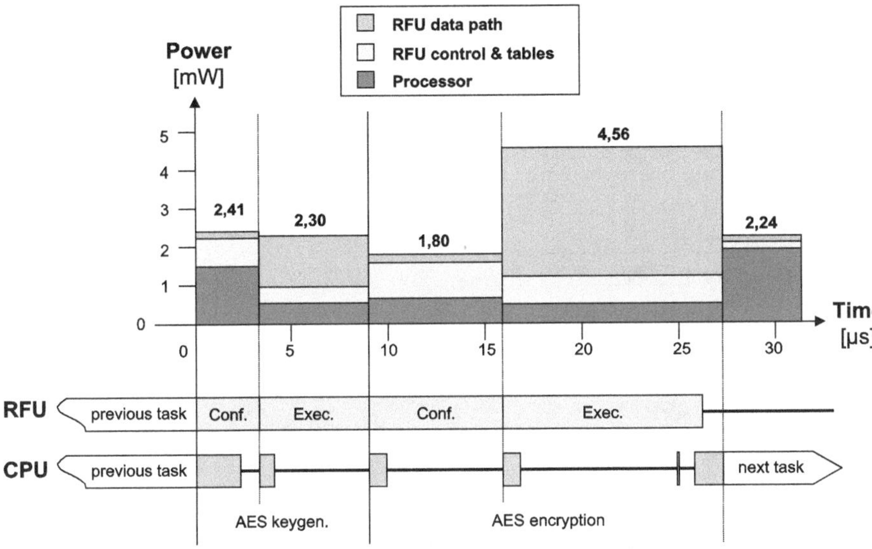

Fig. 15.9 Power consumption of the RFU architecture during the AES task.

The power consumption of the RFU control units represents the costs for inter-task reconfiguration during the configuration phases, and for intra-task reconfiguration during the execution phases. None of them has critical influence. This is quite remarkable, taking into account the high reconfiguration activity denoted in Table 15.6: Some data path components switch configurations up to 100 times between 20 different contexts during the AES encryption (2nd execution phase in Fig. 15.9).

The power consumption of the RFU data path largely depends on how many of its subcomponents are actively used. E.g., only 5 MAC cells are used during key generation, but all 16 cells are used during encryption. When the RFU is completely inactive, it consumes only negligible power, as shown in the last phase named 'next task'. This important effect of shutting down the RFU efficiently is achieved through clock gating and operand isolation techniques. During some phases in Fig. 15.9, the processor's power consumption decreases significantly. This is because the processor is not utilised for large fractions of time and hence experiences only little switching activity on average.

15.6.3.4 Architecture Comparison

Finally, we compare the RFU architecture with the conventional reference architectures. For this purpose, the energy and latency results of all benchmark tasks are given in Tables 15.3 and 15.4 for the LEON2 and the AVR systems [16]. For the ASIC architecture, the key generation is included in the AES encryption task and hence not listed separately. It should be noted here that the clock frequency in WSN nodes is usually not driven to the possible maximum (which might be different for each architecture) but chosen rather low in order to save power. In all of our experiments, a 10 MHz clock frequency was used.

Compared to the processor architectures, the RFU achieves shorter execution times of factor 6 to 10 (up to 32 if considering the peak value of the CRC task) with the LEON2, and factor 6 to 14 (up to 22) with the AVR. More importantly, it also achieves significant energy savings of factor 4 to 6 (up to 22 for the CRC task) for

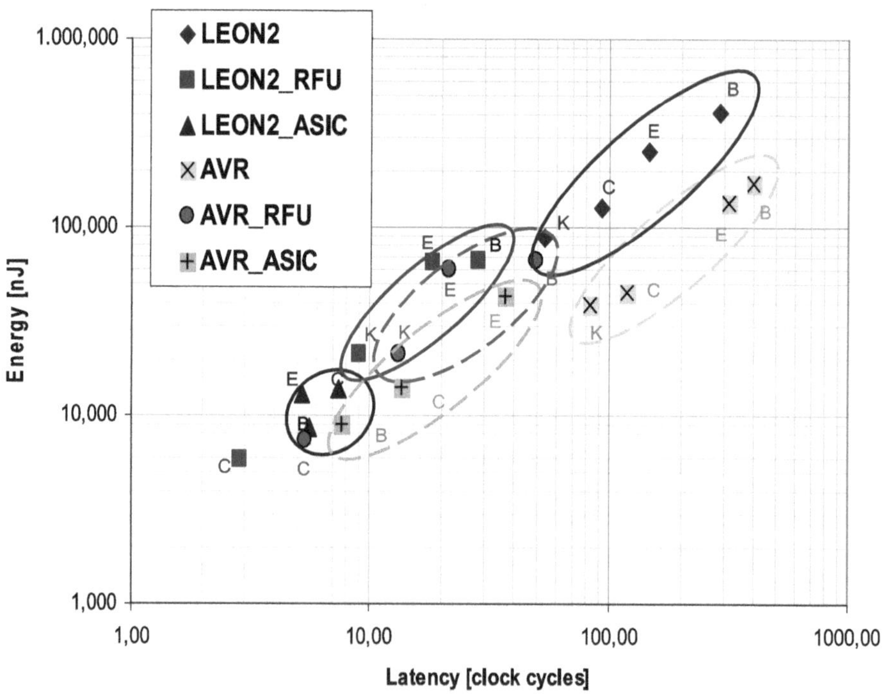

Fig. 15.10 Architecture-characteristic regions in the energy vs. latency design space; C = CRC, B = BCH, E = Enc., K = Keygen [16].

the LEON2, and factor 1.8 to 2.5 (up to 6) for the AVR. This clearly shows the high potential of an RFU integration compared to a simple processor architecture, which is found in most state-of-the-art sensor nodes today. In the case of the CRC task, the RFU achieves even better results than the ASIC, which can be explained simply by the fact that a very light-weight ASIC core has been preferred here (see area values in Table 15.2), whereas the RFU can use its large amount of MAC resources for a fast parallel computation. But in general, the energy consumption of the RFU lies right in the middle between ASICs and processors, while its execution times are more close to ASICs. This can be seen best in Fig. 15.10, which illustrates the typical domains occupied by the different architectures in a latency vs. energy design space.

The 8-bit ATmega AVR is slower than the 32-bit LEON2 but also has lower energy costs. However, the energy consumption of the integrated RFU is independent of the processor—it consumes roughly the same energy with both processors, indicating that its high energy efficiency can be achieved invariably with any RISC processor architecture. The ASIC implementations for the AVR are slower than those for the LEON2, and also have higher energy requirements. At the same time they differ much more than those of the LEON2. This is because the AVR version suffers significantly from the lower data transfer speed of the 8-bit architecture and becomes less efficient the more data is required for a task computation.

15.7 Prototyping of the Sensor Node System

For functional verification under real conditions and for demonstrations of sensor network applications, the proposed sensor node architecture shown in Fig. 15.8 has been prototyped on a FPGA-based platform. A new rapid prototyping platform specifically dedicated to wireless sensor networks has been developed for this purpose [13, 12]. It features a Spartan3-2000 FPGA providing sufficient logic resources for hardware emulation and interfacing, a flexible low-power 868 MHz radio transceiver, and an autonomous power supply using four rechargeable AA batteries. The platform can be equipped with various different sensor types, hence being flexibly usable for a wide spectrum of WSN applications. The platform is not intended for power analysis due to the high power consumption of the FPGA, but highly suitable for functional verification. It was successfully used to show that the RFU-based architecture concept is well applicable to real wireless sensor networks.

Additionally to the prototype platform, a base board for supporting real-time debugging and deployment was developed [12]. It connects the nodes by a secondary Ethernet-based deployment support network, thus providing strong aid of online monitoring and debugging of several nodes without influencing the primary wireless network. Specific debugging interfaces can be implemented together with the sensor node architecture inside the FPGA, thus granting deep insight into internal processes in the monitored sensor node during runtime.

A demonstration application has been developed on the prototype platforms, realising the localisation of a sound source by a wireless sensor network of five nodes, each being equipped with a small acoustic sensor. The RFU is utilised to encrypt and decrypt the transmitted packets in this scenario. The application has been demonstrated successfully at the SPP1148 booth at the FPL conference 2008 [15].

15.8 Generalisation of the Results

The presented results have shown that the large energy savings achieved by the RFU in comparison to processors are an architecture-characteristic feature and not depending on specific functionality. This supports the assumption that similar improvements in energy efficiency can be achieved also for other functions than the analysed benchmark tasks. The same architecture concepts can be applied to construct other RFUs with different function modules, e.g. an RFU supporting integer arithmetic for sensor data processing, or one that merges both arithmetic classes (see Sect. 15.3.1 and [17]). In general, our proposed architecture concept is not even restricted to the WSN domain, but could also be imagined for the design of energy efficient yet flexible embedded systems in other application areas.

Particularly the newly developed reconfiguration mechanism could be adopted well for other reconfigurable architectures, due to its modular structure, good scalability and easy adaptability [14]. It is well suited for coarse-grain architectures featuring fast dynamic reconfiguration, and could also be integrated into design automation tools for such systems. In cooperation with the research group of Prof. Merker (see Chap. 8), this was investigated also for FPGAs. Possible FPGA extensions for realising our reconfiguration mechanism efficiently with fine-grain fabrics as well as a complete tool chain for the synthesis of runtime reconfigurable systems on such FPGAs were proposed. Our results show a promising potential for the automated realisation of efficient dynamically reconfigurable systems on future FPGA structures [27].

15.9 Conclusion

The architecture concept of frequent dynamic reconfiguration of a small heterogeneous data path was analysed in this research work, and a novel mechanism for efficient dynamic reconfiguration was proposed. The coarse granularity and heterogeneity of the RFU provided the desired high energy efficiency, while the dynamic reconfigurability kept the architecture small and flexible. The way how fast dynamic reconfiguration was efficiently utilised for extensively reusing few heterogeneous data path resources distinguishes our RFU from other architectures. Different to common experience, the extensive use of dynamic reconfiguration causes no critical energy and performance overhead for our RFU, even in a worst case scenario.

Our experiments comparing the RFU to ASICs and processors show that the RFU occupies a characteristic design space region (see Fig. 15.10), which is located right in the middle between processors and ASIC implementations. Since ASICs are usually too costly in the WSN domain, programmable platforms like the RFU or processors are often the only economically feasible options. Compared to the LEON2 processor, the RFU achieves large energy savings of factor 4 to 6. Even when compared to a low power 8-bit processor like the ATmega, the RFU reduces the energy consumption by factor 2 on average. Overall, our results demonstrate that the frequent dynamic reconfiguration of small heterogeneous data paths is a highly suitable architecture concept for energy-constraint embedded systems like WSN nodes.

Acknowledgements This work was supported by DFG grants GL 144/25 and GL 144/29, as part of the Priority Programme 1148, *Reconfigurable Computing Systems*.

References

1. Akyildiz, I., Su, W., Sankarasubramaniam, Y., Cayirci, E.: A survey on sensor networks. IEEE Commun. Mag. **40**, 102–114 (2002)
2. Atmel: ATmega103 Datasheet Rev.0945I-AVR-02/07. http://www.atmel.com
3. Blume, H., Feldkaemper, H.T., Noll, T.G.: Model-based exploration of the design space for heterogeneous systems on chip. J. VLSI Signal Process. Syst. **40**(1), 19–34 (2005)
4. Chou, Y., Pillai, P., Schmit, H., Shen, J.: PipeRench implementation of the instruction path coprocessor. In: Proc. Intl. Symp. on Microarchitecture, pp. 147–158 (2000)
5. Cronquist, D., Franklin, P., Fisher, C., Figueroa, M., Ebeling, C.: Architecture design of reconfigurable pipelined datapaths. In: Conf. on Advanced Research in VLSI, pp. 23–40 (1999)
6. Daemen, J., Rijmen, V.: The Design of Rijndael: AES—the Advanced Encryption Standard. Springer, Berlin (2002)
7. Estrin, D., Girod, L., Srivastava, G.P.M.: Instrumenting the world with wireless sensor networks. In: Proc. Intl. Conf. on Acoustics, Speech and Signal Processing (2001)
8. Gaisler, J.: LEON2 Processor User's Manual, Version 1.0.21 XST Edition, Gaisler Research. http://www.gaisler.com (2003)
9. Hauck, S., Li, Z., Schwabe, E.: Configuration compression for the Xilinx XC6200 FPGA. IEEE Trans. Comput.-Aided Des. Integr. Circuits Syst. **18**, 1107–1113 (1999)
10. Hill, J., Culler, D.: Mica: a wireless platform for deeply embedded networks. MICRO, IEEE **22**, 12–24 (2002)
11. Hinkelmann, H., Zipf, P., Glesner, M.: A domain-specific dynamically reconfigurable hardware platform for wireless sensor networks. In: Proc. Intl. Conf. on Field-Programmable Technology, pp. 313–316 (2007)
12. Hinkelmann, H., Reinhardt, A., Glesner, M.: A methodology for wireless sensor network prototyping with sophisticated debugging support. In: Proc. 19th IEEE/IFIP Intl. Symp. on Rapid System Prototyping, pp. 82–88 (2008)
13. Hinkelmann, H., Reinhardt, A., Varyani, S., Glesner, M.: A reconfigurable prototyping platform for smart sensor networks. In: Proc. IV Southern Conference on Programmable Logic, pp. 125–130 (2008)
14. Hinkelmann, H., Zipf, P., Glesner, M.: A scalable reconfiguration mechanism for fast dynamic reconfiguration. In: Proc. Intl. Conf. on Field-Programmable Technology, pp. 145–152 (2008)

15. Hinkelmann, H., Zipf, P., Glesner, M., Alles, M., Vogt, T., Wehn, N., Kappen, G., Noll, T.G.: SPP1148 booth: Application-specific reconfigurable processors. In: Proc. 18th Intl. Conf. on Field Programmable Logic and Applications, p. 350 (2008)
16. Hinkelmann, H., Zipf, P., Glesner, M.: Design and evaluation of an energy-efficient dynamically reconfigurable architecture for wireless sensor nodes. In: Proc. 19th Intl. Conf. on Field Programmable Logic and Applications (2009)
17. Hinkelmann, H., Zipf, P., Li, J., Liu, G., Glesner, M.: On the design of reconfigurable multipliers for integer and Galois field multiplication. Microprocess. Microsyst. **33**, 2–12 (2009)
18. IEEE Std: 802.15.4 Part 15.4: Wireless medium access control (MAC) and physical layer (PHY) specifications for low-rate wireless personal are networks (LR-WPANs) (2003)
19. Lodi, A., Toma, M., Campi, F., Capelli, A., Canegallo, R., Guerrieri, R.: A VLIW processor with reconfigurable instruction set for embedded applications. IEEE J. Solid-State Circuits **38**(11), 1876–1886 (2003)
20. Mei, B., Vernalde, S., Verkest, D., De Man, H., Lauwereins, R.: ADRES: an architecture with tightly coupled VLIW processor and coarse-grained reconfigurable matrix. In: Proc. 13th Intl. Conf. on Field Programmable Logic and Applications (FPL) (2003)
21. Moteiv Corporation: Tmote Sky Datasheet: Ultra low power IEEE802.15.4 compliant wireless sensor module. http://www.moteiv.com (2006)
22. PACT XPP Technologies: XPP-III Processor Overview—White Paper, Version 2.0 (2006)
23. Pionteck, T.: Dynamisch rekonfigurierbare Architekturen für das digitale Basisband und für die Sicherungsschicht drahtloser Netzwerke. Dissertation, Technische Universität Darmstadt (2005)
24. Pionteck, T., Staake, T., Stiefmeier, T., Kabulepa, L.D., Glesner, M.: Design of a reconfigurable AES encryption/decryption engine for mobile terminals. In: Proc. IEEE Intl. Symp. on Circuits and Systems (2004)
25. Rabaey, J., Ammer, M., da Silva Jr., J., Patel, D., Roundy, S.: PicoRadio supports ad hoc ultra-low power wireless networking. Comput. Mag. **33**, 42–48 (2000)
26. Römer, K., Mattern, F.: The design space of wireless sensor networks. IEEE Wirel. Commun. **11**, 54–61 (2004)
27. Rullmann, M., Merker, R., Hinkelmann, H., Zipf, P., Glesner, M.: An integrated tool flow to realize runtime-reconfigurable applications on a new class of partial multi-context FPGAs. In: Proc. 19th Intl. Conf. on Field Programmable Logic and Applications (2009)
28. Schmidt, D., Berning, M., Wehn, N.: Error correction in single-hop wireless sensor networks—a case study. In: IEEE Conf. Design, Automation and Test in Europe (2009)
29. Sweeney, P.: Error Control Coding: From Theory to Practice. Wiley, New York (2002)
30. Trimberger, S., Carberry, D., Johnson, A., Wong, J.: A time-multiplexed FPGA. In: Proc. IEEE Symp. on FPGAs for Custom Computing Machines, pp. 22–28 (1997)
31. Vassiliadis, S., Wong, S., Gaydadjiev, G., Bertels, K., Kuzmanov, G., Panainte, E.: The MOLEN polymorphic processor. IEEE Trans. Comput. **53**(11), 1363–1375 (2004)
32. Virtual Silicon: 2003 Single-Port SRAM Compiler UMC 0.13 μm Datasheet
33. Wittig, R.D., Chow, P.: OneChip: An FPGA processor with reconfigurable logic. In: IEEE Symp. on FPGAs For Custom Computing Machines, pp. 126–135 (1996)

Chapter 16
DynaCORE—Dynamically Reconfigurable Coprocessor for Network Processors

Carsten Albrecht, Jürgen Foag, Roman Koch, Erik Maehle, and Thilo Pionteck

Abstract This chapter presents DynaCORE, a dynamically reconfigurable coprocessor architecture for network processors. The architecture takes over computationally intensive tasks from network processors and provides dynamically exchangeable hardware assists for complex payload processing. According to the actual traffic profile, DynaCORE autonomously determines an optimal set of hardware assists. The hardware assists reside in a grid of exchangeable tiles which are connected by a runtime adaptable network-on-chip. System simulation, hardware architecture as well as reconfiguration management strategies are presented within this chapter.

16.1 Introduction

Network applications substantially demand high computational power. As network link performance already achieves about 10 Gb/s (OC-192) and 40 Gb/s (OC-768), only a few ten nanoseconds per packet are available to process incoming packets in real-time. Typically, systems requiring such high computational power are implemented as ASICs, yet the flexibility required in network applications hamper such solutions. Frequently changing protocols and standards as well as changing traffic profiles ask for more flexible solutions than typically provided by hardwired ASICs. For such constrained design spaces, runtime reconfigurable architectures offer a good trade-off between performance and flexibility. On the one side, tasks

Carsten Albrecht · Roman Koch · Erik Maehle · Thilo Pionteck
Institute of Computer Engineering, University of Lübeck, Ratzeburger Allee 160, 23538 Lübeck, Germany, e-mails: albrecht@iti.uni-luebeck.de, koch@iti.uni-luebeck.de, maehle@iti.uni-luebeck.de, pionteck@iti.uni-luebeck.de

Jürgen Foag
Rohde & Schwarz, Radiocommunications Systems Division, Mühldorfstraße 15, 81674 Munich, Germany, e-mail: juergen.foag@rohde-schwarz.com

M. Platzner et al. (eds.), *Dynamically Reconfigurable Systems*,
DOI 10.1007/978-90-481-3485-4_16, © Springer Science+Business Media B.V. 2010

implemented as hardware assists (HAs) onto FPGA resources provide much higher performance than software solutions, on the other side HAs can be exchanged at runtime, providing the flexibility required for the application area.

Based on this idea, the DynaCORE project was set up. DynaCORE is a runtime reconfigurable coprocessor for network processors (NPs) and allows computationally intensive tasks to be offloaded from the NP. The NP still performs header processing while payload processing like encryption/decryption, compression or network intrusion detection are done within DynaCORE. By exploiting the runtime reconfigurability of modern FPGAs such as Xilinx Virtex-4, DynaCORE autonomously adapts the types and number of HAs to the requirements given by actual traffic characteristics. An adaptable, fault tolerant runtime reconfigurable network-on-chip (NoC) is employed to provide the communication infrastructure for a varying number and different sizes of HAs within the FPGA. A detailed description of these hardware features, the rule-based reconfiguration control logic and an application example are given in this chapter which is organised as follows. Section 16.2 provides an overview of NPs and system requirements in the area of network applications. The overall system architecture of DynaCORE is introduced in Sect. 16.3 and the corresponding SystemC simulation model is given in Sect. 16.4. Sections 16.5 through 16.7 highlight key aspects of DynaCORE which are the adaptable NoC, the reconfiguration control logic and technical realisation of the runtime reconfiguration. A system demonstrator using the FlexPath NP architecture designed at the TU München as an NP is presented in Sect. 16.8, before, finally the chapter closes with concluding remarks in Sect. 16.9.

16.2 Network Processors

Over the years, a variety of hardware devices have been created to satisfy the increasing demand for bandwidth and computational power. Current implementations combine ASICs, application-specific instruction processors (ASIPs), hard-wired but weakly programmable co-processors, field programmable gate arrays (FPGAs), and general purpose processors (GPPs).

The generic NP architecture comprises two parts: a control and a data plane. The control plane forms the slow path. It is responsible for protocol management and exception handling. The data plane is the core of an NP. The control plane can be sufficiently executed by a GPP whereas the data plane realises the fast path by combining basic hardware systems as mentioned above. Generally, an NP is a hybrid device consisting of software programmable parts and special hardware units for packet processing. Its basic design feature is a chip-level multiprocessor block. Two programming models are used for high performance system software: Run-to-completion and pipelining. The first one replicates a single task on all processors, the latter distributes stages of processing over different processors. The level of parallelism is increased by multi-threaded processors. Helper units are applied in two ways. Some perform pre- or post processing such as classification, transmission scheduling, or buffer management independently, others support the multiprocessor block by providing frequent and computational-intensive tasks such as table lookups (search engine), checksum computation, or semaphore management.

High-end NP devices are e.g. Hifn 5NP4G [10] or Intel IXP2400 [12]. These NPs differ in their principle programming models. While the Hifn 5NP4G implements run-to-completion, the IXP2400 is based on functional pipelining. The Hifn 5NP4G supports its packet engines by a set of co-processors whereas only a few co-processors and hardware units are added to the pipelining approach of the IXP2400.

16.2.1 Requirements

For the last ten years, the composition of Internet backbone traffic has been rather stable in respect of the protocol layers 1 to 4 of the OSI layer model. In contrast to lower layers, higher layer protocols vary more frequently. In general, there is a trend towards applying secure connections established on any layer. On the top layers the secure variants of FTP and remote shell applications form an appreciable share of traffic, on the lower layers the integration of secure variants such as IPsec become more and more important [7].

A considerable number of publications deals with the analysis of Internet Protocol (IP)-based backbone traffic. It consists of about 85.1% of TCP packets or 95.7% of TCP-based data volume. The residual traffic is mainly dominated by UDP. Only about 2% of packets and less than 1% of the data volume make up the non-TCP/UDP part [24]. Most secure connections of higher-layer protocols are established on top of TCP. Nevertheless, web applications using the HTTP(S) protocol generate the lion's share [9]. NPs as mentioned above provide bandwidths of 4 Gb/s. So, the IXP2400 for example applied to the backbone would have to deal with 3.8 Gb/s TCP traffic. With regard to the statistics this amount includes at least 0.5 Gb/s of traffic requiring payload processing.

In the area of edge routers the traffic composition deviates. Virtual private networks (VPNs) often combine encryption with compression so that the computational effort is high. It has to be assumed that most routers transport similar packet flows without any need for encryption or compression. Conversely, a small number of routers generate a high amount of encrypted traffic. In those routers the ratio of encrypted to unencrypted traffic is several times higher than in a backbone. Reliable numbers, however, can hardly be given because of the lack of publicly available statistics on edge-router traffic.

Packets demanding high computational effort such as encrypted or compressed data usually use just part of the maximum transfer unit. Design parameters such as buffer sizes and time-out values are to be adjusted to best ratio of performance and resource utilisation. Therefore, DynaCORE is designed to deal with average-sized packets. Worst-case assumptions base on 1 Gb/s incoming traffic stream and a packet size of 507 bytes. Thus, DynaCORE has about 4 ms to process a packet as well as for a dynamic exchange of hardware assists, if applicable.

16.3 System Architecture

DynaCORE (Dynamically adaptable COprocessor based on REconfiguration) is a loosely-coupled coprocessor architecture for NPs. It is based on a runtime reconfigurable FPGA and targeted to time consuming payload processing tasks. A typical application scenario for DynaCORE as an accelerator within an edge router is depicted in Fig. 16.1. The NP performs standard jobs based on header processing such as bridging, IP forwarding, and quality-of-service (QoS) protocols. Payload processing tasks such as encryption/decryption, compression, or network-intrusion detection are off-loaded from the NP via standard Gigabit Ethernet network interfaces to DynaCORE. By making use of standard network interfaces, the design of DynaCORE is to a large extent independent from a specific NP. The NPs described in Sect. 16.2 can be used as well as the FlexPath architecture designed at the TU München (see Chap. 17). The NP performs header-based pre- and post-processing as necessary in order to off-load payload processing to DynaCORE.

The internal architecture of DynaCORE is shown in the upper part of Fig. 16.1. The integral parts are a *dispatcher*, a *reconfiguration manager* (RM), reconfigurable

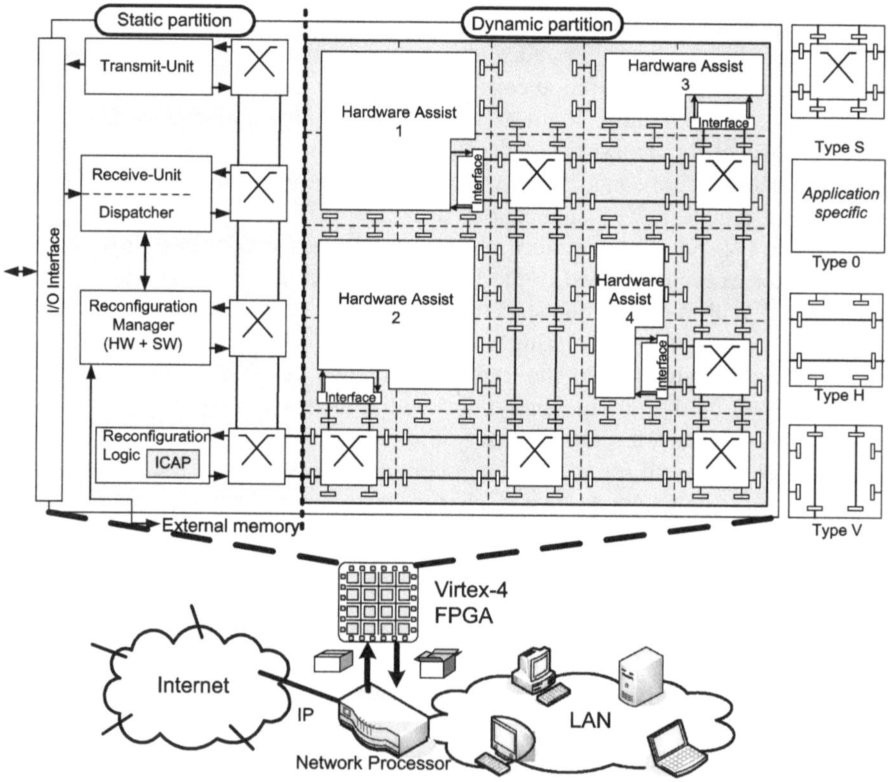

Fig. 16.1 DynaCORE system integration and internal system architecture.

hardware assists (HAs), and a partially reconfigurable *internal interconnect*. The system is completed by *receive* and *transmit* units converting data between the frame format used for communication with the NP on the one side and the DynaCORE-internal protocol on the other side.

The *dispatcher* [1] recognises the type of processing requested for a particular packet and dispatches the packet to an appropriate HA. The rules on the basis of which the dispatcher decides to which of the HAs a particular packet shall be routed are supplied by the RM. In case there is no HA providing the requested functionality the RM is notified and the packet is passed to a software-based fallback unit referred to as *software-based hardware assist* (see below).

The *reconfiguration manager* (RM) is the central control component within DynaCORE and realised as a processor-based subsystem. Its behaviour is thus mostly determined by software running on an embedded processor core. The RM is supplied with information from the dispatcher as well as with status information from monitors within the HAs. It maintains a global view on the system state and cyclically computes a suitable system configuration for the current and expected near-future request pattern. If necessary, the FPGA is reconfigured. The actual process of reconfiguring the FPGA is also controlled by the RM.

The *hardware assists* (HAs) are the actual data processing units and comprise the processing core, data counters for monitoring purposes, and a generic interface which makes the HA compatible with the communication protocol used within DynaCORE. The processing core is generally an off-the-shelf IP core for a certain algorithm. By means of the counter values which are periodically sent to the RM, the degree of utilisation of the processing core can be determined.

The *software-based hardware assist* is a particular HA which includes a processor core and allows for a variety of algorithms to be executed in software. This HA makes use of a hard-wired PowerPC 405 processor core as available in Virtex-4 FX FPGAs. In order to save configurable logic resources, the processor is used standalone, i.e. without any peripherals or memories attached to it. It merely uses its caches for storage of instructions and data. Software is loaded into the caches by means of an FPGA-internal JTAG connection under control of the RM [14].

The *internal interconnect* consists of static and reconfigurable parts. While there are static point-to-point connections between the external I/O interface and the transmit/receive units as well as between these units and the dispatcher, a reconfigurable network-on-chip (see Sect. 16.5) is used for connecting the static parts of the system to the reconfigurable HAs. The reconfigurable part of the interconnect together with the HAs resides in the reconfigurable area as depicted in Fig. 16.1.

16.4 Model

For testing and analysis purposes of DynaCORE, a formal model and simulation opportunities are necessary to support design decisions in an early stage. The application of reconfiguration capabilities in particular makes performance analysis difficult and demands additional tests to verify the system behaviour. Current hardware can be modelled as discrete event-based systems. In combination with dynamic reconfiguration the modelling concept moves to dynamic structure systems [26]. In the following the basic theory details are introduced and applied to the model of DynaCORE which leads to a tile-based, dynamically reconfigurable SystemC model with constant complexity.

16.4.1 Principles of Theory

Usual hardware models base on discrete event-based system (DEVS) description. Dynamically reconfigurable systems are systems which cannot intuitively be described by a single system model. Instead, multiple models and a transfer function to switch between them are needed to describe the entire system. The dynamically reconfigurable system presented here has a core module which is responsible for the system behaviour. It executes any controlling and management functionality.

Barros [5, 26] introduced a controller-based formalism to describe dynamically structured discrete event-based systems (DSDEVS). It extends the formalisms for standard discrete event systems and hierarchical setups of DEVS, called discrete event-based system networks (DEVN). In [5] the proof of closure under coupling is given. DEVS simulators are able to combine models for execution or even use parallel simulator kernels by assigning an own simulator kernel to each model [6].

The formalism to describe DSDEVS applies the DEVS and DEVN relations. The core model is the controller χ of the DSDEVS. The controller itself has different states and two transition functions to react on external and internal events. A reaction can lead to a simple state transition within the current controller model. It may also lead to a state transition that switches to a state of another controller model. So, this recursive definition yields a controller model the states of which are again models of the system. Barros proved that DSDEVS are a subset of atomar DEVS and that DSDEVN and DSDEVS are also closed under coupling [5]. Thus, there is no need for additional effort to model and simulate those systems if there is a DEVS simulator.

16.4.2 Modelling DynaCORE

The model of DynaCORE following the formalism of Barros is described by the following 4-tuple: $\text{DSDEVN}_\Delta = \langle X_\Delta, Y_\Delta, \chi, M_\chi \rangle$. Δ is the identifier variable of the system. X_Δ is the domain of all inputs and Y_Δ is the range of all valid output. In this case, $X_\Delta = Y_\Delta$ input and output of the system both have to be well-formed in the sense of the exchange protocol with the NP. Finally, M_χ contains the model description of the controller as a DEVS [26]: $M_\chi = \langle X_\chi, S_\chi, Y_\chi, \delta^\chi_{int}, \delta^\chi_{ext}, \lambda_\chi, \tau_\chi \rangle$. The sets X_χ and Y_χ describe the inputs and outputs of the controller. S_χ is the set of states, the controller may take on. $\delta_{int} : Q \times S_\chi \rightarrow S_\chi$ and $\delta_{ext} : S_\chi \rightarrow S_\chi$ are the transition functions. δ_{ext} reacts on external events, δ_{int} is dedicated to internal events. The domain Q extends the state space by time information so that an internal event arises if the hold time of a state expires. The hold time is determined by $\tau : S_\chi \rightarrow \mathbb{R}^+_{0,\infty}$. The output of a state is determined by $\lambda_\chi : S_\chi \rightarrow Y_\chi$.

The different configurations of the system are included in the state definition. Each system architecture requires its own representative state. A state $s_\chi \in S_\chi$ of the controller model M_χ is a 6-tuple: $s_\chi = \langle D^\chi, \{M_i^\chi\}, \{I_i^\chi\}, \{Z_{i,j}^\chi\}, \xi^\chi, V^\chi \rangle$. D^χ contains all modules except the controller itself applied in this state, $\{M_i^\chi\}$ consists of all models of components in D^χ, and I_i^χ includes all modules on which module $i \in D^\chi$ has impact on. $Z_{i,j}^\chi$ is the chaining function of output and input of concatenated modules given in I_i^χ. ξ^χ is the serialisation of the events in a non-parallel model. Applying SystemC [11] for simulation, the SystemC kernel provides and executes ξ^χ. Finally, V^χ contains all decision parameters for χ to raise an event handled by δ_{int}^χ.

The most crucial point is the definition of the transition functions δ^χ. Their quality dominates the success of the system. δ_{ext}^χ provides an interface for the NP to raise events from outside. This option is neglected in the following to the benefit of focusing on internal adaptation and parametrisation of δ_{int}^χ.

For a system with three exchangeable HAs A, B, C and a static part providing the basic modules for operation Fig. 16.7 shows a finite state machine describing the states and their interdependencies. This automaton is set up from a complete graph by removing all edges that would lead to a reconfiguration of more than one HA, i.e. only edges with limited costs are kept [4]. Because of the HA placement these transitions allow exchanging only a single HA per reconfiguration. Based on this state graph δ_{int}^χ is defined by attributing each edge with rules for decision making. Each rule has the form:

$$\text{IF} \quad \forall h \in \{A, B, C\} : eval(s_u, s_v, l_h, T_h) \quad \text{THEN} \quad s_u \rightarrow s_v.$$

So, the rule evaluates the current load l_h of each HA h and compares it to a threshold given by T_h. Obviously the evaluation depends on the edge introduced by (s_u, s_v). In case the premise becomes true δ_{int}^χ performs the transition $s_u \rightarrow s_v$. The application of a rule-based description with the semantics of RERAL [8] for the transition function allows for easy extensibility and a compact form.

16.4.3 Simulation

The crux of designing high bandwidth applications is the internal interconnect. The analysis of its performance demands sufficiently detailed models of the interconnect itself as well as of its interfaces. Here, the bus model is a simplified model of the processor local bus (PLB, CoreConnect library) [13]. Its complexity is kept simple but the timing and arbitration characteristics, including QoS features, are implemented. The bus throughput is limited by 2 Gb/s. The examined interconnects are established by one and two buses, respectively. The bus arbitration of the dual-bus model is marginally modified. Bus requests are performed on both buses simultaneously and grant is taken with priority from the first bus. Thus, the bus utilisation in the dual-bus model is irregular [4]. In case both buses are occupied a grant request is queued.

A complete system model applying a bus-based interconnect is shown in Fig. 16.2. The HAs are modelled at a high abstraction level. They consist of a bus-interface and a core model which is generic and time accurate. Only the compression core manipulates the packet size in a deterministic way based on statistical results. The interface executes all needed communication functionalities such as sending, receiving, buffering as well as forwarding for pipelined processing. Moreover, monitoring to collect system state information such as load distribution is incorporated in the interfaces. Driven by counters collected data is sent to the RM frequently. The static part models are basically similar to the HA models. For the software instance used as a fallback HA the performance parameters are scaled down to meet the degraded performance of embedded software compared to hardware.

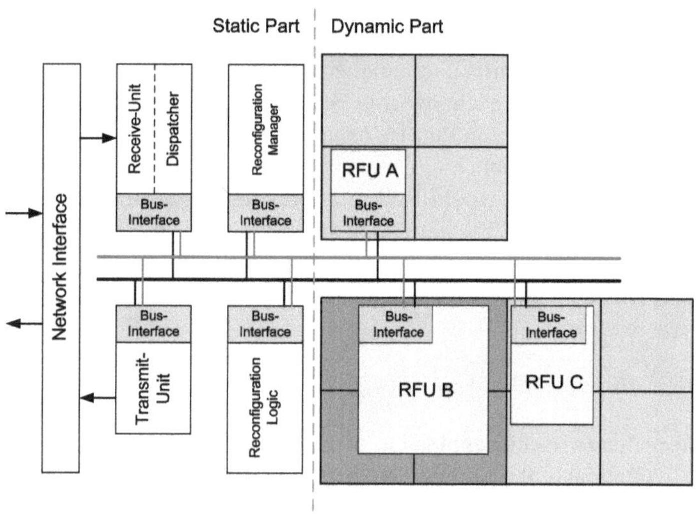

Fig. 16.2 Bus-based model applying one or two PLB-like bus models.

The principle to perform runtime reconfiguration in a SystemC simulation model is basically similar to OSSS+R described in Chap. 7 or ReChannel [23]. The reconfigurable parts are broken into tiles which have a SystemC framework model [2]. This framework allows the use of any method and provides generic interfaces. The tile-based modelling method keeps the complexity of a model constant and independent of the number of states of χ.

As introduced in Sect. 16.2, the workload for DynaCORE is a fraction of (edge) router traffic. Internet traffic as well as file transfers or other streaming applications have certain characteristics. Two main qualities are self-similarity and burstiness [15, 16]. In [3], an application-oriented traffic generator applying stochastic models for generating synthetic traffic to simulate and test network devices is presented. It is used to model and generate test streams for the DynaCORE simulation model. The simulation results of the bus models [4] are based on a traffic stream multiplexed of four b-modelled streams [25]. Each stream represents a different class of DynaCORE processing. The application classes used here are Deflate (A), AES (B), and DES (C). A pipelined process is also established by Deflate and DES. All four streams are multiplexed adapted to the bandwidth limit of 1 Gb/s.

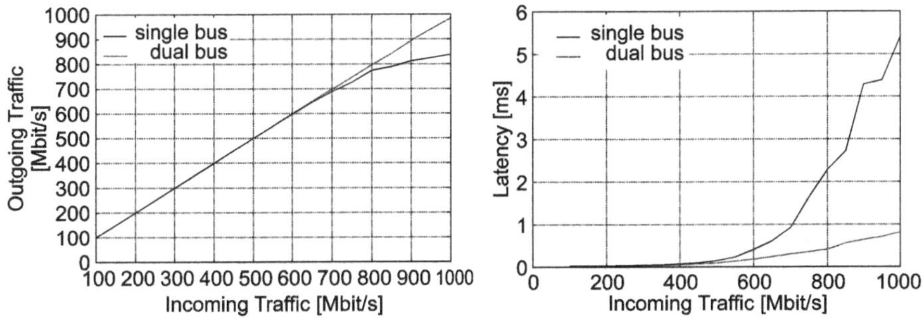

Fig. 16.3 Throughput and latency of the bus-based model for single and dual-bus mode [4].

The single- and dual-bus models are evaluated for certain performance issues. An excerpt of the complete examination [4] is given in Fig. 16.3. It depicts the most requested parameters of network devices: throughput and latency. The throughput shown in Fig. 16.3(a) unveils that the single bus is not capable to deal with the complete input bandwidth. The latency approves that there is a performance break down at about 600 Mb/s. Higher data rates are well processed by the dual-bus model. The latency is limited on an acceptable level and the output achieves the data rates of the input side. Internally, though, a high amount of buffer memory is necessary when bursts occur or a reconfiguration is executed. Reconfigurations are simulated with a configuration data size of up to 520 KByte for four tiles at once. In the experiments the configuration data size is assumed to one, two, and four tiles. So the bus is occupied for 480 μs at maximum to transfer the data to the reconfiguration interface. In these cases, even the dual-bus system becomes a bottleneck because of a limited degree of parallelism. Taking all simulations into account, the system has to be designed for a performance of at least $1.5\times$ input bandwidth. Additionally, the bus system does not scale. The PLB, e.g., is just capable of connecting eight instances. Even in this small sample model seven interfaces are consumed. Thus, a more scalable and adaptive communication infrastructure is required.

16.5 Runtime Adaptive Network-on-Chip

Network-on-Chips offer high degrees of flexibility, scalability and parallelism. These characteristics prove NoCs as an ideal communication infrastructure for flexible systems such as DyanCORE. By combing NoC principles with runtime reconfigurable FPGAs, a runtime adaptable NoC called CoNoChi (Configurable Network-on-Chip) was set up. CoNoChi [20, 17] reuses the infrastructure for performing a dynamic exchange of HAs and applies this to NoC switches and links as well. These can be inserted or removed from the communication network at runtime, so that only the required number of switches and links have to be instantiated. The network protocol supports the dynamic relocation of HAs, the generation of processing chains consisting of multiple HAs, and mechanisms to guarantee QoS. CoNoChi switches operate in virtual cut-through mode and are equipped with four full-duplex 16-bit links. Routing bases on dynamic routing tables provided by the RM of DynaCORE.

16.5.1 Architecture

A detailed view of the modular NoC architecture is given in the right part of Fig. 16.1. CoNoChi resides within a grid of equally-sized and individually reconfigurable tiles. Each tile may contain one of four tile types: a network switch (type *S*), a horizontal link (type *H*), a vertical link (type *V*) or part of an HA (type *0*). HAs may span over multiple adjacent tiles and are connected to the network by means of a generic interface. Each tile provides special communication units at fixed positions at the edges of the tile. These units, called bus macros, allow communication across the border of adjacent tiles. In Fig. 16.1, bus macros are shown as two small white rectangles connected by a line.

CoNoChi distinguishes physical and logical addresses. Physical addresses are bound to switches and are used for routing. Logical addresses are evaluated by HA interfaces and switches. They enable the support of several tasks within one HA and provide smooth processing in case HAs are relocated. The complete protocol stack is depicted in Fig. 16.4. It is organised in three layers: a data link layer, a network layer and an application layer. The data link layer is processed by switches while the link layer and the *AT* field of the application layer are analysed by the interfaces. The *Context ID* and *Type of Proc.* fields of the application layer are used by the DynaCORE dispatcher to handle and identify data streams belonging together.

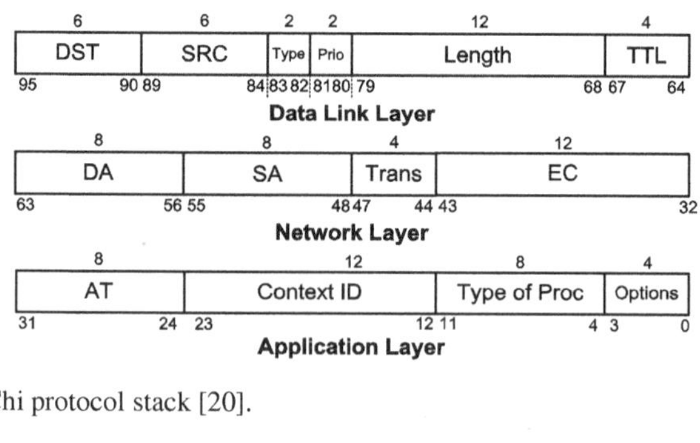

Fig. 16.4 CoNoChi protocol stack [20].

Within CoNoChi, several message classes are defined. For distinction, the *Type* and priority (*Prio*) fields of the data link layer are used in combination with the *AT* field. A *payload class* is used to transport user data to and from HAs. If state information of HAs is required by the RM a *monitor class* is used. A *Log Addr* class provides HA interfaces with their logical addresses as well as with physical and logical addresses for packet forwarding. Packet forwarding allows the setup of processing chains when packets processed by one HA have to be sent to another HA for further processing. A *NoC Msg* class is used for NoC internal messages, e.g. to enable/disable links of a switch. Reconfiguration data can also be transmitted via the NoC, using a *Rec Data* class.

16.5.2 Runtime Adaptation

An example of a reconfiguration scenario including network topology adaptation is presented in Fig. 16.5. Here, HA 1 will be replaced by two smaller HAs 2 and 3. As for the new HA 2 no free connection to the network exists, the network structure has to be adapted, i.e. a third switch is inserted. To guarantee full connectivity, special precautions have to be taken not to isolate parts of the NoC during reconfiguration. Representing the network topology as a hierarchical graph with switches as nodes and links as edges, only inner nodes of the NoC graph can be removed so that the NoC graph is not separated in two subgraphs. Additionally, a link between two switches can only be removed if there is another path between them. These preconditions have to be checked by the RM in advance. When these conditions are fulfilled, the routing tables of switches adjacent to the network tiles to be reconfigured are updated by the RM so that no packets are sent along the link. In addition, links of switches to the tile under reconfiguration are disabled by special NoC messages so that potential glitches caused by the reconfiguration process are blocked. After reconfiguration, the new switch is introduced to all switches by updating their routing tables, and the disabled links are reactivated.

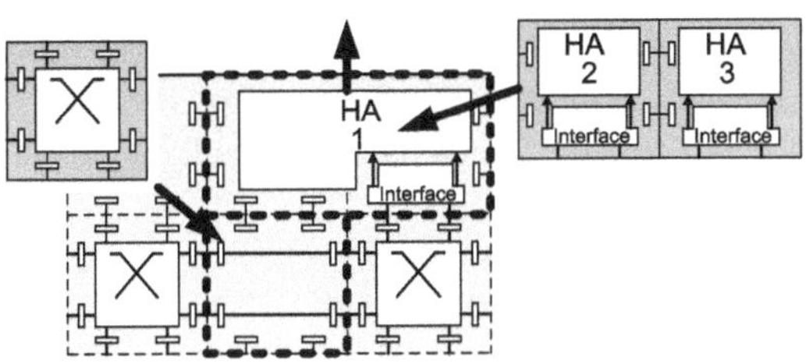

Fig. 16.5 Dynamic exchange of processing modules with network adaptation [18].

16.5.3 Fault Tolerance

CoNoChi is prepared to cope with different kind of faults [19]: single event upsets (SEUs), single event functional interrupts (SEFIs), and permanent hardware faults. SEUs and SEFIs are mainly caused by radiation and describe the inversion of a memory cell. Depending on the semantics of the memory cell, processed data or configuration data is changed. In the context of CoNoChi, the first one leads to faulty NoC packages while the latter impacts functionality. The problem of bit flips in the packet header is addressed by the *EC* field of the protocol stack. Payload faults are not considered as the application area tolerates temporary faults. SEFIs and permanent hardware faults both would lead to a malfunction of CoNoChi. While SEFIs can be corrected by rewriting configuration data, permanent hardware faults require a rearrangement of CoNoChi. Thus, the type of a fault has to be determined in order to apply the proper compensation mechanism. As hardware overhead for error detection and compensation has to be kept low, a stepwise mechanism is applied to determine the type of faults.

A failure exists if a network packet is corrupted, if it is not sent to the correct destination, or if it is delayed too long. All these cases can be detected by means of test packets sent in regular intervals to all switches by the RM. The last switch on each path under test returns the test packet to the RM. If a test packet returns corrupted or does not return at all, a fault in the system is detected. Yet, its location is unknown. Before the defect tile is localised the test is repeated so that SEUs can be excluded.

The procedure of detecting the defect tile starts with dividing the path under test into overlapping segments from the RM to each switch along the path. For each segments a test packet is sent which has to be rebounced by the last switch of each segment. When a faulty segment is identified, the routing tables of the surrounding switches are updated to bypass this segment. After this, the fault has to be corrected. Therefor the exact tile showing the defect does not need to be determined. The procedure starts with the tiles containing the last two switches of the path under test. At first, these switches are reset and new routing tables are sent. After this, the test is repeated. If the test packets return correctly, then an SEU in the switches occurred. If the tests still fail, a SEFI or a permanent hardware error is assumed. In this case, the configuration data of each tile is read back and compared with the configuration data originally written in. In case of a mismatch a SEFI occurred and the original configuration can be rewritten. Then, the test of the corresponding tile has to be repeated, as hard errors may also affect the readback of configuration data. If a mismatch cannot be detected, then a hardware fault occurred.

16.6 Reconfiguration Management

For a successful application of the reconfiguration management the point of reconfiguration has to be well determined and the overhead should be limited for avoidance of degrading effects. During reconfiguration the system is only partially available and, thus, provides reduced performance. The analysis is set up on a comparison of two variants utilising the decision graph given in Fig. 16.7. The first variant is the basic method whereas the second one reflects an optimisation of the basic method taking its analysis results into account.

16.6.1 Basic Method

The parameter set for the simulation runs are determined by test runs. The dual-bus model meets the processing requirements and is not too complex to be repeated for a statistical coverage. The parameters include HAs of 400 Mb/s, 512 KByte overall buffer size, a hardware/software processing ratio of $1 : 10$, and a deflate compression of $1 : 0.7$. System parameters are evaluated every 50 µs and the reconfiguration data size is 520 KByte. The interval is rather short and the amount of data is very large so that these values form a very ambitious environment. The load thresholds for each HA function is set on 10% as a low level and 90% as a high level. Exceeding the high level will lead to an additional HA if another deceeds the low limit. The overall simulated time is about 3 s and the simulation run is repeated 34 times. The data streams consist of two to six IPsec/IPcomp-streams and are processed as described in Sect. 16.4.3. For standardised analysis the ratio of average output stream to average input stream is investigated. The average of all simulation runs of this ratio is $\bar{x} = 0.8837$. The 95% confidence interval of the actuarial expectation is $0.8517 \leq \mu_X \leq 0.9156$. The reasons for a ratio less than 1 are at least two-fold: part of the stream is compressed, and during reconfiguration packet loss may occur.

16.6.2 Optimised Approach

The benchmark of the reconfigurations is drawn by investigating 3597 reconfigurations out of ten simulation runs. A norm is needed to classify the quality of each reconfiguration. The norm for δ^X_{int} is computed off-line after a run and is set-up by three criteria:

- The buffer contents and its behaviour over time indicates the load of a module. If there are almost exhausted buffers all over the system, the system has to deal with a high load at all whereas exhausted buffers at certain modules point out hot spots in the system. Increasing or decreasing buffer load shows significant changes in the load profile. An increasing load in particular indicates a too low performance of the module for the offered load.
- Another feature is the change in the output data rate. A short term success is an increasing data rate during the first three observations after a reconfiguration. The decision is made by comparing the input data rate shifted by the latency of the system with the current output. If the output is greater, the system draws from stock and frees buffer capacity which is a positive feedback. Otherwise a negative feedback is drawn.
- Finally, a long term interval is observed. The comparison is performed analogously. The long term interval aligns itself with the short term one lasts until the next reconfiguration starts. Occassionally, the interval does not exist because of an immediately launched reconfiguration reaching a non-adjacent state.

Applying these criteria to the detected reconfigurations almost two-thirds are rated positive. The distribution of positive and negative transitions is not uniform. Negative transitions mainly occur at the centre $A - B - C$ of the decision graph of Fig. 16.7 whereas most positive ones are on the scope of the graph. The main reason for the negative rates seem to be the fixed thresholds. States are kept as long as there is not any HA with a load less than the low level.

Applying these criteria to the simulation results of Sect. 16.6.1 each transition has positive and negative rates. Based on the ratio of these rates, the transitions are extended by a moment of inertia Θ. So, the set of variables V^{χ} evaluated to raise an internal event is extended by $\forall u, v : \{\Theta_{u,v}\}$. The integration of the moment of inertia into the transition rules is as follows:

$$\begin{aligned} &\text{IF} \quad \theta_{u,v} < \Theta_{u,v} - 1 \wedge \forall f \in \{A, B, C\} : eval(u, v, l_f, t_{f,u,v}) \\ &\quad \text{THEN} \quad \theta_{u,v} \leftarrow \theta_{u,v} + 1,\ \forall (u, v'),\ v \neq v' : \theta_{u,v'} \leftarrow 0 \\ &\text{IF} \quad \theta_{u,v} \geq \Theta_{u,v} - 1 \wedge \forall f \in \{A, B, C\} : \ eval(u, v, l_f, t_{f,u,v}) \\ &\quad \text{THEN} \quad u \rightarrow v,\ \theta_{u,v} \leftarrow 0 \end{aligned}$$

θ is a counter for keeping the multiple sequential decision for the same transition. If the decision counter exceeds a certain level Θ, the usual transition rule is executed. If the decision sequence is interrupted by another one, the counter θ is initialised.

Figure 16.6 shows the effect of Θ. The top bar shows the basic method that switches the state instantly if the threshold is exceeded. The bottom bar resembles the optimised approach. A state is kept after matching the threshold for a certain period to avoid reactions on short term phenomena. The determination of individually adapted $\Theta_{u,v}$ utilises the ratio of positive and negative rates. Based on the test runs all transitions are rated. Depending on the ratio a moment $\Theta \in [3, 8] \subset \mathbb{N}$ is added. This interval is determined by test runs applying a single Θ for all transitions. A high percentage of positive rates per transition assigns a small Θ. In Fig. 16.7 each transition is attributed with a transition index, the ratio of rates and its individual Θ.

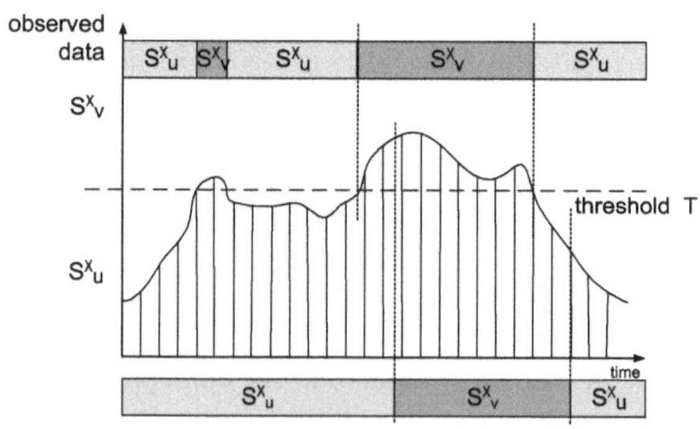

Fig. 16.6 Basic and optimised method to determine point of reconfiguration.

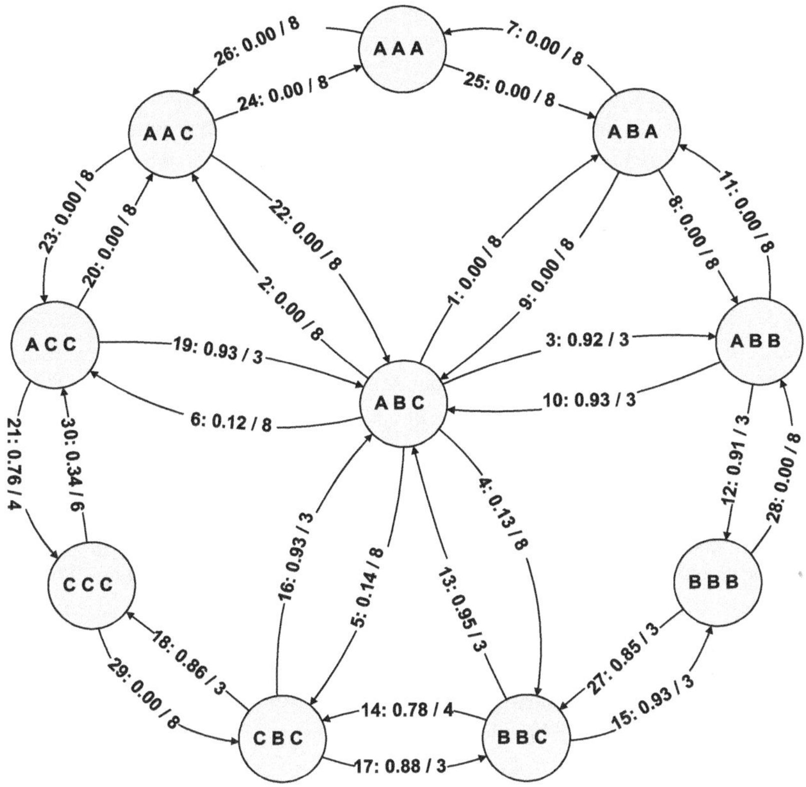

Fig. 16.7 Decision graph attributed by rate ratios and individual Θ's.

The repetition of the 34 simulation runs with absolutely the same parameters except the modified δ_{int}^{X} yields a new average $\bar{y} = 0.9023$. Its 95% confidence interval for the actuarial expectation is $0.8738 \leq \mu_Y \leq 0.9309$. Because of the identical input data and parameters the simulation runs can be statistically tested on a significant difference by a paired test. The null hypothesis H_0 for a sign test assumes that $\bar{x} = \bar{y}$, its counterpart H_1 assumes $\bar{x} \neq \bar{y}$. On a significance level of 95% the null hypothesis has to be refused. It has to be assumed that the difference of the averages $\bar{x}$ and $\bar{y}$ is significant and, thus, the optimised method achieves a better result. A stronger statistical test assuming a normal distribution for the paired differences which cannot be refused is a paired t-test. It is also successful on a significance level of 95% for the data compared here.

Another benefit of the optimised method is given by absolute values for again the same ten runs evaluated for the determination of Θ. The total number of reconfigurations is reduced to 2339 instead of 3597. A reconfiguration occupies the bus system for about 480 μs to transfer the reconfiguration data. So, the effort to adapt the system is strongly decreased and resources for processing are freed.

16.7 Technical Aspects

Technical properties and constraints of the target platform influence several design decisions or required special consideration also at higher abstraction levels. In the tiled grid of CoNoChi the size of the tiles must be consistent with the granularity at which the underlying FPGA can be reconfigured. The smallest unit of reconfiguration in a Xilinx Virtex-4 FPGA is a configuration frame which affects 16 CLB rows, i.e. 32 slices, along in a column. Therefore, a CoNoChi tile vertically extends over a multiple of 32 slices. Horizontally, the overhead for communication with neighbouring tiles gives a realistic lower limit of 12 CLB columns, i.e. 24 slices, for the width of a tile. For 16-bit bidirectional communication with four neighbours 96 slices are occupied by the portions of bus macros located within the tile. Thus, in a minimum-sized tile 672 slices remain for the implementation of actual functionality. For a number of HAs, however, this amount is not sufficient. For example, the AES decryption core used in the demonstrator described in Sect. 16.8 requires more than twice as many slices. On the other hand, choosing a larger tile size beforehand would result in a waste of logic resources. Therefore, CoNoChi is designed to support adjacent tiles to be merged as well as to be split up again later. The FPGA development tools, including Xilinx' *early access partial reconfiguration* (EAPR) tools, however, do not provide any support for scenarios with reconfigurable regions of varying number or shape. Therefore, the custom solution described below had to be developed. A detailed description of the technique is given in [22].

Technically, each CoNoChi tile is represented by a partially reconfigurable region (PR region) in the floorplan of the system. Adjacent PR regions as well as PR regions and static parts of the system are connected by bus macros. In order to realise a tile merge, the floorplan must be adapted so that it contains one fewer PR region while one PR region extends over the area previously occupied by two. Moreover, the adapted floorplan must not include bus macros located entirely within the enlarged PR region. Followingly, the top-level design which instantiates reconfigurable modules and bus macros also has to change. It is thus necessary to run the EAPR implementation tools all anew as if there was an entirely new design. The resulting partial bitstream for the enlarged, merged tile, though, will eventually be configured into an active FPGA containing the original design. It is therefore crucial that the tools as they are run for the adapted floorplan produce a result which is compatible with this original design. At first sight, this does not seem to be a serious issue as only the clearly contoured area of the merged tile will be reconfigured and as the boundary of that area corresponds exactly to the outer boundary of the two tiles in the original design. The affected PR regions, however, may also contain resources used by static parts of the design. For example, I/O buffers required for external communication may reside within the area and also have signals routed to them. The EAPR tools take care of such situations by first preventing PR modules to use resources already used by static parts of a design and then merging the native circuit descriptions (NCD) of PR modules and static parts. This EAPR feature, though, cannot be used across different floorplans and different top-level designs as the static parts, including bus macro instantiations, have to be re-implemented and may thus use resources differently. The solution is to force the *common* static part, i.e. almost everything excluding bus macros, to be implemented in exactly the same way in any scenario of merged or non-merged tiles. This can be accomplished by building a hard macro block from the initial implementation of the static part, a task which is not trivial, but technically feasible as shown in [22]. Derived scenarios with merged tiles then have to use an adapted top-level design which instantiates the static hard macro "as is". Figure 16.8 shows FPGA editor views of a simple example of a system comprising four tiles (left) and of the same system with two tiles merged (right). Using this technique, the resulting partial bitstreams of different scenarios can safely be intermixed.

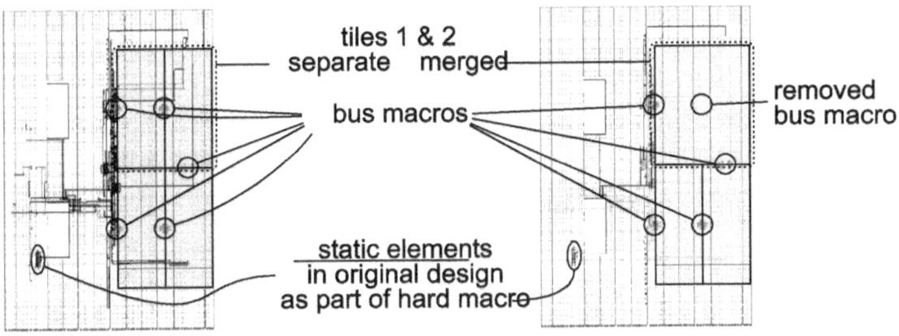

Fig. 16.8 FPGA editor views of separate and merged tiles.

For the actual reconfiguration of the FPGA, the internal configuration access port (ICAP) is used. In Xilinx Virtex-4 FPGAs the ICAP features a 32-bit data interface which can be clocked at up to 100 MHz. In case of a Virtex-4 FX60 FPGA, the configuration data for the entire device which amounts to 2.5 MB can be written within 6.6 ms. A minimum-sized tile as described above can be reconfigured within at most 200 μs. This is the worst-case parameter which has to be taken into account by the RM when scheduling the reconfiguration.

16.8 Evaluation

The applicability of the DynaCORE approach and the technical feasibility of its architecture were shown by means of proof-of-concept implementations. Based on the individual concepts presented in the previous sections of this chapter, a demonstrator was realised using a Xilinx Virtex-4 FX60 (XC4VFX60) FPGA as a platform. This FPGA particularly features an ICAP interface, two hard-wired PowerPC 405 processor cores, and two Gigabit Ethernet MAC components enabling the use of up to four Gigabit Ethernet interfaces with only little extra logic. The architecture of the demonstrator is depicted in Fig. 16.9. It comprises receive, dispatch and transmit units, a reconfiguration management component based on a PowerPC-based subsystem, and three reconfigurable hardware assists. Two distinct Gigabit Ethernet interfaces are used for communication with an NP-alike testing environment and for status information, respectively.

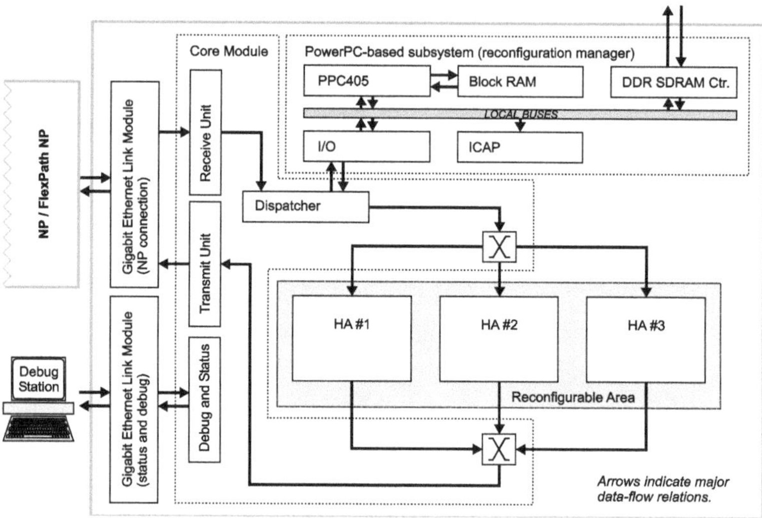

Fig. 16.9 Demonstrator architecture.

The resource requirements of the individual parts of the demonstrator are shown in Table 16.1(a). The figures indicate that about 50% of the total number of slices available in an XC4VFX60 FPGA were utilised for the demonstrator implementation. Table 16.1(b), in addition, shows the amounts of FPGA slices required by the individual functionalities supported in the setup.

Table 16.1 Resource requirements.

(a) Demonstrator

Part	Slices
Top-level logic	69
Base system	3,421
Reserved area for HA #1, #2, #3	3 × 2,944
Total	12,322

(b) Individual HAs

HA functionality	Slices
AES encryption	1,326
AES decryption	1,567
3DES encryption/decryption	1,343
XOR encryption/decryption	69

A complete system of NP and DynaCORE was set up with the *FlexPath NP* demonstrator (see Chap. 17) as the NP. This setup was first established with the individual demonstrators running in Lübeck and Munich, respectively, utilising an Ethernet-over-TCP tunnel connecting both demonstrators through the DFN network. This configuration allowed to run non-timecritical interoperability tests. In a more realistic scenario, both demonstrators were directly connected using their Gigabit Ethernet interfaces. This system of FlexPath NP and DynaCORE demonstrators was presented at the SPP1148 booth of FPL'08 [21]. Artificial network traffic was generated by a hardware traffic generator, processed by the FlexPath NP part and partially forwarded to the DynaCORE part. The latter determined whether or not it was necessary to reconfigure the FPGA in order to fulfill the request, processed the packet using an appropriate HA, and sent the result back to the FlexPath NP part for post-processing. The requested type of processing as well as the load of the HAs could be observed at an additional PC serving as the debug station as shown in Fig. 16.9.

16.9 Summary

The increasing gap between computational power and line speed combined with the short-term and long-term requirements of traffic profiles, i.e. a varying traffic mixture over time and changes in the protocol norms and applications, demand high flexibility, adaptibility, and performance of network devices. The solution proposed here is a dynamically reconfigurable coprocessor DynaCORE as a loosely coupled add-on for NPs. Beside device-dependent components its architecture bases on IP-cores employed as HAs for time-critical deep packet processing tasks, a RM to adapt the system, and a topology-adaptive NoC called CoNoChi. The application of a NoC is motivated by a performance analysis based on a formally derived simulation model. It is used for evaluating performance and functional issues. For the implementation of the runtime reconfigurable system a tile-based design method is introduced. Each tile may comprise a single module or forms part of a module in a tile compound. The technical design issues are demonstrated in combination with the FlexPath NP. The RM algorithm is analysed based on the simulation model of DynaCORE and traffic pattern models. Applying the results for optimising the algorithm leads to a significant improvement of the reconfiguration management and, thus, system performance.

Acknowledgements This work was supported by DFG grant MA 1412/5, project *DynaCORE*, as part of the Priority Programme 1148, *Reconfigurable Computing Systems*.

References

1. Albrecht, C., Koch, R., Pionteck, T.: On the design of a loosely-coupled run-time reconfigurable network coprocessor. In: Proc. of the Workshop on Media and Streaming Processors 2005 (MSP-7) (2005)
2. Albrecht, C., Pionteck, T., Koch, R., Maehle, E.: Modelling tile-based run-time reconfigurable systems using systemC. In: 21st European Conference on Modelling and Simulation, pp. 509–514. European Council for Modelling and Simulation (2007)
3. Albrecht, C., Osterloh, C., Koch, R., Pionteck, T.: An application-oriented synthetic network traffic generator. In: 22nd European Conference on Modelling and Simulation. European Council on Modelling and Simulation, Nicosia, Cyprus (2008)

4. Albrecht, C., Ross, P., Koch, R., Pionteck, T., Maehle, E.: Performance analysis of bus-based interconnects for a run-time reconfigurable co-processor platform. In: Proceedings of the 16th Euromicro Conference on Parallel, Distributed and Network-Based Processing (PDP 2008), pp. 200–205. IEEE Computer Society, Los Alamitos (2008)
5. Barros, F.J.: Dynamic structure discrete event system specification: a new formalism for dynamic structure modeling and simulation. In: Proceedings of the 27th Conference on Winter Simulation, pp. 781–785. IEEE Computer Society, Los Alamitos (1995)
6. Barros, F.J.: Abstract simulators for the DSDE formalism. In: Proceedings of the 30th Conference on Winter Simulation, pp. 407–412. IEEE Computer Society Press, Los Alamitos (1998)
7. Cooperative Association for Internet Data Analysis: Network Monitors of www.caida.org, Online Realtime Statistics (2008)
8. Döring, A.C., Lustig, G., Albrecht, C., Obelöer, W.: Building a compiler for an application-specific language. In: Proceedings of the First Scottish Functional Programming Workshop (2000)
9. Fei, Y., Jones, J., Lakkas, K., Zheng, Y.: Measurement of the usage of several secure Internet protocols from Internet traces. Student Project, Department of Computer Science and Engineering, University of California, San Diego, CA, USA (2002)
10. Hi/fn Inc.: Hifn 5NP4G Network Processor, Product Brief (2004)
11. IEEE Computer Society: IEEE Standard 1666-2005 Systemc Language Reference Manual. 3 Park Avenue, New York, NY 10016-5997, USA (2006)
12. Intel Corporation: Intel IXP2400 Network Processor, Datasheet (2004). Document Number: 301164-011
13. International Business Machines Corporation: The CoreConnect Bus Architecture, Whitepaper (1999)
14. Koch, R., Pionteck, T., Albrecht, C., Maehle, E.: A lightweight framework for runtime reconfigurable system prototyping. In: 18th IEEE/IFIP International Workshop on Rapid System Prototyping, pp. 61–64. IEEE Computer Society, Los Alamitos (2007)
15. Leland, W.E., Taqqu, M.S., Willinger, W., Wilson, D.V.: On the self-similar nature of Ethernet traffic (extended version). IEEE/ACM Trans. Netw. **2**(1), 1–15 (1994)
16. Luo, S., Marin, G.A.: Realistic Internet traffic simulation through mixture modeling and a case study. In: Proceedings of the 37th Conference on Winter Simulation, pp. 2408–2416 (2005)
17. Pionteck, T., Koch, R., Albrecht, C.: Applying partial reconfiguration to networks-on-chips. In: International Conference on Field Programmable Logic and Applications (FPL), Madrid, Spain (2006). doi:10.1109/FPL.2006.311208
18. Pionteck, T., Koch, R., Albrecht, C.: A dynamically reconfigurable packet-switched network-on-chip. In: Proceedings of Design Automation and Test in Europe, pp. 136–137 (2006)
19. Pionteck, T., Albrecht, C., Koch, R., Maehle, E.: Performance and reliability monitoring in network-on-chips. In: Workshop on Diagnostic Services in Network-on-Chips (DSNOC) (2008)
20. Pionteck, T., Albrecht, C., Koch, R., Maehle, E.: Adaptive communication architectures for runtime reconfigurable system-on-chips. Parall. Process. Lett. (PPL) **18**, 275–289 (2008)
21. Pionteck, T., Koch, R., Albrecht, C., Maehle, E., Meitinger, M., Ohlendorf, R., Wild, T., Herkersdorf, A.: SPP1148 booth: Network processors. In: FPL 2008, Heidelberg, Germany (2008)
22. Pionteck, T., Koch, R., Albrecht, C., Maehle, E.: A design technique for adapting number and boundaries of reconfigurable modules at runtime. Int. J. Reconfigurable Comput. (2009, to appear)
23. Raabe, A., Hartmann, P.A., Anlauf, J.K.: ReChannel: describing and simulating reconfigurable hardware in systemC. ACM Trans. Des. Autom. Electron. Syst. **13**(1), 1–18 (2008)
24. Sprint Corporation: IP Monitoring Project, URL: http://ipmon.sprint.com/ (2001–2004)
25. Wang, M., Chan, N.H., Papadimitriou, S., Faloutsos, C., Madhyastha, T.M.: Data mining meets performance evaluation: fast algorithms for modeling bursty traffic. In: Proceedings. 18th International Conference on Data Engineering, pp. 507–516 (2002)
26. Zeigler, B.P., Praehofer, H., Kim, T.G.: Theory of Modeling and Simulation, 2nd edn. Academic Press, San Diego (2000)

Chapter 17
FlexPath NP—Flexible, Dynamically Reconfigurable Processing Paths in Network Processors

Rainer Ohlendorf, Michael Meitinger, Thomas Wild, and Andreas Herkersdorf

Abstract This chapter presents the results of the FlexPath network processor (NP) project. Based on observations on current NP implementations and relevant Internet traffic scenarios, a new NP architecture is defined that makes use of reconfigurable packet processing paths in order to improve the system performance. We propose to extend state-of-the-art processor-centric NP architectures with specific hardware units in order to classify the incoming traffic into separate processing classes. For each traffic class, we can provide an optimized processing path, i.e. a functional unit traversal sequence within the NP. In addition, we propose to offload significant shares of the traffic to a dedicated hardware path in order to bypass the CPU cluster and save precious programmable processing resources. We also address the problem of multi-processor load balancing in the context of multi-core network processors. The concepts have been evaluated on an analytical and simulative level, and finally a demonstrator has been implemented on an FPGA in order to prove the claimed performance advantage by measurements.

17.1 Introduction

FlexPath is a network processor (NP) concept with flexible, dynamically reconfigurable packet processing paths. The basic idea of a FlexPath NP is that the way of a packet through the system is not fixed, but may vary depending on packet content or system status. We want to obtain an optimal utilization of the provided system resources, high data throughput and low packet processing latency along with defined quality-of-service (QoS) criteria.

Rainer Ohlendorf · Michael Meitinger · Thomas Wild · Andreas Herkersdorf
Institute for Integrated Systems, Technische Universität München, Munich, Germany,
e-mails: rainer.ohlendorf@tum.de, michael.meitinger@tum.de,
thomas.wild@tum.de, herkersdorf@tum.de

M. Platzner et al. (eds.), *Dynamically Reconfigurable Systems*,
DOI 10.1007/978-90-481-3485-4_17, © Springer Science+Business Media B.V. 2010

Current NP architectures consist of a heterogeneous mix of processing resources (e.g. general-purpose or networking ASIP processors, hardware accelerators, etc.). In addition, the great variety of application protocols in the data networks impose different processing requirements on the packet processing infrastructure. This observation led us to the question, whether it would not be beneficial to differentiate the incoming traffic into different application classes according to specific processing requirements, and assign them to architectural entities, which are optimized for the respective application domains. In the field of parallel processor cluster NPs, all packets have to be processed by the CPU cluster. If hardware acceleration is needed to achieve complex tasks in real-time, the accelerator is also controlled by a CPU and the interrupt rate seen by the processor cluster may even increase in contrast to a pure software solution (run-to-completion model) [10]. The FlexPath NP strives to achieve a better overall performance by two basic measures:

- Relieving general-purpose processor resources by moving recurring standard tasks to hardware units that achieve better performance at lower cost (chip area), thus saving the flexibility of programmable processors for applications that require intelligent software processing.
- Providing different resource traversal patterns (processing paths) that are optimized for the individual application, and access them directly after packet reception and classification on the receive interfaces. As the processing requirements change over time, a run-time reconfigurable solution is desired that allows adaption towards the current load situation during system runtime.

In contrast to most other projects inside this book, reconfiguration in our project is not achieved by partial dynamic reconfiguration of FPGA resources. The timing requirements faced in FlexPath NP are much harder than in the other applications presented in this book. The partial dynamic reconfiguration of an FPGA takes about 10–100 ms; however, the reconfiguration of the packet classification rule base must be achieved within a packet inter-arrival time. We do not intend to reconfigure our system after each packet, but when a reconfiguration becomes necessary, it has to be fast enough not to interrupt the incoming packet stream. Assuming a minimum frame size of 40 Bytes at Gigabit speed the maximum allowed reconfiguration time is in the order of 200–300 ns (see also [6]). Therefore, we switch between different processing paths for the various packet types by (re-)writing memory contents in the packet classification unit. Another advantage of this approach is that we are thus not constrained to using a reconfigurable hardware platform (FPGA), but a FlexPath NP can be implemented in conventional ASIC technology.

The rest of the chapter is structured as follows: In Sect. 17.2 we introduce the FlexPath NP concept. The concept will be evaluated in Sects. 17.3 and 17.4 using formal analysis and simulation results. In Sect. 17.5 we present our FPGA demonstration platform. After describing the architecture, several experiments demonstrate the advantages of our concept. Finally, we conclude the chapter by summarizing the main benefits of a reconfigurable architecture in the high-performance network processing field.

Due to space limitations, we reference mainly our own publications in this chapter and only the most important related work. For a full coverage of related work we refer the reader to the respective references listed in our own publications.

17.2 FlexPath NP Concept

In most commercial NP architectures a cluster of programmable processing elements (PEs) is responsible for packet processing. Usually hardware accelerators speed up the processing for special tasks like packet storage or cryptographic algorithms. The cluster itself may be programmed as pipeline, i.e. each PE performs only a small task on each packet, or as a run-to-completion system. In the latter case, each PE performs the full processing for one packet. Nevertheless, each packet traverses the PE cluster. Hardware accelerator management or hand-over of control packets to the control plane is done by a data plane PE.

Regarding the protocol mix of a typical backbone router traffic (see [18]) the processing requirements are very heterogeneous. For the majority of packets (75–85% TCP, roughly 10% UDP) IP forwarding is sufficient. With the increasing share of real-time applications in the Internet, also quality-of-service (QoS) techniques such as DiffServ become important features in network nodes. Here, packets are marked with different forwarding priorities, in order to give high-priority traffic an expedited forwarding compared to best effort traffic. In addition, modern router environments may also deploy new services like virus-scanning or other forms of deep packet processing in the future. In this case, only the pure TCP acknowledgment packets without payload (roughly 25% of the TCP packets) are limited to IP forwarding, whereas for all other packets deep packet processing may be performed. Another 5–8% of the traffic consists of cryptographic packets, i.e. at least in an edge router environment these packets are potentially candidates for en- or decryption. Most of the remaining packets contain control protocols (less than 2–3%) that are processed by a control plane CPU. Control protocols are usually complex to implement but have relaxed timing requirements.

In contrast to most other commercial architectures we want to take care of the different processing requirements of each packet already at the input and find an optimal way through the system for each packet. We think that it is not efficient to send each packet to a processor first, that decides afterwards about the further path (e.g. hardware accelerator or control plane). In such a system, the interrupt rate for the PE cluster is higher than the packet rate. In a FlexPath NP packets are classified in hardware already at the input. A dispatcher decides about the further path of the packets. Since packets are directly sent to an appropriate processing instance, the interrupt rate of the PE cluster can be reduced and headroom for processing is created.

The path decision is influenced by two important characteristics:

- *Application dependent path decision*: The protocol type of each packet is taken into account. Control packets are sent directly to the control plane, encrypted packets directly to a hardware crypto core for decryption. Packets that must be processed by a processor are sent to the PE cluster.

- *Load dependent path decision* (*Load Balancing*): This is important when having multiple instances of functionally identical processing elements, like the PEs in the PE cluster. So, not only a decision about the processing element type (like PE) is important, but also a decision to which specific PE. This decision is dependent on the current load situation in the NP.

In order to classify the incoming packets according to their specific application class, it is necessary to inspect the protocol headers in hardware. In addition to extracting relevant header fields, we can also use hardware to perform some basic recurring tasks of network processing. In this way, we even achieve IP forwarding capability without the need of PE intervention, as we will show later. When we are able to bypass the PE cluster and process certain packet types fully in hardware, we call this the AutoRoute path.

17.2.1 Application Dependent Path Decision

The decisions about the most suitable processing path have to be made on a per-packet basis. In the FlexPath NP, this decision is taken by a hardware assist called Path Dispatcher [12, 14] (see Fig. 17.1) that contains a reconfigurable rule base. This rule base determines the particular processing path depending on the incoming packet type. Based on the packet types we see a variety of alternative processing paths in the FlexPath NP:

- Packets may be sent to a number of co-processors (e.g. authentication or decryption) before being handed over to the PE along with the obtained results. Computing amount and packet event rate of the PE cluster are reduced compared to a processor-centric architecture.
- Packets with relatively simple and regular processing requirements, like plain IP forwarding, can be processed completely in hardware (AutoRoute). Additional FlexPath-specific hardware is needed, that will be introduced in the following.
- All other packets are sent directly to the PE cluster as in a processor-centric NP.

The Pre-Processor unit performs basic functions like packet header parsing, extraction of pertinent header information (destination/source addresses, L4 port numbers, priority bits, protocol IDs), packet prioritization and classification and calls the next-hop lookup engine. Thus, the Pre-Processor retrieves sufficient information about the incoming packet which enables the Path Dispatcher to take decisions about the further handling of the packet. Packets traversing the PE cluster also benefit from the pre-processing, since basic integrity checks are already performed.

The Post-Processor [7] performs basic egress packet modification operations like MAC address exchange, TTL decrement, and IP checksum calculation. Together with the functionality of the Pre-Processor and the next-hop lookup engine, these operations are enough to perform IP forwarding and finish the packets in hardware on the AutoRoute path. The Post-Processor may also perform basic packet modifications like IPsec header insertion instead of a PE.

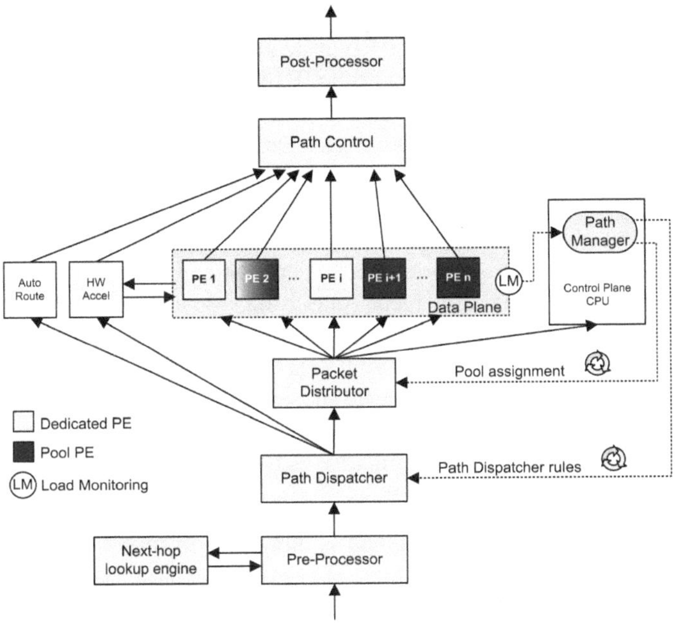

Fig. 17.1 Packet flow within the FlexPath NP.

Alternating path decisions for packets from the same connection (e.g. TCP packets with payload vs. TCP acknowledgment packets) may lead to packet reordering. As shown in [5], packet reordering can have a significant performance penalty on network throughput, especially when considering the dominating TCP protocol. Therefore we propose the Path Control unit to re-sequence the packets at the egress side on a per-flow basis (see [8]). This also creates additional freedom in processor assignment, which is important for the load balancing schemes discussed in the next section.

17.2.2 Load Dependent Path Decision

The path assignment performed in the Path Dispatcher not only differentiates between software and hardware processing, but is also used for load balancing within the processor cluster. In contrast to proposals from prior art publications [4, 17, 16], we differentiate the incoming traffic in our Path Dispatcher unit, and can—in cooperation with the Packet Distributor—implement different load balancing strategies for different traffic classes. The Packet Distributor provides several queues for the different traffic classes and features a multi-processor interrupt controller that associates the queues with the PEs. In addition, our Path Control unit solves the problem of packet reordering, which is specifically addressed by every scheme found in prior publications and that limits the efficiency of the presented load assignment schemes. We provide a scheme that optimizes the NP performance with respect to QoS requirements and maximizes the PE utilization.

For stateless processing applications like IP forwarding, we propose a prioritized spraying of the packets over all PEs in the processor cluster. For each traffic class we provide a single queue, into which packets are assigned by the Path Dispatcher. The packets can then be fetched from any free PE for processing according to predefined priority levels that correspond to the QoS requirements of each traffic class. In contrast to the spraying described in [4], where packets are assigned to different queues with a one-to-one association between queue and PE, our spraying is effectively performed after a common queue (one-to-n association). By spraying packets over a multitude of PEs from a single queue we can exploit a pooling gain effect and insure that the packets suffer minimal delay before being processed.

Packet spraying is not suitable for stateful networking applications, where a flow-specific state has to be maintained and accessed for every packet. The problems associated with distributed state information is twofold:

- Packets from the same flow might be processed by two different PEs at the same time. In order to preserve integrity, the state information would have to be secured by locking semaphores, thus PEs would idle during conflicting accesses and effectively waste time.
- Even if only one packet is processed at a time, the state information has to be kept in a centralized shared memory to provide access for all PEs in the processor cluster. The simulations performed in [11] have shown that shared memory architectures do not scale well to a larger number of cores. Alternatively, a local copy of the state information might be generated, but complex cache coherency measures would have to be installed to insure data consistency over several parallel copies.

Therefore, we propose to basically maintain a fixed flow-to-PE mapping for those application classes. We periodically monitor the load on the individual processors and reconfigure some flow-to-PE assignments when mandated by a large load imbalance within the processor cluster.

Configuration and updating of Path Dispatcher and Packet Distributor is the task of a software process (called Path Manager) running on the NP control plane CPU. This software process has to monitor the system status like processor loads, register IPsec connection setups etc. in order to decide if a system reconfiguration improves the system behavior.

17.3 Formal Analysis

We have performed analytical investigations on the effect of hardware offloading for the overall NP performance [10]. We choose a multi-processor NP with six parallel microengines as a reference architecture. We assume a clock frequency of 232 MHz and with a CPI of 1.0 we can derive a processing budget of 573 instructions per packet for minimum sized packets at 1.2 Gbit/s (2× OC-12 Sonet/SDH speed). This architecture is then compared in terms of the available instructions per packet processing budget to a FlexPath NP with one or two RISC CPUs and AutoRoute traffic shares varying between 0% and 100% (see Fig. 17.2).

The FlexPath NP has fewer PEs than the reference architecture and the dual-core FlexPath NP has 30% lower nominal processing performance (MIPS). Consequently, directing about 30% of the traffic towards AutoRoute would compensate this difference. In addition, it can be seen in Fig. 17.2 that offloading higher percentages of the traffic towards AutoRoute significantly increases the available instruction budget for the remaining traffic that traverses the CPU cluster.

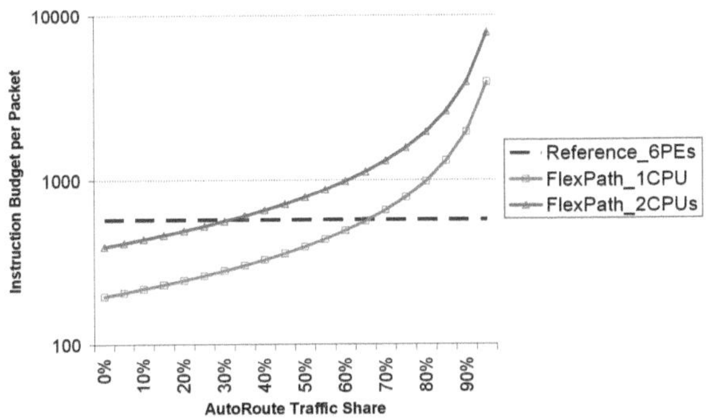

Fig. 17.2 Instruction Budget in Reference Scenario vs. FlexPath NP for Varying AutoRoute Traffic Shares.

Roughly a quarter of the Internet traffic consists of TCP acknowledgment packets without further payload. These packets are prime candidates for AutoRoute, as they only need to be forwarded and no payload processing is applied to them. In addition, those packets are of minimum length and thus impose the highest interrupt rates onto a processor cluster.

In mobile GPRS networks only traffic going to the local base stations is subject to intensive protocol processing. The majority of packets are part of route-through traffic to and from neighboring radio network controllers (RNC). Here, we are able to offload up to 90% of the traffic, assuming that up to 10 chained RNCs aggregate traffic towards a single serving GPRS support node (SGSN). Exploiting such high traffic offload means that we are able to address the CPU-assigned applications with relatively high computational power and do not waste precious programmable resources for doing standard tasks that can be more effectively performed in hardware.

17.4 Simulative Exploration

As the analytical investigation of the FlexPath NP architecture has shown encouraging results, we have refined our investigations on a simulative level. We have first investigated the expected performance improvement of FlexPath in comparison to processor-centric architectures. Then, we investigated the proposed load balancing strategy. As we have not yet implemented load balancing on our FPGA demonstrator platform (Sect. 17.5.2) we show the simulation results in detail.

17.4.1 FlexPath NP Architecture Evaluation

We investigated the expectable system performance by a set of simulations on a trace-based SystemC architecture model [11]. The goal of these simulations is to estimate the achievable NP throughput on a System-on-Chip (SoC) platform with standard RISC cores and the FlexPath-specific hardware offload and AutoRoute features. We realize the functional entities presented in Fig. 17.1 on SoC execution resources with a central on-chip bus system (see Fig. 17.3). In order to calibrate the simulation model with real-world figures, we have profiled an open IP stack that we adapted to run on a preliminary hardware system with readily available IP cores. We implemented this initial prototype on a Xilinx Virtex-II Pro development board that consists only of the Memory Management unit (see Sect. 17.5), Pre-Processor, memories and a single PowerPC. Based on the extracted execution traces of the IP forwarding software and the access patterns of the BlockRAM and SDRAM memories, we have investigated potential system bottlenecks and scalability behavior with respect to clock frequencies, number of parallel processors and different memory technologies. The simulation model features a cycle-accurate model of IBM's CoreConnect PLB bus, thus the simulator is well suited to cover the effects of multiple masters accessing the shared media over a common communication infrastructure.

Without being depicted in separate figures, three main results of the performance simulations are summarized in the following.

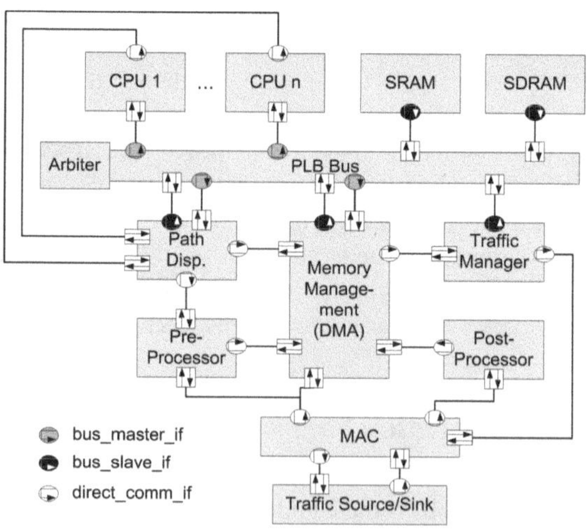

Fig. 17.3 SystemC Performance Model of the FlexPath SoC Architecture.

The first result of the simulations is that in order to scale up the processing performance of the regarded NP, it is more beneficial to increase the number of RISC processors rather than increasing the processors' clock frequencies. It also became clear that scaling the system to more than four CPU cores requires higher capacity NoC interconnect structures and distributed memories, in order to avoid running into either a communication or memory access bandwidth bottleneck.

Providing a separate fast SRAM for the software increases the system performance, as reads from the SDRAM with the used Xilinx memory controller have quite a long access latency and the code does not fit into the caches of the PowerPCs on the FPGA.

Finally, we investigate the performance of a FlexPath NP architecture with a single CPU, using the hardware offload possibilities of Pre-Processor and Post-Processor and AutoRoute forwarding. The individual performance figures for the different scenarios had to be omitted from this chapter for space reasons, but can be found in [11] and [13].

17.4.2 Load Balancing in FlexPath NP

In order to evaluate the proposed load balancing schemes, we have performed a set of simulations using real traffic traces obtained by CAIDA [2, 3] that are processed by a FlexPath NP with QoS-aware IP forwarding and IPsec encryption. The traffic is split up into high-priority forwarding and best effort forwarding when spraying the stateless traffic. The packets for IPsec encryption are assigned to dedicated CPUs. We developed a functional simulation model of our FlexPath NP that exactly resembles the packet pre-processing, Path Dispatcher, queuing and Path Control functions. In order to obtain performance figures, we have considered processing delays from measurements with a combined IP forwarding/IPsec stack, but this time omitted the memory access and DMA effects. A model of the Path Manager process monitors the CPU cluster load every 50 ms and optimizes the flow-to-CPU assignment by reconfiguring the Path Dispatcher rule base if necessary.

Figure 17.4 shows the packet loss rate of our combined scheme compared to the most recent load balancing scheme proposed by Kencl and Shi in 2006 [16]. As

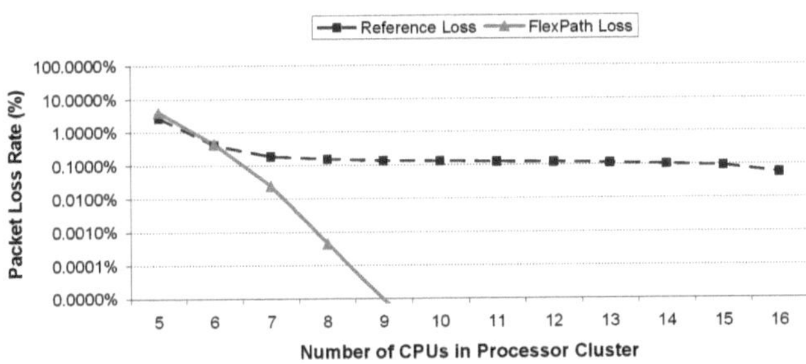

Fig. 17.4 Packet Loss Rate in FlexPath NP versus Reference Implementation.

Kencl and Shi assign the entire traffic to CPU-specific queues, head-of-line blocking effects can be observed for the forwarding packets. Forwarding packets that arrive behind an IPsec packet are affected, as the processing latency for IPsec is up to a factor of 300 larger compared to plain forwarding. Consequently, a certain number of packets are lost, even when provisioning a CPU cluster that has enough processing power to handle the entire traffic. In the FlexPath scenario, we reach a lossless operation beyond nine processors in the cluster.

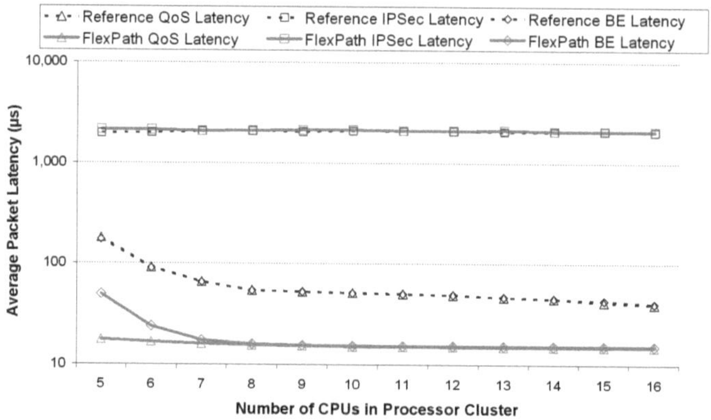

Fig. 17.5 Packet Latencies per Traffic Class.

Figure 17.5 illustrates packet latencies for the before mentioned setup. As FlexPath differentiates the traffic classes in the Path Dispatcher before the input queues of the processor cluster, we can insure proper prioritization of the QoS packets. It can be seen that the best effort packets suffer a longer latency (and a significant packet loss) while the processor cluster is overloaded with the IPsec and QoS traffic for a smaller number of CPUs. As more CPUs are added, the latencies of best effort and QoS converge as the processing consists of exactly the same forwarding function for both traffic types. In contrast, since [16] does not consider traffic classes, and with the head-of-line blocking effects caused by the IPsec packets, significantly higher processing latencies for both the QoS and best effort traffic can be observed.

17.5 FPGA Demonstrator

17.5.1 Architecture

In order to verify our FlexPath concept with real measurements rather than with simulation results only, we have developed a Multi-Processor System-on-Chip implementation of a FlexPath NP on a single Xilinx FPGA [9]. The block diagram of our system is depicted in Fig. 17.6. As interconnect we are using the CoreConnect Processor Local Bus (PLB) with two PowerPC 405 CPUs connected that are used for packet processing. One CPU always runs the data plane stack, whereas the second CPU can be configured to perform control and/or data plane functions. The stack supports IPv4 forwarding and features an IPsec implementation, which is limited to manual keying. In the following we describe in short the architecture of our system:

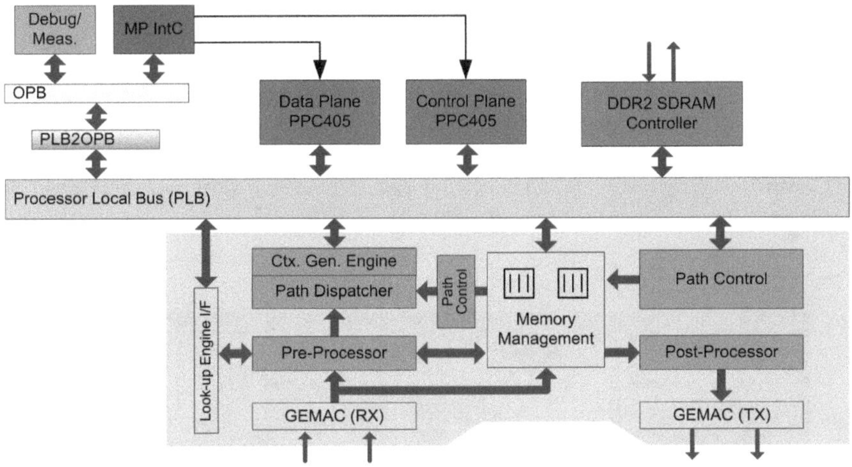

Fig. 17.6 Block diagram of FlexPath NP architecture.

- *Memory Management*: The Memory Management unit is an autonomously working DMA engine, which stores incoming packets in the main memory. A packet descriptor is created and handed over to Path Control and Path Dispatcher. The packet descriptor contains pointers to the packet data in the memory. When receiving a packet descriptor at the transmit side (e.g. from the Path Control), the packet is read out again and sent to the Gigabit Ethernet Media Access Control (GEMAC).
- *Pre-Processor*: The Pre-Processor parses the incoming packets, performs integrity checks (e.g. IPv4 header checksum) and extracts a set of important header fields depending on the protocol stack of the incoming packet. All this information is passed on to the Path Dispatcher for processing path selection. The Pre-Processor may also initiate a hardware next-hop lookup.
- *Path Dispatcher*: Based on the extracted fields from the Pre-Processor, the Path Dispatcher evaluates the runtime reconfigurable rule base for every incoming packet (see [12, 14]) and makes a path decision.
- *Context Generation Engine*: Depending on the outcome of the packet classification, the Context Generation Engine either stores the extracted header fields from the Pre-Processor for later usage within a CPU or hardware accelerator. If the packet is eligible for AutoRoute, the correct instructions for the Post-Processor have to be generated and stored in memory. In this case the packet descriptor is directly sent to the egress Path Control.

- *Path Control*: The architecture of the Path Control is divided into two modules. The ingress part adds a flow ID and a consecutive sequence number to each packet descriptor to preserve the original input sequence. The egress part checks the packet sequence and performs the re-sequencing if needed (see [8, 19]).
- *Post-Processor*: The Post-Processor performs the egress packet manipulations. It is designed as a pipeline engine working at line speed (see [7]). It is programmed by specific assembler like instructions sent along with the packet. These instructions either have to be created by a CPU or the Context Generation Engine in case of AutoRoute.
- *Packet Distributor*: We have not yet implemented a fully functional version of the Packet Distributor as described in Sect. 17.4. In the current state, the function of the Packet Distributor is implemented by the Multi-Processor Interrupt Controller (MP-IntC) and a set of registers that store a single packet descriptor for each provisioned processing path.

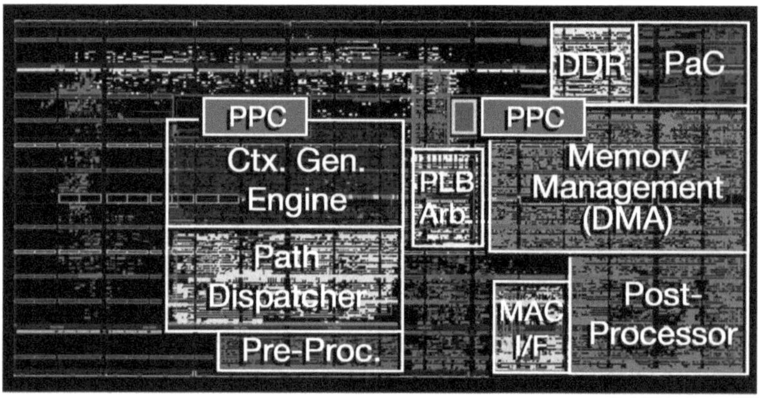

Fig. 17.7 Floorplan of FlexPath NP Demonstrator.

We built up our demonstrator on a Xilinx ML410 prototyping board with a Virtex-4 FX 60 FPGA. The board contains an external DDR2-SDRAM and two Gigabit Ethernet PHYs. The operating frequency of our system is 100 MHz. Besides the two PowerPCs, the FlexPath NP consumes about 16,000 slices out of the 25,280 available slices on the Virtex-4 (63%). Additionally, we are using 77 (33%) built-in SRAM-blocks with 2 kbit each. Figure 17.7 shows the floorplan of our system. Most of the modules are arranged around the central PLB. The most slice consuming modules are the Memory Management, Context Generation Engine, Post-Processor and the Path Dispatcher (cf. Table 17.1).

The goal of this demonstrator is to validate the benefits of a FlexPath NP compared to a processor-centric architecture. As we are limited to using the two PowerPC hard macros provided by the FPGA (soft core CPUs would offer a worse area-performance tradeoff), we can not achieve the same throughput performance

Table 17.1 Resource usage on a Xilinx Virtex-4 FX60.

Module	Slices	BlockRAMs
Memory Management	3285	5
Path Dispatcher	1446	14
Ctx. Generation Engine	1978	4
Pre-Processor	682	1
Post-Processor	1615	3
MP Interrupt-Ctrl.	187	0
Path Control	1288	14
Others/Glue Logic	5525	36
Σ	16006 (63%)	77 (33%)

of commercially available NPs. However, the FlexPath-specific hardware modules have been designed to work at 3.2 Gbit/s (32 bit data path running at 100 MHz in our FPGA design). Both the processing performance as well as the hardware modules could be scaled towards a higher performance, if the design was moved from an FPGA to standard cell ASIC technology.

17.5.2 Experiments

In the following section we show measurement results obtained from the FPGA demonstrator. We especially focus on the following four aspects of a FlexPath NP architecture:

- Pre-classification of packets in hardware (i.e. Path Dispatcher) is helpful to enhance the performance of a NP.
- The FlexPath NP allows to combine the packet distribution techniques of dedicated assignment and packet spraying and exploit the benefits according to the simulations performed in Sect. 17.4.2.
- Additional hardware support by Pre-Processor and Post-Processor and making use of the AutoRoute processing path further enhances the NP performance and relieves the processor complex in accordance with the results obtained in [11, 13].
- We have also built up a common demonstrator with the DynaCORE group from the University of Lübeck (see Chap. 16 and [15]). The DynaCORE was used as an off-chip hardware accelerator for the AES encryption function within our IPsec scenario. We could demonstrate the direct calling of co-processor resources from the CPU and forwarding the encrypted packet back from the accelerator via the AutoRoute path to the actual transmit interface.

17.5.2.1 Demonstrator Setup

We imposed the following traffic mix consisting of two different TCP flows on our NP system. The first flow is a traffic according to the simple IMIX specification [1] with a data rate of 100 Mbps and a resulting packet rate of 34.5 kpps. This traffic can be entirely forwarded on the IP layer, which is an example of a stateless networking application that may be performed either in software or AutoRoute. A second flow consists of a sequence of 512 byte packets with constant packet inter-arrival time, that can be configured to various data rates. This flow is considered to enter an IPsec tunnel at the router, such that the payload has to be AES encrypted and tunnel headers added before the packets can be forwarded towards the egress interface. The IPsec traffic belongs to the class of stateful networking applications, for which we have to insure processing by the same processing element.

The Path Dispatcher can separate the two connections by their IP addresses and assigns different paths for them. Since we differentiate the incoming traffic only based on the IP addresses—and not the full IP 5 tuple—we have to perform a full Security Policy Database (SPD) check in software for all potential IPsec packets. The database is implemented in the IPsec stack and tells whether a packet must be encrypted, forwarded without encryption or discarded at the router.

In our configuration, the data plane CPU 1 can process IP forwarding as well as IPsec packets. CPU 2 is also used as data plane CPU, but only has IP forwarding capabilities. In the following, we examine six configurations:

1. All packets are processed by CPU 1. This is the simplest case without using the FlexPath-specific capabilities and will be used as a reference. CPU 2 is used as control plane processor.
2. All packets are processed by CPU 1 with packet pre-classification by the Path Dispatcher resulting in an assignment to different input queues for IP forwarding and IPsec traffic. There are two different software entry points for packets fetched from either queue. Again, CPU 2 is used as control plane processor.
3. The forwarding traffic is assigned exclusively to CPU 2, while the IPsec traffic remains exclusively assigned to CPU 1.
4. The forwarding traffic is sprayed among CPU 1 and CPU 2, while IPsec traffic is only assigned to CPU 1.
5. CPU 1 is processing IPsec traffic, while the forwarding traffic is routed via AutoRoute.
6. IPsec traffic is processed by both DynaCORE and CPU 1, while forwarding traffic is sprayed among CPU 1 and CPU 2.

17.5.2.2 Single Data Plane CPU (Scenarios 1 and 2)

The results of the first two scenarios are shown in Fig. 17.8. Without any IPsec traffic traversing the NP, the CPU load is 70% in scenario 1, while the CPU load is reduced to only 37% in scenario 2 due to partially offloading the SPD check to the Path Dispatcher. The available processing performance of the system can thus almost be doubled by exploiting the hardware offload possibilities offered by FlexPath. When increasing the IPsec traffic data rate, the CPU load increases rapidly. An additional traffic of 1.6 Mbps (0.4 kpps) in scenario 2 already doubles the load

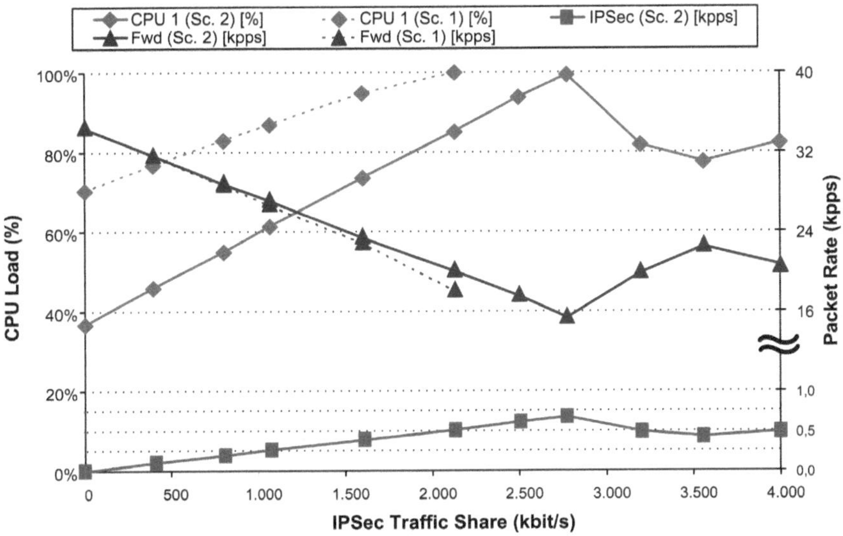

Fig. 17.8 Measurement Results for Scenarios 1 and 2.

of the CPU, which proves that encryption in software is very computationally expensive. At the same time, the output packet rate of the forwarding traffic decreases. All IMIX packets can be served when there is no IPsec traffic in the system. With an IPsec data rate of 400 kbps we already observe a loss rate of 8% for the IMIX packets—even though the CPU load is still below 80%. This effect was traced back to the long processing latency of IPsec packets of more than 1 ms. During this time, the ingress path is blocked. As soon as all input queues are filled, incoming packets are discarded at the system input. This leads to a packet loss even though there is enough processing power left between two IPsec packets.

The CPU load reaches its maximum of almost 100% with an IPsec data rate of 2.8 Mbps in scenario 2. Whereas for smaller IPsec shares there is no IPsec packet loss, we see an increasing loss rate for higher IPsec data rates. Since the processing effort per IPsec packet is high, each lost packet reduces the CPU load significantly, leaving more room for IP forwarding. For higher IPsec input data rates the CPU load and overall forwarding rate vary depending on the IPsec data rate and thus IPsec packet inter-arrival time.

17.5.2.3 Two Data Plane CPUs (Scenarios 3 and 4)

In the following two scenarios, we use the second PowerPC as an additional data plane CPU. In scenario 3 all IPsec packets are assigned to CPU 1 and the entire IP forwarding traffic is assigned to CPU 2. We see a constant processing load of 30% for CPU 2 in Fig. 17.9 independent from the IPsec data rate. No IMIX packets are lost in the NP. The load of CPU 1 starts from zero and increases linearly with the IPsec data rate.

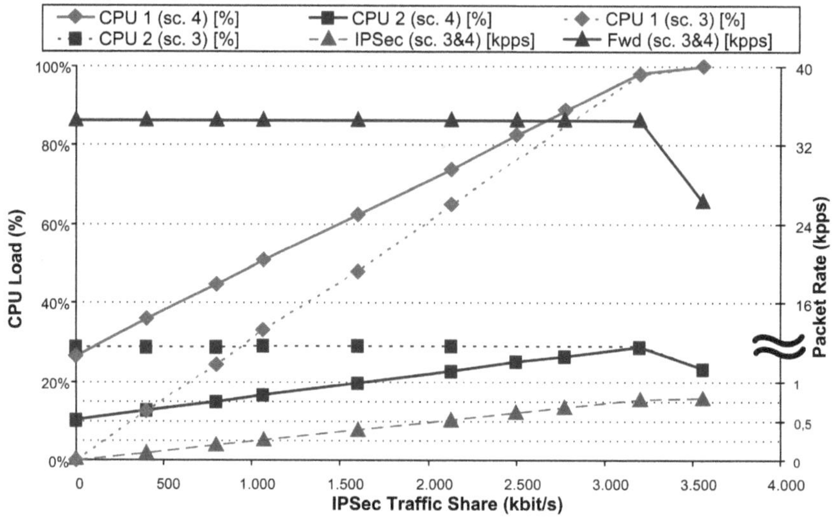

Fig. 17.9 Measurement Results for Scenarios 3 and 4.

In the spraying scenario for the forwarding traffic (scenario 4), the IMIX packets are processed by both CPUs. Without IPsec packets, one would expect the same load for both CPUs. However, we observe a load of 26% on CPU 1 and 10% on CPU 2. The reason for this imbalanced load distribution is due to the specific implementation of the spraying mechanism in our demonstrator. When there is a packet in the spraying queue, the Interrupt Controller will notify both CPUs. The CPU that reacts first by reading out the Interrupt Service Register will get the packet. As long as both CPUs are idle, this will always be CPU 1, since it has the higher bus priority. When increasing the IPsec data rate, CPU 1 will mainly process IPsec packets, while CPU 2 will take over the forwarding packets. The curves for the CPU loads finally approach the values observed in scenario 3 for high IPsec loads.

The cumulative CPU load in scenario 4 with IMIX packets sprayed on both CPUs $(26\% + 10\%)$ is larger compared to the forwarding load in scenario 3 (30%). In contrast to CPU 2, the stack on CPU 1 also includes code for the IPsec implementation. This leads to a larger executable and more cache line replacements, although the forwarding functionality is the same. In addition, interrupting both CPUs in the spraying scenario leads to the execution of the interrupt service routine in both CPUs, although processing is performed only on one CPU. This increases the overall CPU load.

A decline of the IMIX output data rate can be observed in both scenarios, when reaching 100% processor utilization from IPsec packets. At this point, the second

CPU still has sufficient headroom for processing additional traffic. The effect can be explained with head-of-line blocking in the ingress data path pipeline of our NP demonstrator. In the simulation scenario investigated in Sect. 17.4.2, packets arriving at the Packet Distributor with a full queue destination are discarded, keeping the ingress pipeline free for non-blocked destinations. Since the current implementation lacks this feature, packets are discarded at the MAC buffer as soon as one CPU path is blocked. This leads to a discard of packets with arbitrary destination. With a full-fledged Packet Distributor we would not expect a packet loss for the IMIX packets here.

17.5.2.4 AutoRoute (Scenario 5)

In this scenario we use the full hardware AutoRoute path for IP forwarding packets. All IPsec packets are guaranteed to be processed by the CPU. The results shown in Fig. 17.10 are similar to those seen in scenario 3 (Fig. 17.9). The load of CPU 1 is increasing linearly with the IPsec data rate and all IMIX packets are processed in a lossless fashion. When the load of CPU 1 reaches 100%, we see the same head-of-line blocking effect as discussed before. The same throughput as before can be achieved with only a single CPU now. The compute performance of the second CPU is saved by the AutoRoute packet path, which now remains available for another IPsec connection, for example. In this scenario, the AutoRoute path is also not fully loaded; it could handle much more forwarding traffic than the 100 Mbps.

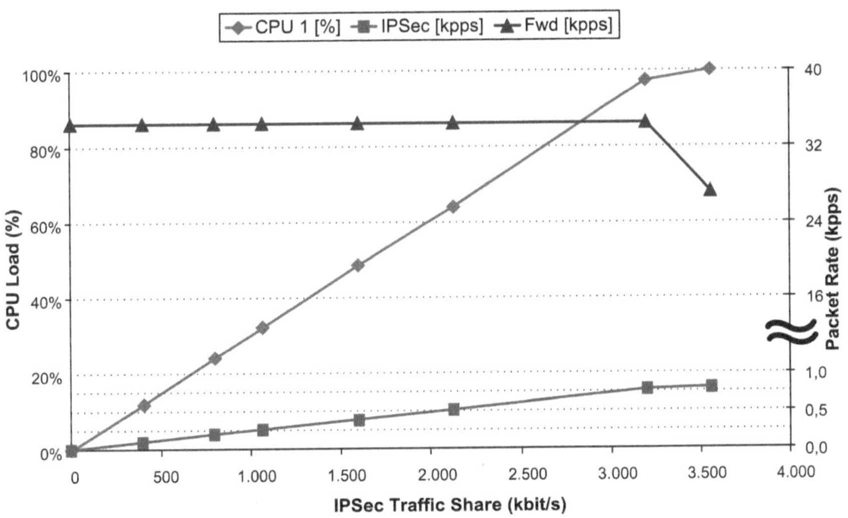

Fig. 17.10 Measurement Results for Scenario 5.

17.5.2.5 DynaCORE (Scenario 6)

In a last step, the FlexPath NP demonstrator was coupled with the DynaCORE system, developed at the University of Lübeck (see Chap. 16 and [15]). As could be seen in the results discussed before, IPsec encryption requires a processing effort, that is about a factor of 300 greater than that of pure IP forwarding. Therefore, it is desirable to move the AES encryption function into a dedicated hardware assist. The remaining parts of the processing, such as header processing, IPsec database checks and tunnel header insertion are still performed on the data plane CPU. Both systems communicate via a standard Gigabit Ethernet link. A proprietary tunneling protocol on top of the Ethernet layer was defined for packet exchange.

An incoming IPsec packet is first sent towards CPU 1 in the FlexPath NP. A modified and smaller stack performs the SPD check and constructs the tunnel header just like in the previous scenarios. Instead of applying the AES encryption, the packet is sent to DynaCORE with the tunneling protocol that also contains information about the connection parameters (i.e. cryptographic key). The two tunnel headers are stored in the packet context and are inserted into the packet by the Post-Processor in order to avoid costly copy operations from the CPU. After transmission on a separate Ethernet port to DynaCORE, the packet payload is encrypted according to the parameters communicated in the DynaCORE header. The processed packet is then sent back towards the FlexPath NP. At this stage, processing of the packet is almost complete. If the Pre-Processor finds an appropriate lookup result for the address in the outer IP header, the packet can actually be AutoRouted towards the final interface. The Post-Processor only removes the DynaCORE header—the remaining header fields can be left unchanged, since they have been prepared in the correct form by the CPU before sending the packet over to DynaCORE. Consequently, the packet has to traverse the CPU only once.

Although the communication between network processor and accelerator via Ethernet is certainly not the most efficient way, it has proven very helpful to couple the two demonstrator systems, that have been developed at two remote sites without much problems.

17.6 Conclusion

In this chapter, we summarized the main contributions of the FlexPath NP project. We have proposed a novel network processor architecture, that uses run-time reconfigurable processing paths to improve the performance. In contrast to most works in the field of reconfigurable computing, we do not reconfigure functional units during system runtime, but change the flow of packets through the system by modifying memory contents of a reconfigurable classification rule base. We can thus extend the scope of reconfigurable systems also to ASIC implementations. We further show how exploiting reconfigurability improves the performance in a field with very high computational and timing requirements. Extensive hardware offload that allows bypassing the CPU cluster for significant shares of the network traffic has been demonstrated. Performing the hardware offload with a range of generic operations leads to a solution that is not restricted to a specific application. We thus preserve the flexibility associated with programmable NP solutions. For our demonstrator we used standard RISC processors instead of proprietary PEs, in order to simplify software development and having access to standard tool chains. Finally, a novel load balancing scheme was proposed that considers the different requirements of several networking traffic classes in the load assignment process.

The FlexPath NP architecture was derived based on an analysis of current Internet protocols and applications. The concept has been evaluated and refined using both formal analysis and SystemC-based simulation platforms. Finally, we have implemented a complete NP demonstrator on an FPGA development board and were able to show by measurements that the expected performance benefits can be achieved in an actual implementation.

Based on the results of this project, we would suggest further research towards reconfiguring CPU resources with dedicated hardware assists (e.g. crypto cores). This has to go along with an appropriate path reconfiguration scheme implemented in the Path Manager. In addition, the concept of reconfigurable processing paths and specific hardware offload could also be extended to NPs with proprietary PEs rather than standard RISC CPUs.

Acknowledgements We would like to thank the following parties for their contributions to the success of this project:

- The German research foundation (DFG) for funding the FlexPath NP project for four years and associating it with the SPP1148 "Reconfigurable Computing".
- Daniel Llorente, who has developed the Memory Manager IP core during the SmartFlow project, funded by the Bavarian State Ministry of Economy, Infrastructure, Transport and Technologies under grant reference IUK 178/001.
- The Xilinx University Partnership program for giving us access to the FPGA development tool chain and Xilinx CoreAlliance IP cores.
- The team of the DynaCORE project from the University of Lübeck, for the cooperation within the SPP 1448 program, that culminated in the common demonstration of our two platforms at the FPL 2008 in Heidelberg.
- CAIDA, for giving us access to real Internet traffic traces used in the NP load balancing simulations. Support for CAIDA's Internet traces is provided by the National Science Foundation, the US Department of Homeland Security, and CAIDA members.

References

1. Agilent: Mixed Packet Size Throughput. http://advanced.comms.agilent.com/n2x/docs/insight/2001-08/TestingTips/1MxdPktSzThroughput.pdf (2001)
2. CAIDA: Anonymized OC-48 Traces. https://data.caida.org/datasets/oc48/oc48-original (2002)
3. CAIDA: Anonymized 2008 Internet Traces (OC-192), https://data.caida.org/datasets/passive-2008 (2008)

4. Dittmann, G., Herkersdorf, A.: Network processor load balancing for high-speed links. In: SPECTS 2002, San Diego, CA, USA (2002)
5. Govind, S., Govindarajan, R., Kuri, J.: Packet reordering in network processors. In: IPDPS 2007, Long Beach, CA, USA (2007)
6. Herkersdorf, A., Claus, C., Meitinger, M., Ohlendorf, R.: Reconfigurable processing units vs. reconfigurable interconnects. In: Dagstuhl Seminar on Dynamically Reconfigurable Architectures, Dagstuhl Seminar Proceedings 06141, Dagstuhl, Germany (2006)
7. Meitinger, M., Ohlendorf, R., Wild, T., Herkersdorf, A.: A programmable stream processing engine for packet manipulation in network processors. In: ISVLSI 2007, Porto Alegre, Brazil (2007)
8. Meitinger, M., Ohlendorf, R., Wild, T., Herkersdorf, A.: A hardware packet resequencer unit for network processors. In: ARCS 2008, Dresden, Germany (2008)
9. Meitinger, M., Ohlendorf, R., Wild, T., Herkersdorf, A.: FlexPath NP—a network processor architecture with flexible processing paths. In: SoC 2008, Tampere, Finland (2008)
10. Ohlendorf, R., Herkersdorf, A., Wild, T.: FlexPath NP—a network processor concept with application-driven flexible processing paths. In: CODES + ISSS 2005, Jersey City, NJ, USA (2005). http://doi.acm.org/10.1145/1084834.1084904
11. Ohlendorf, R., Wild, T., Meitinger, M., Rauchfuss, H., Herkersdorf, A.: Performance evaluation of RISC-based SoC platforms in network processing applications. In: IC-SAMOS 2006, Samos, Greece (2006)
12. Ohlendorf, R., Meitinger, M., Wild, T., Herkersdorf, A.: A packet classification technique for on-chip processing path selection. In: WASP 2007, Salzburg, Austria (2007)
13. Ohlendorf, R., Wild, T., Meitinger, M., Rauchfuss, H., Herkersdorf, A.: Simulated and measured performance evaluation of RISC-based SoC platforms in network processing applications. J. Syst. Architect. **53**(10), 703–718 (2007). doi:10.1016/j.sysarc.2007.01.009
14. Ohlendorf, R., Meitinger, M., Wild, T., Herkersdorf, A.: A processing path dispatcher in network processor MPSoCs. IEEE Trans. VLSI Systems **16**(10), 1335–1345 (2008). doi:10.1109/TVLSI.2008.2002048
15. Pionteck, T., Koch, R., Albrecht, C., Maehle, E., Meitinger, M., Ohlendorf, R., Wild, T., Herkersdorf, A.: SPP1148 booth: network processors. In: FPL 2008, Heidelberg, Germany (2008)
16. Shi, W., Kencl, L.: Sequence-preserving adaptive load balancers. In: ANCS 2006, San Jose, CA, USA (2006)
17. Shi, W., MacGregor, M., Gburzynski, P.: Load balancing for parallel forwarding. IEEE Trans. Networking **13**(4), 790–801 (2005)
18. Sprint Academic Research Group: IP Data Analysis (2009). https://research.sprintlabs.com/packstat/packetoverview.php
19. Traboulsi, S., Meitinger, M., Ohlendorf, R., Herkersdorf, A.: An efficient hardware architecture for packet re-sequencing in network processor MPSoCs. In: DSD 2009, Patras, Greece (2009)

Chapter 18
AutoVision—Reconfigurable Hardware Acceleration for Video-Based Driver Assistance

Christopher Claus and Walter Stechele

Abstract Using dynamically reconfigurable systems makes sense especially when a high degree of flexibility is demanded and the application requires inherent parallelism to achieve real time constraints. The AutoVision architecture is a new Multi Processor System-on-Chip (MPSoC) architecture for video-based driver assistance systems, using run-time reconfigurable hardware accelerator engines for video processing. According to various driving conditions (highway, city, sunlight, darkness, tunnel entrance), different algorithms have to be used for video processing. These different algorithms require different hardware accelerator engines, which are loaded into the AutoVision chip at run-time of the system. The aim of this project was to find out how to use fast dynamic partial reconfiguration to load and operate the right hardware accelerator engines in time (without loosing a single video frame), while removing unused engines in order to save precious chip area.

18.1 Introduction

In future automotive systems, video-based driver assistance will improve safety. Video processing for driver assistance requires real time implementation of complex algorithms. A pure software implementation does not offer the required real time processing, based on available hardware in automotive environments. Therefore hardware acceleration is necessary. Dedicated hardware circuits (ASICs) can offer the required real time processing, but they do not offer the necessary flexibility. Video algorithms for driver assistance are not standardized, and might never be. Algorithmic research is expected to continue in future years. So a flexible, programmable hardware acceleration is required. Specific driving conditions (e.g. highway, country side, urban traffic, tunnel) require specific optimized algorithms. Re-

Christopher Claus · Walter Stechele
Institute for Integrated Systems, Technische Universität München, Munich, Germany,
e-mails: `Christopher.Claus@tum.de`, `Walter.Stechele@tum.de`

M. Platzner et al. (eds.), *Dynamically Reconfigurable Systems*,
DOI 10.1007/978-90-481-3485-4_18,

configurable hardware offers high potential for real time video processing, and is adaptable to various driving conditions and future algorithms.

Today's systems for driver assistance offer features such as adaptive cruise control and lane departure warning. Both video cameras and radar sensors are used. On highways and two-way primary roads a safe distance to previous cars can be kept automatically over a broad speed range. However, for complex driving situations and complex environments, e.g. urban traffic, there are no established and reliable algorithms. This is a topic for future research.

Algorithms for video processing can be grouped into high level application code and low level pixel operations. High level application code requires a high degree of flexibility, and thus is good for a standard processor implementation. Pixel manipulation on the other hand requires applying the same operation on many pixels, and thus seems to be a good candidate for hardware acceleration.

Hardware acceleration for image processing algorithms is beneficial:

- if reading/writing of consecutive areas in memory is demanded. This allows for burst transfers of pixel data.
- if the processing of huge pixel neighborhoods using regular scans over the image is required. The central pixel and its complete neighborhood could be processed in one hardware clock cycle due to inherent parallelism.
- if the algorithm requires multiple scans over input and intermediate images, which can be done using a deeply pipelined structure.
- if the processing in hardware leads to a significant data reduction. Then embedded CPUs can continue to work on this intermediate result.

In the AutoVision project only the performance intense parts are accelerated by coprocessor engines, which usually leads to a significant data reduction. The rest of the algorithm, the high level application code, is implemented fully programmable on one standard PowerPC (PPC) CPU cores to remain easily updateable and allow flexibility for new algorithms. In addition, the application code running on one of two available PPCs is able to trigger a reconfiguration process. Hence the coprocessors available on the system can be exchanged during run-time, which allows a much larger set of hardware accelerated functionality than would normally fit onto a device. This process makes use of the partial dynamic partial reconfiguration capabilities of Xilinx Virtex FPGAs. If a coprocessor is exchanged the remainder of the system, containing for example the video-input and video-output, remains fully operational. Beside supplying the reconfiguration controller with the information where to fetch the reconfiguration data (partial bitstream) the second PPC is responsible for verifying and visualizing the reconfiguration.

In Sect. 18.1.1 recent work is surveyed. Section 18.1.2 presents a typical example where reconfigurable video processing can be applied. In addition, the coprocessor engines are introduced. An overview of the AutoVision architecture is given in Sect. 18.2. The concepts and optimizations that led to the utilization of Dynamic Partial Reconfiguration (DPR) in video-based driver assistance systems is presented in Sect. 18.3. The results obtained are presented in Sects. 18.4 and 18.5. Finally this contribution is concluded with an outlook on further plans in Sect. 18.6.

18.1.1 State of the Art & Related Work

The idea of using hardware acceleration as a solution for computationally intensive applications for real-time visual recognition has been developed many years ago. Since then, companies begun to manufacture System on Chips (SoCs) to be used in automotive or robotic environments. One of them is Mobileye [16], who presented their EyeQ chip with two ARM cores and 4 dedicated coprocessors for object classification, tracking, lane recognition and filter applications. Dedicated hardware circuits (ASICs) can offer the required real-time processing, but they do not offer the necessary flexibility. Video algorithms for driver assistance are not standardized, and may never be. Algorithmic research is expected to continue in the future. Thus a flexible, programmable hardware acceleration is needed. In their second generation of the EyeQ chip (EyeQ2), Mobileye introduced three programmable Vector Microcode Processors to enhance the flexibility of the system.

But the problem, that unused coprocessors persist on the device while they are not needed and thus occupying a lot of resources remains. FPGAs can be used to cope with that problem by updating their configuration information when needed as already mentioned in Sect. 18.1.1. Various authors addressed the problem of fast DPR, to be used in the signal and image processing domain.

Manet et al. [15] present an evaluation of DPR for real signal and image processing applications. Although some other applications are mentioned, the focus is on applying DPR for Software Defined Radio (SDR). For fast reconfiguration in their system an Internal Configuration Access Port (ICAP, see Sect. 18.3.3.1) controller for Virtex-4 devices has been implemented, which achieves a throughput of 350 KB/ms at a frequency of 100 MHz. This is close to the theoretical maximum of 400 KB/ms at 100 MHz on Virtex-4, as the input data width of ICAP is 4 bytes.

Another project that requires fast DPR is reported in [17]. Shelburne et. al. present the Metawire approach that uses fast DPR to emulate a Network-on-Chip (NoC). As especially the NoC router nodes consume a lot of resources on an FPGA, the Metawire architecture uses the configuration circuitry as a relatively high-performance NoC. The configuration information of a block RAM (BRAM) is read by the ICAP and written to another BRAM afterwards. To accelerate this process an overclocked Virtex-4 ICAP is used, which is capable of providing a bandwidth in excess of 200 KB/ms.

In [3] Bomel et al. present the fast downloading of partial bitstreams for DPR via a standard Ethernet framework on a Virtex-II Pro device. Similar to the implementation presented in [15], Bomel et al. use an ICAP controller attached to the On-chip Peripheral Bus (OPB). Their measurements show an obtained ICAP throughput of up to 50 KB/ms (400 Kbit/ms).

Contrary to the FlexPath project (Chap. 17) where reconfiguration is performed by switching between different processing paths according to various packet types, in AutoVision complete functional units, namely the coprocessor engines, are physically exchanged on the device. Further differences between the abstraction level concerning the reconfiguration of interconnects and functional modules are described in [12].

18.1.2 Typical Scenario & Hardware Accelerators

In general, a large number of different coprocessors could be implemented on a System-on-Chip (SoC) in parallel. However, this is not resource efficient, as depending on the application, just a subset of all coprocessors will be active at the same time. In this section a typical scenario is presented which shows that it is necessary for the hardware to adapt to a changing environment. The hardware accelerators used for this scenario were implemented for a proof-of-concept study.

With hardware reconfiguration, hardware resources can be used more efficiently. Coprocessor configurations can be loaded into an FPGA whenever needed, depending on the application. This especially makes sense when the driving situations are mutually exclusive (day-time/night-time driving, forward/backward driving, highway/urban environments).

The following typical driving scenario is considered: A car is driving at daytime on a highway, where other cars can be detected by feature points on their silhouette or shape. Then the car is approaching a tunnel. Here it is meaningless to search for feature points as it is more important for the driver to see what is inside the tunnel. Thus an algorithm for contrast enhancement on the dark tunnel entrance is desirable. Inside the tunnel, due to the low luminance level, only the lights of other vehicles are promising features and have to be distinguished from tunnel lights. Therefore another algorithm is used. When the car is leaving the highway and enters an urban environment, a detection of pedestrians according to their motion seems beneficial. This scenario could be arbitrarily extended by other situations such as changing weather conditions (rain, fog, snow etc.). In the AutoVision project this specific typical scenario is used as a proof of concept.

For the image processing in the situations described above several different hardware accelerators for pixel processing, so called Engines, are required. The Engines are attached as bus masters to the Processor Local Bus (PLB). This design allows for direct memory access (DMA) of the Engines without involving the PowerPC (PPC) cores in the pixel transfer, which leads to a great offload of the CPUs. Simultaneous read and write transfers are supported by the two separate read and write data busses of the PLB, each 64-bit wide.

The extraction of feature points on the shape of a car for instance is done by a hardware accelerator called the ShapeEngine. The contrast enhancement near gloomy tunnel entrances is done by the ContrastEngine. The pixel-level processing inside the tunnel can be accelerated by a coprocessor, called the TaillightEngine. Finally in urban environments, Optical Flow, which is calculated using two separate accelerators, can be used to detect moving objects, such as pedestrians or cyclists. A detailed description about the organization and performance of the TaillghtEngine, the ShapeEngine and the OpticalFlow can be found in [1, 10, 11] respectively. An implementation of the TaillightEngine on the Erlangen Slot Machine, which originates from a collaboration between the ReCoNodes (see Chap. 3) and the AutoVision project, can be found in [2]. The input images and the processed output from all the engines described above can be seen in Fig. 18.1.

As can be seen from this scenario, different driving conditions require different algorithms and thus different accelerators for image processing. As most of these situations are mutually exclusive it is meaningful to load the desired functionality only if required instead of implementing all Engines for the image processing algorithms in parallel.

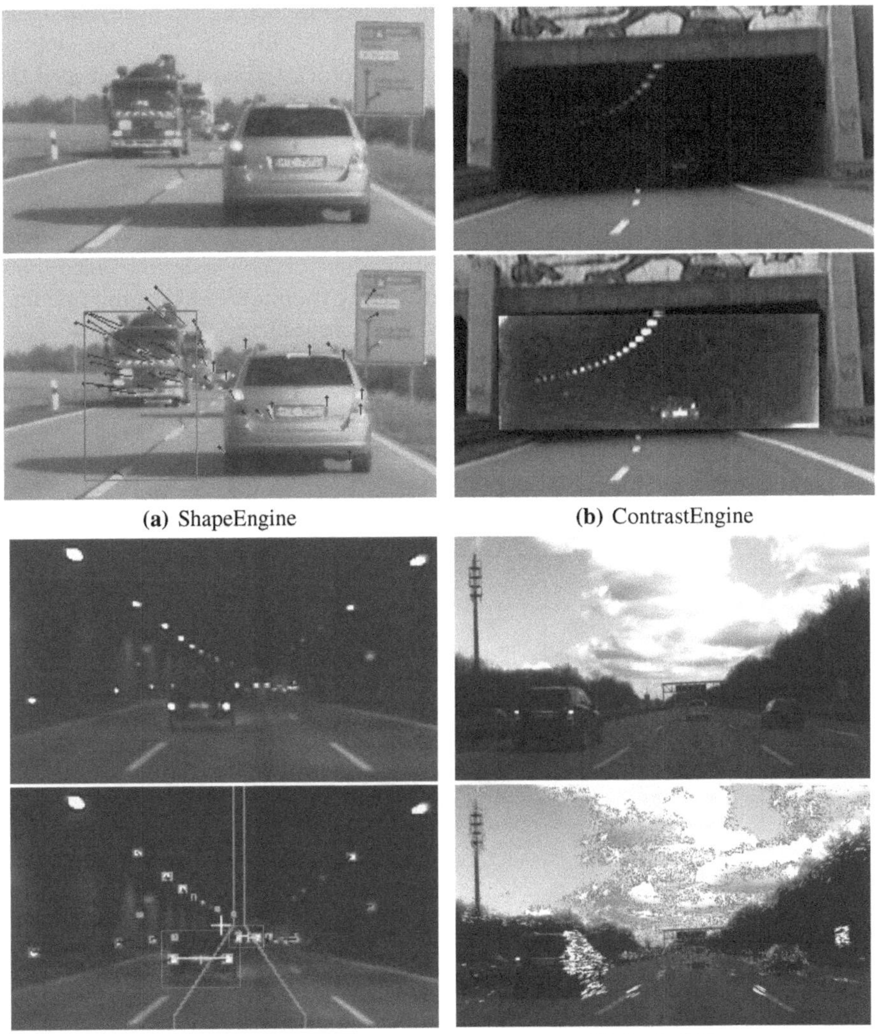

(a) ShapeEngine

(b) ContrastEngine

(c) TaillightEngine

(d) Optical Flow

Fig. 18.1 Unprocessed input and processed output of the ShapeEngine (a), the ContrastEngine (b), the TaillightEngine (c) and the Optical Flow (d).

18.2 AutoVision Architecture

The AutoVision system has been implemented on a Xilinx Virtex-II Pro device with two embedded PPC cores. The incoming pixel data is buffered in on-chip RAM resources (BRAMs) inside the Video-Input core. If enough data is buffered for a burst transfer, the Video-input core transfers the data into the DDR SDRAM, which acts as frame buffer. The DMA-capability of the Video-input core, and using bursts instead of single-pixel-transfers, leads to a reduced PLB utilization. The same applies to the Video-Output core which is used to fetch processed data and send it to the output pins of the FPGA e.g. for displaying it on an external monitor.

In [8], the Erlangen Slot Machine (ESM) (see Chap. 3) was compared against the Virtex-II Pro based system from the AutoVision project in terms of suitability for reconfiguration and embedded video processing. By combining the advantages of both platforms a modified platform for reconfigurable embedded video processing has been developed. In this section the modifications, namely a digital Video input and output, a combination of on-chip BRAM and off-chip SRAM for intermediate pixel storage and the usage of an embedded CPU are summarized.

In the current AutoVision demonstrator, the analog input has been replaced by a fully digital interface, which delivers 25 frames per second in progressive scan mode, and connects a CameraLink camera via a custom accessory board to the FPGA pins. The IP core *Video-Input*, responsible for writing the incoming pixel data into the DDR SRAM, is connected to the PLB, as can be seen in Fig. 18.2.

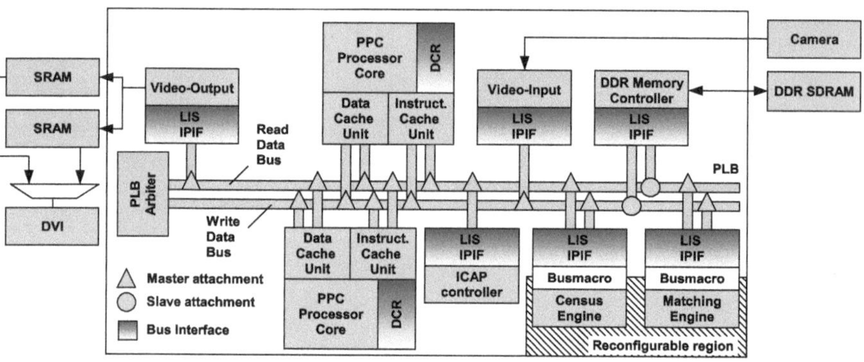

Fig. 18.2 Blockdiagram of the AutoVision Project.

Various tests have proven that it is beneficial to use fast RAM (on-chip block RAM) for convolution processes. Instead of loading the same pixel various times from the main memory, storing a few pixel lines inside a local memory attached to the hardware accelerator represents a resource efficient alternative. Details about the modular concept of the hardware accelerators can be found in Sect. 18.2.1. Processing images in that way is almost as fast as storing the complete image inside the expensive on-chip block RAM.

One important design consideration was to use a random access framebuffer to display image data on a monitor using standard frequencies. This was necessary in order to cope with the problem that the image data has to be transferred multiple times over a central bus or crossbar if the output frequency (60 Hz, 75 Hz etc.) is higher than the input frequency (25 Hz, 30 Hz etc.). Once the image data to be displayed has been sent to another custom accessory board on which two SRAMs for double buffering are used, it can be displayed at any desired frequency. As soon as an image has been processed it can be written into one of the framebuffer SRAMs, while the image stored in the other SRAM is read out and displayed on the monitor.

In [8] it has been concluded that an on-chip CPU is beneficial in order to gain fast access to the image data from both hardware and software. Instead of using a central bus or crossbar which connects the main memory, the CPU and the accelerators, another solution has been implemented. By connecting various HW accelerators, or the ICAP controller directly to a Multiport Memory controller, a tremendous speedup due to longer burst transfers and address pipelining can be achieved.

By utilizing the reconfiguration controller described in Sect. 18.3.3.1 it is possible to ensure that the reconfiguration is fast enough, so that during this process no video frame is dropped.

For additional information on the AutoVision architecture the interested reader is referred to [6] and [7], where the AutoVision concept is presented.

18.2.1 The AddressEngine—A Pixel Processing Pipeline

The AddressEngine (AE) can be seen as a wrapper around a user-defined pixel operation. It is a pixel processing pipeline that can independently initiate transfers from and to the DDR SDRAM for fast pixel access (fetching and storing pixels). The pixel data is transferred from the DDR SDRAM to the AE using direct memory access (DMA). After the AE has received the start address of the picture to be processed, it fetches the pixel data, processes it and writes it back into the main memory. The pixel data to be processed does not necessarily have to be a full image. The processing solely of a region of interest (ROI) within an image is also supported. A maximum of 16 64-bit words when connected to a PLB via the LIS IPIF are grouped in a full burst. By connecting the AE directly to the Multiport Memory Controller (MPMC) via the LIS Native Port Interface (LIS NPI), 32 64-bit words can be transferred in one single burst. The maximum resolution currently supported is an image size of 1024×1024 pixels, independent of the color depth of the pixels.

Figure 18.3(a) shows the architecture of the *AddressEngine*. In the basic configuration it consists of six main modules, namely the *InputFSM*, the *Local Input memory*, the *Matrix*, the *User Logic*, the *Local Output Memory* and the *OutputFSM*. A bus interface, the so called LIS-IPIF [19], can be used to attach the *AddressEngine* to a PLB bus.

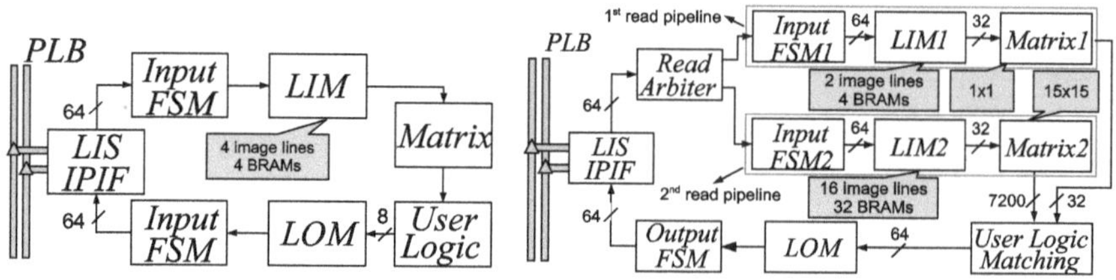

(a) Blockdiagram AddressEngine (b) Blockdiagram MatchingEngine

Fig. 18.3 Blockdiagrams of the AdressEngine in basic configuration and the MatchingEngine.

The module *InputFSM* is mainly responsible for reading the image data and storing it temporarily into the *Local Input Memory* (*LIM*). This includes sending requests and providing the calculated addresses to the main memory.

The *LIM* consists of BRAMs and is used as intermediate storage for all pixels currently required by the processing function. This is beneficial because if the same pixel has to be accessed multiple times from the external memory (e.g. when neighborhood operations are required) the processing time can be reduced dramatically by utilizing a local memory. Thus the number of bus transfers is reduced significantly.

The *Matrix* represents the processing window and is shifted from left to right over the image. It consists of registers in order to access one pixel and its complete neighborhood in one clock cycle. If the *Matrix* jumps into the next row the next complete image line is requested from the main memory. Hence, this avoids stalling the complete pixel processing pipeline.

The *User Logic* can be understood as a user-defined pixel operation on a central pixel and its neighborhood. Depending on the implementation of the *User Logic* the AE is responsible for feature point detection, contrast enhancement or the calculation of the Optical Flow.

To avoid stalling the pixel processing pipeline at the output, a *Local Output Memory* (*LOM*) lying between the *User Logic* and *OutputFSM* is necessary. It is used for collecting the result pixels and preparing a burst transfer to the main memory, which results in a lower bus utilization compared to the transfer of individual pixels.

Finally, the *OutputFSM* is responsible for setting up the process to write the result pixels back into the main memory at the correct position.

By extending the basic configuration described above and by flexibly connecting the main modules with each other more powerful hardware accelerators, such as the ShapeEngine, TaillightEngine or Optical Flow can be generated. Introducing so called *Intermediate Local Memory* (*ILM*) further deepens the image processing pipeline and prevents intermediate pixel data from being written back into the DDR SDRAM when pixel processing is not completed yet.

In Fig. 18.3(b) a block diagram of the MatchingEngine is depicted. The MatchingEngine is one of two building blocks of the Optical Flow core. The main functionality of the MatchingEngine is to search for corresponding pixels in consecutive frames. The accelerator searches within a predefined window (15×15) and is able to deliver one result every clock cycle. One major difference between the original processing pipeline of the AE and the MatchingEngine is that the input path of the latter must read data from two separate images. Thus the input processing path consists of two read pipelines. The read arbiter and the two Input FSMs fill both read pipelines. The User Logic of the MatchingEngine is configured to check whether a correspondence between pixels in consecutive frames within the predefined 15-by-15 window is found. Another elementary difference between the AE and the MatchingEngine is that the former writes back a complete image to the DDR SDRAM, while the latter one exports a list of correspondences.

By analyzing the conceptual differences between the AE in Fig. 18.3(a) and the MatchingEngine in Fig. 18.3(b) it is obvious that both Engines are fundamentally different. Although all of the Engines are based on the AE, their internal structures greatly differ from each other. Therefore a reconfiguration of the complete functional block has to be performed in order to save resources rather than just parameterizing the functionality by updating configuration registers.

18.3 Fast Dynamic Partial Reconfiguration

Dynamic partial reconfiguration (DPR) is the ability to reconfigure a certain portion (partial) during run-time (dynamic) of the device. Currently this feature is only offered by Xilinx Virtex and Spartan devices. Thus some of the approaches (e.g. bitstream modification) described in this section are Xilinx specific but can be applied to other future devices. However, the approaches for higher configuration data throughput and interconnect optimizations are generally applicable.

18.3.1 Motivation for Fast Reconfiguration

A requirement for future video-based driver assistance, especially when used for safety critical applications, is that video frames must not be dropped. This requirement again leads to the fact that the reconfiguration of an Engine has to be as fast as possible , or at least so fast that the real-time requirement (processing of at least 25 fps) is not violated. Depending when and how often the system has to be reconfigured, the reconfiguration process can be separated in Inter Video Frame Reconfiguration and Intra Video Frame Reconfiguration.

18.3.1.1 Inter Video Frame Reconfiguration

The Inter Video Frame Reconfiguration (InterVFR) is defined as exchange of a reconfigurable module *between* two consecutive video frames. InterVFR is required to e.g. exchange the ContrastEngine with the TaillightEngine in the typical scenario described in Sect. 18.1.2. The time to process an image with the ContrastEngine is denoted as T_{i1}. The processing time required by the TaillightEngine is denoted as T_{i2}. The reconfiguration time for the exchange is denoted as T_R in Fig. 18.4(a). If $T_{i1} + T_R < 40$ ms, then interVFR is possible.

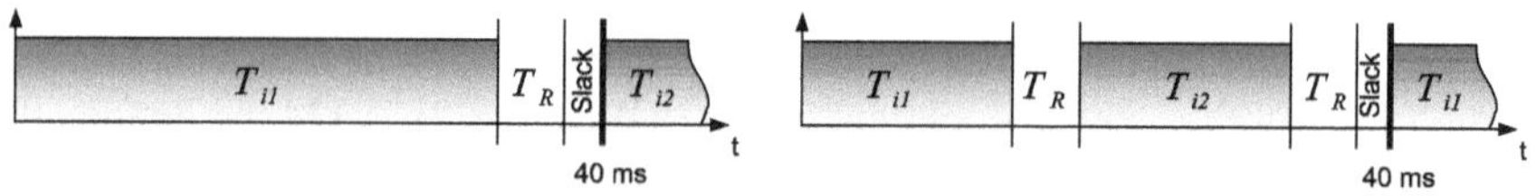

(a) interVFR **(b)** intraVFR

Fig. 18.4 Inter and Intra Video Frame Reconfiguration.

18.3.1.2 Intra Video Frame Reconfiguration

The Intra Video Frame Reconfiguration (IntraVFR) is defined as exchange of reconfigurable modules *within* one video frame. The hardware accelerator for the Optical Flow can serve as an example here. It consists of two engines, namely the CensusEngine and the MatchingEngine, which are executed sequentially. The time to process an image with the CensusEngine is denoted as T_{i1}. The processing time required by the MatchingEngine is denoted as T_{i2}. The size of the partial bitstreams for the CensusEngine and the MatchingEngine is considered to be the same. Thus the reconfiguration time for both Engines is equal and denoted as T_R in Fig. 18.4(b). If $T_{i1} + T_{i1} + 2xT_R < 40$ ms, then intraVFR is possible.

18.3.2 Bitstream Modification

Basically there are two ways to decrease the size of a bitstream: using compression techniques and removing unnecessary data. While almost all approaches in literature use the former (e.g. [13]), the Combitgen tool [4] removes redundant configuration frames. Bistream compression can then be applied to further reduce the bitstream size, if desired.

18.3.2.1 Combitgen

A frame is the smallest accessible entity on the FPGA. It is one bit wide and has a height of the complete device in Virtex-II, a height of 16 CLBs in Virtex-4 and 20 CLBs in Virtex-5. Configuration frames can be grouped into so called columns. These represent a number of consecutive frames, which are used to configure a CLB, a Block-RAM, a Block-RAM Interconnect switch matrix, DSP silces etc. In order to write frames into the configuration memory a pipelined approach is used as shown in Fig. 18.5.

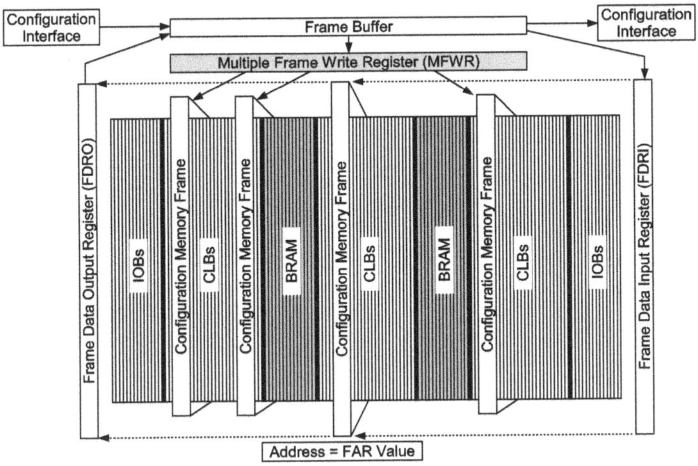

Fig. 18.5 (Re-)configuration process: The bitstream data can either be shifted from the frame buffer into the FDRI or copied into the MFWR. From both registers the bitstream data is written in the configuration memory in one clock cycle.

The first frame of data is written into the frame data input register (FDRI). A second frame is shifted through the frame buffer while the entire first frame is loaded in parallel into the configuration memory latches. To guarantee that the last frame of data is written into the configuration memory, a pad frame is required. Thus, writing a single frame of data in that manner requires at least two frames: the data frame itself, followed by the pad frame to flush the data frame out of the FDRI.

In principle, two flows are available to generate partial bitstreams. Module-based and Difference based [18]. The Module-based flow is used when *more than two* modules should be exchanged during run-time. Depending on the area defined by the user constraints a partial bitstream is created containing all the configuration frame information of this area. The Module-Based Reconfiguration Flow is only able to write whole columns into the bitstream, rather than just frames. In summary this flow can create partial bitstreams for an arbitrary number of modules but might contain large amounts of unnecessary data.

The Difference-based Reconfiguration Flow is recommended by Xilinx only if small design changes between *exactly two designs* are needed. In this flow a bitstream is created containing only the differences of the two different designs. Thus single, often non-continuous configuration frames make up the partial bitstream. Contrary to the module based flow, this flow can create partial bistreams for at most two modules but with almost zero overhead.

Combitgen combines the advantages of these design flows into one program, while attempting to avoid their disadvantages. It reads several toplevel bitstreams, processes them and writes out one partial bitstream for each of the toplevel ones that are able to reconfigure from any other toplevel to this one. The bitstreams are read and executed like the on-chip configuration logic would, as can be seen in Fig. 18.6.

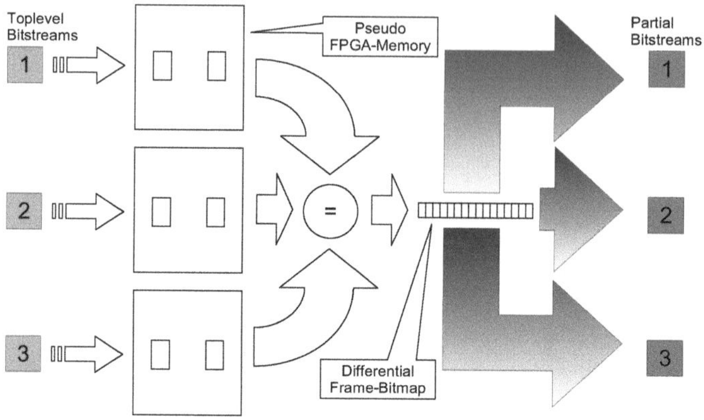

Fig. 18.6 Data Flow of Combitgen with 3 toplevel bitstreams as input and 3 resulting partial bitstreams with unnecessary pad data removed.

The configuration data extracted form every toplevel bitstream is instead written in a separate array, the so called pseudo FPGA configuration memory. After all toplevel memories are filled they are compared frame by frame. If there is a difference found in a frame between any of the numerous toplevel bitstreams it is marked in a differential frame bitmap. This contains one bit for every frame on the FPGA. After the comparison process is complete, all these marked frames are written out into partial bitstreams, one for every toplevel bitstream. Unlike the partial bitstreams created by the Module-based flow, the partial bitstreams created by Combitgen contain single frames instead of columns, which can greatly reduce the amount of configuration data within a partial bitstream. However, it remains possible to be able to configure from any toplevel possible to any other remains.

In order to further decrease the bitstream size, the Multiple Frame Write (MFWR: write only one frame to the MFWR register and copy it to multiple addresses, see Fig. 18.5) command is used whenever there are identical frames on the chip. This feature is provided by the Xilinx tools and can be used to get rid of the problem that when writing a single frame to the configuration memory another one is required to flush the pipeline. Unlike the Xilinx tools for bitstream generation, Combitgen also uses the MFWR command to write single unique frames without the requirement of pad data, and thus can save the extra padding words and achieve both a smaller bitstream size and a shorter reconfiguration time.

18.3.3 Hardware Modification

18.3.3.1 ICAP Controller

To utilize the *Internal Configuration Access Port* (*ICAP*), which allows read and write access to the configuration data, a controller has been implemented to achieve a throughput rate close to the theoretical maximum. This controller, presented in [9], consist of some logic to load the bitstream data, stored in an external memory, and provide the maximum amount of data per clock cycle to the ICAP. The ICAP Input Width is 8 bit in Virtex-II and 32 in Virtex-4 and Virtex-5 devices. In [9], a frequency of 100 MHz was used to send data to the ICAP, although ICAP on VirtexII is specified for 66 MHz. However, higher frequencies are possible by using a simple handshaking protocol. By indicting a *busy* status, the ICAP notifies the controller that it currently is not able to process incoming data. Compared to the implementation in [9] some additional optimizations have been made. The ICAP controller can now be easily connected to any Virtex-II, Virtex-4 or Virtex-5 device. The ICAP input width (IIW), as well as the burst size BS can be configured.

Most of the approaches described in literature use a FIFO as intermediate buffer. In order to obtain higher throughput and thus shorter reconfiguration times many authors concentrate on techniques to keep the FIFO filled all the time. As can be seen in Eq. (18.2) another possibility to obtain higher throughput is to manipulate the frequency. In literature only Shelburne et al. [17] use the ICAP beyond the specified 100 MHz. To safely pass data from one clock domain to another asynchronous clock domain, asynchronous FIFOs are used. To guarantee that the FIFO used for intermediate storage of the bitstream data remains filled all the time, high speed interconnects are necessary. Therefore it is possible to connect the ICAP controller either to a PLB Bus or to a Multiport Memory controller (MPMC) via custom interfaces without any modification (see Sect. 18.3.3.2). To verify that the FIFO remains full all the time, Eq. (18.1) has to be fulfilled. The left side of the equation determines how many bits per second can be written into the FIFO, which of course is dependant on the frequency the FIFO is fed with. The right side of the equation determines how many bits per second can be read from the FIFO and fed into the ICAP.

$$\frac{BS\ [\text{cycles}] * DW\ [\text{bit}]}{BS\ [\text{cycles}] + L\ [\text{cycles}]} * f_1\ [\text{MHz}] > IIW\ [\text{bit}] * f_2\ [\text{MHz}] \tag{18.1}$$

The number of bytes that can be written to the *ICAP* per clock cycle is dependent on the *ICAP Input Width* (*IIW*). As long as the asynchronous FIFO is full, *IIW* bits can be written into the ICAP per clock cycle. The *data width* (*DW*) determines the input and output width of the asynchronous FIFO and is dependent on the incoming data. If a 64-bit wide Processor Local Bus is used, to connect the DDR SDRAM memory and the ICAP controller, the input data width DW of the ICAP is set to 64. The *burst size* (*BS*) determines how many packets with a size of DW are transferred within a burst. If BS is set to 16 and DW is considered to be 64 bit, 128 bytes can be transferred in one burst. The memory access *latency* (L) is used to specify the number of cycles from the point the data is requested until the first word appears at the input of the ICAP controller. This value is strongly dependent on the implementation of the memory controller. Once it has been assured that the FIFO remains full during the whole configuration process, the throughput TP can be calculated using Eq. (18.2).

$$TP\left[\frac{\text{byte}}{\text{s}}\right] = f\ [\text{Hz}] * IIW\ [\text{byte}] * BF[0,1] \tag{18.2}$$

The frequency used to write data to the asynchronous FIFO is the 100 MHz frequency from the PLB. The frequency used to read from the FIFO can be any frequency that can be generated with a Digital Clock Manager (DCM). Results of the throughput obtained by utilizing the ICAP controller on Virtex-II Pro and Virtex-4 devices with an asynchronous FIFO as described above, can be found in Sect. 18.4.

18.3.3.2 Optimizing Throughput Through Modular Design

As mentioned in Sect. 18.3.3.1 the ICAP controller can be attached to either a PLB via the LIS IP interface (IPIF) [19] or to the MPMC via the LIS Native Port Interface (NPI). As the *IIW* in Virtex-4 and Virtex-5 devices has been increased to 32, it is necessary to make sure that the FIFO remains filled and 32 bit per clock cycle can be provided to the ICAP in order to achieve the maximum throughput possible in Virtex-4 devices. For these tests a Virtex-4 FX device on an ML405 development board from Xilinx was used. This device contains an embedded PPC core and the bitstreams are stored in an DDR2 SDRAM which is accessed via the MPMC. Two major changes in the MPMC allow for faster data transfer from the memory into the ICAP's FIFO. The MPMC allows burst sizes of up to 32 64-bit words. In addition, address pipelining is supported which can be used to reduce the latency on the bus by overlapping a new request with an ongoing transfer. Longer burst transfers and address pipelining help ensure a proper filled FIFO, so that the maximum amount of data can be fed into the ICAP every clock cycle.

In Fig. 18.7, three possible connections of the ICAP controller to a DDR or DDR2 SDRAM are depicted. In Fig. 18.7(a) the ICAP controller is connected to the PLB via the LIS IPIF. On the ML405, the DDR2 SDRAM controller from Xilinx is used to interface the memory. As depicted in Fig. 18.7(b) the DDR2 SDRAM controller has been replaced by the MPMC. This setup was not further considered because significantly higher throughput rates were obtained by using the setup shown

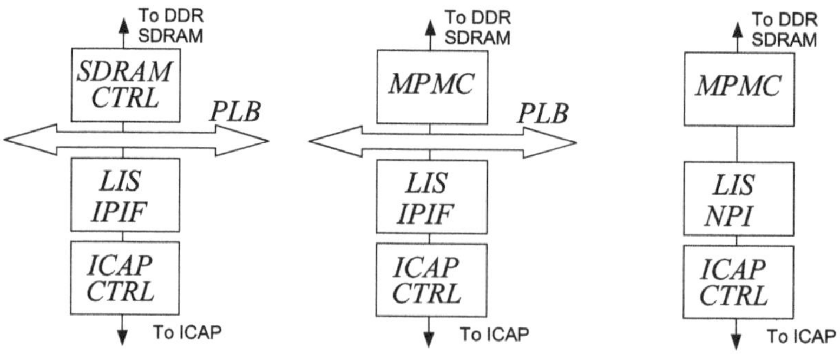

Fig. 18.7 Different interconnects used to connect ICAP and DDR SDRAM.

in Fig. 18.7(c). Here the ICAP controller is connected directly to the MPMC via the LIS NPI. The throughput of each of these setups by using different burst sizes was measured and is shown in Table 18.1.

Table 18.1 Different connections of the ICAP controller and their performance at 100 MHz on an ML405.

Setup	Interface	BS	AP	L	f	T_P
Figure 18.7(a)	LIS IPIF	16	No	31	100	281.93
Figure 18.7(c)	LIS NPI	16	No	19	100	350.68
Figure 18.7(c)	LIS NPI	16	Yes	19	100	400
Figure 18.7(c)	LIS NPI	32	No	19	100	400
Figure 18.7(c)	LIS NPI	32	Yes	1	100	400

Due to the modular design of the ICAP controller and the Engines no changes are necessary when attaching them either to the LIS IPIF or the LIS NPI. Different burst sizes for instance are supported by synthesis attributes (generics).

18.3.3.3 Bitstream Verification

If a device is reconfigured twice within 40 ms (intraVFR) a visual verification such as the approach presented in [14] is not possible anymore. Thus another method to verify the correctness of the reconfiguration process is used. After every reconfiguration, the configuration memory is read back and compared against the bitstream that was used to reconfigure the device. As some primitives, such as registers or BRAMS change their state, this information has to be masked out. The read back is done by the ICAP controller which writes the complete or parts of the configuration memory to the DDR SDRAM. Afterwards a CPU can compare the read back data with the initial bitstream used to configure the device.

As one PowerPC is already busy with the image processing the second PowerPC will take over the task for bitstream comparison and thus extending the AutoVision architecture to an MPSoC. Beside the comparison of bitstreams the second PowerPC is responsible for the reconfiguration management and can also be used to decrypt the bitstream in order to visualize the updated FPGA configuration as shown in [14] during interVFR. The configuration visualization and the novel ICAP controller have been designed and implemented in an interdisciplinary project between professor J. Becker's and our group.

To verify that the device was configured correctly three test were used. The first test was used to verify that the functionality is correct. In another test the on-chip logic analyzer (Chipscope ILA) was used to monitor the correct command and data sequences. Finally a readback of the configuration data via the JTAG Port has been performed. The configuration data read back is used to compare against the initial bitstream. Utilizing this verification method, changes on bitlevel can be easily detected. The frequency on the read side has been increased until the configuration process was not stable anymore.

18.4 Results

A framework to speed up dynamic partial reconfiguration, by utilizing Combitgen and the PLB ICAP controller has been presented in [5]. Further information about how Combitgen and the PLB ICAP controller reduce the total reconfiguration time can be found in this publication. I has been shown that the maximum throughput for Virtex-II devices has been achieved. In order to achieve the same for Virtex-4 devices, the transfer from memory to ICAP has been optimized. As described in Sect. 18.3.3.2 various test to obtain the maximum throughput by connecting the ICAP controller either to the PLB or directly to the MPMC have been performed. As depicted in Table 18.1 the reconfiguration throughput T_P has been measured for the setups shown in Fig. 18.7.

As can be seen in Table 18.1 the maximum T_P could not be achieved without address pipelining AP or increasing the burst size to 32. Only by connecting the ICAP through the LIS NPI 400 KB/ms at a frequency of 100 MHz are possible. As reported in Sect. 18.3.3.1, beside optimizing the transfer from the memory to the ICAP, the frequency used to read from an asynchronous FIFO and provide the data to the ICAP can be increased.

As the ICAP on Virtex-II Pro is specified only up to 66 MHz but no upper boundary is provided, the maximum frequency for ICAP on Virtex-II has been determined. Table 18.2 shows the result of an ICAP controller tested on different Virtex-II devices. In all designs the same size for the reconfigurable region was used, which results in the same bitstream size of 72064 bytes for all the designs. A hardware counter was used to measure the number of cycles for each reconfiguration process. The busy factor BF has been determined by several measurements. Up to a frequency of 140 MHz the reconfiguration on various platforms was successful. Due to production tolerances some of the tested devices achieved a frequency of 150 MHz. But the universally valid maximum frequency for the Virtex-II Pro devices that have been tested is 140 MHz. This has been verified by reading back the configuration data and comparing it against the initial bitstream, as described in Sect. 18.3.3.3. The results from the tests which have been performed to determine the maximum frequency on Virtex-II Pro are depicted in Table 18.2.

At a frequency of 140 MHz a throughput T_P of 129.97 MB/s has been achieved. The same tests are going to be performed on Virtex-4 and Virtex-5 devices. Measurements on Virtex-4 have shown that reconfiguration at frequencies of 140 MHz is possible. In Table 18.3 the ICAP throughput reported by different authors is depicted and compared with our achievements.

As can be seen in Table 18.3, the maximum throughput has been achieved. On Virtex-II Pro and Virtex-4 the maximum throughput for frequencies of 100 and 140 MHz has been achieved. As already mentioned, on Virtex-II Pro 140 Mhz is considered to be maximum frequency that can be used to reconfigure the device safely. The throughput of the Virtex-II device is strongly dependant on the *busy* signal of the ICAP. The measurements indicate a maximum value between 96 and 90 KB/ms due to the *busy* status of the ICAP. The theoretical throughput at 100 MHz (which is 100 KB/ms) can never be achieved due to delays internal to the ICAP.

Table 18.2 Performance measurements.

Frequency [MHz]	Cycle count	T_R [µs]	BF	T_P [MB/s]
80	76656	958.2	0.94	75.21
90	76803	853.37	0.94	84.45
100	76711	767.11	0.94	93.94
110	76249	693.17	0.95	103.96
120	76020	633.5	0.95	113.76
130	76214	586.26	0.95	122.92
140	77624	554.46	0.93	129.97

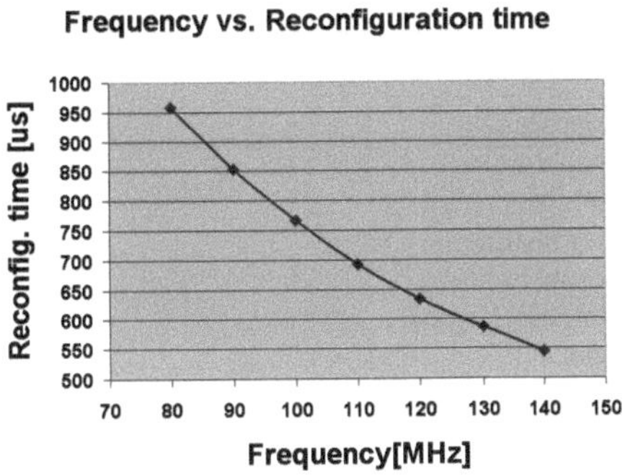

Fig. 18.8 Performance chart.

Table 18.3 Throughput of ICAP controller on Virtex-II and Virtex-4 devices.

Source	Device type	IIW [bit]	Frequency [MHz]	Interconnect	Meas. throughput [KB/ms]	Theor. throughput [KB/ms]	Maximum reached
[17]	V4	32	144	custom Link	219.31	576	No
[15]	V4	32	100	OPB	350	400	No
[3]	V2	8	100	Ethernet	50	90–100	No
Claus et al.	V2	8	100	PLB	90	90–100	Yes
Claus et al.	V2	8	140	PLB	133	133	Yes
Claus et al.	V4	32	100	MPMC	400	400	Yes
Claus et al.	V4	32	140	MPMC	560	560	Yes

Thus the measured throughput is considered as the maximum achievable, since every clock cycle that the ICAP is not busy, the maximum amount of data possible is provided to ICAP. None of the authors in [3, 15] or [17] mention why the maximum is not achieved, but it is most likely due to problems with transferring bitstream data to the ICAP, which happens when a high speed peripheral (ICAP) is connected to an interconnect for low speed peripherals, such as the OPB.

18.5 Performance of the Engines

All Engines in the AutoVision system are capable of processing one pixel plus its complete neighborhood in one clock cycle. The execution time of an image with a resolution of 640×480 pixels (VGA resolution) processed by an Engine running at 100 Mhz can be estimated by $640 * 480 * \frac{1}{100\text{ Mhz}} = 3.072$ ms. The measured processing time of almost all the Engines (depicted in Table 18.4) is close to that value.

An exception constitute the execution time of the CensusEngine and the Matching-Engine. Due to a deep pipeline (see [11]) the processing time of the CensusEngine is a little longer than the execution time of the other Shape-, Taillight- and ContrastEngine, and is nearly 4 ms. Beside the latency added by additional

Table 18.4 Execution times and resource utilization of the coprocessors running at 100 MHz on a VirtexII-Pro (XC2VP30) for an image resolution of 640×480 pixels.

Coprocessor	Execution time [ms]	# of LUTs	# of Flip-Flops	# of BRAMs
ShapeEngine	3.142	8450 (30%)	2281 (8%)	16 (11%)
TaillightEngine	3.18	3695 (14%)	2546 (9%)	12 (8%)
ContrastEngine	3.103	1002 (4%)	504 (2%)	6 (4%)
CensusEngine	3.95	3140 (11%)	1029 (3%)	14 (10%)
MatchingEngine	5.92	10787 (39%)	8171 (29%)	36 (26%)
Optical Flow	9.89	13927 (51%)	9200 (34%)	50 (34%)

pipeline stages, the additional $3.95 - 3.072 = 0.878$ ms of the ContrastEngine are spent for read and write data transfers. As can be seen in Fig. 18.3(b) the MatchingEngine consists of two processing pipelines, which run simultaneous. As the MatchingEngine is also capable of processing one pixel per clock cycle a theoretical execution time around 3.072 ms can be expected. The time difference is caused by a bottleneck which was identified as the DDR SDRAM memory controller in the current Virtex-II Pro based system. Measurements show that around 30 cycles pass from the first data request until the last data chunk of a 128 byte burst appears at the input of the MatchingEngine. If two complete VGA resolution images, 32-bit color depth each, have to be transferred form the DDR SRAM to the MatchingEngine in 128 byte bursts, $\frac{2*640*480*32 \text{ bits}}{128*8 \text{ bit}} = 19{,}200$ transfers are required. The fact that a single transfer takes 30 cycles leads to a total execution time of $19{,}200 * 30 * \frac{1}{100 \text{ MHz}} = 5.76$ ms, which is close to the measured 5.92 ms. By replacing the DDR SDRAM controller with an optimized controller, such as the MPMC, would result in an additional decrease of the Engine's execution time.

Additionally it can be seen that implementing all the coprocessors on the device in parallel would exceed the available resources. Finally, the results in Table 18.4 show that even if the execution time of the MatchingEngine could not be improved, the image processing time of the CensusEngine and the MatchingEngine plus the reconfiguration time (around 4 ms on Virtex-II Pro and 1 ms on Virtex-4 per Engine) is fast enough to perform IntraVFR. This results are going to be shown in our final demonstrator.

18.6 Conclusion and Outlook

The AutoVision project has shown that fast DPR can be utilized in a real time environment, such as video based driver assistance, without violating the real time constraints. This has been achieved through bitstream optimization, modifications of the ICAP controller, optimizations in memory transfer and minimization of the hardware accelerators' resource utilization. In addition, an architectural concept for these high performance hardware accelerators for video processing has been developed, which led to the implementation of various engines. As the whole system is implemented using a modular design approach, adaptions to other devices, platforms and IP cores can be done easily.

The results obtained are going to be shown in a final demonstrator. With this demonstrator for IntraVFR (see Sect. 18.3.1.2) it can be shown that fast dynamic reconfiguration can be applied in real-time systems in order to save precious on-chip resources. Using the example of the Optical Flow, where two hardware accelerators work sequentially, it should be demonstrated that two partial reconfigurations (from CensusEngine to MatchingEngine and vice versa) within 40 ms are possible without the loss of video frames. In addition, after each reconfiguration process the configuration memory should be read back to verify its correctness. Finally, the fast DPR is going to be shown on Virtex-5, in order to prove that the presented approaches are also valid for state of the art devices.

Acknowledgements We would like to thank the following parties for their contributions to the success of this project:

- The German research foundation (DFG) for funding the AutoVision project for four years within the focus program 1148 "Reconfigurable Computing".
- The team of Andreas Herkersdorf, especially Johannes Zeppenfeld and Florian H. Müller who contributed substantially to the success of AutoVision.
- Jürgen Becker, Michael Hübner and Lars Braun from the ITIV in Karlsruhe for their technical support in the field of DPR primarily during the beginning of the project and for the fruitful cooperation within the focus program 1148.
- The Xilinx Research Labs in San Jose and the Xilinx Automotive Division in Detroit and Munich. P. Lysaght, A. Donlin, B. Blodget, P. Patel, J. Mason, N. DiFiore, P. Zoratti and R. Stewart provided insightful comments and feedback as well as development platforms for our project.
- Klaus Gresser from BMW for providing test video sequences for our research.
- All the students who worked in the AutoVision project. Without their generous and valuable input this work would not have been possible.
- Everybody who was not mentioned here, but had contributed direct or indirectly to perform this work. Our sincerely acknowledgments.

References

1. Alt, N., Claus, C., Stechele, W.: Hardware/software architecture of an algorithm for vision-based real-time vehicle detection in dark environments. In: Proceedings of DATE '08, pp. 176–181. ACM, New York (2008)
2. Angermeier, J., Batzer, U., Majer, M., Teich, J., Claus, C., Stechele, W.: Reconfigurable HW/SW architecture of a real-time driver assistance system. In: Proceedings of ARC'08, London, UK, pp. 149–159. Springer, Berlin (2008)
3. Bomel, P., Crenne, J., Ye, L., Diguet, J.P., Gogniat, G.: Ultra-fast downloading of partial bitstreams through Ethernet. In: Proceedings of ARCS '09, Delft, The Netherlands. Lecture Notes in Computer Science, pp. 72–83. Springer, Berlin (2009)
4. Claus, C., Müller, F.H., Stechele, W.: Combitgen: A new approach for creating partial bitstreams in Virtex-II Pro devices. In: Proceedings of the Workshop on Dynamically Reconfigurable Systems (ARCS'06), Frankfurt a. M., Germany, pp. 122–131 (2006)
5. Claus, C., Müller, F.H., Zeppenfeld, J., Stechele, W.: A new framework to accelerate Virtex-II Pro dynamic partial self-reconfiguration. In: Proceedings of IPDPS '07, Long Beach, California, pp. 1–7 (2007)

6. Claus, C., Stechele, W., Herkersdorf, A.: AutoVision—a run-time reconfigurable MPSoC architecture for future driver assistance systems. Inf. Technol. **49**(3), 181–187 (2007)
7. Claus, C., Zeppenfeld, J., Müller, F., Stechele, W.: Using partial-run-time reconfigurable hardware to accelerate video processing in driver assistance systems. In: Proceedings of DATE '07, Nice, France, pp. 498–503 (2007)
8. Claus, C., Stechele, W., Kovatsch, M., Angermeier, J., Teich, J.: A comparison of embedded reconfigurable video-processing architectures. In: Proceedings of FPL '08, Heidelberg, Germany, pp. 587–590 (2008)
9. Claus, C., Zhang, B., Stechele, W., Braun, L., Hübner, M., Becker, J.: A multi-platform controller allowing for maximum dynamic partial reconfiguration throughput. In: Proceedings FPL '08, Heidelberg, Germany, pp. 535–538 (2008)
10. Claus, C., Huitl, R., Rausch, J., Stechele, W.: Optimizing the SUSAN corner detection algorithm for a high speed FPGA implementation. In: Proceedings of FPL '09, Prague, Czech Republic (2009)
11. Claus, C., Laika, A., Jia, L., Stechele, W.: High performance FPGA based optical flow calculation using the census transformation. In: Proceedings of IV '09, Xi'an, China (2009)
12. Herkersdorf, A., Claus, C., Meitinger, M., Ohlendorf, R., Wild, T.: Reconfigurable processing units vs. reconfigurable interconnects. In: Dynamically Reconfigurable Architectures, no. 06141 in Dagstuhl Seminar Proceedings (2006)
13. Hübner, M., Ullmann, M., Weissel, F., Becker, J.: Real-time configuration code decompression for dynamic FPGA self-reconfiguration. In: Proceedings of IPDPS '04, Santa Fe, New Mexico, vol. 4, p. 138b (2004)
14. Hübner, M., Braun, L., Becker, J., Claus, C., Stechele, W.: Physical configuration on-line visualization of Xilinx Virtex-II FPGAs. In: Proceedings of ISVLSI '07, Porto Alegre, Brazil pp. 41–46 (2007)
15. Manet, P., Maufroid, D., Tosi, L., Gailliard, G., Mulertt, O., Ciano, M.D., Legat, J.D., Aulagnier, D., Gamrat, C., Liberati, R., Barba, V.L., Cuvelier, P., Rousseau, B., Gelineau, P.: An evaluation of dynamic partial reconfiguration for signal and image processing in professional electronics applications. EURASIP J. Embed. Syst. **2008** (2008)
16. Mobileye: http://www.mobileye-vision.com
17. Shelburne, M., Patterson, C., Athanas, P., Jones, M., Martin, B., Fong, R.: Metawire: using FPGA configuration circuitry to emulate a network-on-chip. In: Proceedings of FPL '08, Heidelberg, Germany, pp. 257–262 (2008)
18. Xilinx, Inc.: XAPP290: Two Flows for Partial Reconfiguration: Module Based or Difference Based. 2100 Logic Drive, San Jose, CA, USA (2004)
19. Zeppenfeld, J.: LIS-IPIF Specification. Lehrstuhl für Integrierte Systeme (LIS), Technische Universität München, Arcisstr. 21, 80280 München, v1.5 edn. (2007). http://www.lis.ei.tum.de/?lisipif

Chapter 19
Procedures for Securing ECC Implementations Against Differential Power Analysis Using Reconfigurable Architectures

Marc Stöttinger, Felix Madlener, and Sorin A. Huss

Abstract Side channel attacks have changed the design of secure cryptosystems dramatically. Today a reasonable designed cryptosystem has not only to be cryptographically secure, but to be resistant against side channel attacks as well. Therefore many countermeasure techniques have been developed in the last years to avoid exploitable information leaking. We introduce new concepts of countermeasure approaches against differential power analysis attacks to an essential operation of elliptic curve cryptography in $\mathbb{GF}(2^n)$. Similar to many other published countermeasures we are focusing on the architecture layer to secure the cryptographic operations. This type of countermeasures is geared to the well-known hiding methods in this research field, but we apply them on a different implementation layer. For securing the multiplication over $\mathbb{GF}(2^n)$, an essential operation in elliptic curve cryptography, we propose a countermeasure, which is highly scalable and thus allows to select arbitrary trade-offs between performance and side channel resistance.

19.1 Introduction

Nowadays the demand for effective public key cryptosystems is clearly increasing. Because of its low requirements and high performance the Elliptic Curve Cryptography (ECC) scheme has been established as a new standard next to the RSA factorization scheme for public key cryptography.

From a mathematical point of view ECC is regarded as secure. However, real-world hardware implementations introduce additional properties besides the functionality. These properties may lead to some information leakage in so-called side-channels such as power consumption history, which is not considered at all

Marc Stöttinger · Felix Madlener · Sorin A. Huss
Integrated Circuits and Systems Lab, Computer Science Dept., Technische Universität Darmstadt, Darmstadt, Germany, e-mails: stoettinger@iss.tu-darmstadt.de, madlener@iss.tu-darmstadt.de, huss@iss.tu-darmstadt.de

M. Platzner et al. (eds.), *Dynamically Reconfigurable Systems*,
DOI 10.1007/978-90-481-3485-4_19,

by cryptographic algorithms. Thus, an unsecured implementation can lead to the exposure of the secret key by using well-known side channel attack techniques [4]. We present a shuffling-based approach to secure ECC against side channel attacks approaches such as the differential power analysis (DPA) [12].

The approach is applied to the most computing intensive part of ECC, the $\mathbb{GF}(2^n)$ arithmetic. We demonstrate that this application provides additional synergistic effects, which clearly distinguish this work from other shuffling approaches. We also introduce new concepts for a dedicated consideration of parallelism and dynamic resource binding to avoid the exploitation of their specific power consumption.

At the beginning of this chapter we will give a short introduction into the mathematics of ECC. We will also introduce the novel *enhanced Multi-Segment-Karatsuba* (eMSK) multiplication scheme, which is a main element of our approach to secure ECC implementations. In Sect. 19.3 we will describe the basic concept of side-channel attacks and classify the existing countermeasure approaches against such attacks. Our approach to secure ECC implementations is then presented in Sects. 19.4 and 19.5. Here, we describe the application of shuffling-based countermeasures to the eMSK multiplication and give some experimental results that demonstrate the feasibility of our approach. We also show, how this concept can be extended to other operations like Elliptic Curve multiplication.

19.2 Elliptic Curve Cryptography

Beside the widely known *RSA* scheme the *Elliptic Curve Cryptography* is the most established scheme for asymmetric cryptography. The security of ECC is based on the difficulty of calculating the discrete logarithm on elliptic curves. It achieves the same security level as a similar RSA-based solution with significantly shorter keys. This is of special interest in the domain of embedded systems, where the available memory is an important issue.

The mathematical theory of elliptic curve cryptosystems exhibits a strictly layered architecture as depicted in Fig. 19.1. The top layer forms the Elliptic Curve

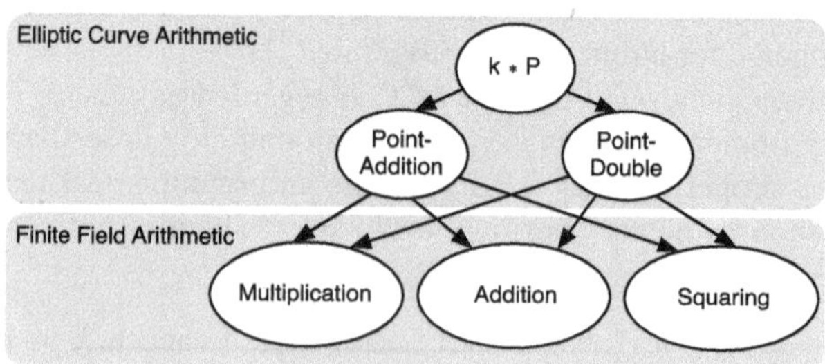

Fig. 19.1 Scheme of the mathematical layer hierarchy for elliptic curve arithmetic.

Arithmetic and exhibits the operations *Point-Multiplication* ($k * P$), *Point-Addition*, and *Point-Double* on the elliptic curve. These operations are based on the arithmetic operations in the underlying Finite Field Arithmetic. Each Elliptic Curve Arithmetic operation requires the execution of multiple lower-level operations. Thus, an cryptographic operation like a digital signature requires just two $k * P$ operations, but many thousands of finite field operations.

19.2.1 Elliptic Curve Arithmetic

The elliptic curves are represented over a finite field $\mathbb{F}$. They form an additive, Abelian group with a well-defined addition operation and a neutral Element $\mathcal{O}$, which is called the point at infinity. They are defined by the simplified Weierstrass equation, a cubic equation for a non-supersingular elliptic curves, defined by:

$$E : y^2 + xy = x^3 + ax^2 + b \tag{19.1}$$

with $a, b, x, y \in \mathbb{F}$ and $b \neq 0$. The points of the curve E are a set of solution to solve the cubic in (19.1). In this chapter we concentrate on elliptic curves over the finite field $\mathbb{GF}(2^n)$ and their arithmetic implementation for hardware architectures. For further more detailed information about elliptic curves we refer to [7].

The EC point multiplication $k * P$ as depicted in Fig. 19.1 is an accumulation of points on the curve E:

$$\underbrace{P + P + \cdots + P}_{k\ times} = k * P = R. \tag{19.2}$$

To perform the $k * P$ computation the two elliptic curve arithmetic operations *Point-Addition* and *Point-Double* are applied. The operation *Point-Addition* $P + Q = R$ is used to calculate the addition of two different points on an elliptic curve. The operation *Point-Double* $P + P = R$ is applied when P and Q are identical. Both operations are composed of operations on the underlying finite field arithmetic as shown in Fig. 19.1. It is possible to define them in different ways, varying in the number of utilized finite field operations. They should be selected with respect to the available system components to obtain the optimal performance. Most ECC solutions are based on projective coordinates instead of affine coordinates. This allows to minimize the extremely expensive finite field inversion, required otherwise.

19.2.2 Finite Field Arithmetic

The EC-level algorithms require a set of operations on the underlying finite field $\mathbb{GF}(2^n)$. Multiplication, squaring, and addition are the most common operations, which have to be executed on a large scale. All these operations are based on logical shift and XOR operations in $\mathbb{GF}(2^n)$.

Squaring and addition scale linearly with their bitwidth and thus can be implemented efficiently as completely combinational logic in digital hardware. The $\mathbb{GF}(2^n)$ multiplication, however, may be prohibitively expensive for bitwidths of cryptographic relevance. Thus, a computation scheme is needed, which combines a combinational multiplier of reasonable size with a sequential algorithm to determine the final result.

Such a computation scheme should be flexible enough to allow for an arbitrary partitioning between the faster combinational and the slower, but more resource efficient sequential part of the multiplication scheme.

The partitioning can be performed asymmetrically for both operands. The most common approach for partitioning is the bit serial implementation with an $1 * n$ bit wide combinational multiplier. Another approach aims at a symmetric partitioning for both operands. In this case an $m * m$ combinational multiplier has to be utilized. For the implementation of DPA resistant implementations the symmetrical approach is more promising. This is because power consumption of each operation depends on as a many bits of both operands as possible.

19.2.2.1 Karatsuba Multiplication

In 1962 Karatsuba and Ofman [10] reported on a sequential multiplication scheme denoted as Karatsuba-Ofman-Algorithm (KOA). It computes the product of two m-digit numbers by dividing the operands into two $\frac{m}{2}$-digit segments and by performing multiple operations on these segments. Until today this scheme is the fastest known sequential multiplication algorithm without precomputations. In addition to its efficiency, the KOA is well-suited for our intended countermeasure approach as we will explain in Sect. 19.4.1

The KOA computes the product of two finite field elements A and B by:

$$\begin{aligned} C = A \cdot B &= (A_1 x^{m/2} \oplus A_0) \cdot (B_1 x^{m/2} \oplus B_0) \\ &= T_1 x^m \oplus [T_1 \oplus T_2 \oplus T_3] x^{m/2} \oplus T_3 \end{aligned} \tag{19.3}$$

with T_1, T_2, and T_3 given by

$$T_1 := A_1 B_1 \tag{19.4}$$

$$T_2 := (A_1 \oplus A_0) \cdot (B_1 \oplus B_0) \tag{19.5}$$

$$T_3 := A_0 B_0. \tag{19.6}$$

It can be observed that one full multiplication is now realized by 3 multiplications of half bitwidth and some addition operations. Obviously, a recursive KOA approach is possible in order to derive a multiplication algorithm that splits the operands into four, eight or, any number of segments being a power of 2. The KOA-based sequential multiplication scheme requires 3^n multiplications for a partitioning into 2^n segments. The maximal supported bitwidth of the combined multiplication scheme is then determined by the product of the combinational bitwidth and the number of segments.

E.g., a system featuring an 192 bit cryptographic bitwidth and an available finite field multiplication of 18 bit would then require at least $\frac{192}{18} = 11$ segments. However, a design with 16 segments has to be implemented as the KOA scheme only supports powers of 2. This design takes 81 finite field multiplications.

19.2.2.2 Multi-Segment-Karatsuba Multiplication

During our studies we observed that the restriction to powers of 2 is too limiting. In the given example, a partitioning into 11 segments would be sufficient. While the realization with 16 segments supports $16 * 18 = 288$ bit instead of the required 192 bit, this does not help in most cryptographic applications aimed to standardized bitwidths. A smaller combinational multiplier may provide a better bitwidth utilization, however this comes at the cost of a larger overall execution time.

To address this limitation, we proposed a novel modified multiplication scheme, called Multi-Segment-Karatsuba (MSK) [5]. The MSK enhances the idea behind the KOA and exploits special properties of the mathematical field $\mathbb{GF}(2^n)$. It defines arbitrary MSK_k schemes, which partition the multiplication operands A and B into k segments.

$$\begin{aligned}\text{MSK}_k(A, B) = & \left(\bigoplus_{i=1}^{k} S_{i,0}(A, B) \cdot x^{i-1}\right) \\ & \oplus \left(\bigoplus_{i=1}^{k-1} S_{k-i,i}(A, B) \cdot x^{i-1+k}\right)\end{aligned} \tag{19.7}$$

whereas

$$\begin{aligned}S_{m,l}(A, B) &= \left(\bigoplus_{i=1}^{m-1} S_{i,l}(A, B)\right) \oplus \left(\bigoplus_{i=1}^{m-1} S_{i,l+m-i}(A, B)\right) \oplus M_{m,l}(A, B), \\ S_{1,l}(A, B) &= M_{1,l}(A, B) \quad \text{and} \\ M_{m,l}(A, B) &= \left(\bigoplus_{i=l}^{l+m-1} A_i\right) \cdot \left(\bigoplus_{i=l}^{l+m-1} B_i\right).\end{aligned} \tag{19.8}$$

The MSK_k scheme requires $\sum_{i=1}^{k} i = \frac{(k+1)\cdot k}{2}$ multiplications. For $k = 2$ both, the MSK_2 and the KOA, are identical. While the MSK supports the subdivision into any number of segments, the KOA still offers better scaling for a larger number of segments. For any power of two, the classical KOA scheme is clearly more efficient. A multiplication with 16 segments, e.g., would require a total of 81 multiplications with the KOA, but 136 multiplications with the MSK_{16}.

The advantages of the MSK come to place when the desired system parameters (i.e., the number of segments) is not reachable by powers of 2. For the given 192 bit example, the MSK scheme allows for an implementation with 11 segments. This requires just 66 multiplications, compared to 81 multiplications of the classical KOA scheme.

19.2.2.3 Enhanced Multi-Segment-Karatsuba Multiplication

To benefit from the advantages of both, KOA and MSK, we propose in the sequel the *enhanced Multi-Segment-Karatsuba* (eMSK). The eMSK combines the efficiency of the KOA scheme with the flexibility of arbitrary MSK_n schemes. This is achieved by applying multiple multiplication schemes for a smaller number of segments in a sequential order. In most cases this combination of smaller MSK schemes will lead to better results than one larger MSK scheme as we will demonstrate.

In case of our example it would be possible to create a design with 12 segments for the multiplication of 192 bit. This segmentation is obtained by the application of two KOA schemes and one MSK_3 scheme and is denoted as $eMSK_{2*2*3}$. Albeit this sequential algorithm does support 12 segments, it requires only 54 multiplications, which is 20% less than the corresponding MSK_{11} scheme.

The performance of the different sequential algorithms is compared in Fig. 19.2. The graph denotes the number of required multiplications for an increasing number of segments. It can be observed that the KOA curve forms a step-function whenever the next power of 2 is reached. The MSK is only partially better than the KOA-scheme. In contrast, the novel eMSK shows the best overall performance. In the worst case it is identical to the KOA scheme, which is an upper bound for the eMSK performance.

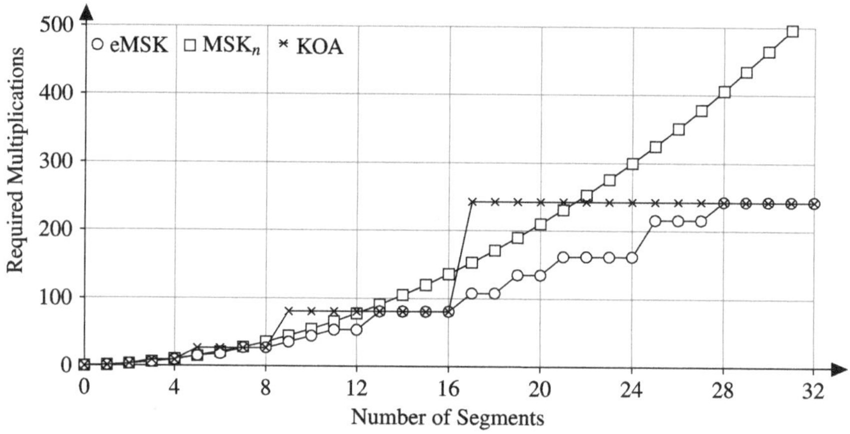

Fig. 19.2 Performance comparison of the Karatsuba-based sequential multiplication schemes.

To obtain the correct eMSK scheme for any set of bitwidths and the related number of segments one needs a dedicated algorithm. This algorithm has to identify the appropriate schemes for the recursive application that will finally lead to the smallest number of multiplications.

This algorithm is given in Algorithm 19.1. It derives the eMSK from two design constraints, the overall bitwidth bw and the maximal available combinational bitwidth $comb$. The overall bitwidth characterizes the security level of the design, whereas the combinational bitwidth is limited by the available hardware resources. The algorithm starts with a prime factorization of bw. A certain part of this prime set will form the combinational multiplication and the remaining primes factors form the set $seqPrimes$. For each of these prime numbers the corresponding sequential multiplication scheme is applied. In two nested optimization loops the supported bitwidth bw is decreased to the next power of 2 and the combinational bitwidth $comb$ is decreased down to 1. For each feasible combination the resulting sequential multiplication scheme is evaluated in order to find the optimal solution.

Algorithm 19.1 Generate eMSK Scheme

```
Require: Cryptographic bitwidth bw, maximal combinational bitwidth comb.
  for all i in bw, bw + 1, ..., NextPowerOfTwo(bw) do
    optNumMults := ∞
    BitwidthPrimes := PrimesOf(i)
    for all j in comb, comb − 1, ..., 1 do
      CombPrimes := PrimesOf(j)
      if CombPrimes ⊂ BitwidthPrimes then
        seqPrimes = BitwidthPrimes \ CombPrimes
        if NumMults(seqPrimes) < optNumMults then
          optNumMults := NumMults(seqPrimes)
          optSequence := seqPrimes
          optCombWidth := j
        end if
      end if
    end for
  end for
  return (optNumMults, optSequence, optCombWidth)
```

As a result of the eMSK scheme a large number of small basic multiplications is obtained. The operands of these basic operations are formed from a sum of distinct words taken from the input operands. The partial results have to be added at varying positions of an accumulation register, which finally contains the overall result. All basic multiplications are data-independent to each other, which means that they can be executed in any order. This property stems from the commutative and associative nature of $\mathbb{GF}(2^n)$.

19.3 Side Channel Attacks

Side channel attacks (SCA) are a very efficient and elegant method to attack cryptographic system implementations. These attacks are taking place in the scenario of either chosen plaintext or chosen chiphertext attacks. Even mathematically secure algorithms may be susceptible to side channel methods. SCA exploit the properties of the *implemented* cryptographic algorithm to access information on the processed data and the secrets of the cryptographic device. The different behaviors of the implementation while processing data generate a specific signature in different physical domains such as time or energy. This leaking information like the processing duration, the power consumption, or the electro magnetic radiation, can be misused to gain knowledge about the secrets of the cryptosystem.

In this chapter we focus on side channel attacks, which exploit the power consumption of the target device, i.e. we address, the power analysis attacks. Using only information leaks in the power consumption of a circuit, no further knowledge about the implementation of a cryptographic algorithm is necessarily required. Power analysis attacks can be ordered into two groups: the simple power analysis attack (SPA) and the differential power analysis attack (DPA). Especially DPA has been established as a common power attack method onto cryptographic implementations. During a SPA the attacker records the power consumption of the attacked cryptosystem for different cryptographic operations. Afterwards, the attacker directly tries to interpret the measured power consumption and map this information to the secret key of the cryptosystem. In general, the attacker tries to discover specific behavior properties depending on the cryptographic operation sequence. The more powerful DPA exploits correlation methods to extract more information from the measured trace about the cryptographic secret. This is achieved by monitoring the power consumption P_{t_0} of a device at a certain point in time t_0. Therefore, the power consumption is measured many times for the same cryptographic operation and hereby varying publicly accessible input parameters. The variance of the publicly input data is needed to determine deterministic properties of the implementation.

Equation (19.9) shows the composition of the power consumption figure P_{t_0}. It stems from the sum of processing consumption $P_{process}$ and the disturbance of the environment modeled as P_{noise}. Note that only $P_{process}$ depends on the actual input parameters. Equation (19.10) shows the variance behavior of the power measurement for this single point in time. The behavior of the disturbance power P_{noise} is assumed to be a noise process with the distribution $\mathcal{N}(0, \sigma)$. $P_{process}$, in turn, can be separated into two parts. The power consumption P_{data} represents the input-dependent arithmetic components, whereas P_{op} addresses the data-independent part of the design. Therefore, Eq. (19.10) can be mapped to the more detailed description of the variance behavior given in Eq. (19.11). Considering the fact of monitoring the same point in time t_0 for a large number of traces, it can be assumed that we will always measure the same, data-independent power consumption P_{op}. Thus, the variance behavior of the measured power consumption simplifies as given in Eq. (19.12).

Assuming an unsecured implementation, $Var(P_{t_0})$ is now strongly related to the variance behavior of P_{data}. With this knowledge an attacker can use a set of power measurements to estimate the unaccessible and secret parameter (i.e., key) of the cryptosystem. Similar to [1] and [13] the attacker can derive the key through correlating a consumption hypotheses with the measured traces.

If the number of traces is sufficient, then the highest correlation will refer to the actual key. Establishing a sophisticated power model is nontrivial. In addition the efficiency of a DPA strongly depends on the chosen power model. A power model characterizes the relationship between the possible intermidiate values of the cryptographic operations and certain power levels. Adequate models require a deep knowledge about the actual implementation. Two well-known power models for DPA attacks are the Hamming distance power model (HDPM) and the Hamming weight power model (HWPM) [16]. Both models are based on possible intermidiate values. Figure 19.3 illustrates how DPA exploits one of the common power models for calculating the correlation between the hypothesis and the measured power traces. The HWPM is a rather simple power model. For its exploitation an attacker needs no knowledge of the architecture or the implementation of the cryptographic device. She or he only has to know which cryptosystem has been implemented on the device, because this model is based on the binary ones of a value written into a register. From the plaintext or chiphertext scenarios the attacker can compute an key-depended intermediate value and then use the HWPM and correlation methods to analyze the measured power traces. The more advanced HDPM estimates the power consumption by guessing the changes of ones and zeros between two computational steps. In order to use this power model for an exploitation of the information contained in the power trace the attacker needs some knowledge about the architecture of the implemented cryptosystem. The advantage of analyzing the different states of a register is a more precise power model for CMOS technology based architectures. When considering the very large system integration (VLSI) layer for CMOS technology-based devices, the dynamic current during the switching process is still higher than the static leakage current of the device. The more information an

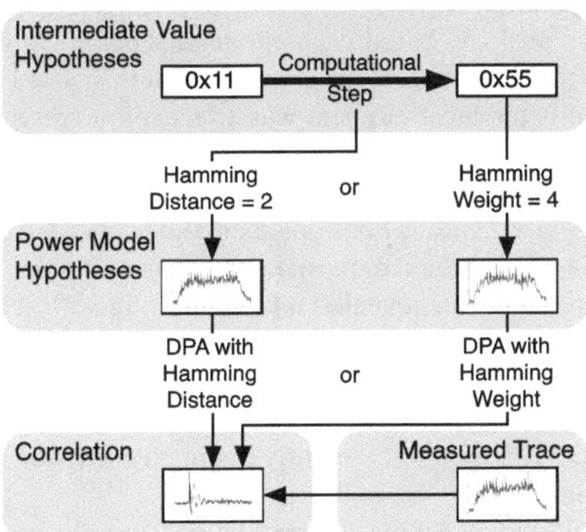

Fig. 19.3 Workflow of a differential power analyze attack.

attacker has about the implemented architecture the better the HWPM maps the intermediate, precalculated values to estimated power levels.

$$P_{t_0} = P_{process} + P_{noise} \tag{19.9}$$

$$Var(P_{t_0}) = Var(P_{process}) + \sigma \tag{19.10}$$

$$Var(P_{t_0}) = Var(P_{data}) + Var(P_{op}) + \sigma \tag{19.11}$$

$$Var(P_{t_0}) = Var(P_{data}) + \sigma \tag{19.12}$$

19.3.1 Information Leaking of ECC Hardware Architectures

As already mentioned the more knowledge an attacker has of the cryptosystem implementation, the more detailed power models can be established. In principle, each cryptosystem has some weak points, where confidential information may leak from the device. The amount of the possible leaking information depends on the implementation, the platform architecture, and the implemented cryptographic algorithm. Any data dependend branches during cryptographic operations are vulnerable to side channel attacks. Significant value changes are also susceptible to the differential power attack, like calculating and storing secret intermediate results.

The cryptographic security of an ECC cryptosystem is based on the hard mathematical problem of solving the discrete logarithm on an elliptic curve. The result of this problem is a point on the elliptic curve. This point is rather simple to allocate with the knowledge of a base point and an order of *Point-Multiplications* of different points. The base point is handled as the public key and the order of *Point-Multiplications* as the secret key. The bits of the key value controls the number of *Point-Double* or a *Point-Addition* executions during the *Point-Multiplication*. A *Point-Double* is performed if a key bit is zero, otherwise a *Point-Addition* is performed for the computation of $k * P$.

In order to attack an ECC based cryptographic scheme, an attacker may identify several vulnerable properties of the cryptosystem to start a DPA. Attacking the *Point-Multiplication* is the most efficient way to a exploit secret information about the ECC implementation. Even when running a SPA the secret can be interpreted directly from the power trace, because of the different computation durations of *Point-Addition* and *Point-Double*. For implementations of *Point-Addition* and *Point-Double* with equal computation length, the scheme is still vulnerable to a DPA. In that case, the secret key can be revealed bitwise starting with the second most significant bit and ending with the least significant one. To estimate the current bit the attacker has to hypothesize the state of the previous bit. The binary decision between two possible values for each bit is the advantage of this iterative attack. Under these circumstances the DPA is used to efficiently distinguish between the correlation of two different hypotheses.

A second possibility to reveal the secret of an ECC cryptosystem is by analyzing the intermediate values of the processed data on the Finite Field arithmetic. Tracing and analyzing the power consumption of this arithmetic is more expensive than the previous one. The attacker has to evaluate more than two basic hypothesis correlation results and thus cannot determine the secret by a binary decision. Based on the bit length of the possible intermediate values the attacker usually has to correlate $2^{bit\ length} - 1$ hypotheses with the recorded power traces. When attacking the finite field arithmetic it is suitable to focus on the multiplying and the squaring in $\mathbb{GF}(2^n)$. Based on the mathematical properties of $\mathbb{GF}(2^n)$ only these two operations require a reduction. The need of performing a reduction depends on the current operand values of these operations. This data dependency can be exploited to estimate the processed data. Weak implementations of the arithmetic units, such as modules featuring different processing times for multiplication and squaring, clearly may be vulnerable to SPA. The attacker can thus distinguish between a multiplication and a squaring operation, he or she can then make sensible assumptions of the performed operation in the EC arithmetic. Based on the different compositions a *Point-Addition* or a *Point-Double* may be identified from the number of performed multiplications and squaring over $\mathbb{GF}(2^n)$. Based on this knowledge it is rather simple to extract the order of the performed operations in the EC arithmetic and finally the secret k may be estimated.

19.3.2 General Countermeasure Techniques

Countermeasures against side channel attacks are one of the hot topics in cryptographic engineering since their discovery by Kocher et al. [12]. Securing a cryptosystem against side channel attacks by countermeasures does not lead to a perfectly secure cryptosystem. Instead, in general countermeasures increase the number of side channel measurements, which are required to extract the correct key by stochastic methods like analyzing the Pearson correlation coefficient. Adding a countermeasure to an unsecured design will not lead to an absolute SCA resistant cryptosystem,but it increases the effort of attacking the system. However, the number of needed power traces to successfully reveal the correct key is usually taken as a factor of quality how resistant the implementation is against DPA. As the Table 19.1 depicts, the concepts of counter measures can be divided into two main groups namely, masking and hiding.

Table 19.1 Known countermeasures against power analysis attacks.

		Circuit level	Register level	Algorithm level
Masking		–	Gate masking	Operand blinding
Hiding	Time based	–	Clockless logic	Shuffling Dummy operations
	Amplitude based	Noise generator	Dual-rail logic Triple-rail logic	Random precharge

The countermeasure concept of masking combines random values with the original values to randomize the power consumption. This technique can be used on different implementation levels. A masking countermeasure on algorithmic level adds a random additive or multiplicative value to the input data and thus misleads the estimation of Hamming weight or distance. Applying this technique on the register level it is also possible to implement arithmetic components, which contain masked logic structures. In contrast, the hiding method disturbs the typical power consumption of the cryptographic process by means of an artificially generated power consumption. The two parameter characteristics of a power consumption value in a trace, i.e., time point and amplitude, are the degrees of freedom to falsify the statistic analysis of the DPA. Therefore, the hiding countermeasure can be catalogized in two groups. The time based hiding methods disturbs the DPA by a significant reordering the power consumption values over time. The amplitude based techniques try to increase the *Signal-to-noise* (*SNR*) to hide the leaking information or to balance the power consumption. Each of these methods is suitable for an application on different implementation layers. A very common balancing technique on the circuit level layer is the dual-rail logic [18]. This technique tries to balance the power consumption by implementing complementary design logic. However, designing such balanced circuits is very difficult and negatively affects both, the resource requirements and the performance of an implementation.

Most of the published countermeasure techniques against power attacks are designed to harden the private key AES algorithm [6, 2, 17, 16]. The research work in the field of securing public key ECC against power attacks mostly focuses on masking or dummy operations. The interesting operation of the EC cryptosystem is the *Point-Multiplication* because its processing depends directly on the secret. One obvious method is to add dummy operations to the *Point-Multiplication* [4, 9] another approach is to mask this operation with a random value [4, 11, 3, 15]. Securing the *Point-Multiplication* via a randomized projective representation of the point coordinates is also possible as published in [4, 8].

19.4 Countermeasure Through Reconfiguration

Most known countermeasures for hardware try to handle dynamic algorithms on top of a static architecture. Existing countermeasures like masking and dummy operations as described in Sect. 19.3.2 either require additional resources and/or delay the computation considerably. Adapting the dynamic randomness of the algorithm to the architecture opens a new dimension of data independent power consumption behavior. With FPGA-based hardware implementation designers have a very large degree of freedom to develop dynamic hardware structures through reconfiguration, which are extremely well-suited for masking and hiding purposes. We want to give an overview on some new concepts to use the basic concepts of masking and hiding in a more architecture-centric context. Additional composition of a mutating control flow or a datapath with an algorithmic countermeasure will improve the resistance against SCA too.

We discuss new countermeasures which exploit a dynamic architecture in order to hide the data dependent specific power consumption without the performance drawbacks of other countermeasure approaches. The proposed countermeasures either change dynamically the control flow or the data flow of a cryptographic operation to secure the implementation. In the first part we demonstrate how to apply this new techniques on the eMSK multiplier to secure the finite field arithmetic. This is followed by a description of mapping this hiding concept to the next abstract layer, the Elliptic Curve Arithmetic, as an important base for our further research work.

19.4.1 Securing ECC on Finite Field Arithmetic Level

An effective way to harden an implementation against SCA is the addition of noise to P_{t_0}. To achieve this, existing countermeasures either manipulate the time or the amplitude domain. Our approach combines both domains to obtain optimal results. Regarding the time domain, we perform a randomized rescheduling, where the execution order is determined during runtime. This technique sounds similar to the shuffling technique for AES in [16], but it is in fact to a large extent different to plain shuffling. In presence of this randomized execution order, an attacker does no longer know which operation will take place at a given point in time t_0. Thus, the data-independent power consumption $Var(P_{op})$ can no longer be assumed to remain constant at t_0 and Eq. (19.12) is no longer valid. Instead of Eq. (19.12) we have to reconsider Eq. (19.11). The amplitude domain can be used to add noise by exploiting the accumulation process of the eMSK. As the eMSK algorithm sums up all n intermediate results r_n in one common accumulation register, the results of previous basic operations $op(\cdot,\cdot)$ affect the actual power consumption $P_{data,i}$ too:

$$\begin{aligned} \bigoplus_{i=0}^{i=n} r_i &= \bigoplus_{i=1}^{i=n} op(data\ segment_i, r_{i-1}) \rightarrow P_{data} \\ &= \bigoplus_{i=1}^{i=n} P_{data,i} + P_{data,i-1} \end{aligned} \tag{19.13}$$

Due to the randomized execution order, the influence of previous intermediate results r_{i-1} is hidden and the noise of P_{data} is increased. This property does not hold for existing shuffling approaches, which target other cryptographic operations (as detailed in [16]). Thus, the application of shuffling techniques to the amplitude domain is a novel contribution of this research work.

By combining benefits from this shuffling method for the time domain and the hiding method for the amplitude domain, our countermeasure improves the overall resistance with nearly no drawbacks on the performance. We will detail this in the next sections.

As stated before, the eMSK scheme allows to reschedule the execution order of basic multiplications without changing the composite $\mathbb{GF}(2^n)$ multiplication and all superior ECC algorithms. Thus, we can apply the proposed countermeasure to our unsecured MSK multiplier, by rescheduling the basic multiplications. In order to apply the proposed countermeasure, the control flow has to be changed from a static schedule to a random-based dynamic schedule of the basic operations. We perform the rescheduling by swapping random elements of the execution sequence. By swapping operations each basic multiplication will be performed only once without additional management control. Algorithm 19.2 denotes the complete eMSK multiplication with rescheduling. For comparison purposes we implemented the unsecured version as well. Figure 19.4 depicts the implemented design with the data path on the left and the control section on the right hand. The data path is identical for both variants.

Algorithm 19.2 Multiplication with eMSK $z \leftarrow x \cdot y$

Require: (x, y) Bitvectors
  $n :=$ Number of Basic Multiplications
  $z :=$ Accumulation Register
  $N[\,] :=$ Array of all Basic Multiplications

  if Execute first time **then**
    $N[\,] := \{0, 1, 2, \ldots, n\}$
  end if
  for $i = 0$ to n **do**
    $r \leftarrow \text{rand}(1, n)$
    $N_{temp} \leftarrow N[r]$
    $N[r] \leftarrow N[0]$
    $N[0] \leftarrow N_{temp}$
  end for
  for $i = 0$ to n **do**
    $z_{temp} \leftarrow$ Execute $N[i]$-th Basic Operation
    $z \leftarrow z + z_{temp}$
  end for
  $z \leftarrow \text{reduce}(z)$
  return z

The secured and the unsecured version only differ in their control unit. A runtime replacement of the control unit by means of partial reconfiguration thus becomes possible. The unsecured control unit simply consists of a program RAM for the instructions and a corresponding program counter. The counter can be reseted by an external trigger to start a new multiplication. The control unit of the secured implementation is extended by a swapping unit and by a random number generator. This random value is being used as an address to swap the corresponding RAM entry with the first entry. As this simple swapping step only affects two program entries at the same time, it has to be performed multiple times to obtain a global randomization. To guarantee a sufficiently randomized shuffle effect, the swapping operation is performed in every idle clock cycle of the hardened multiplier. During our measurements we delayed the time between two multiplications to guarantee a certain minimum amount of shuffling.

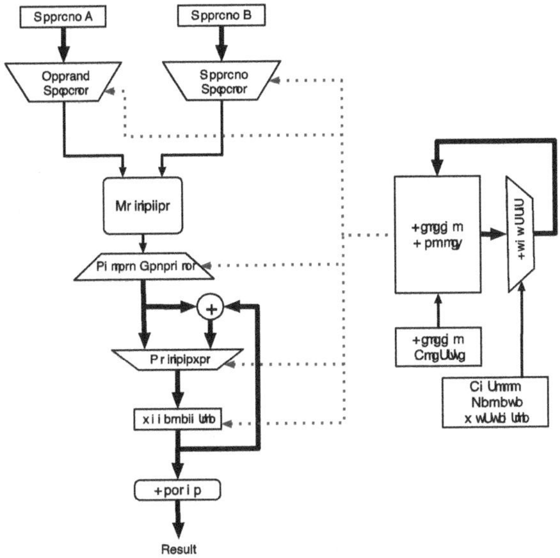

Fig. 19.4 eMSK implementation with DPA countermeasure.

We implemented the 192-bit wide $\mathbb{GF}(2^n)$ multiplications both with 6 segments and with 24 segments to illustrate the scalability of our approach. For the $eMSK_6$ this leads to 18 basic 32 bit multiplications, the $eMSK_{24}$ needs 162 basic multiplications to execute the composed multiplication. The operand selector combines the input segments required for the computation of each basic multiplication. The multiplication result is then set up such, that it can be accumulated to arbitrary segments of the 384-bit word. The accumulator can switch between a direct load, which is required at the start of a new field multiplication, and the accumulator path for subsequent execution steps. The accumulated result is then finally reduced to comply to the mathematical field $\mathbb{GF}(2^n)$.

While the Operand Selector and the Pattern Generator resources grow in size depending on the number of segments, this is compensated by the smaller combinational multiplier unit. Therefore, the MSK_{24} requires less resources than the MSK_6 design and features a higher processing time.

19.4.2 Securing ECC on Elliptic Curve Arithmetic Level

SCA on the Elliptic Curve Arithmetic level exploits the sequential calculation of the scalar multiplication on the curve. Our concept to secure the elliptic curve arithmetic is based on the hiding method as detailed before. Adapting the technique to harden the $\mathbb{GF}(2^n)$ arithmetic on the more abstract layer seems to be a viable approach. For a successful attack the variance of a data dependend power consumption at a certain time point has to be determined. Varying the time point of the operation execution or the processing duration will cause the need for more data samples needed to estimate the variance behavior of the overall power consumption.

We exploit parallelized processes to add noise to the power consumption. Based on the composed algorithms to compute either a *Point-Double* or a *Point-Addition* some multiplications over $\mathbb{GF}(2^n)$ can be parallelized. A scheme of different parallel processing $\mathbb{GF}(2^n)$ multipliers for each basic operation of the elliptic arithmetic clearly affect the computation time of the *Point-Multiplication* for each execution. Utilizing this effect we use different schedules, based on different resource bindings and control flows, for computing $k * P$. Therefore, the interesting time point of storing the intermidiate result of a *Point-Double* or a *Point-Addition* into a register has a variance on its own and will add misleading traces to the correlation test. For a suitably chosen projective representation, like the López-Daham projective [14] coordinates with $Z = 1$, both basic operations of the elliptic curve arithmetic can be parallelized very efficiently.

Figure 19.5 depicts a schematic of an elliptic curve arithmetic architecture with autonomous multiplier units. Each multiplier has its own local memory for the processing data. A bus connects the distributed memory architecture to provide the data exchange between the autonomous processing units.

With five autonomous eMSK multipliers this architecture is massively parallel thus providing a fast computation of *Point-Multiplication*. Each *Point-Double* and *Point-Addition* is kind of a macro featuring a specific instruction order of operations

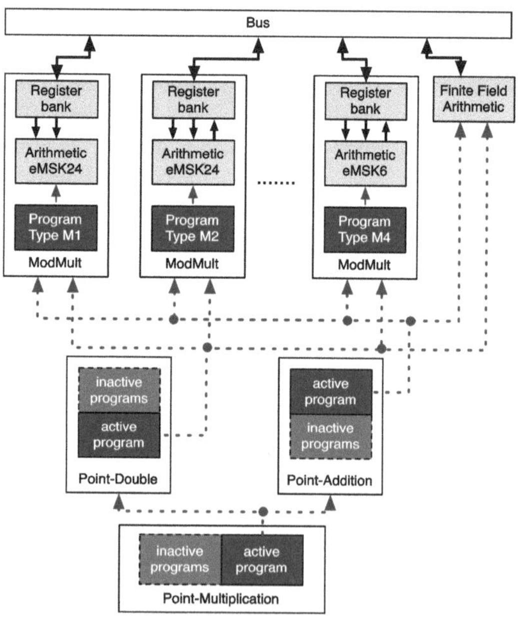

Fig. 19.5 Architecture of an arithmetic logic unit for *ECC Point-Multiplication*.

on the finite field arithmetic. Therefore, the data path and the control flow can be strictly separated. The autonomous multiplier units, the unit for the finite field addition, and the bus handle the dataflow for a *Point-Multiplication*. The control flow of the operations on the elliptic curve arithmetic is then implemented on a primitive memory and program counter structure.

By extending the control unit structure with a more intelligent program counter and offset addressing it is possible to map different control flows with different resource utilizations within the program memory. Now it is possible to call via an offset address different instruction orders for each *Point-Double* or *Point-Addition*. To archive more randomness for the calculation $k * P$, the binding of multipliers for the operation is chosen randomly and will thus affect the processing time for each execution. In case that a designer wants to increase the entropy of the runtime to perform a whole *Point-Multiplication*, the architecture can be extended by more eMSK multipliers each with a different number of segments.

As stated before, the number of segments of eMSK clearly affects both the power consumption for each basic operation and the number of needed clock cycles. An architecture with a heterogeneous multiplier structure introduces an additional degree of freedom to distribute the execution time needed for storing the sensitive intermediated results of the *Point-Multiplication*. The varying power consumption for the different multiplier types additionally makes it more difficult to find similar structures to determine the begin or the end of performed cryptographic operations. One part of the future work will therefore concentrate on implementing an optimized architecture that will fit on our target device to attack and thus to quantify this novel scheme.

19.5 DPA Experiments on Countermeasures

We have set up a Side Channel Attack Lab to attack our own designs in order to quantify the proposed countermeasures. One measurement setup inside this lab has been built up to perform DPA attacks on FPGA implementations. Attacking our secured implementations under real conditions gives a good feedback on the countermeasures. Due to the size of our target FPGA we are not able to attack hardened designs with a high resource consumption, e.g., the dynamic binding architecture presented in Sect. 19.4.2, but we can evaluate the basic concepts on smaller designs and abstract their impact on SCA attacks. Therefore, we will demonstrate in an empirical experiment the practical resistance of our hiding method applied on the eMSK.

The setup consists of a FPGA development board, an Agilent DSO6052A oscilloscope, and a computer running Matlab. As the target for our attacks we use a SASEBO-G FPGA development board, which is the side-channel experimental board developed by Advanced Industrial Science and Technology (AIST) and the Tohoku University, Japan, in the METI project. Due to the structure and fabrication characteristics of the board the physical errors in measurements are very small.

19.5.1 Resistance of the Secured eMSK-Multiplier

In order to test the quality of our approach we attacked both the unsecured and secured MSK implementation in our DPA lab. We set up two different scenarios for the attack. First, we imagined an attacker without any knowledge about the implementation and used the HWPM. In the second scenario we performed a white box attack, now using the HDPM. But for both scenarios we assume the attacker already knows the time points where some vulnerable side channel information leaks from the design during the execution of a multiplication. He or she has not to perform the difficult estimation to find the leaking points in the power trace plot.

During the attack we regard the second operand as the inaccessible key, which has to remain secret. To measure the different traces we generated 400,000 random values for the first operand. The second operand was set to one of the 100 equally distributed random values. After capturing the data sets for each scenario, we generated the key hypotheses with both the HWPM and the HDPM and correlated the measured traces with the hypotheses.

First, we attacked the unsecured $eMSK_6$ with both power models and 5,000 measured traces. The results of this attack are presented in Figs. 19.6 and 19.7. Each figure shows the correlation plots over the monitored time window for all 100 key hypotheses. As described in [1] and [13] an attacker can use these correlation plots to get knowledge about the correct key. The correlation trace with the highest value over all correlation traces refers to the best fitting key. This key is assumed to be the single correct key, because the power consumption between the hypothesis and the measured trace matches best. The black marked trace in both figures is the correct key hypothesis and shows the highest correlation value. In this case an attacker would have found the correct key hypothesis and, thus, the attack would have succeeded. A deeper analysis shows that the HDPM attack succeeds after 80 traces, the HWPM needs about 570 traces. Figures 19.8 and 19.9 illustrate the DPA results of the secured $eMSK_6$ implementation at 50,000 traces. In both figures the black printed, correct hypotheses can not be distinguished from the other gray printed hypotheses. For this design, 50,000 traces are not enough to determine the correct hypotheses.

In the next step we attacked the secured implementation of the $eMSK_6$ and the $eMSK_{24}$ with both the HWPM and the HDPM. We performed multiple attacks in which we increased and the number of measured power traces by 25,000. Each attack step was performed four times and an attack was regarded as successful, if at least one of this four attacks revealed the secret operand. For both power models a successful attack requires significantly more measured traces compared to the corresponding unsecured variants. Table 19.2 documents the required traces to succeed with the attack. While the $eMSK_6$ can successfully be attacked with 75,000 traces, the $eMSK_{24}$ design is not successfully attackable, even when running 100,000 traces. This clearly demonstrates the scalability of our proposed countermeasure approach.

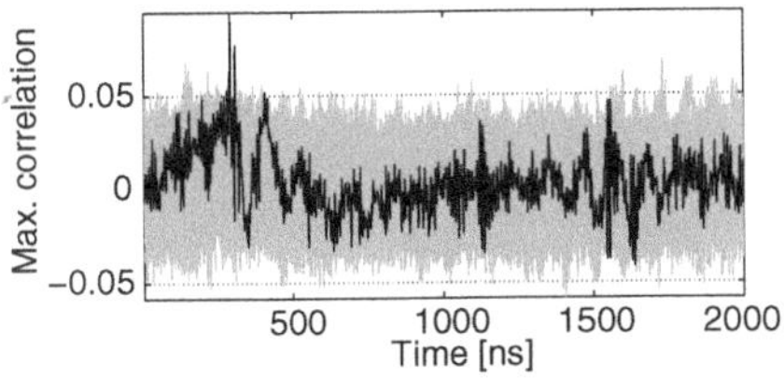

Fig. 19.6 HWPM on unsecured eMSK with 5,000 traces.

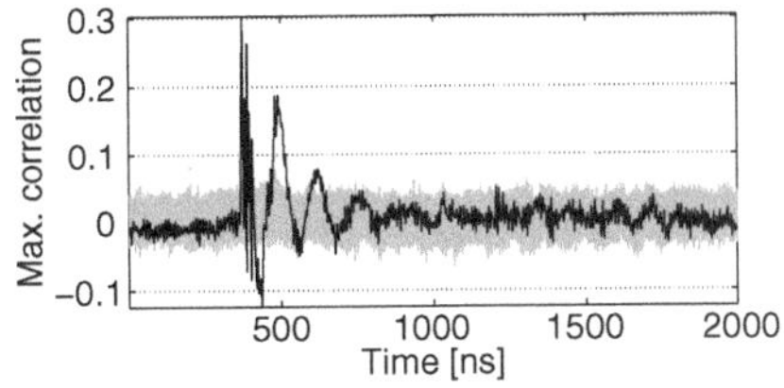

Fig. 19.7 HDPM on unsecured eMSK with 5,000 traces.

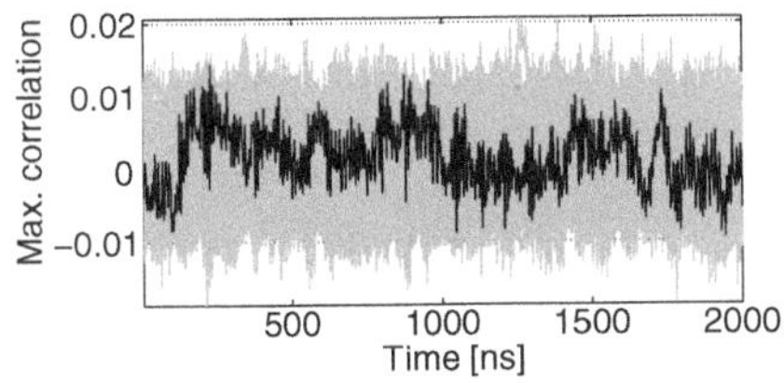

Fig. 19.8 HWPM on secured eMSK with 50,000 traces.

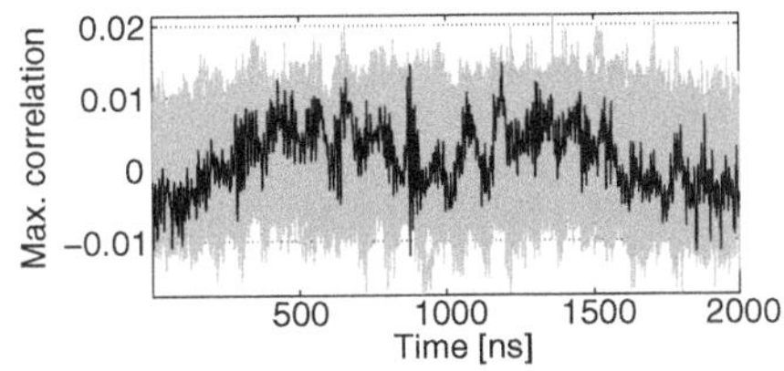

Fig. 19.9 HDPM on secured eMSK with 50,000 traces.

Table 19.2 Success of DPA attacks.

	Traces		80	570	50 k	75 k	100 k
Unsecured	$eMSK_6$	HDPM	Yes	Yes	Yes	Yes	Yes
	$eMSK_6$	HWPM	No	Yes	Yes	Yes	Yes
Secured	$eMSK_6$	HDPM	No	No	No	Yes	Yes
	$eMSK_6$	HWPM	No	No	No	Yes	Yes
	$eMSK_{24}$	HDPM	No	No	No	No	No
	$eMSK_{24}$	HWPM	No	No	No	No	No

19.6 Conclusion and Future Work

We presented a novel technique to secure a fundamental operation of Elliptic Curve Cryptography, namely the multiplication in the finite field $\mathbb{GF}(2^n)$, against DPA and similar power attacks. Until now, this shuffling-based technique has not yet been applied to $\mathbb{GF}(2^n)$. We have shown that we can take advantage of the amplitude domain to gain additional resistance against side channel attacks, which is not the case for other existing shuffling approaches. We presented a new algorithm for efficient finite field multiplication and a generic architecture for its implementation aimed to DPA countermeasures. Experiments have shown that the secured eMSK multiplier requires significantly more power traces (625 times in the case of MSK_6 and HDPM) to successfully attack an implementation. By choosing an appropriate segmentation for the eMSK scheme the DPA resistance can be adjusted in a wide range. This unique scalability property trades off side channel security vs. performance. The required additional logic resources are quite small.

In future works the introduction of parallelism and dynamic binding will be analyzed in detail. Therefore, we are going to develop a target platform able to drive DPA attacks on the architecture introduced in Sect.19.4.2. By using the mapping concepts on the architectural layer we will provide an additional degree of flexibility for dynamic execution rescheduling on the elliptic curve arithmetic. We will investigate in how to exploit partial reconfiguration concepts to harden the architectures against DPA attacks. In combination with dynamic reconfiguration features of FPGAs this additional flexibility will lead to further substantial improvements against power attacks.

Acknowledgements This work was supported by DFG grant HU 620/12, project *Procedures for securing ECC implementations against differential power analysis using reconfigurable architectures*, as part of the Priority Programme 1148, *Reconfigurable Computing Systems*, in cooperation with CASED (www.cased.de).

References

1. Aigner, M., Oswald, E.: Power analysis tutorial. Technical report, Institute for Applied Information Processing and Communication, University of Technology Graz (2000)
2. Blomer, J., Merchan, J.G., Krummel, V.: Provably secure masking of AES. In: SAC, pp. 69–83. Springer, Berlin (2004)
3. Ciet, M., Joye, M.: (Virtually) free randomization techniques for elliptic curve cryptography. In: Information and Communications Security (ICICS 2003). Springer, Berlin (2003)
4. Coron, J.S.: Resistance against differential power analysis for elliptic curve cryptosystems. In: Workshop on Cryptographic Hardware and Embedded Systems (CHES'99), pp. 292–302. Springer, Berlin (1999)
5. Ernst, M., Jung, M., Madlener, F., Huss, S.A., Blümel, R.: A reconfigurable system on chip implementation for elliptic curve cryptography over GF(2^n). In: Workshop on Cryptographic Hardware and Embedded Systems, Redwood City, USA (2002)
6. Golic, J.D., Tymen, C.: Multiplicative masking and power analysis of AES. In: Workshop on Cryptographic Hardware and Embedded Systems (CHES'02), pp. 198–212. Springer, Berlin (2003)
7. Hankerson, D., Menezes, A.J., Vanstone, S.: Guide to elliptic curve cryptography. Springer, Secaucus (2003)
8. Joye, M., Tymen, C.: Protections against differential analysis for elliptic curve cryptography—an algebraic approach. In: Workshop on Cryptographic Hardware and Embedded Systems (CHES'01), pp. 377–390. Springer, Berlin (2001)
9. Joye, M., Yen, S.M.: The Montgomery powering ladder. In: Workshop on Cryptographic Hardware and Embedded Systems (CHES'02), pp. 291–302. Springer, Berlin (2002)
10. Karatsuba, A., Ofman, Y.: Multiplication of multidigit numbers on automata. In: Sov. Phys. Dokl. 7(7), 595–596 (1963)
11. Koç, P.I.Ç.K., Naccache, D., Paar, C., Clavier, C., Joye, M.: Universal exponentiation algorithm—a first step towards provable SPA-resistance (2001)
12. Kocher, P.C., Jaffe, J., Jun, B.: Differential power analysis. In: CRYPTO '99: Proceedings of the 19th Annual International Cryptology Conference on Advances in Cryptology, pp. 388–397. Springer, Berlin (1999)
13. Le, T.H., Canovas, C., Clédière, J.: An overview of side channel analysis attacks. In: ASIACCS '08: Proceedings of the 2008 ACM Symposium on Information, Computer and Communications Security, pp. 33–43. ACM, New York (2008). doi:10.1145/1368310.1368319

14. López, J., Dahab, R.: Improved algorithms for elliptic curve arithmetic in gf(2n). In: SAC '98: Proceedings of the Selected Areas in Cryptography, pp. 201–212. Springer, Berlin (1999)
15. Mamiya, H., Miyaji, A., Morimoto, H.: Efficient countermeasures against rpa, dpa, and spa. In: Workshop on Cryptographic Hardware and Embedded Systems (CHES '04), pp. 343–356. Springer, Berlin (2004)
16. Mangard, S., Popp, T., Oswald, M.E.: Power Analysis Attacks—Revealing the Secrets of Smart Cards. Springer, Berlin (2007)
17. Oswald, M.E., Schramm, K.: An efficient masking scheme for AES software implementations. In: WISA 2005. Lecture Notes in Computer Science, pp. 292–305. Springer, Berlin (2006)
18. Razafindraibe, A., Robert, M., Maurine, P.: Improvement of dual rail logic as a countermeasure against DPA. In: Lecture Notes in Computer Science, pp. 270–275. Springer, Berlin (2007)

Chapter 20
Reconfigurable Controllers—A Mechatronic Systems Approach

Roland Kasper and Steffen Toscher

Abstract This chapter covers the design and implementation of reconfigurable controllers based on modern reconfigurable FPGAs for mechatronic systems. The partial reconfiguration of FPGAs enables a dynamic adaptation of the designed controller functionalities to varying hardware requirements. The chapter introduces a design methodology for partially reconfigurable systems with a specific mechatronic design and implementation approach. Moreover, the design methodology comprises an implicit and distributed reconfiguration management with hard real-time capabilities. The use of reconfigurable hardware in mechatronic systems is also supplemented by infrastructural components and functional groups included in the design and implementation flow. The applicability and performance of the design and implementation methodology are verified with a selected application of a mechatronic control system.

20.1 Introduction

Mechatronic systems integrate mechanical, electrical and software components into innovative products offering heretofore unmatched functionality, performance and cost. Important markets like automotive industry, robotics or home electronics document the advantage of this integration as most of the products developed in the past or presently under development are based on a mechatronic approach. A big challenge in this integration process is the adaption of all necessary single functions to perform the specified function of the complete system. The shift of former mechanical implementations to electric and electronic hard- and software solutions is a primary trend of this evolution. The flexibility gained by this movement together with the rapid enhancements of microelectronics offering steadily increasing

Roland Kasper · Steffen Toscher
Institute of Mobile Systems, Otto-von-Guericke University Magdeburg, Magdeburg, Germany,
e-mail: `roland.kasper@ovgu.de`

M. Platzner et al. (eds.), *Dynamically Reconfigurable Systems*,
DOI 10.1007/978-90-481-3485-4_20, © Springer Science+Business Media B.V. 2010

resources at decreasing costs, are key elements for the success of mechatronic products. On the other hand mechatronic controllers are embedded systems with a close interaction between soft- and hardware and the physical environment that have to do their work within the narrow and strict limitations of a product and have to fulfill high demands on hard real-time behavior. And last but not least, mechatronic products have to compete in very cost sensitive markets which forces all hard- and software resources to be at a minimum.

To meet these demands especially designed microcontrollers or signal processors were used in mechatronic systems from the beginning. Besides analog and digital process and communication interfaces these processors offer optimized architectures to implement algorithms for digital feedback and feedforward controllers, digital filters, observers or control and supervision logic necessary to realize the complex behavior of a mechatronic system. Fast and sophisticated interrupt systems or even hardware support for multitasking assist the processors to fulfill hard real-time conditions. In spite of these features microcontrollers and signal processors follow the same technological roadmap and face the same technological borders than other processor architectures. As noted in [5] the semiconductor industry has turned its attention to new chip architectures that can still take advantage of shrinking transistor size but avoid the increase of power as a consequence of steadily increasing clock frequencies. In typical mechatronic applications this race of frequency never took place as passive cooling is the only way to get rid of heat. Further problems associated with electromagnetic compatibility and increasing costs limitted the operation frequency in mechatronics computing to very moderate values. On the other hand there is a growing number of economic attractive products e.g. in automotive applications, in robotics or in medical technology that are as hungry for computing power than modern games.

Actually computer industry tries to solve these problems with multicore chips. In the simplest case homogeneous multicore chips are used. This result in a relatively simple architecture, as the chip manufacturers only has to place multiple copies of the same microprocessor core on the same silicon die. For mechatronic applications homogeneous multicore designs are very critical as chip size and cost grow and it is very difficult to keep all cores running on a high rate of usage. As a consequence a significant part of hardware investment is not used permanently. A fact that will not be tolerated in very cost sensitive markets as mechatronics. The more complex inhomogeneous approach uses several processor cores with different functionality and capabilities integrated together to form system-on-chip (SoC) architectures. This offers the opportunity to design a chip for specific application domains. Of course this feature is very attractive for all mechatronic high volume markets, as very special demands in I/O-capabilities or specialized computation tasks can be met with the minimum of needed chip size. On the other hand developing software for these architectures that meets hard real-time conditions is very critical task.

The main problem of software development for mechatronic and other control or signal processing focused application is the inherent parallelism of these systems. This becomes evident with a look on usual and wide spread specification methods in these fields. Data flow driven designs very often defined by a signal flow or a block diagram dominate great parts of the systems. Control flow is defined using state machines whose states contain data flow, as shown in Fig. 20.1. To implement such designs on a processor the intrinsically parallel signal flow has to be serialized on the processor's operation code level, while synchronization and real-time conditions have to be fulfilled. Today this step is one of the most critical ones in mechatronic system development because it cannot be automated efficiently, e.g. by automatic code generation tools like Matlab Realtime Workshop or others. In many cases real-time operation systems have to be used to implement code in a safe way, whereby computational load and cost of the processor is increased further. This serialization process is also responsible for the need of processor clocks in range up to several hundreds of Megahertz, while implementing systems that need sampling rates of some tens of kilohertz or lower.

To avoid these difficulties true parallel implementations of the embedded system direct in hardware can be used. Flexible architectures that allow dynamic programming of the underlying hardware are offered e.g. by FPGAs. They allow for matching the complete data flow and all parallel computing elements of a design directly to the parallel signal lines/buses and logical elements of an FPGA (Fig. 20.1). Very high computation speed and data transfer rates are possible in this way as the high clock frequencies of FPGAs can be exploited. The drawback of this direct approach is a very huge consumption of resources e.g. for wide busses or a large number of hardware multipliers to operate with 16 or 32 bit numbers. Nevertheless, many applications are documented where FPGAs are used to implement very high bandwidth digital filters [7] or as a kind of numeric processor [3] assisting a standard processor, microcontroller or DSP to solve very computation- or I/O-intensive tasks, which would exceed the processor's capabilities. One possibility to reduce this big resource consumption is the bit serial implementation of the complete data flow and all numeric operations [4]. It shows that the gap between operation frequency offered by an FPGA and the sampling frequency demanded by a typical mechatronic application is big enough to serialize 32 or even 64 bit values, which is more than enough for most applications. On the other hand bit serialization of data flow and computational elements is a non-trivial task. A domain specific solution for mechatronic systems is presented in [13], but up to now there are no tools available that support this step in an automated way.

A significant advantage of completely parallelized or partially serialized implementation in hardware is exact timing behavior. All sampling times and delays of the implemented embedded system depend only on well defined characteristic timing values of underlying hardware elements. Using established synchronous design methods, where all activities are related to one or several clocks, timing accuracy of few nanoseconds can be guaranteed independent from the behavior of the implemented system. This is far away from the actual values offered by processors, where typical values are in the range of several microseconds. And usually software solutions do not ensure these times. For systems with many branches or threads there only estimations or measurements of these times are available. Unfortunately key elements of speed up of modern processor architectures like pipelining and caches duplicate the inaccuracies of estimations and measurements as they modify actual runtimes significantly. With increasing complexity and for safety critical applications the above mentioned excellent real-time behavior of embedded mechatronic systems implemented direct in hardware is indispensable.

With the arrival of dynamically reconfigurable FPGAs a new level of flexibility in consumption of FPGA resources was attained. The static programming model of classical FPGAs forced the user to implement all parts of an algorithm on the hardware as a whole, because each part must be available when needed. With dynamic reconfiguration, algorithms can be split. Thus only the parts actually at work have to be implemented in hardware and consume resources. Actually inactive parts that will become active can be loaded dynamically at runtime. Evidently dynamic reconfiguration is well suited for applications like software radio [1], where algorithms only change at times the frequency band or some coding scheme changes. Also for very fast applications like video processing there are well defined points in time for dynamic reconfiguration of the computing pipeline as images are processed with a defined frame rate [10]. In many applications mechatronic systems are designed to work at different situations or at different working points, which can be specified using a combination of state machines and block diagrams as already mentioned and shown in Fig. 20.1. In a static solution the complete specification is synthesized and loaded on the FPGA. In case of dynamic reconfiguration only the active state and its associated block diagram must be present on the FPGA, which reduces the need of resources significantly, if there are a larger number of states. This relies on the fact that there is only one active state in a state machine at a time. On the other hand the time a new bitstream has to be loaded is not known because the transition from one state to the other depends on events generated by conditions of dynamically varying signals defined by the block diagram. Thus loading of new bitstreams must be very fast and with an exact timing to guarantee hard real-time conditions

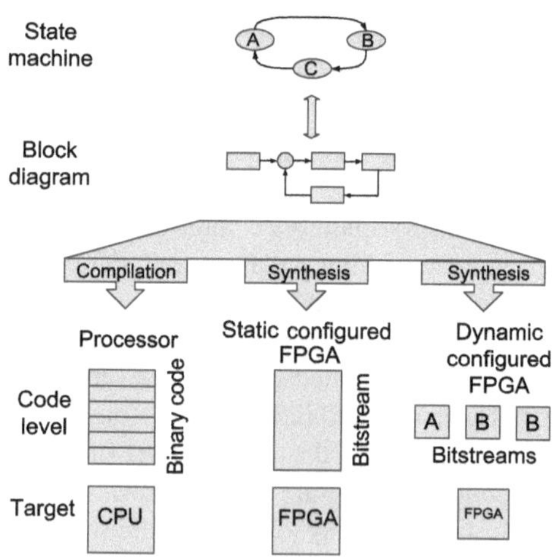

Fig. 20.1 Comparison of implementation strategies of mechatronic systems including processors, static and dynamically reconfigurable FPGAs.

also in the situation of dynamic reconfiguration and to keep the advantage of strict timing of direct hardware implementations.

Starting from these considerations this chapter first describes a design methodology for reconfigurable mechatronic controllers, which is based on high level design flows and design tools commonly used in the domain of mechatronics and control systems. Implicit definition and integration of a sophisticated reconfiguration management together with a distributed implementation of the reconfiguration management and a hard real-time loading and activation strategy for partially reconfigurable controller functionality are key features of the presented methodology. Focusing on partially reconfigurable FPGAs as target hardware a technical structure for reconfigurable mechatronic control systems is introduced, keeping in mind important implementation issues such as resources and interfaces. Special focus is given to a fast, reliable and resource-effective implementation of a partial reconfiguration solution being supported by a dedicated reconfiguration management in hardware and thus enabling self-reconfiguration and hard real-time operation. To reduce to need of FPGA resources bit serial signal processing and transmission for the implementation of parts of the design. To prove the superiority of dynamically reconfigurable controllers in mechatronics, this chapter presents the design and application of a reconfigurable controller for piezo-electric actuators in automotive applications, which uses hard real-time reconfiguration in the sub millisecond range.

20.2 Design Methodology

The design methodology comprises system-level design flows and tools being commonly used in the domain of mechatronics and control systems. This approach allows an easy integration of partially reconfigurable controller functionality in existing design flows and systems. A distinctive feature of the design methodology is the implicit definition and integration of a sophisticated, hard real-time, distributed reconfiguration management.

20.2.1 Logical Controller Structure and Partitioning

Mechatronic control systems in general are specified as a combination of *data flow* and *control flow* parts that interact closely. Operation control signals generated by finite state machines (FSMs) are used to affect the data flow that typically is defined by signal diagrams or other well known specification methods such as truth tables, hierarchical blocks and the like.

In a first step towards partitioning, the design methodology uses FSMs to map controller functionalities to a logical controller structure. The logical controller structure defines operating states, process-dependent control functions and the interaction between control flow and data flow. Employing state machines and signal diagrams, a logical controller structure is defined by describing single process-dependent functions with signal diagrams and other known specification tools and mapping these functions to the states of an FSM.

As an example, the run-up and operation of a typical mechatronic system such as an electric drive system can be mapped to an FSM consisting of four states such as a *static test*, a *dynamic test*, a main *operation* state and a *fail-safe* state. First, the static test takes place followed by the dynamic test before the system becomes operational. If an error occurs during the tests or operation, the system switches to fail-safe and puts the outputs in known safe states.

Thus, by applying this approach to reconfigurable hardware, loading and activation sequences of reconfigurable controller functionality may be directly specified with state machines. A single operation state within such a logical structure describes a specific controller *configuration* comprising several sub-functions. The transition to another operating state changes the configuration by employing new sub-functions. However, a subset of these functions is crucial to all operating states and thus always active. With regard to reconfigurable systems, these sub-functions may be referred to as *static*, while sub-functions being dependent on specific operation states may be referred to as *partially reconfigurable*. Thus, the FSM-based specification approach provides an implicit partitioning of controller functionality into static and partially reconfigurable components. Static sub-functions will be communication and process interfaces, memory and reconfiguration management as well as other high-level controller functions. The sub-set of partially reconfigurable and thus run-time exchangeable functions will rather include specific operation state dependent control algorithms, test and fail-safe functions.

In a next step, function subsets are mapped to a generic reconfigurable platform. This platform independent partitioning on hardware level follows well known partial reconfiguration design flows such as by Xilinx [6]. According to this design flow, static sub-functions are mapped to a static, i.e. non-reconfigurable part of the target hardware called *static base system*. Besides static functionality, the base system comprises I/O and other global hardware resources. Run-time exchangeable functionality is mapped to partially reconfigurable areas of the target hardware. In addition to Xilinx design flow terms, reconfigurable hardware areas are referred to as *reconfigurable modules* whereas run-time exchangeable functions are referred to as *hardware tasks*. Thus, reconfigurable modules represent well-defined hardware areas serving as placeholders for hardware tasks. Hardware tasks may be activated and loaded into reconfigurable modules at run-time by means of partial reconfiguration. Although a controller may comprise several reconfigurable modules, there is only one active task per module at a particular time.

In a further hardware-level partitioning step, hardware tasks are assigned to reconfigurable modules depending on the logical controller structure. Operating state dependent functions being loaded/activated subsequently may be assigned to a single module, whereas tasks being loaded/activated at the same time require separate modules. The state machine representation of logical controller structure facilitates this assignment, since operating states and their specific functionality may be directly mapped to reconfigurable modules and hardware tasks. Figure 20.2 illustrates this approach.

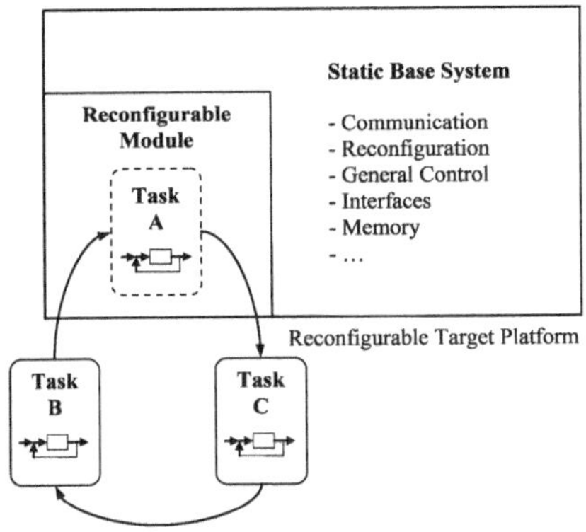

Fig. 20.2 State machine based mapping of hardware tasks to a reconfigurable module.

20.2.2 *Specification of Reconfigurable Controller Functionalities*

The specification of reconfigurable controller functionalities is based on a consistent application of state machines. Thus, reconfigurable modules are specified with state machines comprising single hardware tasks as states. In the context of the specification methodology, every module is described by a particular state machine called *Module FSM*. While the states of a Module FSM represent module and operating state specific hardware tasks, the transitions between those states are typically dependent on process parameters such as the revolution speed of an electric drive system or test and failure conditions.

Figure 20.3 shows a generalized example of a Module FSM with three tasks. Beginning with the start state represented by task 1, the other tasks are subsequently loaded into the module upon their specific transition conditions. A more detailed example of a Module FSM is given in Sect. 20.4 with regard to a reconfigurable controller for piezo-electric actuators.

The functional specification of hardware tasks involves block diagrams and the like and may comprise a code generation step towards a hardware description lan-

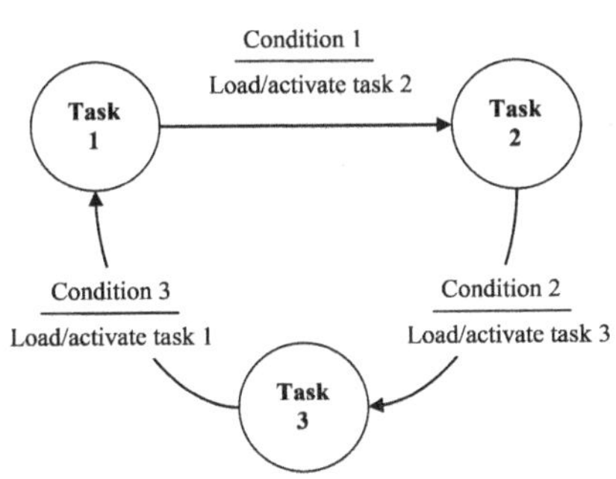

Fig. 20.3 Generic example of a Module FSM.

guage. However, in the context of a reconfigurable controller a hardware task also has to reflect the transitions of its assigned higher-level Module FSM. This is accomplished by an additional state machine implemented in each task. This *Task FSM* controls the sub-states of a single task and is identical for all hardware tasks. Moreover, the concept involves a global identifier for each task, the *Task Marker*, denoting the active task and enabling activation/loading sequences for subsequently active tasks.

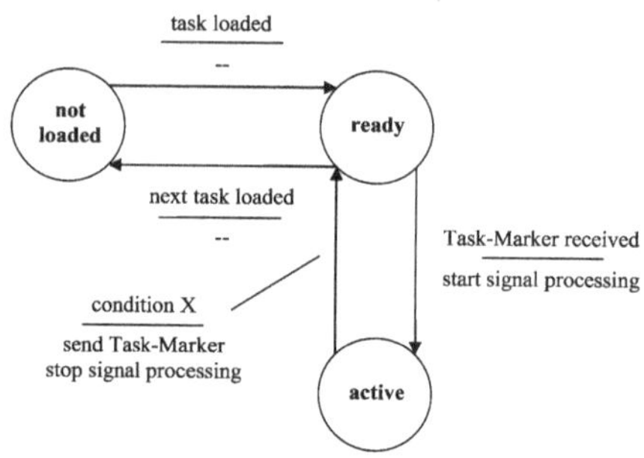

Fig. 20.4 Task FSM.

The structure of the Task FSM is derived from a basic state model of software tasks in automotive real-time operating systems [8]. As shown in Fig. 20.4, the Task FSM comprises three states representing the potential status of a task: **not loaded**, **ready** and **active**. The transition from not **loaded** to **ready** takes place after the task has been loaded from memory and the local reset signal has been deasserted. Upon receiving its corresponding Task Marker, the Task FSM switches to **active** and enables the task's signal processing and local outputs. The Task FSM switches back to **ready** when a certain external condition is met and a transition in the higher-level Module FSM occurs. This causes the deactivation of the task's local outputs and a Task Marker is sent to activate the next task. Replacing the task with another one causes a transition to the virtual state **not loaded**.

20.2.3 Distributed Reconfiguration Control and Activation Strategies

The state machine based specification of reconfigurable controller functionalities supports a generic distributed reconfiguration management that can be directly applied to hardware. Being based on a functional and hardware independent partitioning and specification approach, the reconfiguration management is implicitly incorporated in the controller design at an early overall design stage. Moreover, the reconfiguration management is self-contained, requires no external components and thus enables self-reconfiguration features.

On module level, the Module FSM may be broken down into its single states and implemented directly in the single hardware tasks of the corresponding module. Thus, a Module FSM will at no time be completely implemented on the target platform but only partially in a distributed way through the active task. Therefore, hardware tasks represent not only states of a Module FSM, but additionally implement the out-bound transition of the Module FSM and initiate loading and activation sequences for subsequent tasks. On hardware task level, the Task FSM accomplishes this by means of the Task Marker. A task may activate another task by sending the corresponding Task Marker. Optionally, the static base system may send a Task Marker to activate a specific task, e.g. in order to initiate signal processing.

With regard to hard real-time requirements of mechatronic control systems, the loading and activation time for a single hardware task is of particular importance. In the context of partially reconfigurable hardware, this time is called reconfiguration time T_R and is dependent on module dimension, task complexity, implementation methods and characteristics of the target hardware. As for modern reconfigurable hardware platforms such as FPGAs, reconfiguration times may be in the range of milliseconds.

The influence of the reconfiguration time T_R on the reconfiguration control results from the relation between the sampling time T_S and the processing time T_P. T_S is given by the real-time needs of the mechatronic control problem, while T_P is determined by the logic inside a hardware task and its clock frequency. The relation between T_R, T_P and T_S separates two cases. In the first case, T_S is equal or larger than $T_P + T_R$. In this situation, a loading and activation sequence, i.e. a reconfiguration, may take place within one sampling period without loss of any data. Since typical sample times of mechatronic systems are in the range of 100 kHz and fast reconfiguration solutions are available (see Sect. 20.3), this may be applicable for a majority of controllers. However, in a second case, $T_P + T_R$ may exceed T_S. Therefore, one or more samples could not be processed during reconfiguration. This is obviously not acceptable for hard real-time systems, but may be avoided by task pre-loading.

Task pre-loading on hardware level is based on the enlargement of hardware resources assigned to a single module. Thus, required hardware resources per module increase proportionally to the number of tasks to be pre-loaded. The resulting multitask module is specified by an extended Module FSM. In this context, task activation is a simple switching matter whereas the actual reconfiguration procedure is safely accomplished beforehand. However, a pre-loading strategy is only feasible if the total number of Module FSM states N_S is greater than the maximum number of accessible states $N_R + 1$.

20.3 Structure and Implementation

This section introduces a technical structure for reconfigurable mechatronic control systems. The technical structure has been designed with regard to partially reconfigurable FPGAs as target hardware. Thus, it is closely connected to implementation issues such as resources, interfaces and in particular the hardware-dependent implementation of partial reconfiguration. Furthermore, the technical controller structure complements the reconfigurable design parts with static infrastructure components. The overall controller implementation is accomplished in a very resource-effective way by a direct hardware implementation of all controller functionality and the use of bit serial signal processing and transmission for parts of the design.

20.3.1 Structure and Components

The technical controller structure targets partially reconfigurable FPGAs and maps reconfigurable modules and hardware tasks as well as the static base system to a generic FPGA hardware. On a higher level, the controller structure comprises the target platform FPGA and a memory unit storing the configuration data that represents the hardware tasks. The memory unit may be realized with external memory elements or with dedicated memory blocks on an FPGA depending on the availability and number of such blocks.

On the FPGA level, the technical controller structure comprises the reconfigurable modules, the static base system and a dedicated communication system transferring the Task Marker. While reconfigurable modules contain the hardware tasks, the static base system comprises infrastructure components such as a reconfiguration/memory management, communication and process interfaces and the reconfiguration interface including the actual hardware-level reconfiguration control. Figure 20.5 illustrates this approach with an exemplary technical structure of a reconfigurable mechatronic controller comprising two reconfigurable modules.

As can be seen from Fig. 20.5, the dedicated communication system connects the reconfigurable modules and the static base system of the controller. It ensures a failure-free transfer of the Task Markers and other crucial reconfiguration data between the static base system and the Task FSMs in the reconfigurable modules. The communication system consists of a persistent hard macro and features subscriber connections for reconfigurable modules and their tasks as well as for the static base system. The latter includes additional functionalities to initiate and monitor the run-up and re-run-up of communication. The overall structure of the communication system is a shift register being formed by all subscribers each storing one bit of the message frame and shifting it to the next subscriber on every clock cycle. Moreover, local multiplexers in the subscriber connections prevent any interaction during a reconfiguration. This ensures a failure-free operation of the communication system while a module is being reconfigured.

The reconfiguration and memory management in the static base system includes a hardware-based Reconfiguration FSM, a also hardware-based Memory FSM, a resolution function for pre-loading distinction and a task address table (see Fig. 20.6). The task address table is a specific memory block containing size, memory locations, and type of all task related configuration data. Configuration data may be dynamically received through the communication interface of the controller and is written to the memory by the Memory FSM that simultaneously updates the task address table.

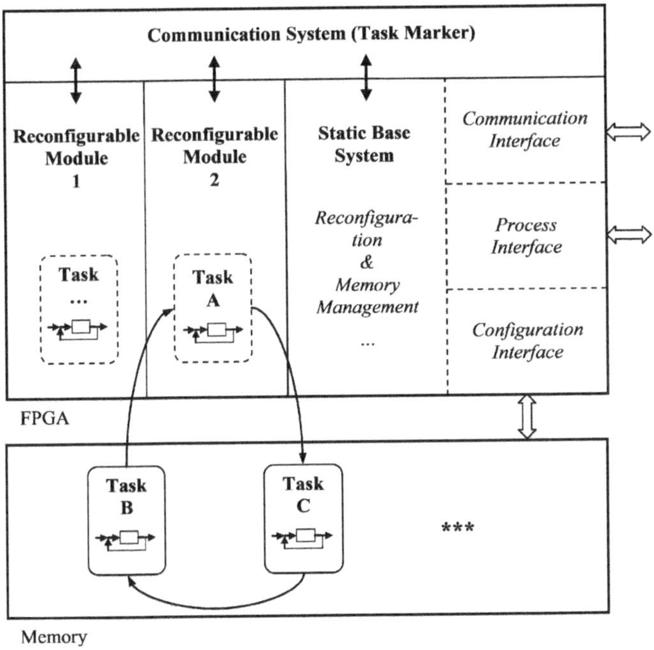

Fig. 20.5 Exemplary controller structure.

The loading and reconfiguration sequence for a specific task and the corresponding reconfigurable module is initiated upon receiving a request and a Task Marker via the dedicated communication system. Subsequently, the reconfiguration state machine accesses the task address table and obtains the specific memory location of the task that is to be loaded. Next, this task's configuration data is read from its memory location and is sent to the actual configuration interface accomplishing the partial reconfiguration.

The communication and process interfaces have been designed as statically, i.e. off-line, reconfigurable components and may implement different functionality depending on the actual application. An exemplary USB 2.0 based communication interface has been implemented employing the external EZ-USB FX2 USB microcontroller by Cypress. Thus, the communication interface on the reconfigurable controller implements a Master/Slave communication with the FX2 based on a further hardware state machine. With regard to process interfaces, exemplary implementations comprise pulse width modulation (PWM) and single line switching outputs and inputs as well as a sophisticated Delta Sigma ADC. The Delta Sigma ADC consists of a small external delta sigma modulator that generates a one bit pulse code modulated signal representing the analog input and an extensive FPGA based bit serial multi-stage filtering and decimation part [14]. Resolution and sampling frequency of the Delta Sigma ADC are adjustable by adaptation of the digital filter stages without any changes in the analog off-chip design. Other feasible process

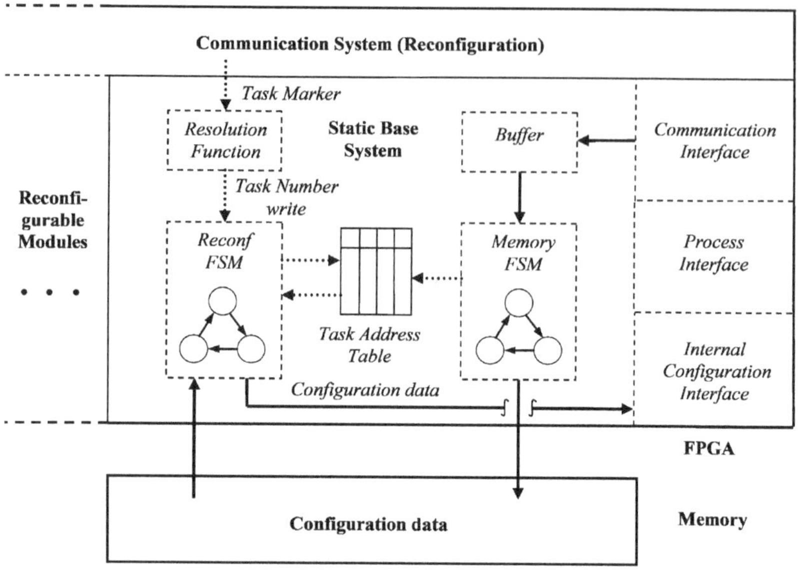

Fig. 20.6 Reconfiguration and memory management.

interfaces for reconfigurable mechatronic controllers include solutions based on external ADC and DAC components.

Moreover, the static base system implements an advanced state saving/backup and state recovery system for reconfigurable control functionalities in hardware tasks such as integral and derivative parts of PI/PID closed-loop control algorithms. State saving/backup is based on an automated backup structure continuously saving actual state values in a dedicated memory component in the static base system, while state recovery after a reconfiguration is based on an automated task specific write-back procedure.

20.3.2 Implementation and Target Hardware

The implementation of reconfigurable mechatronic controllers maps the technical structure to an actual hardware device. Due to the availability of a working design flow supporting partial reconfiguration, Xilinx FPGAs are chosen as target hardware.

The Xilinx partial reconfiguration design flow may be applied to reconfigurable mechatronic controllers in a straight forward way based on the well-defined technical controller structure. However, being a non-standard hard macro, the dedicated communication system for the transfer of the Task Marker must be constrained additionally and assigned to adjacent locations of the reconfigurable modules and the static base system in a the same manner as standard bus macros. As a result of the Xilinx partial reconfiguration design flow, configuration data for the complete controller comprising the static base system and first tasks as well as for the single hardware tasks will be generated. A target FPGA by Xilinx may then be configured using the fully implemented controller data. The configuration data of single hardware tasks may be send to the communication interface of the controller after initial configuration of the device or later. Upon being received by the communication interface, the hardware task configuration data is stored in the controller's memory and the task address table is updated accordingly.

In contrast to most scenarios and common implementations of FPGA-based partially reconfigurable systems, the reconfigurable mechatronic controllers are implemented following a direct hardware approach rather than employing an embedded microprocessor. All functionality of a controller is directly mapped to hardware and thus facilitates hard real-time operation. This is of particular importance for the reconfiguration management as well as for the actual control functionality such as closed-loop control algorithms. Moreover, the multi-level state machine based controller design can be mapped to hardware in a very resource efficient way and supports the portability of controller functionality.

Another distinctive feature of the reconfigurable mechatronic controllers is the application of bit serial signal processing and transmission means based on the approach presented in [9]. Bit serial hardware task functionality as well as bit serial data transmission between the hardware tasks and the base system support a very resource efficient controller implementation. Furthermore, bit serial algorithms support the overall scalability of the controller since only one line per signal is needed regardless of the bit width.

20.3.3 Partial Reconfiguration Solution

The implementation of actual partial reconfiguration processes follows the overall direct hardware implementation approach and represents the final component of the reconfiguration and memory management. As can be seen from Fig. 20.6, a dedicated hardware state machine called the Reconfiguration FSM realizes a direct memory access (DMA) and reads the configuration data of the hardware task to be loaded from the controllers memory. Simultaneously, the Reconfiguration FSM writes the data read from the memory to an configuration interface that is part of the target device. Thus, the distributed reconfiguration management, the static reconfiguration and memory management and the Reconfiguration FSM enable self-reconfiguration features since both the decision to load a task and the actual performing of the loading/reconfiguration sequence are accomplished locally. Moreover, this approach facilitates task loading/reconfiguration sequences under real-time conditions.

With regard to Xilinx FPGAs, the Reconfiguration FSM implements the SelectMAP configuration protocol and is directly connected to the internal configuration access port ICAP, but may also employ I/O resources to access an external configuration interface. Since direct access to ICAP can be realized by a single hardware based state machine, this reconfiguration solution is very efficient, fast and requires only a small amount of logic resources. No embedded microprocessor reconfiguration control system is used.

Due to the direct hardware implementation of the Reconfiguration FSM, the reconfiguration time, i.e. the loading time, for a hardwaretask is directly dependent on the ICAP configuration clock (max. device dependent) and the size of the task that is to be loaded. Thus, the reconfiguration time T_R of a hardware task comprising x_{bytes} configuration bytes at a given ICAP clock frequency f_{ICAP} may be computed for the 32 Bit configuration mode as

$$T_R = \frac{1}{f_{ICAP}} * \frac{x_{bytes}}{4}. \tag{20.1}$$

20.4 Application

This section covers the application of partially reconfigurable controllers in mechatronic systems. The section in particular introduces a reconfigurable controller for piezo-electric actuators in automotive applications. The application of partially reconfigurable hardware to rapid control prototyping in the field of automotive controller design is covered by a further paper by the authors [12]. Another application of partially reconfigurable controllers to mechatronic control problems by the authors can be found in [11]. This paper describes a run-time reconfigurable FPGA-based drive controller for electrical drive systems that is specified, structured and implemented using the here-presented methodology.

20.4.1 A Reconfigurable Controller for Piezo-Electric Actuators

The reconfigurable controller for piezo-electric actuators is based on a power drive circuit and control algorithms presented in [2] and targets automotive fuel injection systems. Figure 20.7 shows the basic power amplifier topology. It comprises two switching elements, i.e. MOSFETs or IGBTs, including drivers, an inductor for current limiting purposes, the actual piezo-electric actuator, a small resistor for current measurement and the voltage supply part with storage capacitors and a DC/DC-converter. With a supply voltage being greater than the positive maximum actuator voltage or smaller than the negative maximum actuator voltage, the actuator can be charged and discharged in a very fast and highly dynamic way.

Control functions for the power drive circuit include a closed-loop voltage and current control based on voltage and current measurements. The here-presented controller employs two-level control algorithms for this purpose. On the voltage control level, the algorithms generate a charge/discharge signal, while the current control algorithms generate on/off signals for the power-electronic switches. Both control

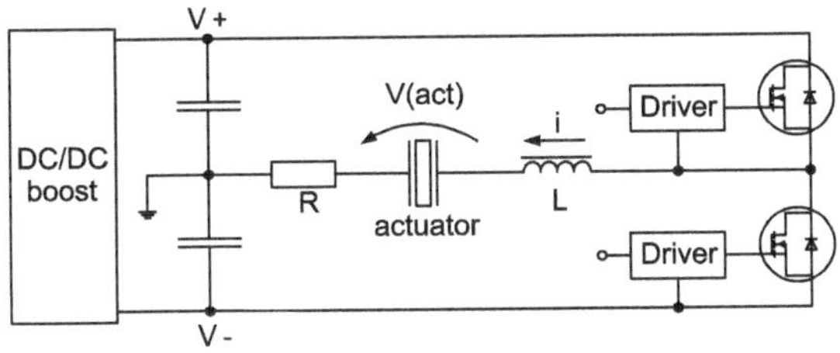

Fig. 20.7 Power drive circuit topology for piezo-electric actuators [2].

functions include a hysteresis with regard to the controlled variables. Additionally, an FPGA-based reconfigurable controller may implement control functions for the voltage supply part, i.e. the storage capacitor voltage control through the DC/DC-converter.

The reconfigurable controller for piezo-electric actuators is specified using a logical controller structure for tests, run-up and operation of the system. A subsequent partitioning leads to reconfigurable and static functionalities that are represented by a Module FSM, hardware tasks and a technical controller structure. The Module FSM comprises four hardware tasks and implements the pre-loading strategy according to Sect. 20.2 in order to enable hard real-time task switching. Thus, the controller requires two single reconfigurable modules on the target platform. However, a controller implementation may also be accomplished with one single module if the achievable loading/reconfiguration times for the hardware tasks are acceptable for the envisaged actuator control. Figure 20.8 shows the Module FSM of the reconfigurable controller for piezo-electric actuators.

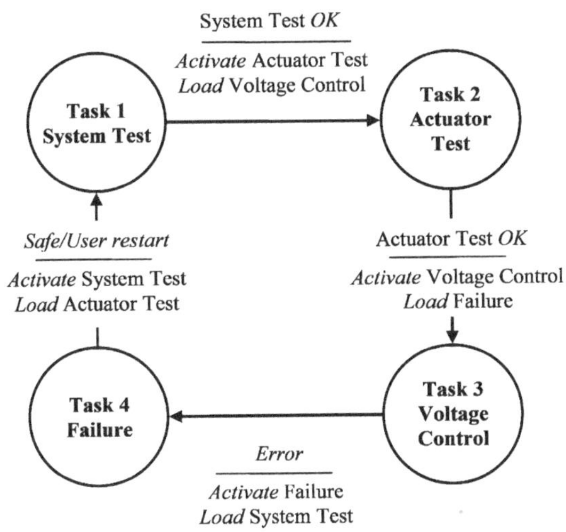

Fig. 20.8 Module FSM of the reconfigurable controller for piezo-electric actuators.

The task set of the controller comprises the following hardware tasks:

- **Task 1** implements simple system tests upon the system's start. The tests comprise initial actuator voltage checks regarding maximum and minimum values. The actuator is not being charged or discharged at that stage. Other potential system test algorithms may include internal or external communication and interface tests as well as tests regarding the DC/DC-converter and the storage capacitors.
- **Task 2** is being activated after the system tests are completed and comprises basic actuators tests. Using a simple test function, the actuator is charged and discharged by an open-loop voltage and closed-loop current control algorithm. The resulting actuator voltage is captured and compared to given limit values. Other potential actuator tests include a parametrization of the voltage and current control hysteresis as well as the determination of voltage limits of the attached actuator by mechanical impact detection and thus an adaptation of set value datasets.
- **Task 3** represents the main operating state of the controller and implements a closed-loop voltage control for the attached actuator. The closed-loop voltage control is based on a comparison of digitized actual voltage values with set values being generated or received by the static base system of the controller. The actual voltage values are subject to a constant hysteresis in order to avoid oscillation. The direct comparison of voltage values is implemented with digital comparators acting as two-level controllers and generating charge (*load*) and discharge (*unload*) signals for the actuator. In addition, switching between loading and unloading the actuator may be subject to an adjustable delay. Other potential main operation functions may be added.
- **Task 4** of the task set is being activated in case of a failure or under erroneous system conditions and transfers the system in a fail-safe state. A basic error detection algorithm is implemented in the main operating state and exemplarily detects any voltages exceeding pre-set limits. Following an error detection, the fail-safe algorithm uses voltage and current control functions to bring the actuator in a neutral voltage state. The dynamics of this process are adjustable through parametrization of the voltage and current control functions. Other potential functions for a fail-safe mode include a precautionary modification of voltage set values as well as a precautionary adaptation of hysteresis values and switching times for following tasks thus preventing future system failures.

The technical structure of the controller comprises two reconfigurable modules for the hardware tasks, the dedicated communication system and the static base system. Figure 20.9 shows a graphical representation of the controller structure and supplemental external interface components. The controller generates the binary signals *Gate_1* and *Gate_2* for the external power drive circuit and thus controls the energy flow into the actuator. The resulting actuator voltage U_{act_actual} is fed to the controller by a voltage divider and subsequent analog-digital conversion. The actuator current is measured using the voltage drop of the small resistor and comparing it to external reference values with analog comparators and voltage dividers. Thus, the comparator circuits generate four logical signals representing the state of the actuator current: I_{up} for a current higher than and I_{down} for current lower than the reference value. I_{up} and I_{down} are generated for the positive (I_{up}/I_{down} $Gate$ 1) and negative (I_{up}/I_{down} $Gate$ 2) actuator current, respectively.

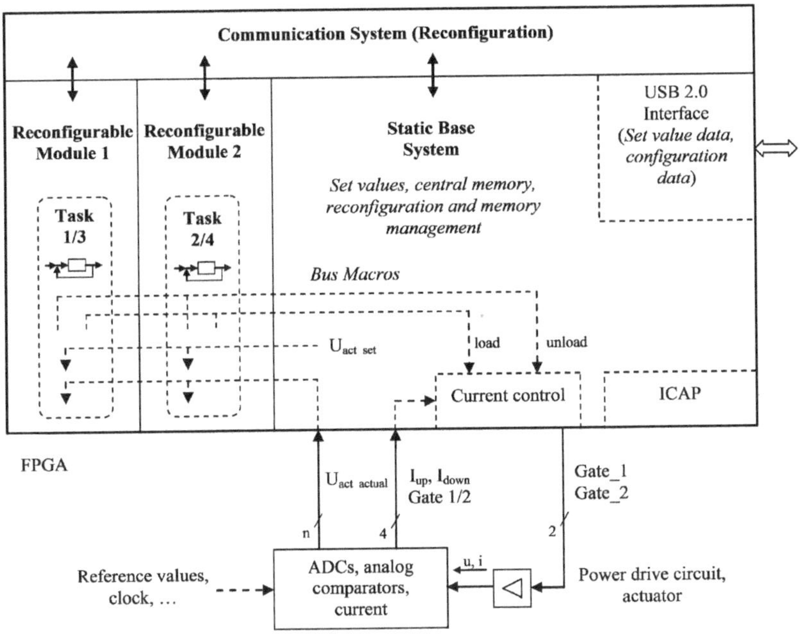

Fig. 20.9 Structure of the reconfigurable controller for piezo-electric actuators.

The static base system implements interfaces, infrastructure components and the central memory of the controller. Communication is accomplished with the USB 2.0 interface and includes receiving of configuration data, of set value data and transfer of diagnostic data. Actuator voltage values are read in from external ADCs, while the actuator current is monitored by the binary comparator circuit signals. Further infrastructure components include the reconfiguration and memory management with the Reconfiguration FSM and ICAP as well as set value generation/alteration functions. Moreover, the static base system implements the actuator-level current control. The two-level control algorithm uses the load/unload signals generated by the voltage control to set the control signals for the power drive circuit depending on the actuator current. The binary current-state signals from the analog comparators may be digitally filtered in the FPGA in order to minimize the influence of high-frequency noise. Switching operations are conducted if the actuator current exceeds or falls below the reference values of the analog comparators while the load/unload signal is set. The adjustable switching frequency is dependent on the dynamics of the actual piezo-electric actuator and the power drive circuit.

Controller implementation targets a Xilinx Virtex-4 XC4VLX60 FPGA and is accomplished with the Xilinx partial reconfiguration design flow. The USB 2.0 interface employs the external EZ-USB FX2 microcontroller by Cypress. The reconfigurable modules on the FPGA are 4 by 20 Configurable Logic Blocks (CLBs) wide and thus provide each 80 CLBs (equaling 320 CLB slices). The static base system may use the remainder of the FPGA and additionally implements 256 KB of on-chip BlockRAM as the controller's central memory. Memory and ICAP are employed in a 32 Bit wide data mode. The controller is generally clocked with 40 MHz and additionally uses a 200 MHz clock for the Reconfiguration FSM thus facilitating a reconfiguration (ICAP) clock frequency of 100 MHz. Furthermore, signal processing for the piezo-control functions is implemented with an exemplary data width of 8 Bit. However, the signal processing data width can be easily adapted offline due to the direct hardware implementation of all controller functionality.

With regard to the implementation results, the static base system allocates about 915 CLB slices (460 DFFs, 1520 LUTs) and 118 BlockRAM resources. The hardware tasks require about 70 to 90 CLB slices (60 DFFs, 105 LUTs to 80 DFFs, 145 LUTs). The achievable reconfiguration times are in the range of 91 to 111 μs.

20.5 Conclusion

The design methodology and hardware platform for dynamically reconfigurable mechatronic controllers presented in this chapter currently build a powerful link between advanced mechatronic system design and actual research in reconfigurable computing focused to the need of the mechatronic system engineer. Starting from common specification tools like block diagrams and state machines the reconfigurable structure of the mechatronic controller can be defined in a natural problem oriented way, which embeds very fine in the usual design flow. Design tools familiar to a mechatronic system engineer can be used and combined with standard tools for FPGA synthesis and reconfiguration. Adapting the structure of the reconfigurable system to the functional structure of the mechatronic controller preserves original physical structure, e.g. data flow and functional blocks, and facilitates analysis, design, test and implementation of the reconfigurable system. Hard real-time operation specific to parallel hardware implementation is guaranteed as the physical parallel data flow is not modified. Only temporarily inactive parts are swapped out.

This strong commitment to hard real-time operation, essential to all mechatronic systems, is continued for the hardware platform and the reconfiguration scheme developed. The combination of a relatively small static infrastructure always available on the FPGA and dynamically loadable functional parts guarantees exact points of reconfiguration and the restoration of actual dynamic internal states of a reconfigured component. The hardware implementation of the reconfiguration manager without the need of any processor is very fast allowing reconfiguration times of only a few microseconds which are met exactly. In the sum, design methodology and hardware platform fit perfectly and allow the design of hard real-time systems starting from specification up to the hardware implementation of the reconfigurable system. This fact was shown in this chapter using the example of a reconfigurable controller for piezo-electric actuators.

Besides the benefits of resource minimization offered by dynamic reconfiguration a special bit-serial implementation of data flow structures is used to reduce resource needs to a very low level. From the viewpoint of resources and implementation cost, dynamically reconfigurable systems could be a very attractive alternative to microcontroller based solutions in the near future, in particular if hard real-time conditions have to be met. Unfortunately the tool chain available for specification and implementation of dynamically reconfigurable mechatronic controllers is a hard obstacle in the usual mechatronic design process which today delays a wide spread usage of this very interesting technology significantly. Without integration in the standard design process and a support by standard tools like Matlab/Simulink or others even the first steps directed to use reconfigurable controllers in mechatronic systems are too expensive. At the other end, reconfigurable FPGA architectures are not optimally suited for mechatronic applications. Typical I/O-Peripherals, a standard for microcontrollers, are not available. External devices connected via very fast serial interfaces may meet the speed demands but will not be accepted in harsh environments or with respect to costs. From these practical points of view a rather slow stepwise move to dynamic reconfiguration concentrated on specific mechatronic application fields will be a probable scenario. On the other hand, from a more technological point of view these obstacles could be lifted away with not too much effort. In this situation not only motor controllers or piezo-controllers as part of an engine ECU can be implemented as shown in this chapter. In fact complete ECU systems for vehicle or robot control can be implemented using dynamically reconfigurable hardware thus avoiding today's problems of software development and testing associated mainly to real-time operation. Against this background we encourage mechatronic system engineers as well as decision makers to keep in touch with this promising technology of hardware implementation and dynamical reconfiguration of mechatronic controllers. As with all new technologies it is not easy to find the best point in time to switch horses. With increasing complexity and real-time demands of mechatronic controllers, the increasing effort and cost to manage, test and maintain associated software projects, dynamically reconfigurable computing offers a path to realize a significant advantage in competition in the development of new mechatronic products.

Acknowledgements This work was supported by DFG grant KA 1186/13-1, project *Rekonfigurierbare Hardware zur Steuerung und Regelung mechatronischer Systeme unter harten Echtzeitbedingungen*, as part of the Priority Programme 1148, *Reconfigurable Computing Systems*.

References

1. Delahaye, J.P., Gogniat, G., Roland, C., Bomel, P.: Software radio and dynamic reconfiguration on a dsp/fpga platform. Freq., J. Telecommun. **58**, 152–159 (2004)

2. Gnad, G., Kasper, R.: Power drive circuits for piezo-electric actuators in automotive applications. In: Proc. IEEE International Conference on Industrial Technology ICIT 2006, pp. 1597–1600 (2006)
3. He, P., Jin, M., Yang, L.: High performance dsp/fpga controller for implementation of hit/dlr dexterous robot hand. In: IEEE International Conference on Robotics and Automation ICRA '04, pp. 3397–3402 (2004)
4. Kasper, R., Reinemann, T.: Gate level implementation of high speed controllers and filters. In: FAC Conference on New Technologies for Computer Control NTCC 2001, pp. 170–175 (2001)
5. Kindratenko, V.V.: Novel computing architectures. In: Computing in Science and Engineering, pp. 54–57 (2009)
6. Lysaght, P., Blodget, B., Mason, J., Young, J., Bridgford, B.: Invited paper: Enhanced architectures, design methodologies and cad tools for dynamic reconfiguration of xilinx fpgas. In: Proc. International Conference on Field Programmable Logic and Applications FPL '06, pp. 1–6 (2006). doi:10.1109/FPL.2006.311188
7. Meyer-Baese, U.: Digital signal processing with field programmable gate arrays, 2nd edn. Springer, Berlin (2004)
8. OSEK/VDX Operating System Specification 2.2.3 (2004)
9. Reinemann, T.: Verfahren zur direkten implementierung von algorithmen auf gatterebene (method for direct gate level implementation of algorithms). PhD thesis, Otto-von-Guericke-Universität Magdeburg (2003)
10. Sedcole, N.P.: Reconfigurable platform-based design in fpgas for video image processing. PhD thesis, Imperial College of Science, Technology and Medicine, University of London (2006)
11. Toscher, S., Kasper, R.: A run-time reconfigurable fpga-based drive controller for electrical drive systems. In: Proc. International Conference on Computer Engineering & Systems ICCES '07 (2007). doi:10.1109/ICCES.2007.4447074
12. Toscher, S., Gnad, G., Kasper, R.: Reconfigurable hardware in automotive systems—a rapid prototyping approach. In: Proc. 4th International Signal Processing Conference GSPx 2006 (2006)
13. Toscher, S., Reinemann, T., Kasper, R.: Implementation of a reconfigurable hard real-time control system for mechatronic and automotive applications. In: Proceedings of the 20th IEEE International Parallel and Distributed Processing Symposium (IPDPS'06) (2006)
14. Toscher, S., Reinemann, T., Kasper, R., Hartmann, M.: A reconfigurable delta-sigma adc. In: Proc. IEEE International Symposium on Industrial Electronics, vol. 1, pp. 495–499 (2006)

Index

M. Platzner et al. (eds.), *Dynamically Reconfigurable Systems*,
DOI 10.1007/978-90-481-3485-4, © Springer Science+Business Media B.V. 2010

FSC
www.fsc.org
MIX
Papier | Fördert
gute Waldnutzung
FSC® C083411

Kolibri 360 GmbH
Ferdinand-Jühlke-Straße 7
99095 Erfurt, Deutschland
produktsicherheit@kolibri360.de